EUKARYOTIC TRANSCRIPTION FACTORS

Fifth Edition

Cover illustration: a stylized image depicting the nutlin-2 inhibitor and its similarity to the p53 transcription factor, which allows it to block the interaction of p53 with MDM2. (Based on an image kindly provided by Dr B Groves and Dr L Vassilev)

EUKARYOTIC TRANSCRIPTION FACTORS

Fifth Edition

David S. Latchman

Master, Birkbeck, University of London and Professor of Genetics, Birkbeck and University College London

AMSTERDAM • BOSTON • HEIDELBERG • LONDON • NEW YORK • OXFORD
PARIS • SAN DIEGO • SAN FRANCISCO • SINGAPORE • SYDNEY • TOKYO
Academic Press is an imprint of Elsevier

Academic Press is an imprint of Elsevier
360 Park Avenue South, New York, NY 10010-1710, USA
84 Theobald's Road, London WC1X 8RR, UK
30 Corporate Drive, Suite 400, Burlington, MA 01803, USA
525 B Street, Suite 1900, San Diego, CA 92101-4495, USA

First edition 1991
Second edition 1995
Third edition 1998
Fourth edition 2004
Fifth edition 2008

Notice
No responsibility is assumed by the publisher for any injury and/or damage to persons
or property as a matter of products liability, negligence or otherwise, or from any use or
operation of any methods, products, instructions or ideas contained in the material
herein. Because of rapid advances in the medical sciences, in particular, independent
verification of diagnoses and drug dosages should be made

British Library Cataloguing in Publication Data
A catalogue record for this book is available from the British Library

Library of Congress Cataloging in Publication Data
A catalog record for this book is available from the Library of Congress

ISBN: 978-0-12-373983-4

For information on all Academic Press publications
visit our web site at books.elsevier.com

Typeset by Charon Tec Ltd
(A Macmillan Company), Chennai, India
www.charontec.com
Printed and bound in Great Britain
07 08 09 10 11 10 9 8 7 6 5 4 3 2 1

Working together to grow
libraries in developing countries

www.elsevier.com | www.bookaid.org | www.sabre.org

ELSEVIER   BOOK AID
International   Sabre Foundation

To Hannah
With love

CONTENTS

Colour Plates are located between pages 244 and 245

LIST OF TABLES

ABOUT THE AUTHOR

David S. Latchman graduated with a BA (First Class Honours) in Natural Sciences from Cambridge University and subsequently obtained a PhD in Genetics from Cambridge University.

Following a period of post-doctoral research at Imperial College London, he was appointed to a lectureship at University College London in 1984. Subsequent appointments at UCL included Director of the Medical Molecular Biology Unit (1988), Professor of Molecular Pathology and Head of the Department of Molecular Pathology (1991), Director of the Windeyer Institute of Medical Sciences (1996) and in 1999, Dean of the Institute of Child Health and Professor of Human Genetics.

At the beginning of 2003, he took up a new appointment as Master of Birkbeck, University of London, leading this multi-faculty College comprising over 1,000 staff and 16,000 students studying at levels from short courses to PhDs. He continues to retain an active research laboratory at the Institute of Child Health, UCL and in recognition of this he is now Professor of Genetics at Birkbeck and UCL.

Since he set up his own laboratory in 1984, Professor Latchman's interests have focused particularly on the regulation of gene expression in mammalian cells by specific transcription factors. Initial studies, involved the eukaryotic virus herpes simplex virus (HSV) and were aimed at elucidating why this virus was able to replicate in epithelial cells, whilst establishing silent latent infections in neuronal cells.

The identification of a cellular POU family transcription factor, Oct-2, which was expressed in neuronal cells and blocked the virus-lytic cycle, led to an interest in the role of specific POU family transcription factors in regulating cellular as well as viral gene expression in neuronal cells. As well as studies of the role of Oct-2, these investigations resulted in the characterization of a small sub-family of POU family transcription factors which comprised three members: Brn-3a, Brn-3b and Brn-3c. In particular, the key role of Brn-3a in promoting neuronal differentiation and enhancing the survival of neuronal cells was defined by Professor Latchman's laboratory and has important therapeutic potential for the treatment of human neurodegenerative diseases involving losses of neuronal cells.

More recently, Professor Latchman's group has also studied the role of specific transcription factors in the heart and in particular, in its response to damage caused by ischaemia, as in a heart attack. They have demonstrated that a specific transcription factor STAT-1 becomes activated during ischaemia and induces a number of genes which enhance the programmed cell death of cardiac cells, resulting in the loss of these irreplaceable cells. Thus, inhibition of STAT-1 represents a potential therapeutic mechanism for decreasing the damage caused by interruption of blood flow to the heart.

On the basis of his scientific research, Professor Latchman was awarded a DSc by London University in 1994 and was appointed a Fellow of the Royal College of Pathologists (FRCPath) in 1999. He serves on a number of Committees, including the Department of Health, Genetics and Insurance Committee, the National Biological Standards Board and its Scientific Policy Advisory Committee (Chairman), the Health Protection Agency and the Research Strategy Committee of Universities UK. He is also Chairman of London Higher, the umbrella organization which represents all London's Universities.

PREFACE

The fourth edition of *Eukaryotic Transcription Factors* differed radically from its predecessors in the manner in which the immense amount of information about transcription factors was organized. Thus, for the first time, the book adopted a single approach of dealing in turn with the specific properties of transcription factors, using a range of examples. This contrasted with the dual approach of previous editions in which the roles of individual transcription factors were discussed in the initial chapters, followed by chapters on the more mechanistic aspects of their function.

This single approach has proved successful and is therefore continued in this new edition. However, as in all previous editions, the immense amount of progress that has been made since the publication of the fourth edition has necessitated updating throughout the volume.

The most important changes, however, reflect a fundamental change in our view of transcription factor function, which has taken place progressively over a number of years since the initial editions of this book were published. Thus, for a considerable period it was believed that transcription factors acted primarily at the level of transcriptional initiation by interacting directly or indirectly with the basal transcriptional complex to enhance or reduce its activity. This clearly remains a central function of transcription factors. However, it is now clear that transcription factors also play a key role both before and after the process of transcriptional initiation by the basal transcriptional complex.

In particular, transcription factors have a vital role in organizing the chromatin structure of a particular gene so that it is either in an open chromatin structure compatible with transcription or in a more tightly packed closed chromatin structure that does not allow transcription to occur. The basic processes that modify chromatin structure and the key role of histone modifications in this effect now merit a more extensive discussion in Chapter 1 (section 1.2). Moreover, the effect of transcriptional activators on chromatin structure now merits a separate section of Chapter 5 (section 5.5), and similarly, the effect of transcriptional repressors on such chromatin structure is also now discussed in a separate section of Chapter 6 (section 6.4). This new section also includes a subsection on the important topic of small inhibitory RNAs since, although these inhibitory molecules act mainly at stages subsequent to transcription, they can also modulate transcription by regulating chromatin structure (section 6.4.2).

As well as having an effect on chromatin structure prior to transcriptional initiation, transcription factors can also affect the process of transcriptional elongation that occurs subsequent to transcriptional initiation. For this reason, Chapter 3 now contains a new section dealing with the basic process of transcriptional elongation, whilst Chapters 5 and 6 now contain additional sections that deal respectively with the effect on transcriptional elongation of activating (section 5.6) and inhibitory (section 6.5) transcription factors.

The significant advances that have been made in understanding the manner in which transcription factors regulate gene expression have been paralleled by an enhanced understanding of the manner in which these processes can go wrong to produce human disease. As more and more information on this topic accumulates, it becomes increasingly clear that transcription factors are important targets for therapies aimed at treating a wide range of human diseases. Accordingly, a new section has been added to Chapter 9 dealing specifically with the topic of transcription factors and the treatment of human disease (section 9.5).

It is hoped that these changes will allow the book to continue to provide an effective guide to the enormous complexity of transcription factors and the manner in which a relatively small number of factors control highly complex processes, such as embryonic development and functioning of the adult organism.

Finally, I would like to thank Miss Maruschka Malacos for her continued efficiency in typing the additional text and making the very large number of modifications to the existing text, which were required to keep the work up-to-date. I am also most grateful to Dr Luna Han and the staff of Elsevier for commissioning this new edition of the book and for the efficient manner in which they have produced it.

David S. Latchman

PREFACE TO THE FOURTH EDITION

It is now over ten years since the first edition of Eukaryotic Transcription Factors was published. It is obvious that in that time an enormous amount of information about transcription factors has accumulated and this has been reflected in subsequent editions of the book. However, over the past years, we have moved from a situation where only a few transcription factors had been characterized in any detail, to a situation where a very large number of transcription factors have been extensively characterized. This has led to the decision in this new edition to abandon the dual structure of previous editions in which the role of a few transcription factors in inducible, cell-type specific and developmental gene regulation was extensively discussed, followed by chapters dealing with the mechanistic aspects of transcription factors.

In the new edition therefore, the book adopts a single approach of dealing in turn with the specific properties of transcription factors, using a range of examples including those which were extensively discussed in previous editions but also others as appropriate. This has allowed a much more detailed analysis of various mechanistic aspects which have become of increasing importance in recent years.

As before, the work begins with a chapter on DNA sequences and chromatin structure in which the section on the modulation of chromatin structure by chromatin remodelling complexes and histone modifying enzymes has been considerably expanded to reflect recent work. This is followed, as before, by a chapter describing the methods used to analyse the properties of transcription factors which now has an additional section dealing with the methods of identifying target genes for previously uncharacterized transcription factors. As before, this is followed by a chapter dealing with RNA polymerase enzymes and the basal transcriptional complex.

Following these three initial chapters, however, the format of the book has dramatically changed. Thus, Chapter 4 now deals extensively with specific transcription factor families. Moreover, since these families are defined primarily on the basis of their DNA binding domain, this chapter also deals with the features that allow these various factors to bind to DNA. Subsequently, separate chapters deal with activation and repression of transcription respectively, replacing the single chapter which previously dealt with both these processes. This has allowed a considerable expansion of the discussion of these topics, allowing subjects such as the mediator complex, co-activators and the activation or repression of transcription by alterations in chromatin structure, to be discussed in much greater depth.

Similarly, the single chapter in the previous edition dealing with the regulation of transcription factor synthesis and activity, has now been split into two chapters dealing respectively with the regulation of transcription factor synthesis and the regulation of transcription factor activity. Again, this has allowed a number of topics, such as the regulation of transcription factor activity by a variety of different post-translational modifications, to be discussed in greater depth. As part of these changes, the chapter on transcription factors and human disease has been moved to the end of the work and is followed by a final conclusion chapter.

It is hoped that these changes will avoid the increasing duplication that would have been necessary if the initial approach had been maintained and will allow the work to build on the success of its predecessors, by providing an up-to-date account of this critically important topic.

Finally, I would like to thank Miss Maruschka Malacos for typing the text and coping with the necessity to move around large and small sections, to reflect the change in the structure of the book. I am also most grateful to Dr Tessa Picknett and the staff at Elsevier Science (Academic Press) for commissioning this new edition and producing it with their customary efficiency.

David S. Latchman

PREFACE TO THE THIRD EDITION

As in previous years, the period between the publication of the second and third editions of this book has been marked by a considerable further accumulation of information about individual transcription factors and the manner in which they act. This new edition has therefore been extensively updated to reflect this and several sections have been completely rewritten.

As well as such increased general understanding of transcription factors, a major new theme unifying much of this information has emerged. This involves the role of co-activator molecules such as CBP in the action of a number of different activating transcription factors as well as the finding that such co-activators frequently possess histone acetyltransferase activity indicating that they may act by modulating chromatin structure. In addition to discussion of co-activators in the appropriate sections on individual transcription factors, the new edition of this work now includes specific new sections dealing with this important topic. Thus the role of chromatin structure and histone acetylation in the regulation of gene expression is now introduced in Chapter 1 (section 1.4), the role of CBP in cyclic AMP mediated gene activation where it was originally discovered is discussed in Chapter 4 together with other aspects of this signalling pathway (section 4.3) and the interaction of transcriptional activators with co-activators is discussed in a separate section of Chapter 9 (section 9.2.4).

In addition to these new sections on this aspect, other new sections have been added describing topics which are now of sufficient importance to merit a separate section. These are the methods used to determine the DNA binding specificity of an uncharacterized transcription factor (Chapter 2, section 2.3.4), the Pax family transcription factors (Chapter 6, section 6.3.2), anti-oncogenic transcription factors other than p53 or Rb (Chapter 7, section 7.3.4) and the regulation of transcription factor activity by protein degradation and processing (Chapter 10, section 10.3.5). Similarly, Chapter 7 now includes an extensive discussion of the role of transcription factors in diseases other than cancer and its title has therefore been changed to 'Transcription factors and human disease' (from 'Transcription factors and cancer').

As well as these changes in the text, we have been able to include, for the first time, a special section of colour illustrations illustrating various aspects of transcription factor structure which are being progressively elucidated. It is hoped that all these changes will allow this new edition, like its predecessors, to provide an up-to-date overview of the important area of transcription factors and their vital role in regulating transcription in different cell types, during development and in disease.

Finally, I would like to thank Mrs Sarah Franklin for her efficiency in producing the text and dealing with the need to make numerous changes from the previous edition, as well as Mrs Jane Templeman for continuing to use her outstanding skills in the preparation of the numerous new illustrations in this edition. Thanks are also due to Tessa Picknett and the staff at Academic Press for producing this new edition with their customary efficiency.

David S. Latchman

PREFACE TO THE SECOND EDITION

In the four years since the first edition of this work was published, the explosion of information about transcription factors has continued. The genes encoding many more transcription factors have been cloned and this information used to analyse their structure and function culminating, in many cases, with the use of inactivating mutations to prepare so called 'knock out' mice, thereby testing directly the role of these factors in development. Nonetheless, the examples used in the first part of this book to illustrate the role of transcription factors in processes as diverse as inducible gene expression and development still remain amongst the best understood. The discussion of these factors has therefore been considerably updated to reflect the progress made in the last few years. In addition, new sections have been added on topics such as TBP; the *myc* oncogene and anti-oncogenes where the degree of additional information now warrants a separate section.

Even greater changes have been necessary in the second part of the book, which deals with the mechanisms by which transcription factors act. Thus, for example, the sections in Chapter 9 on the mechanisms of transcriptional activation and on transcriptional repression have been completely rewritten. In addition the increasing emphasis on transcriptional repression discussed in Chapter 9 has led to a change in the title of Chapter 10 to 'What regulates the regulators?' (from 'What activates the activators?'). Moreover, this chapter now includes a much more extensive section on the interaction between different factors, which is another major theme to have emerged in the past few years. It is hoped that these changes will allow the new edition to build on the success of the first edition in providing an overview of these vital factors and the role they play in gene regulation.

Finally, I would like to thank Jane Templeman, who has prepared a large number of new illustrations to complement the excellent ones she provided for the first edition, and Sarah Chinn for coping with the necessity of adding, deleting or amending large sections of the first edition. I am also grateful to Tessa Picknett and the staff at Academic Press for commissioning this new edition and their efficiency in producing it.

David S. Latchman

PREFACE TO THE FIRST EDITION

In my previous book, *Gene Regulation: A Eukaryotic Perspective* (Unwin–Hyman Ltd., 1990), I described the mechanisms by which the expression of eukaryotic genes is regulated during processes as diverse as steroid treatment and embryonic development. Although some of this regulation occurs at the post-transcriptional level, it is clear that the process of gene transcription itself is the major point at which gene expression is regulated. In turn this has focused attention on the protein factors, known as transcription factors, which control both the basal processes of transcription and its regulation in response to specific stimuli or developmental processes. The characterization of many of these factors and in particular the cloning of the genes encoding them has resulted in the availability of a bewildering array of information on these factors, their mechanism of action and their relationship to each other. Despite its evident interest and importance, however, this information could be discussed only relatively briefly in *Gene Regulation*, whose primary purpose was to provide an overview of the process of gene regulation and the various mechanisms by which this is achieved.

It is the purpose of this book therefore to discuss in detail the available information on transcription factors, emphasizing common themes and mechanisms to which new information can be related as it becomes available. As such, it is hoped the work will appeal to final year undergraduates and postgraduate students entering the field as well as to those moving into the area from other scientific or clinical fields who wish to know how transcription factors may regulate the gene in which they are interested.

In order to provide a basis for the discussion of transcription factors, the first two chapters focus respectively on the DNA sequences with which the factors interact and on the experimental methods which are used to study these factors and obtain the information about them provided in subsequent chapters. The remainder of the work is divided into two distinct portions. Thus Chapters 3 to 7 focus on the role of transcription factors in particular processes. These include constitutive and inducible gene expression, cell type-specific and developmentally regulated gene expression and the role of transcription factors in cancer. Subsequently, Chapters 8 to 10 adopt a more mechanistic approach and consider the features of transcription factors which allow them to fulfil their function. These include the ability to bind to DNA and modulate transcription either positively or negatively as well as the ability to respond to specific stimuli and thereby activate gene expression in a regulated manner.

Although this dual approach to transcription factors from both a process-oriented and mechanistic point of view may lead to some duplication, it is the most efficient means of providing the necessary overview both of the nature of transcription factors and the manner in which they achieve their role of modulating gene expression in many diverse situations.

Finally, I would like to thank Mrs Rose Lang for typing the text and coping with the continual additions necessary in this fast moving field and Mrs Jane Templeman for her outstanding skill in preparing the illustrations.

David S. Latchman

ACKNOWLEDGEMENTS

I would like to thank all the colleagues, listed below, who have given permission for material from their papers to be reproduced in this book and have provided prints suitable for reproduction.

Figures 4.1 and **7.11**, photographs kindly provided by Professor W.J. Gehring from Gehring, Science 236, 1245 (1987) by permission of the American Association for the Advancement of Science.

Figure 4.15, photograph kindly provided by Dr P. Holland from Holland and Hogan, Nature 321, 251 (1986) by permission of Macmillan Magazines Ltd.

Figure 4.27 redrawn from Redemann *et al.,* Nature 332, 90 (1988) by kind permission of Dr H. Jackle and Macmillan Magazines Ltd.

Figures 4.32 and **4.36** redrawn from Schwabe *et al.*, Nature 348, 458 (1990) by kind permission of Dr D. Rhodes and Macmillan Magazines Ltd.

Figure 4.41 redrawn from Abel and Maniatis, Nature 341, 24 (1989) by kind permission of Professor T. Maniatis and Macmillan Magazines Ltd.

Figures 7.5 and **7.9**, photographs kindly provided by Dr R.L. Davis from Davis *et al.*, Cell 51, 987 (1987) by permission of Cell Press.

Figures 7.14 and **7.15,** photographs kindly provided by Dr R. Krumlauf from Graham *et al.*, Cell 57, 367 (1989) by permission of Cell Press.

Figure 8.6, photograph kindly provided by Professor M. Beato from Willmann and Beato, Nature 324, 688 (1986) by permission of Macmillan Magazines Ltd.

Figure 8.13, photograph kindly provided by Dr C. Wu from Zimarino and Wu, Nature 327, 727 (1987) by permission of Macmillan Magazines Ltd.

I am also especially grateful to the colleagues who have provided colour prints of transcription factor structures, allowing us to include this feature.

Plate 1 kindly provided by Dr J. H. Geiger.

Plate 2 kindly provided by Dr T. Li and Professor C. Wolberger.

Plate 3 kindly provided by Professor P. E. Wright from Lee *et al.*, Science 245, 635 (1989) by permission of the American Association for the Advancement of Science.

Plate 4 kindly provided by Professor R. Kaptein from Hard *et al.*, Science 249, 157 (1990) by permission of the American Association for the Advancement of Science.

Plate 5 kindly provided by Dr D. Rhodes from Schwabe *et al.*, Cell 75, 567 (1993) by kind permission of Cell Press.

Plate 6 kindly provided by Professor D. Moras.

Plate 7 kindly provided by Dr B. Groves and Dr L. Vassilev.

CHAPTER 1

DNA SEQUENCES, TRANSCRIPTION FACTORS AND CHROMATIN STRUCTURE

1.1 THE IMPORTANCE OF TRANSCRIPTION

The fundamental dogma of molecular biology is that DNA produces RNA which in turn produces protein. Hence, if the genetic information which each individual inherits as DNA (the genotype) is to be converted into the proteins that produce the corresponding characteristics of the individual (the phenotype), it must first be converted into an RNA product. The process of transcription whereby an RNA product is produced from the DNA is therefore an essential element in gene expression. The failure of this process to occur will obviously render redundant all the other steps that follow the production of the initial RNA transcript in eukaryotes, such as RNA splicing, transport to the cytoplasm or translation into protein (for review of these stages see Nevins, 1983; Latchman, 2005).

The central role of transcription in the process of gene expression also renders it an attractive control point for regulating the expression of genes in particular cell types or in response to a particular signal. Indeed it is now clear that in the vast majority of cases where a particular protein is produced only in a particular tissue or in response to a particular signal this is achieved by control processes which ensure that its corresponding gene is transcribed only in that tissue or in response to such a signal (for reviews see Darnell, 1982; Latchman, 2005). For example, the genes encoding the immunoglobulin heavy and light chains of the antibody molecule are transcribed at high level only in the antibody producing B cells whilst the increase in somatostatin production in response to treatment of cells with cyclic AMP is mediated by increased transcription of the corresponding gene. Therefore, while post-transcriptional regulation affecting, for example, RNA splicing or stability plays some role in the regulation of gene expression (for reviews see Mata *et al.*, 2005; Stetefeld and Ruegg, 2005), the major control point lies at the level of transcription.

1.2 CHROMATIN STRUCTURE AND ITS REMODELLING

1.2.1 CHROMATIN STRUCTURE AND GENE REGULATION

The central role of transcription both in the basic process of gene expression and its regulation in particular tissues has led to considerable study of this process. Initially such studies focused on the nature of the DNA sequences within individual genes which were essential for either basal or regulated gene expression. These sequences will be discussed in section 1.3. It is now clear, however, that the accessibility of these DNA sequences and hence their ability to regulate gene expression is controlled by the manner in which they are packaged in the cell. The packaging of DNA will therefore be discussed in this section.

It has been known for some time that the DNA in eukaryotic cells is packaged by association with specific proteins such as the histones into a structure known as chromatin (for reviews see Wolffe, 1995; Felsenfeld and Groudine, 2003; Latchman, 2005). The fundamental unit of this structure is the nucleosome in which the DNA is wrapped twice around a unit of eight histone molecules (two each of histones H2A, H2B, H3 and H4) (for reviews see Kornberg and Lorch, 1999; Khorasanizadeh, 2004). This structure is compacted further into the so-called solenoid structure in genes that are not transcriptionally active or about to become active (for review see Mohd-Sarip and Verrijzer, 2004). In contrast, active or potentially active genes exist in the simple nucleosomal structure. Moreover, in the regulatory regions of these genes, nucleosomes are either removed altogether or undergo a structural alteration which facilitates the binding of specific transcription factors to their binding sites in these regions (Fig. 1.1).

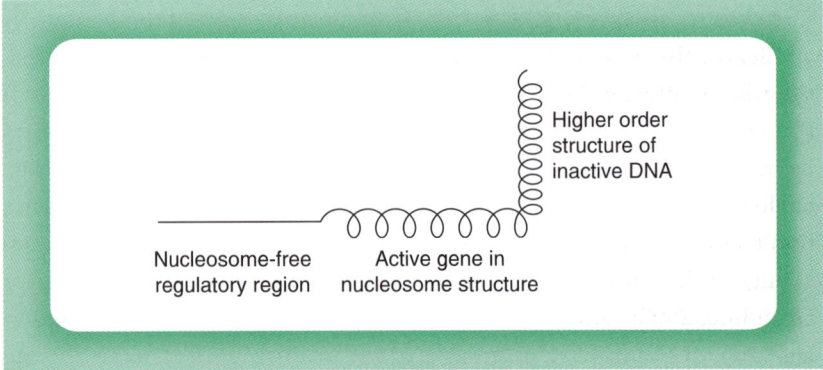

Figure 1.1

Levels of chromatin structure in active or inactive DNA.

Interestingly, the tightly packed solenoid structure can be compacted even further, by extensive looping, to form the chromosomes that are visible during cell division. These loops are linked at their bases to a protein scaffold known as the nuclear matrix, with such linkage occurring via specific DNA sequences, known as matrix attachment regions (MARs) (Fig. 1.2; for review see Horn and Peterson, 2002).

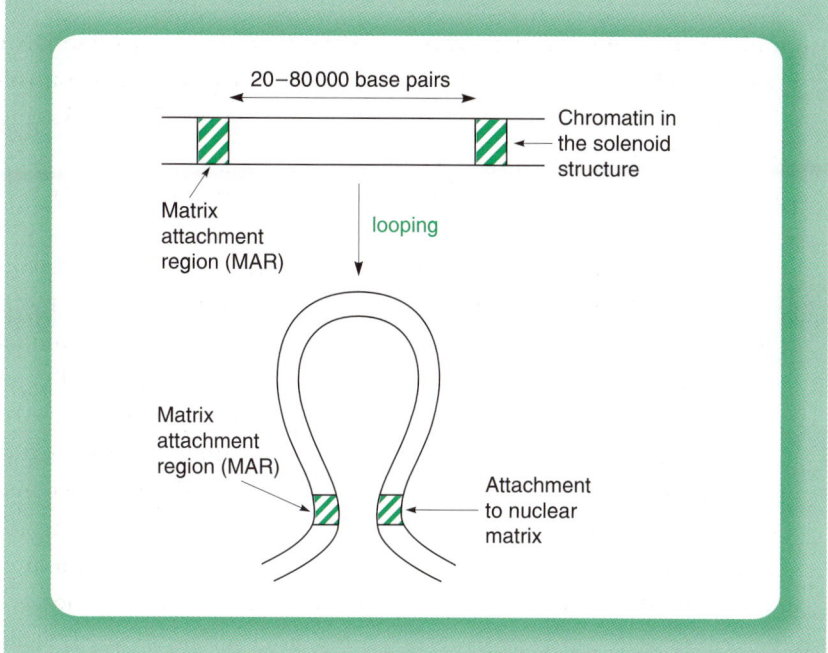

Figure 1.2
The tightly packed solenoid structure can be further compacted by the formation of loops. These loops (which contain approximately 20 000–80 000 bases of DNA) are attached to the nuclear matrix via specific DNA elements known as matrix attachment regions.

Clearly, the access of a transcription factor to its appropriate binding site will be affected by the manner in which that site is packaged within the chromatin structure. Evidently, therefore, genes that are about to be transcribed must undergo changes in chromatin structure which facilitate such transcription by allowing access of activating transcription factors to their binding sites. Although a detailed discussion of these changes is beyond the scope of this book (for reviews see Felsenfeld and Groudine, 2003; Khorasanizadeh, 2004; Carey, 2005; Latchman, 2005), at least two mechanisms that can alter chromatin structure are of particular importance in terms of transcription factor regulation and these will be discussed in turn.

1.2.2 CHROMATIN REMODELLING FACTORS

A number of studies have identified protein complexes which are capable of binding to DNA, hydrolysing ATP and using the energy generated to disrupt the nucleosomal structure and it is clear that chromatin remodelling by such complexes plays a critical role in the regulation of gene expression (for reviews see Mellor, 2005; Saha *et al.*, 2006; Serna *et al.*, 2006). The best characterized of these remodelling complexes is the SWI/SNF complex which contains a number of different polypeptides. It was originally defined in yeast but has now been identified in a range of organisms including humans. The critical role of this complex in regulating gene expression is indicated by the phenotype of the brahma mutation in *Drosophila* which inactivates the SWI2 component of the complex. Thus, in this mutant the genes encoding several homeobox-containing proteins, which control the correct patterning of the body (see Chapter 4, section 4.2), remain in an inactive chromatin structure and are hence not transcribed. This results in a mutant fly with a grossly abnormal body structure (for review see Simon, 1995).

It is likely that SWI/SNF and other chromatin remodelling complexes can act by, at least, three different methods to alter the accessibility of the DNA. Thus, they may act by altering the association of the histone molecules within the nucleosome so that the nucleosome structure is changed in such a way as to allow other factors to bind to DNA (nucleosome remodelling: Fig.1.3a). Secondly, they may act by causing the nucleosome to move along the DNA, so exposing a particular DNA sequence (nucleosome sliding: Fig.1.3b). Finally, they may act by displacing a nucleosome so that it leaves the target DNA and binds to another DNA molecule (nucleosome displacement: Fig.1.3c). All these methods have in common, the use of ATP hydrolysis to alter the nucleosome in some way so as to allow a particular region of DNA to become more accessible and hence bind specific regulatory factors.

Evidently, these mechanisms beg the question of how the SWI/SNF complex is itself recruited to the genes that need to be activated. This can occur via its association with the RNA polymerase complex or by its association with other transcription factors which can bind to their specific DNA binding sites even in tightly packed, non-remodelled chromatin. These processes are discussed in Chapter 5 (section 5.5.1).

Interestingly, it has been shown that chromatin remodelling complexes can also be recruited to the DNA by the SATBI protein which is involved in the looping of the chromatin into a highly compact structure (Yasui *et al.*, 2002) (see section 1.2.1). This provides a link between the looping process and chromatin remodelling/gene regulation and

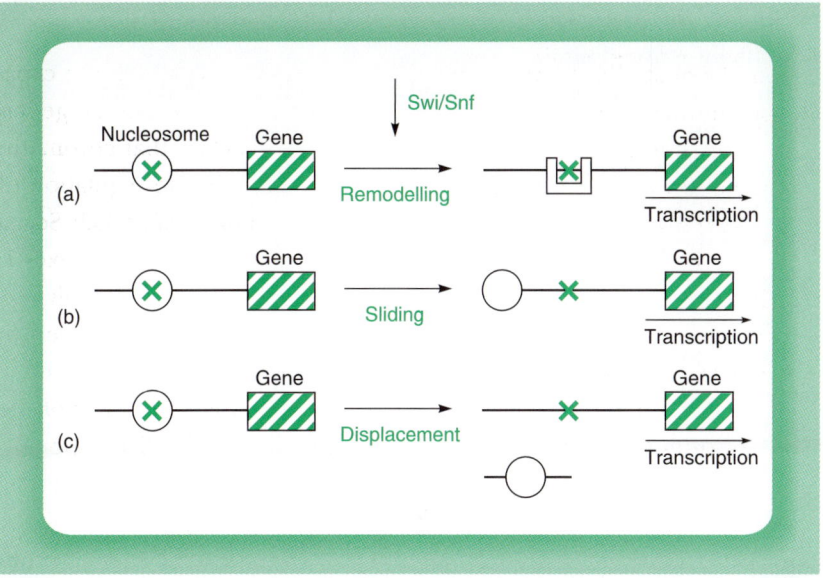

Figure 1.3

The Swi/Snf complex can allow a regulatory protein access to its binding site (X) by (a) producing an altered structure of the nucleosome in a process known as nucleosome remodelling; (b) inducing nucleosome sliding to a different position on the DNA; or (c) displacing the nucleosome onto another DNA molecule.

suggests that such remodelling processes can target the large regions of DNA (20 000 to 80 000 bases of DNA) contained in individual loops (for review see Li *et al.*, 2006).

1.2.3 HISTONE MODIFICATIONS

The histone molecules which play a key role in chromatin structure are subject to a number of post-translational modifications such as phosphorylation, ubiquitination or acetylation (for reviews see Felsenfeld and Groudine, 2003; Jaskelioff and Peterson, 2003; Khorasanizadeh, 2004; Clayton *et al.*, 2006; Millar and Grunstein, 2006; Turner, 2007). In particular, the addition of an acetyl group to a free amino group in lysine residues in the histone molecule reduces its net positive charge. Such acetylated forms of the histones have been found preferentially in active or potentially active genes where the chromatin is less tightly packed. Moreover, treatments that enhance histone acetylation such as addition of sodium butyrate to cultured cells result in a less tightly packed chromatin structure and the activation of previously silent cellular genes. This suggests that hyperacetylation of histones could play a causal role in producing the more open chromatin structure characteristic of active or potentially active genes.

These ideas have been confirmed by a recent study that examined the distribution of acetylated histone H3 across the entire genome of human T lymphocytes (Roh *et al.*, 2005). This showed that acetylated histones were located at known regulatory sequences in the DNA. Moreover, activation of the T cells resulted in the appearance of histone H3 acetylation at new sites in the DNA, coinciding with opening up of the chromatin and gene activation.

Hence, activation of gene expression could be achieved by factors with histone acetyltransferase activity which were able to acetylate histones and hence open up the chromatin structure, whereas inhibition of gene expression would be achieved by histone deacetylases which would have the opposite effect (Fig. 1.4). Most interestingly, recent studies have identified both components of the basal transcriptional complex and specific activating transcription factors with histone acetyltransferase activity as well as specific inhibitory transcription factors with histone deacetylase activity (for reviews see Brown *et al.*, 2000; Carrozza *et al.*, 2003). These findings, which link studies on modulation of chromatin structure with those on activating and inhibitory transcription factors, are discussed further in Chapters 5 and 6.

It is clear therefore that histone acetylation plays a key role in regulating chromatin structure. However, in the last few years it has become increasingly clear that other histone modifications such as methylation, phosphorylation or the addition of the small protein ubiquitin (ubiquitination) are also involved in this process. Thus, like acetylation of lysine residues, phosphorylation of serine residues in specific histones is associated with a more open chromatin structure and gene activation (for review see Nowak and Corces, 2004). Moreover, such phosphorylation of histone H3 occurs in response to growth factor stimulation of cells, with the resulting phosphorylated histone H3, localizing to genes such as the cellular oncogenes c-*fos* - c-*myc*, which are switched on by growth factor treatment.

In contrast, methylation of lysine residues in the histones can promote either a more open or a more closed chromatin structure depending on the specific amino acid residue involved. For example, methylation on lysine residue 4 of histone H3 promotes a more open chromatin structure, whilst methylation of lysine 9 has the opposite effect (for review see Martin and Zhang, 2005) (Fig. 1.5).

Interestingly, the different histone modifications can interact with one another. Thus, for example, demethylation of the lysine amino acid at position 9 in histone H3 facilitates phosphorylation of serine 10 and

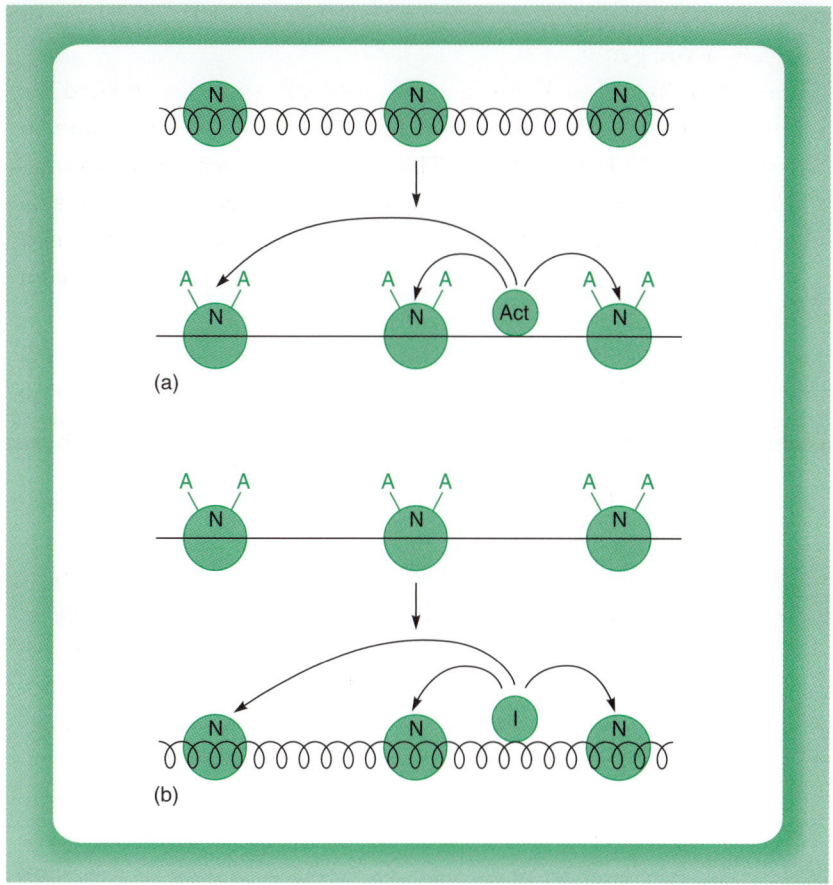

Figure 1.4

(a) An activating molecule (Act) can direct the acetylation of histones in the nucleosome (N) thereby resulting in a change in chromatin structure from a tightly packed (wavy line) to a more open (solid line) configuration. (b) An inhibitory molecule can direct the deacetylation of histones thereby having the opposite effect on chromatin structure.

acetylation of lysine 14 of H3, leading to opening of the chromatin and gene activation (Fig. 1.6). Indeed, it has been suggested that methylation of lysine 9 and phosphorylation of serine 10 on histone H3 serve as a 'binary switch' promoting chromatin closing or opening respectively (for review see Fischle *et al.*, 2003).

   Hence, the histones show a complex pattern of post-translational modifications, which interact with one another and alter chromatin structure. This complex pattern of modification has led to the idea of a 'histone code' in which the chromatin structure of a particular gene is specified by the pattern of different modifications of the histones which package it (for reviews see Goll and Bestor, 2002; Millar and Grunstein,

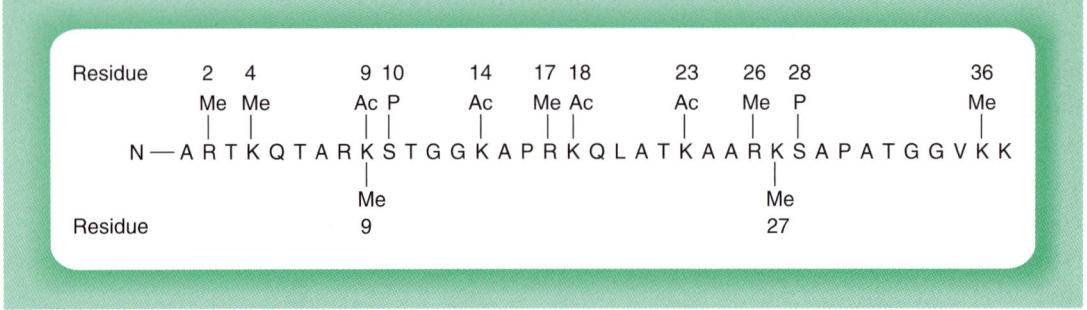

Figure 1.5

Post-translational modifications of the first thirty-seven amino acids of histone H3 by acetylation (Ac) or methylation (Me) of lysine residues (K) methylation (Me) of arginine residues (R) or phosphorylation (P) of serine residues (S). Modifications which produce opening of the chromatin structure are shown above the line and those producing a closed chromatin structure are shown below the line.

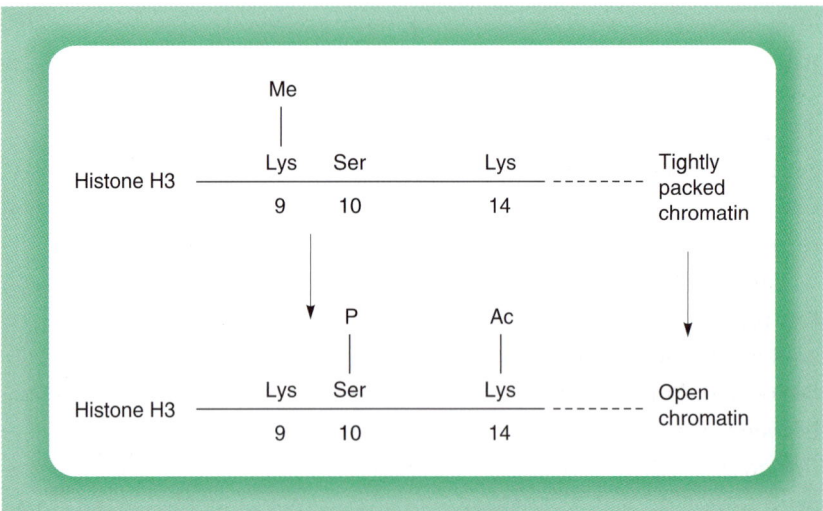

Figure 1.6

Demethylation of the lysine amino acid at position 9 in histone H3 facilitates phosphorylation of serine 10 and acetylation of lysine 14 leading to a more open chromatin structure.

2006; Turner, 2007). The complex pattern of modifications at the N-terminus of histone H3 is illustrated in Figure 1.5.

Hence, both ATP-dependent chromatin remodelling complexes and alterations in histone acetylation/modification play a vital role in regulating the chromatin structure of specific genes. Although these two processes have been discussed separately, it is likely that chromatin remodelling and histone modification enzymes co-operate. Thus, for example, it has been shown that acetylation of histones can allow recruitment of SWI/SNF to a promoter (Agalioti *et al.*, 2002) as well as preventing it from dissociating once it has bound (Hassan *et al.*, 2001). Hence, it appears that these two processes act together to ensure that the DNA sequences involved in transcription control become accessible at the correct time in development or in response to appropriate signals (for review see Narlikor *et al.*, 2002). The nature of these DNA sequences is discussed in the next section.

1.3 DNA SEQUENCE ELEMENTS

1.3.1 THE GENE PROMOTER

The primary aim of chromatin remodelling processes is to expose specific DNA sequences so that these can be targeted by transcription factors involved in the process of gene transcription. In prokaryotes, such sequences are found immediately upstream of the start site of transcription and form part of the promoter directing expression of the genes. Sequences found at this position include both elements found in all genes which are involved in the basic process of transcription itself and those found in a more limited number of genes which mediate their response to a particular signal (for review see Muller-Hill, 1996).

Early studies of cloned eukaryotic genes therefore concentrated on the region immediately upstream of the transcribed region where, by analogy, sequences involved in transcription and its regulation should be located. Putative regulatory sequences were identified by comparison between different genes and the conclusions reached in this way confirmed either by destroying these sequences by deletion or mutation or by transferring them to another gene in an attempt to alter its pattern of regulation.

This work carried out on a number of different genes encoding specific proteins identified many short sequence elements involved in transcriptional control (for reviews see Davidson *et al.*, 1983; Jones *et al.*, 1988). The elements of this type present in two typical examples, the human gene encoding the 70 kd heat-inducible (heat shock) protein (Williams *et al.*, 1989) and the human metallothionein IIA gene (Lee *et al.*, 1987) are illustrated in Figure 1.7.

Comparisons of these and many other genes revealed that, as in bacteria, their upstream regions contain two types of elements: firstly, sequences found in very many genes exhibiting distinct patterns of regulation which are likely to be involved in the basic process of transcription

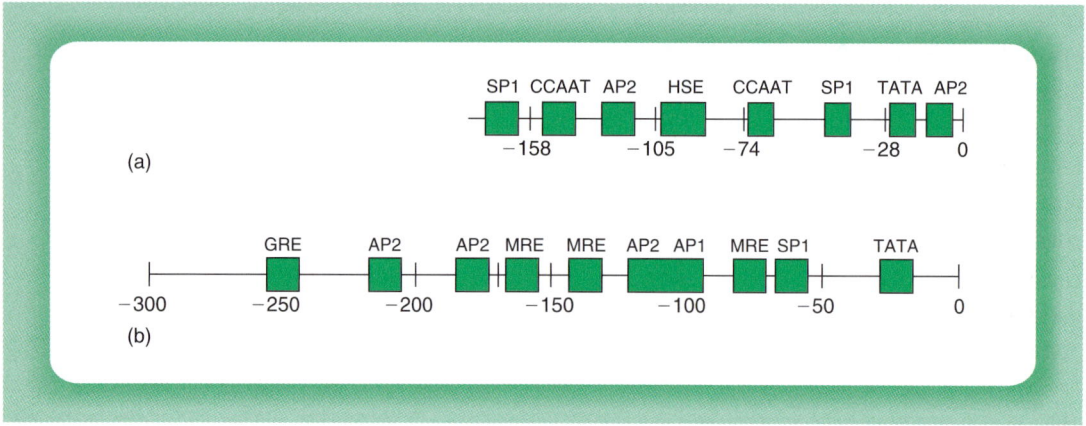

Figure 1.7

Transcriptional control elements upstream of the transcriptional start site in the human genes encoding hsp70 (panel a) and methallothionein IIA (panel b). The TATA, Sp1 and CCAAT boxes bind factors which are involved in constitutive transcription whilst the glucocorticoid response element (GRE), metal response element (MRE), heat shock element (HSE) and the AP1 and AP2 sites bind factors involved in the induction of gene expression in response to specific stimuli.

itself and secondly, those found only in genes transcribed in a particular tissue or in response to a specific signal which are likely to produce this specific pattern of expression. These will be discussed in turn.

1.3.2 SEQUENCES INVOLVED IN THE BASIC PROCESS OF TRANSCRIPTION

Although they are regulated very differently, the hsp70 and metallothionein genes both contain a TATA box. This is an AT rich sequence (consensus TATAA/TAA/T) which is found about thirty base pairs upstream of the transcriptional start site in very many but not all genes. Mutagenesis or relocation of this sequence has shown that it plays an essential role in accurately positioning the start site of transcription (Breathnach and Chambon, 1981). The region of the gene bracketed by the TATA box and the site of transcriptional initiation (the Cap site) has been operationally defined as the gene promoter or core promoter (Goodwin *et al.*, 1990). It is likely that this region binds several proteins essential for transcription, as well as RNA polymerase II itself, which is the enzyme responsible for transcribing protein coding genes.

Although the TATA box is found in most eukaryotic genes, it is absent in some genes, notably housekeeping genes expressed in all tissues and in some tissue specific genes (for reviews of the different classes of core promoters see Smale, 2001; Butler and Kadonaga, 2002). In these promoters, a sequence known as the initiator element which is located over the start site of transcription itself appears to play a critical role in determining the initiation point and acts as a minimal promoter capable of producing basal levels of transcription (see Chapter 3, section 3.6 for a discussion of transcription from promoters containing or lacking a TATA box).

In promoters that contain a TATA box and in those that lack it, the very low activity of the promoter itself is dramatically increased by other elements located upstream of the promoter. These elements are found in a very wide variety of genes with different patterns of expression indicating that they play a role in stimulating the constitutive activity of promoters. Thus inspection of the hsp70 and metallothionein IIA genes reveals that both contain one or more copies of a GC rich sequence (known as the Sp1 box), which is found upstream of the promoter in many genes both with and without TATA boxes (for review see Lania *et al.*, 1997).

In addition, the hsp70 promoter but not the metallothionein promoter contains another sequence, the CCAAT box, which is also found in very many genes with disparate patterns of regulation. Both the CCAAT box and the Sp1 box are typically found upstream of the TATA box, as in the metallothionein and hsp70 genes. Some genes, as in the case of hsp70, may have both of these elements whereas others such as the metallothionein gene have single or multiple copies of one or the other. In every case, however, these elements are essential for transcription of the genes and their elimination by deletion or mutation abolishes transcription. Hence these sequences play an essential role in efficient transcription of the gene and have been termed upstream promoter elements (UPE: Goodwin *et al.*, 1990).

1.3.3 SEQUENCES INVOLVED IN REGULATED TRANSCRIPTION

Inspection of the hsp70 promoter (Fig. 1.7) reveals several other sequence elements that are only shared with a much more limited number of other genes and which are interdigitated with the upstream promoter elements discussed above. Indeed, one of these, which is located approximately ninety bases upstream of the transcriptional start site, is shared only with other heat shock genes whose transcription is increased in response to elevated temperature. This suggests that this heat shock element may be essential for the regulated transcription of the hsp70 gene in response to heat.

To directly prove this, however, it is necessary to transfer this sequence to a non-heat-inducible gene and show that this transfer renders the recipient gene heat inducible. Pelham (1982) successfully achieved this by linking the heat shock element to the non-heat-inducible thymidine kinase gene of the eukaryotic virus herpes simplex. This hybrid gene could be activated following its introduction into mammalian cells by raising the temperature (Fig. 1.8). Hence the heat shock element can confer heat inducibility on another gene, directly proving that its presence in the hsp gene promoters is responsible for their heat inducibility.

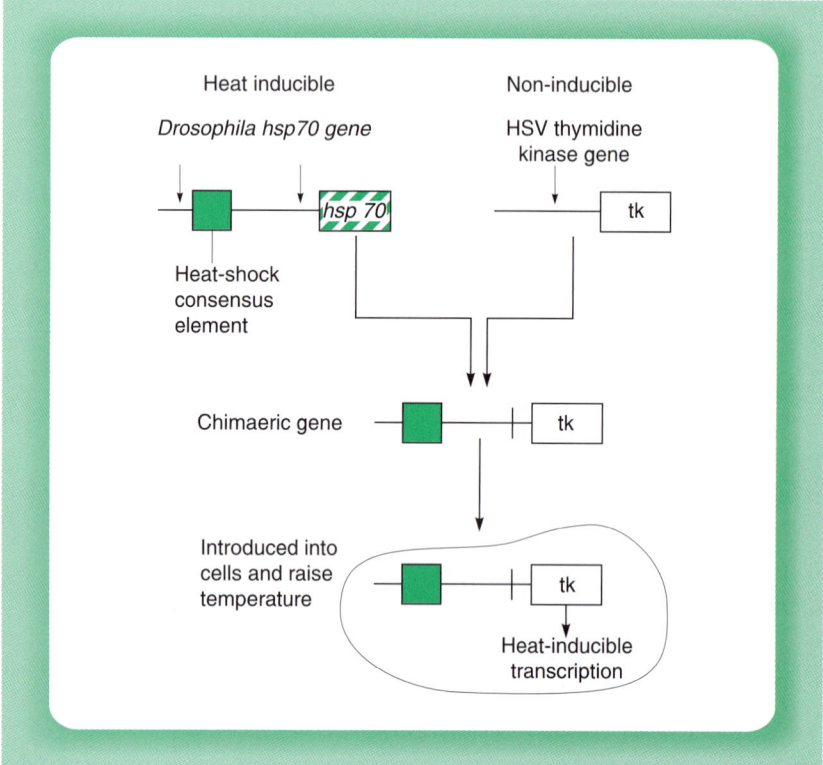

Figure 1.8
Demonstration that the heat shock element mediates heat inducibility. Transfer of this sequence to a gene (thymidine kinase) which is not normally inducible renders this gene heat inducible.

Moreover, although these experiments used a heat shock element taken from the hsp70 gene of the fruit fly *Drosophila melanogaster*, the hybrid gene was introduced into mammalian cells. Not only does the successful functioning of the fly element in mammalian cells indicate that this process is evolutionarily conserved but it permits a further conclusion about the way in which the effect operates. Thus in the cold blooded *Drosophila*, 37°C represents a thermally stressful temperature and the heat shock response would normally be active at this temperature. The hybrid gene was inactive at 37°C in the mammalian cells, however, and was only induced at 42°C, the heat shock temperature characteristic of the cell into which it was introduced. Hence this sequence does not act as a thermostat, set to go off at a particular temperature since this would occur at the *Drosophila* heat shock temperature (Fig. 1.9a). Rather, this sequence must act by being recognized by a cellular protein which, in mammalian cells, is activated only at an elevated temperature characteristic of the mammalian cell heat shock response (Fig. 1.9b).

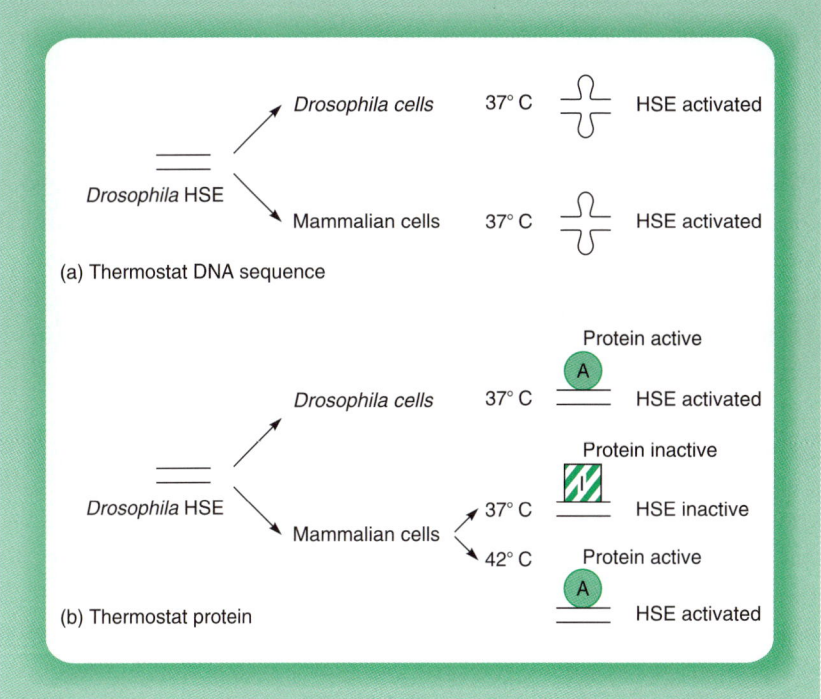

Figure 1.9

Predicted effects of placing the *Drosophila* heat shock element in a mammalian cell if the element acts as a thermostat detecting elevated temperature directly (panel a) or if it acts by binding a protein which is activated by elevated temperature (panel b). Note that only possibility b can account for the observation that the *Drosophila* heat shock element only activates transcription in mammalian cells at the mammalian heat shock temperature of 42°C and not at the *Drosophila* heat shock temperature of 37°C.

This experiment therefore not only directly proves the importance of the heat shock element in producing the heat inducibility of the hsp70 gene but also shows that this sequence acts by binding a cellular protein which is activated in response to elevated temperature. The binding of this transcription factor then activates transcription of the hsp70 gene. The manner in which this factor activates transcription of the hsp70 gene and the other heat shock genes is discussed further in Chapter 8 (section 8.3.1).

The presence of specific DNA sequences that can bind particular proteins will therefore confer on a specific gene the ability to respond to particular stimuli. Thus the lack of a heat shock element in the metallothionein IIA gene (Fig. 1.7) means that this gene is not heat inducible. In contrast however, this gene, unlike the hsp70 gene, contains a glucocorticoid response element (GRE). Hence it can bind the complex of the glucocorticoid receptor and the hormone itself which forms following treatment of cells with glucocorticoid. Its transcription is therefore activated in response to glucocorticoid, whereas that of the hsp70 gene is not (see Chapter 4, section 4.4). Similarly, only the metallothionein gene contains metal response elements (MRE) allowing it to be activated in response to treatment with heavy metals such as zinc and cadmium (Thiele *et al.*, 1992). In contrast, both genes contain binding sites for the transcription factor AP2 which mediates gene activation in response to cyclic AMP and phorbol esters.

Similar DNA sequence elements in the promoters of tissue specific genes play a critical role in producing their tissue specific pattern of expression by binding transcription factors that are present in an active form only in a particular tissue where the gene will be activated. For example, the promoters of the immunoglobulin heavy and light chain genes contain a sequence known as the octamer motif (ATGCAAAT) which can confer B cell specific expression on an unrelated promoter (Wirth *et al.*, 1987). Similarly, the related sequence ATGAATAA/T is found in genes expressed specifically in the anterior pituitary gland such as the prolactin gene and the growth hormone gene and binds a transcription factor known as Pit-1 which is expressed only in the anterior pituitary (for review see Andersen and Rosenfeld, 1994). If this short sequence is inserted upstream of a promoter, the gene is expressed only in pituitary cells. In contrast, the octamer motif, which differs by only two bases, will direct expression only in B cells when inserted upstream of the same promoter (Elsholtz *et al.*, 1990; Fig. 1.10). Hence small differences in control element sequences can produce radically different patterns of gene expression.

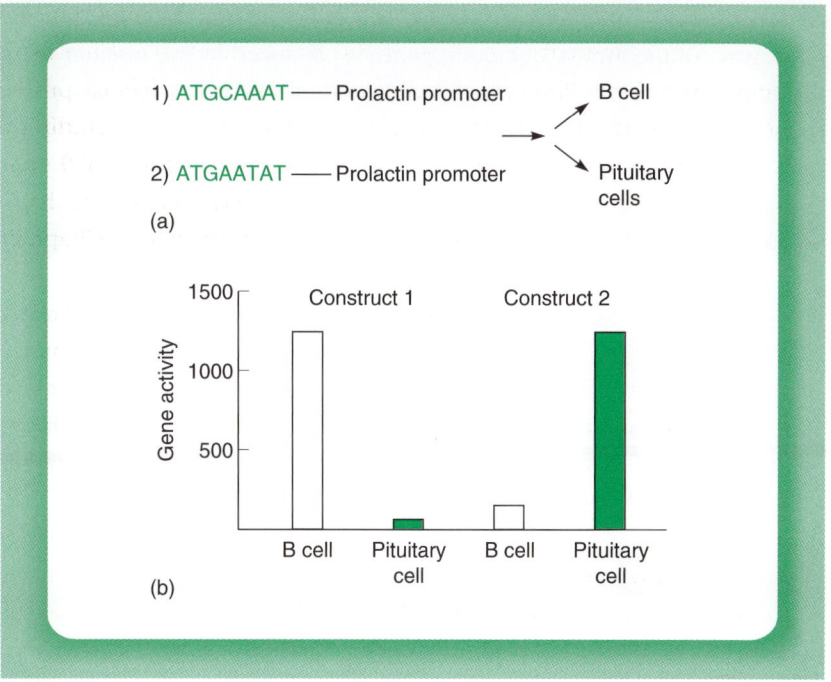

Figure 1.10

Linkage of the octamer binding motif ATGCAAAT (1) and the related Pit-1 binding motif ATGAATAT (2) to the prolactin promoter and introduction into B cells and pituitary cells (panel a). Only the octamer-containing construct 1 directs a high level of activity in B cells, whereas only construct 2 containing the Pit-1 binding site directs a high level of gene activity in pituitary cells (panel b). Data from Elsholtz *et al.* (1990).

1.3.4 SEQUENCES WHICH ACT AT A DISTANCE

(a) Enhancers

One of the characteristic features of eukaryotic gene expression is the existence of sequence elements located at great distances from the start site of transcription, which can influence the level of gene expression. These elements can be located upstream, downstream or within a transcription unit and function in either orientation relative to the start site of transcription (Fig. 1.11). They act by increasing the activity of a promoter, although they lack promoter activity themselves and are hence referred to as enhancers (for reviews see Hatzopoulos *et al.*, 1988; Muller *et al.*, 1988; Pennisi, 2004; Szutorisz *et al.*, 2005; Halfon, 2006). Some enhancers are active in all tissues and increase the activity of a promoter in all cell types whilst others function as tissue specific enhancers which activate a particular promoter only in a specific cell type. Thus the enhancer located in the intervening region of the immunoglobulin genes is active only in B cells and the B cell specific expression of the immunoglobulin genes is produced by the interaction of this enhancer and the immunoglobulin promoter which, as we have previously seen, is also B cell specific (Garcia *et al.*, 1986).

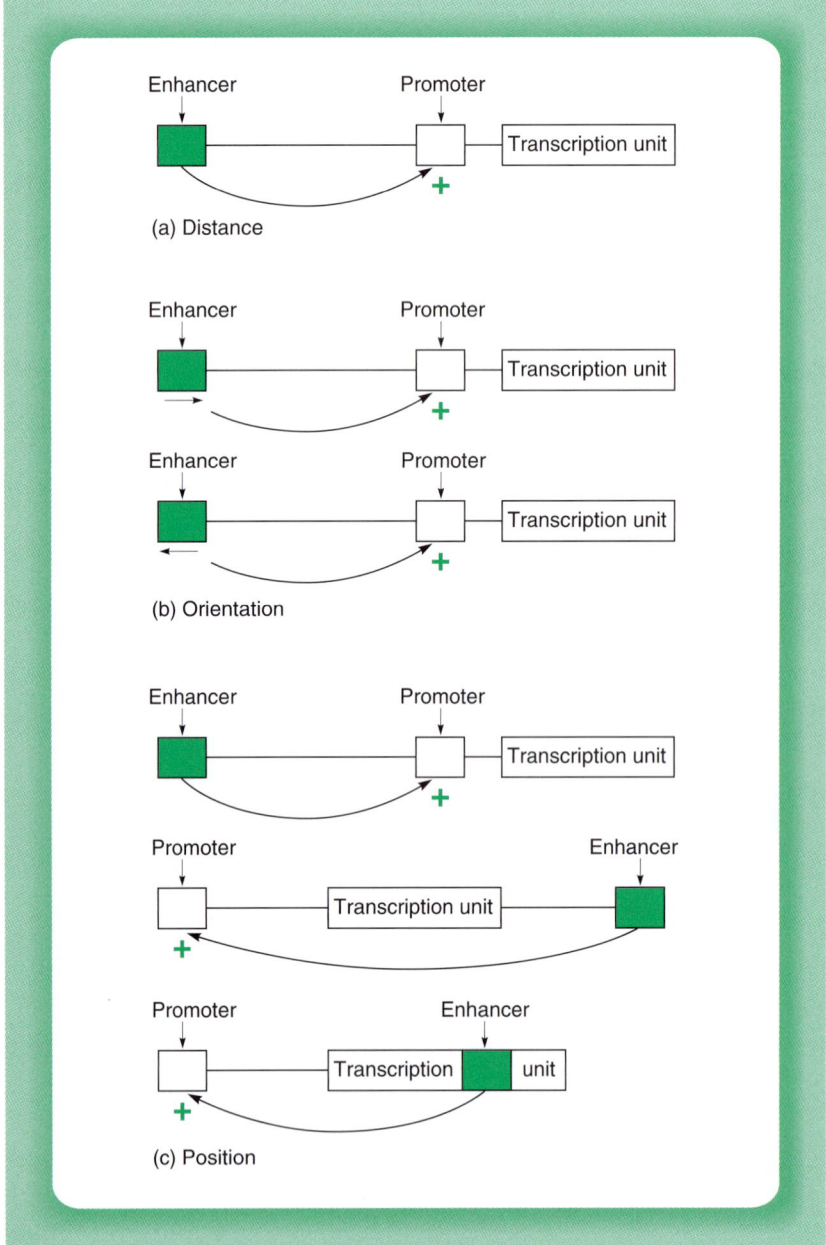

Figure 1.11

Characteristics of an enhancer element which can activate a promoter at a distance (a); in either orientation relative to the promoter (b), and when positioned upstream, downstream, or within a transcription unit (c).

As with promoter elements, enhancers contain multiple binding sites for transcription factors which interact together to mediate enhancer function. In many cases the elements within enhancers are identical to those contained immediately upstream of gene promoters. Thus, the immunoglobulin heavy chain enhancer contains a copy of the octamer sequence (Sen and Baltimore, 1986), which is also found in the immunoglobulin promoters (section 1.3.3). Similarly, multiple copies of the heat shock element are located far upstream of the start site in the *Xenopus* hsp70 gene and function as a heat-inducible enhancer when transferred to another gene (Bienz and Pelham, 1986).

Enhancers therefore consist of sequence elements which are also present in similarly regulated promoters and may be found within the enhancer associated with other control elements and frequently in multiple copies. This has led to the idea that a multi-protein complex known as the enhanceosome assembles on the enhancer and induces transcriptional activation of the target gene. This complex can then recruit proteins such as histone acetylases and the SWI/SNF complex which open up the chromatin (section 1.2) and thereby allow the subsequent binding of activating transcription factors (see Chapter 5, section 5.7) (for reviews see Cosma, 2002; Fry and Peterson, 2002).

One of the roles of the proteins that bind to an enhancer may be to promote looping of the DNA within chromatin (see sections 1.2.1 and 1.2.2). This can result in regulatory proteins bound at the enhancer, interacting with those at the promoter, resulting in stimulation of transcription (for review see Li *et al.*, 2006) (Fig. 1.12).

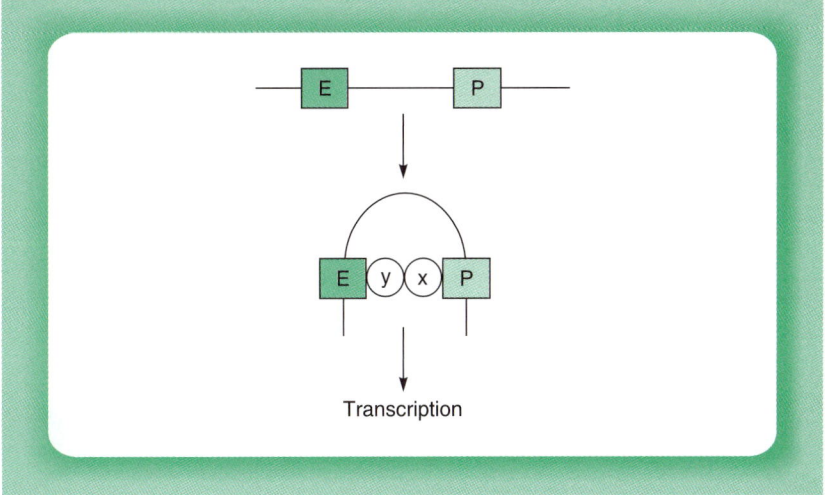

Figure 1.12

Looping of the DNA can bring together regulatory proteins (y) based at an enhancer (E) with proteins (x) bound at the promoter (P).

(b) Locus control regions

The genes encoding the β-globin component of haemoglobin and other related molecules are found clustered together in the genome with five functional genes located adjacent to one another. All of these genes are expressed in erythroid (red blood cell) precursors and not in other cell types and this pattern of expression is dependent on an element located 10–20 kilo-bases upstream of the gene cluster which is known as a locus control region (LCR) (Fig. 1.13). In the absence of this element, none of the genes is expressed in the correct erythroid specific manner (for reviews of LCRs see Bulger and Groudine, 1999; Li *et al.*, 1999; Dean, 2006).

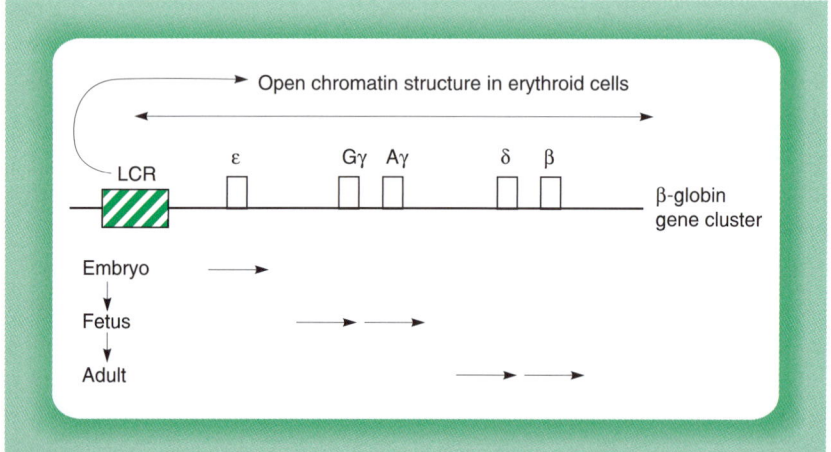

Figure 1.13

The locus control region (LCR) in the β-globin gene cluster directs the correct pattern of chromatin opening in erythroid cells. Regulatory processes acting on each gene in the cluster then allows it to be expressed at the correct time in erythroid development with the ε-globin gene being expressed in the early embryo, the Gγ and Aγ-globin genes in the fetus and the β and δ-globin genes in the adult.

It is likely that the LCR functions by regulating chromatin structure so that the entire region of the genome containing β-globin-like genes is opened up in red blood cell precursors. Each of the genes within the region can then be individually regulated in the red blood cell lineage by their own individual enhancer and promoter elements with, for example, the epsilon globin gene being expressed in the embryo and the β and δ-globin genes in the adult (Fig. 1.13).

Since its original identification in the β-globin locus, LCRs have been found regulating the expression of a number of other gene clusters expressed in different cell types. Interestingly, in several cases, LCRs contain matrix attachment regions (see section 1.2.1). This suggests that a region controlled by an LCR, such as the β-globin cluster, may form a single large loop attached to the nuclear matrix whose chromatin structure is regulated as a single unit.

The typical eukaryotic gene will therefore consist of multiple distinct transcriptional control elements (Fig. 1.14). These are, firstly, the promoter itself, secondly, upstream promoter elements (UPE) located close to it which are required for efficient transcription in any cell type, thirdly, other elements adjacent to the promoter which are interdigitated with the UPEs and which activate the gene in particular tissues or in response to particular stimuli, and lastly, elements such as enhancers or locus control regions which act at a distance to regulate gene expression.

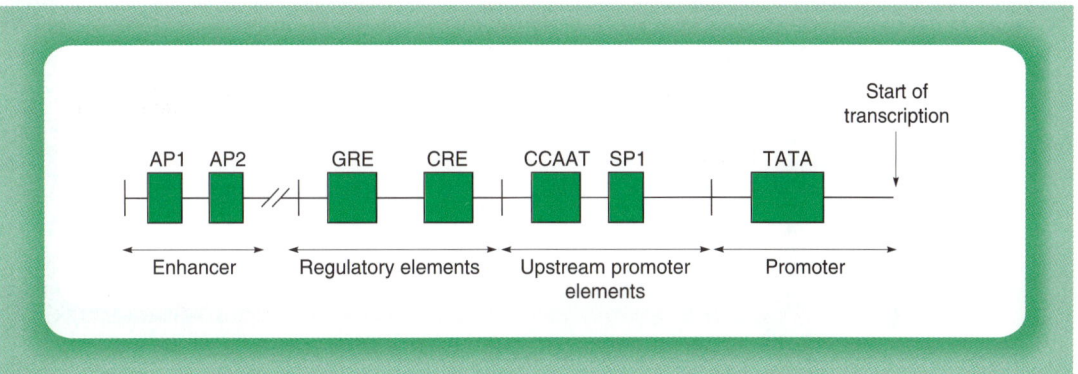

Figure 1.14
Structure of a typical gene with a TATA box-containing promoter, upstream promoter elements such as the CCAAT and Sp1 boxes, regulatory elements inducing expression in response to treatment with substances such as glucocorticoid (GRE) and cyclic AMP (CRE) and other elements within more distant enhancers. Note that as discussed in the text and illustrated in Figure1.6, the upstream promoter elements are often interdigitated with the regulatory elements whilst the same regulatory elements can be found upstream of the promoter and in enhancers.

Such sequences often act by binding positively acting factors that then stimulate transcription (Fig. 1.15a). As will be discussed in later chapters, this could involve the DNA binding protein either altering chromatin structure to make the DNA more accessible to other positively acting regulatory factors or direct stimulation of transcription by the DNA binding protein interacting with RNA polymerase or its associated molecules. Interestingly, however, although most sequences act in such a positive way, some sequences do appear to act in a negative manner to inhibit transcription and these are discussed in the next section.

1.3.5 NEGATIVELY ACTING DNA SEQUENCES

(a) Silencers

Silencer elements, which act to inhibit gene transcription, have been defined in a number of genes including the cellular oncogene c-*myc* (Chapter 9, section 9.3.3) and those encoding proteins such as growth hormone or collagen type II. As with activating sequences, some silencer elements are constitutively active whilst others display cell type specific activity. Thus, for example, the silencer in the gene encoding the T lymphocyte marker CD4 represses its expression in most T cells where CD4 is not expressed but is inactive in a subset of T cells allowing these cells to actively express the CD4 protein (Sawada *et al.*, 1994). In many cases silencer elements have been shown to act by binding regulatory factors which then act to reduce the rate of transcription (Fig. 1.15b), either by promoting a more tightly packed chromatin structure or by interacting with RNA polymerase and its associated molecules in an inhibitory manner.

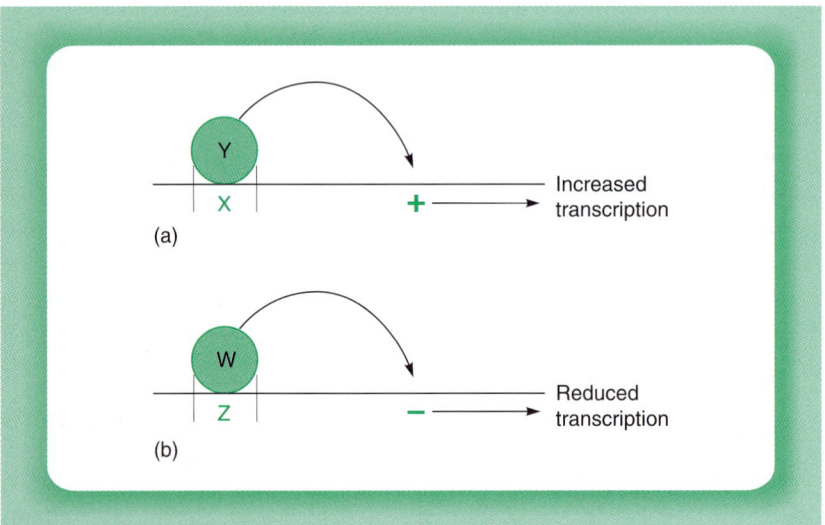

Figure 1.15

Panel (a): A specific DNA sequence (X) can act to stimulate transcription by binding a positively acting factor (Y). Panel (b): In contrast, binding of the negatively acting factor (W) to the DNA sequence Z inhibits transcription.

(b) Insulators

The ability of sequences such as enhancers or LCRs to act over large distances evidently begs the question of how their activity is limited to the genes which they need to regulate and does not affect other genes in adjacent regions. This is achieved by DNA elements known as insulators, which act to block the spread of enhancer or silencer activity (Fig. 1.16) (for reviews see West *et al.*, 2002; Gaszner and Felsenfeld, 2006).

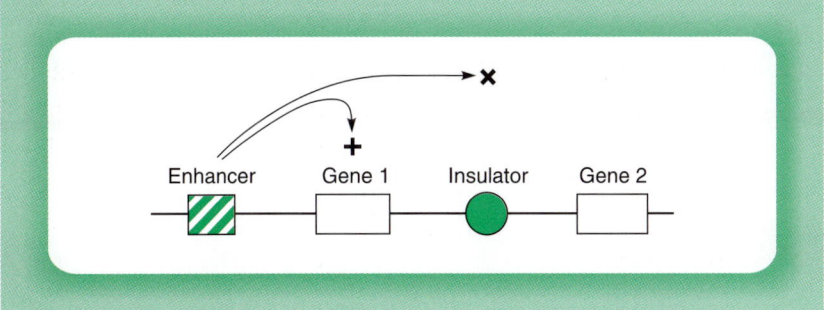

Figure 1.16
An insulator sequence can limit the action of an enhancer to genes located between the enhancer and the insulator.

It is likely that insulators act by blocking the alterations in DNA structure induced by enhancers or silencers. In some cases this involves a direct effect on chromatin structure preventing the opening of chromatin structure induced by enhancers (or the production of a more tightly packed chromatin structure induced by silencers) from spreading to a particular region of chromatin. In other cases an insulator may prevent the looping of DNA, which is required to bring together regulatory proteins bound at the enhancer with their target proteins bound to the promoter (Fig. 1.17).

1.3.6 INTERACTION BETWEEN FACTORS BOUND AT VARIOUS SITES

Obviously the balance between positively and negatively acting transcription factors which bind to the regulatory regions of a particular gene will determine the rate of gene transcription in any particular situation. In some cases binding of the RNA polymerase and associated factors to the promoter and of other positive factors to the UPEs will be sufficient for transcription to occur and the gene will be expressed constitutively. In other cases, however, such interactions will be insufficient and transcription of the gene will occur

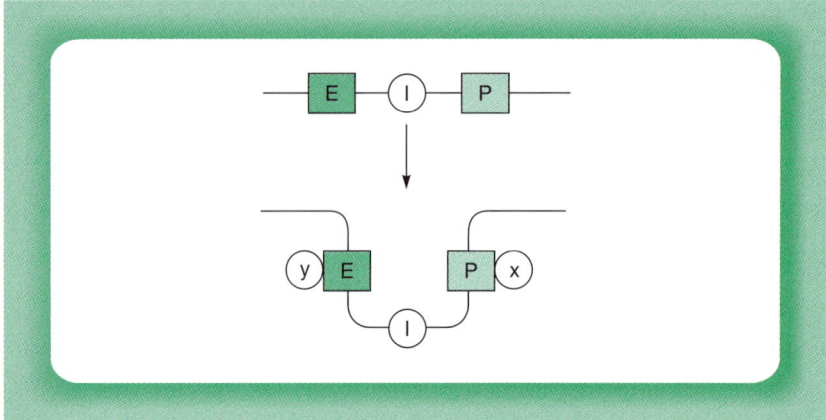

Figure 1.17

An insulator (I) can prevent an enhancer (E) from activating a promoter by altering the looping pattern of the DNA so that proteins (x and y) bound at the promoter and the enhancer cannot interact (compare with Fig. 1.12).

only in response to the binding, to another DNA sequence, of a factor which is activated in response to a particular stimulus or is present only in a particular tissue. These regulatory factors will then interact with the constitutive factors allowing transcription to occur. Hence their binding will result in the observed tissue specific or inducible pattern of gene expression.

Such interaction is well illustrated by the metallothionein IIA gene. As illustrated in Figure 1.7, this gene contains a binding site for the transcription factor AP-1 which produces induction of gene expression in response to phorbol ester treatment. The action of AP-1 on the expression of the metallothionein gene is abolished, however, both by mutations in its binding site and by mutations in the adjacent Sp1 motif which prevent this motif binding its corresponding transcription factor Sp1 (Lee *et al.*, 1987). Although these mutations in the Sp1 motif do not abolish AP1 binding, they do prevent its action, indicating that the inducible AP1 factor interacts with the constitutive Sp1 factor to activate transcription.

Clearly, such interactions between bound transcription factors need not be confined to factors bound to regions adjacent to the promoter but can also involve the similar factors bound to more distant enhancers which are brought together by a looping out of the intervening DNA allowing contact between factors bound at the promoter and those bound at the enhancer (see section 1.3.4).

This need for transcription factors to interact with one another to stimulate transcription means that transcription can also be stimulated by a class of factors which act indirectly by binding to the DNA and bending it so that other DNA bound factors can interact with one another (Fig. 1.18). Thus, the LEF-1 factor which is specifically expressed in T lymphocytes binds to the enhancer of the T cell receptor α gene and bends the DNA so that other constitutively expressed transcription factors can interact with one another thereby allowing them to activate transcription. This results in the T cell specific expression of the gene even though the directly activating factors are not expressed in a T cell specific manner (for review see Werner and Burley, 1997). Similarly, the DNA binding transcription factor HMGI (Y) plays a critical role in the multi-protein enhanceosome which assembles on the interferon β gene enhancer and is essential for the inducibility of this gene following viral infection (for further discussion of the processes involved in the activation of this promoter see Chapter 5, section 5.7).

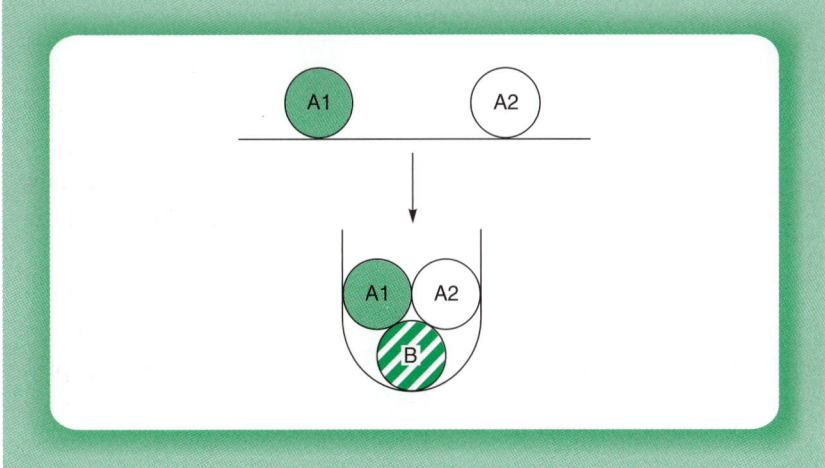

Figure 1.18
A factor which bends the DNA (B) can indirectly activate transcription by facilitating the interaction of two activating transcription factors (A1 and A2).

1.4 CONCLUSIONS

It is clear that both the process of transcription itself and its regulation in particular tissues or in response to particular signals are controlled by short DNA sequence elements located adjacent to the promoter or in enhancers. In turn such sequences act by binding proteins which are either active constitutively or are present in an active form only in a specific tissue or following a specific inducing signal. Such DNA bound transcription factors then interact with each other and the RNA polymerase itself in order to produce constitutive or regulated transcription. The nature of these factors, the manner in which they function and their role in different biological processes form the subject of this book.

REFERENCES

Agalioti, T., Chen, G. and Thanos, D. (2002) Deciphering the transcriptional histone acetylation code for a human gene. Cell 111, 381–392.

Andersen, B. and Rosenfeld, M.G. (1994) Pit-1 determines cell types during the development of the anterior pituitary gland. Journal of Biological Chemistry 269, 29335–29338.

Bell, A.C., West, A.G. and Felsenfeld, G. (2001) Insulators and boundaries: versatile regulatory elements in the eukaryotic genome. Science 291, 447–450.

Bienz, M. and Pelham, H.R.B. (1986) Heat shock regulatory elements function as an inducible enhancer when linked to a heterologous promoter. Cell 45, 753–760.

Breathnach, R. and Chambon, P. (1981) Organization and expression of eukaryotic split genes coding for proteins. Annual Review of Biochemistry 50, 349–383.

Brown, C.E., Lechner, T., Howe, L. and Workman, J.L. (2000) The many HATs of transcription coactivators. Trends in Biochemical Sciences 25, 15–19.

Bulger, M. and Groudine, M. (1999) Looping versus linking: toward a model for long-distance gene activation. Genes and Development 13, 2465–2477.

Butler, J.E.F. and Kadonaga, J.T. (2002) The RNA polymerase II core promoter: a key component in the regulation of gene expression. Genes and Development 16, 2583–2592.

Carey, M. (2005) Chromatin marks and machines, the missing nucleosome is a theme: gene regulation up and downstream. Molecular Cell 17, 323–330.

Carrozza, M.J., Utley, R.T., Workman, J.L. and Cote, J. (2003) The diverse functions of histone acetyltransferase complexes. Trends in Genetics 19, 321–329.

Clayton, A.L., Hazzalin, C.A. and Mahadevan, L.C. (2006) Enhanced histone acetylation and transcription: a dynamic perspective. Molecular Cell 23, 289–296.

Cosma, M.P. (2002) Ordered recruitment: gene-specific mechanism of transcription activation. Molecular Cell 10, 227–236.

Darnell, J.E. (1982) Variety in the level of gene control in eukaryotic cells. Nature 297, 365–371.

Davidson, E.H., Jacobs, H.T. and Britten, R.J. (1983) Very short repeats and co-ordinate induction of genes. Nature 301, 468–470.

Dean, A. (2006) On a chromosome far, far away: LCRs and gene expression. Trends in Genetics 22, 38–45.

Elsholtz, H.P., Albert, V.R., Treacy, M.N. and Rosenfeld, M.G. (1990) A two-base change in a POU factor binding site switches pituitary-specific to lymphoid-specific gene expression. Genes and Development 4, 43–51.

Felsenfeld, G. and Groudine, M. (2003) Controlling the double helix. Nature 421, 448–453.

Fischle, W., Wang, Y. and Allis, C.D. (2003) Binary switches and modification cassettes in histone biology and beyond. Nature 425, 475–479.

Fry, C.J. and Peterson, C.L. (2002) Unlocking the gates to gene expression. Science 295, 1847–1848.

Garcia, J.V., Bich-Thuy, L., Stafford, J. and Queen, C. (1986) Synergism between immunoglobulin enhancers and promoters. Nature 322, 383–385.

Gaszner, M. and Felsenfeld, G. (2006) Insulators: exploiting transcriptional and epigenetic mechanisms. Nature Reviews Genetics 7, 703–713.

Goll, M.G. and Bestor, T.H. (2002) Histone modification and replacement in chromatin activation. Genes and Development 16, 1739–1742.

Goodwin, G.H., Partington, G.A. and Perkins, N.D. (1990) Sequence specific DNA binding proteins involved in gene transcription. In: *Chromosomes: eukaryotic, prokaryotic and viral*, vol. 1 (Adolph, K.W., ed.). CRC Press, Boca Raton, FL, pp. 31–85.

Halfon, M.S. (2006) (Re)modelling the transcriptional enhancer. Nature Genetics 38, 1102–1103.

Hassan, A.H., Neely, K.E. and Workman, J.L. (2001) Histone acetyltransferase complexes stabilise SWI/SNF binding to promoter nucleosomes. Cell 104, 817–827.

Hatzopoulos, A.K., Schlokat, U. and Gruss, P. (1988) Enhancers and other cis-acting sequences In: *Transcription and splicing* (Hames, B.D. and Glover, D.M., eds). IRL Press, New York, pp. 43–96.

Horn, P.J. and Peterson, C.L. (2002) Chromatin higher order folding: wrapping up transcription. Science 297, 1824–1827.

Jaskelioff, M. and Peterson, C.L. (2003) Chromatin and transcription: histones continue to make their marks. Nature Cell Biology 5, 395–399.

Jenuwein, T. and Allis, C.D. (2001) Translating the histone code. Science 293, 1074–1080.

Jones, N.C., Rigby, P.W.J. and Ziff, E.G. (1988) Trans-acting protein factors and the regulation of eukaryotic transcription. Genes and Development 2, 267–281.

Khorasanizadeh, S. (2004) The nucleosome: from genomic organization to genomic regulation. Cell 116, 259–272.

Kornberg, R.D. and Lorch, Y. (1999) Twenty-five years of the nucleosome, fundamental particle of the eukaryote chromosome. Cell 98, 285–294.

Lania, L., Majello, B. and de Luca, P. (1997) Transcriptional regulation by the Sp family proteins. International Journal of Biochemistry and Cell Biology 29, 1313–1323.

Latchman, D.S. (2005) *Gene regulation: a eukaryotic perspective*. Fifth edition. Taylor and Francis, Oxford and New York.

Lee, W., Haslinger, A., Karin, M. and Tjian, R. (1987) Activation of transcription by two factors that bind promoter and enhancer sequences of the human metallothionein gene and SV40. Nature 325, 369–372.

Li, Q., Barkess, G. and Qian, H. (2006) Chromatin looping and the probability of transcription. Trends in Genetics 22, 197–202.

Li, Q., Harju, S. and Peterson, K.R. (1999) Locus control regions coming of age at a decade plus. Trends in Genetics 15, 403–408.

Maniatis, T., Goodboun, S. and Fischer, J.A. (1987) Regulation of inducible and tissue specific gene expression. Science 236, 1237–1245.

Martin, C. and Zhang, Y. (2005) The diverse functions of histone lysine methylation. Nature Reviews Molecular Cell Biology 6, 838–849.

Mata, J., Marguerat, S. and Bahler, J. (2005) Post-transcriptional control of gene expression: a genome-wide perspective. Trends in Biochemical Sciences 30, 506–514.

Millar, C.B. and Grunstein, M. (2006) Genome-wide patterns of histone modifications in yeast. Nature Reviews Molecular Cell Biology 7, 657–666.

Mohd-Sarip, A. and Verrijzer, C.P. (2004) A higher order of silence. Science 306, 1484–1485.

Muller, M.M., Gerster, T. and Schaffner, W. (1988) Enhancer sequences and the regulation of gene transcription. European Journal of Biochemistry 176, 485–495.

Muller-Hill, B.W. (ed.) (1996) *The lac operon: a short history of a genetic paradigm*. de Gruyter, Berlin.

Narlikor, G.J., Fan, H-Y. and Kingston, R.E. (2002) Co-operation between complexes that regulate chromatin structure and transcription. Cell 108, 475–487.

Nevins, J.R. (1983) The pathway of eukaryotic mRNA transcription. Annual Review of Biochemistry 52, 441–446.

Nowak, S.J. and Corces, V.G. (2004) Phosphorylation of histone H3: a balancing act between chromosome condensation and transcriptional activation. Trends in Genetics 20, 214–220.

Pelham, H.R.B. (1982) A regulatory upstream promoter element in the Drosophila hsp70 heat-shock gene. Cell 30, 517–528.

Pennisi, E. (2004) Searching for the genome's second code. Science 306, 632–634.

Roh, T-Y., Cuddapah, S. and Zhao, K. (2005) Active chromatin domains are defined by acetylation islands revealed by genome-wide mapping. Genes and Development 19, 542–552.

Saha, A., Wittmeyer, J. and Cairns, B.R. (2006) Chromatin remodelling: the industrial revolution of DNA around histones. Nature Reviews Molecular Cell Biology 7, 437–447.

Sawada, S., Scarborough, J.D., Kileen, N. and Littman, D.R. (1994) A lineage-specific transcriptional silencer regulates CD4 gene expression during T lymphocyte development. Cell 77, 917–929.

Sen, R. and Baltimore, D. (1986) Multiple nuclear factors interact with the immunoglobulin enhancer sequences. Cell 46, 705–716.

Serna, I., de la, Ohkawa, Y. and Imbalzano, A.N. (2006) Chromatin remodelling in mammalian differentiation: lessons from ATP-dependent remodellers. Nature Reviews Genetics 7, 461–473.

Simon, J. (1995) Locking in stable states of gene expression: transcriptional control during Drosophila development. Current Opinion in Cell Biology 7, 376–385.

Smale, S.T. (2001) Core promoters: active contributors to combinatorial gene regulation. Genes and Development 15, 2503–2508.

Staudt, L.H., Singh, H., Sen, R., Wirth, T., Sharp, P.A. and Baltimore, D. (1986) A lymphoid-specific protein binding to the octamer motif of immunoglobulin genes. Nature 323, 640–643.

Stetefeld, J. and Ruegg, M.A. (2005) Structural and functional diversity generated by alternative mRNA splicing. Trends in Biochemical Sciences 30, 515–521.

Szutorisz, H., Dillon, N. and Tora, L. (2005) The role of enhancers as centres for general transcription factor recruitment. Trends in Biochemical Sciences 30, 593–599.

Thiele, D.J. (1992) Metal regulated transcription in eukaryotes. Nucleic Acids Research 20, 1183–1191.

Turner, B.M. (2007) Defining an epigenetic code. Nature Cell Biology 9, 2–6.

Wang, J. and Manley, J.L. (1997) Regulation of pre-mRNA splicing in metazoa. Current Opinion in Genetics and Development 7, 205–211.

Werner, M.H. and Burley, S.K. (1997) Architectural transcription factors: proteins that remodel DNA. Cell 88, 733–736.

West, A.G., Gaszner, M. and Felsenfeld, G. (2002) Insulators: many functions, many mechanisms. Genes and Development 16, 271–288.

Williams, G.T., McClanahan, T.K. and Morimoto, R.I. (1989) E1a transactivation of the human hsp70 promoter is mediated through the basal transcriptional complex. Molecular and Cellular Biology 9, 2574–2587.

Wirth, T., Staudt, L. and Baltimore, D. (1987) An octamer oligonucleotide upstream of a TATA motif is sufficient for lymphoid specific promoter activity. Nature 329, 174–178.

Wolffe, A. (1995) *Chromatin: structure and function.* Second edition. Academic Press, London and San Diego.

Yasui, D., Miyano, M., Cal, S., Varga-Weisz, P. and Kohwi-Shigematsu, T. (2002) SATB1 targets chromatin remodelling to regulate genes over long distances. Nature 419, 641–645.

CHAPTER 2

METHODS FOR STUDYING TRANSCRIPTION FACTORS

2.1 INTRODUCTION

The explosion in the available information on transcription factors which has occurred in recent years has arisen primarily because of the availability of new or improved methods for studying these factors. Initially such studies may focus on identifying a factor which interacts with a particular DNA sequence, and characterizing this interaction and the methods for doing this are discussed in section 2.2. Subsequently, the protein identified in this way is further characterized and purified and its corresponding gene isolated. The methods involved in the purification and/or cloning of transcription factors are considered in section 2.3. Subsequently, section 2.4 analyses the methods used to characterize such cloned transcription factors, including the methods for determining the DNA binding site or gene targets of a transcription factor which is initially identified by means other than its DNA binding characteristics. (For details of the methodologies involved see Latchman, 1999.)

2.2 METHODS FOR STUDYING DNA–PROTEIN INTERACTIONS

2.2.1 DNA MOBILITY SHIFT ASSAY

As discussed in Chapter 1 (section 1.3), the initial stimulus to identify a transcription factor frequently comes from the identification of a particular DNA sequence that confers a specific pattern of expression on a gene which carries it. The next step therefore, following the identification of such a sequence, will be to define the protein factors which bind to it. This can be readily achieved by the DNA mobility shift or gel retardation assay (Fried and Crothers, 1981; Garner and Revzin, 1981).

This method relies on the obvious principle that a fragment of DNA to which a protein has bound will move more slowly in gel electrophoresis than the same DNA fragment without bound protein. The DNA mobility shift assay is carried out therefore by first radioactively labelling the specific DNA sequence whose protein binding properties are being investigated. The labelled DNA is then incubated with a nuclear (Dignam *et al.*, 1983) or whole cell (Manley *et al.*, 1980) extract of cells prepared in such a way as to contain the DNA binding proteins. In this way DNA–protein complexes are allowed to form. The complexes are then electrophoresed on a non-denaturing polyacrylamide gel and the position of the radioactive DNA visualized by autoradiography. If no protein has bound to the DNA, all the radioactive label will be at the bottom of the gel, whereas if a protein–DNA complex has formed, radioactive DNA to which the protein has bound will migrate more slowly and hence will be visualized near the top of the gel (Fig. 2.1). (For methodological details see Smith *et al.*, 1999.)

This technique can be used therefore to identify proteins which can bind to a particular DNA sequence in extracts prepared from specific cell types. Thus, for example, in the case of the octamer sequence discussed in Chapter 1 (section 1.3.3), a single retarded band is detected when this sequence is mixed, for example, with a fibroblast extract. In contrast, when an extract from immunoglobulin producing B cells is used, two distinct retarded bands are seen (Fig. 2.2). Since each band is produced by a distinct protein binding to the DNA, this indicates that in addition to the ubiquitous octamer binding protein Oct-1 which is present in most cell types, B cells also contain an additional octamer binding protein, Oct-2, which is absent in many other cells.

As well as defining the proteins binding to a particular sequence the DNA mobility shift assay can also be used to investigate the precise sequence specificity of this binding. This can be done by including in the binding reaction, a large excess of a second DNA sequence which has not been labelled. If this DNA sequence can also bind the protein bound by the labelled DNA, it will do so. Moreover, binding to the unlabelled DNA will predominate since it is present in large excess. Hence the retarded band will not appear in the presence of the unlabelled competitor, since only protein-DNA complexes containing labelled DNA are visualized on autoradiography (Fig. 2.3b). In contrast, if the competitor cannot bind the same sequence as the labelled DNA, the complex with the labelled DNA will form and the labelled band will be visualized as before (Fig. 2.3c).

Thus by using competitor DNAs which contain the binding sites for previously described transcription factors, it can be established whether the protein detected in a particular mobility shift experiment is identical or related to any of these factors. Similarly, if competitor DNAs are used which differ in only one or a few bases from the original binding site, the

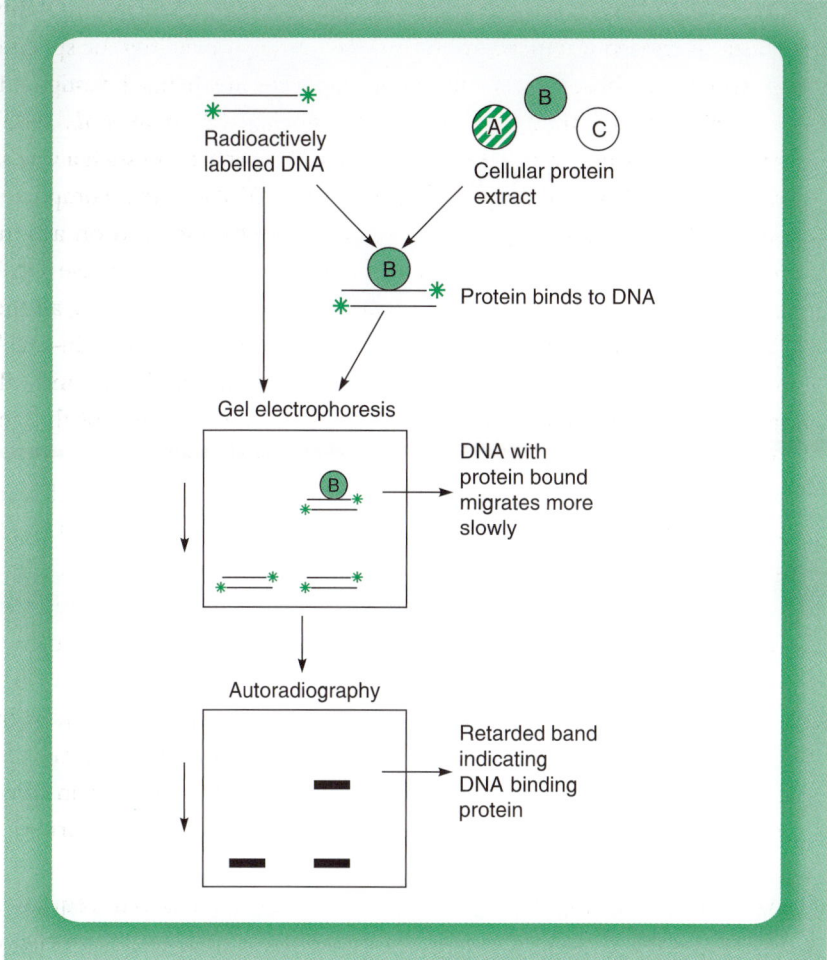

Figure 2.1

DNA mobility shift or gel retardation assay. Binding of a cellular protein (B) to the radioactively labelled DNA causes it to move more slowly upon gel electrophoresis and hence results in the appearance of a retarded band upon autoradiography to detect the radioactive label.

effect of such base changes on the efficiency of the competitor DNA and hence on binding of the transcription factor can be assessed. Figure 2.2 illustrates an example of this type of competition approach, showing that the octamer binding proteins Oct-1 and Oct-2 are efficiently competed away from the labelled octamer probe by an excess of identical unla-belled competitor but not by a competitor containing three base changes in this sequence which prevent binding (ATGCAAAT to ATAATAAT). Similarly, no competition is observed, as expected, when the binding site of an unrelated transcription factor Sp1 is used as the competitor DNA.

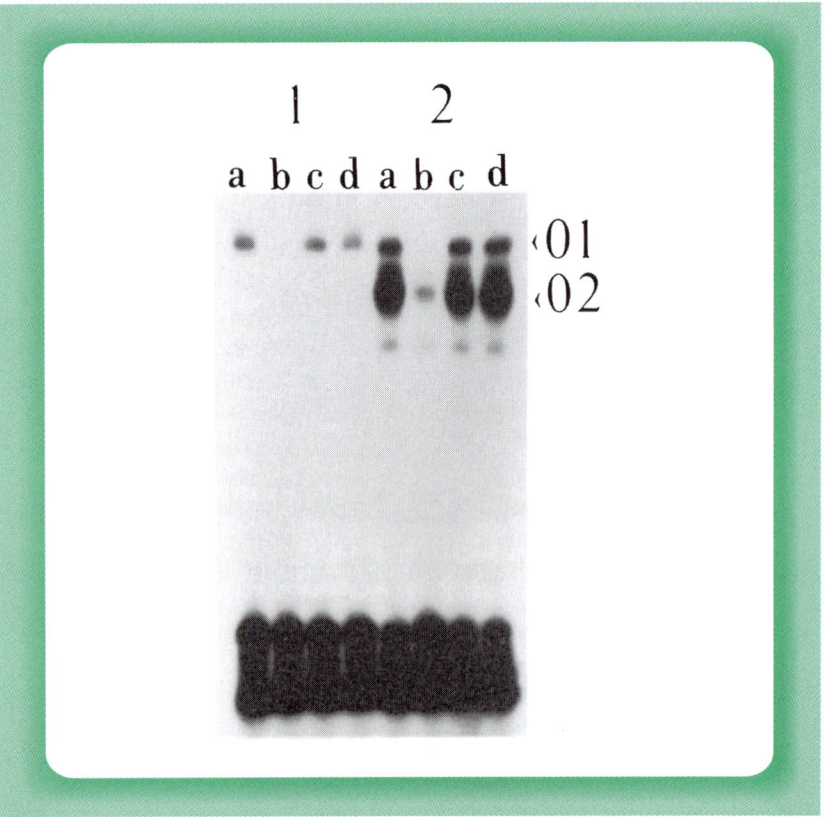

Figure 2.2

DNA mobility shift assay using a radioactively labelled probe containing the binding site for octamer binding proteins (ATGCAAAT) and extracts prepared from fibroblast cells (1) or B cells (2). Note that fibroblast cells contain only one protein Oct-1 (01) capable of producing a retarded band whereas B cells contain both Oct-1 and an additional tissue-specific protein Oct-2 (02). The complexes formed by Oct-1 and Oct-2 on the labelled oligonucleotide in the absence of unlabelled oligonucleotide (track a) are readily removed by a one hundred-fold excess of unlabelled octamer oligonucleotide (track b). They are not removed, however, by a similar excess of a mutant octamer oligonucleotide (ATAATAAT) which is known not to bind octamer binding proteins (track c) or of the binding site for the unrelated transcription factor Sp1 (track d: Dynan and Tjian, 1983). This indicates that the retarded bands are produced by sequence specific DNA binding proteins which bind specifically to the octamer motif and not to mutant or unrelated motifs.

The DNA mobility shift assay therefore provides an excellent means of initially identifying a particular factor binding to a specific sequence and characterizing both its tissue distribution and its sequence specificity.

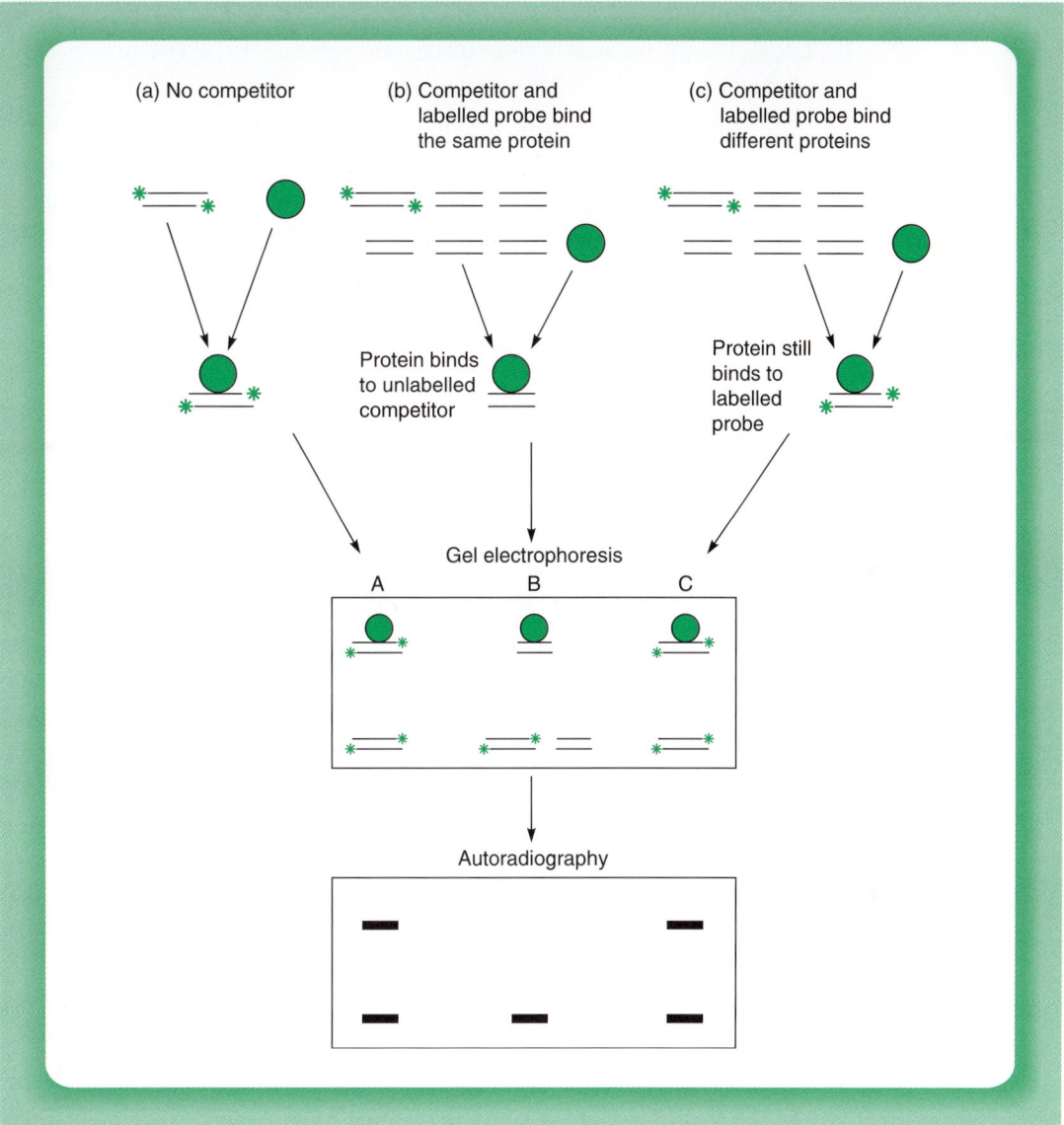

Figure 2.3

Use of unlabelled competitor DNAs in the DNA mobility shift assay. If an unlabelled DNA sequence is capable of binding the same protein as is bound by the labelled probe, it will do so (B) and the radioactive retarded band will not be observed, whereas if it cannot bind the same protein (C), the radioactive retarded band will form exactly as in the absence of competitor (A).

2.2.2 DNAₛₑI FOOTPRINTING ASSAY

Although the mobility shift assay provides a means of obtaining information on DNA–protein interaction, it cannot be used to directly localize the area of the contact between protein and DNA. For this purpose, the DNAseI footprint assay is used (Galas and Schmitz, 1978; Dynan and Tjian, 1983).

In this assay, DNA and protein are mixed as before, the DNA being labelled, however, only at the end of one strand of the double stranded molecule. Following binding, the DNA is treated with a small amount of the enzyme deoxyribonuclease I (DNAseI) which will digest DNA. The digestion conditions are chosen, however, so that each molecule of DNA will be cut once or a very few times by the enzyme. Following digestion the bound protein is removed and the DNA fragments separated by electrophoresis on a polyacrylamide gel capable of resolving DNA fragments differing in size by only one base. This produces a ladder of bands representing the products of DNAseI cutting either one or two or three or four etc. bases from the labelled end. Where a particular piece of the DNA has bound a protein, however, it will be protected from digestion and hence the bands corresponding to cleavage at these points will be absent. This will be visualized on electrophoresis as a blank area on the gel lacking labelled fragments and is referred to as the footprint of the protein (Fig. 2.4). Similar labelling of the other strand of the DNA molecule will allow the interaction of the protein with the other strand of the DNA to be assessed.

The footprinting technique therefore allows a visualization of the interaction of a particular factor with a specific piece of DNA. By using a sufficiently large piece of DNA, the binding of different proteins to different DNA sequences within the same fragment can be assessed. An analysis of this type is shown in Figure 2.5. This shows the footprints (A and B) produced by two cellular proteins binding to two distinct sequences within a region of the human immunodeficiency virus (HIV) control element which has an inhibiting effect on promoter activity (Orchard *et al.*, 1990). Interestingly, some insights into the topology of the DNA–protein interaction are also obtained in this experiment since bands adjacent to the protected region appear more intense in the presence of the protein. These regions of hypersensitivity to cutting are likely to represent a change in the structure of the DNA in this region when the protein has bound, rendering the DNA more susceptible to enzyme cleavage.

As with the mobility shift assay, unlabelled competitor sequences can be used to remove a particular footprint and determine its sequence specificity. In the HIV case illustrated in Figure 2.5, short DNA competitors containing the sequence of one or other of the footprinted areas were used to specifically remove each footprint without affecting the other, indicating that two distinct proteins produce the two footprints.

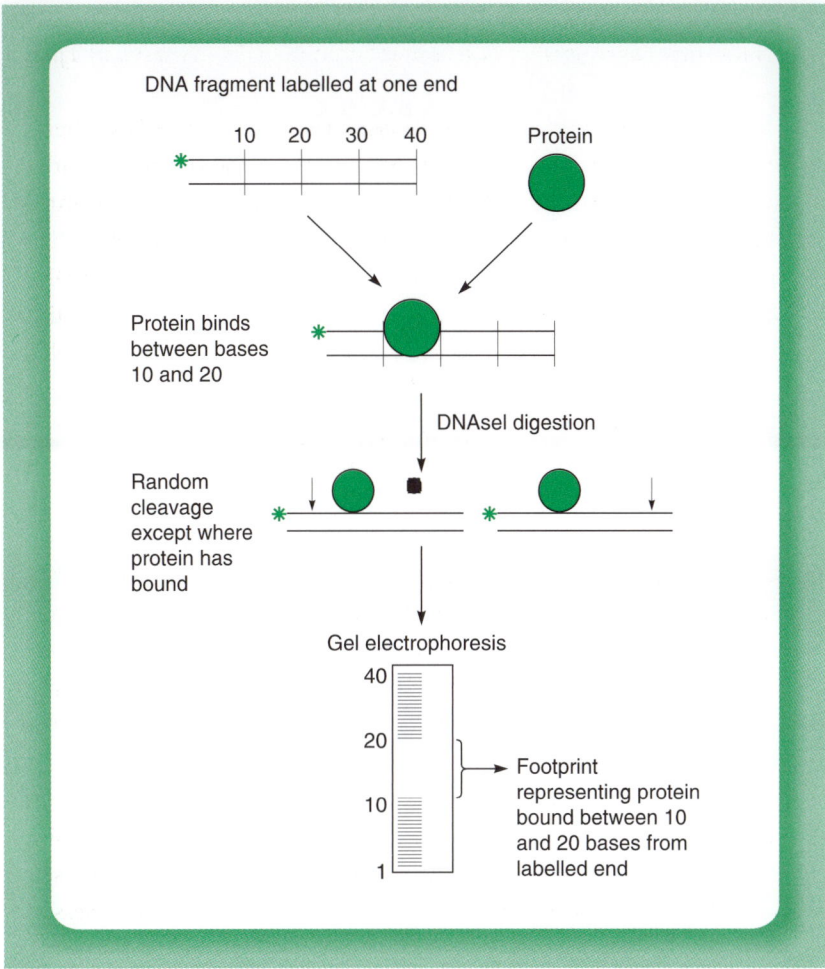

Figure 2.4

DNAseI footprinting assay. If a protein binds at a specific site within a DNA fragment labelled at one end, the region of DNA at which the protein binds will be protected from digestion with DNAseI. Hence this region will appear as a footprint in the ladder of bands produced by the DNA being cut at all other points by DNAseI.

As well as footprinting using DNAseI, other footprinting techniques have been developed which rely on the protection of DNA which has bound protein from cleavage by other reagents that normally cleave the DNA. These include hydroxyl radical footprinting and phenanthroline-copper footprinting which like DNAseI footprinting rely on the ability of the reagents to cleave the DNA in a non-sequence specific manner (for further details see Kreale, 1994; Papavassilou, 1995).

Of greater interest, however, is the technique of dimethyl sulphate (DMS) protection footprinting since it can provide information on the

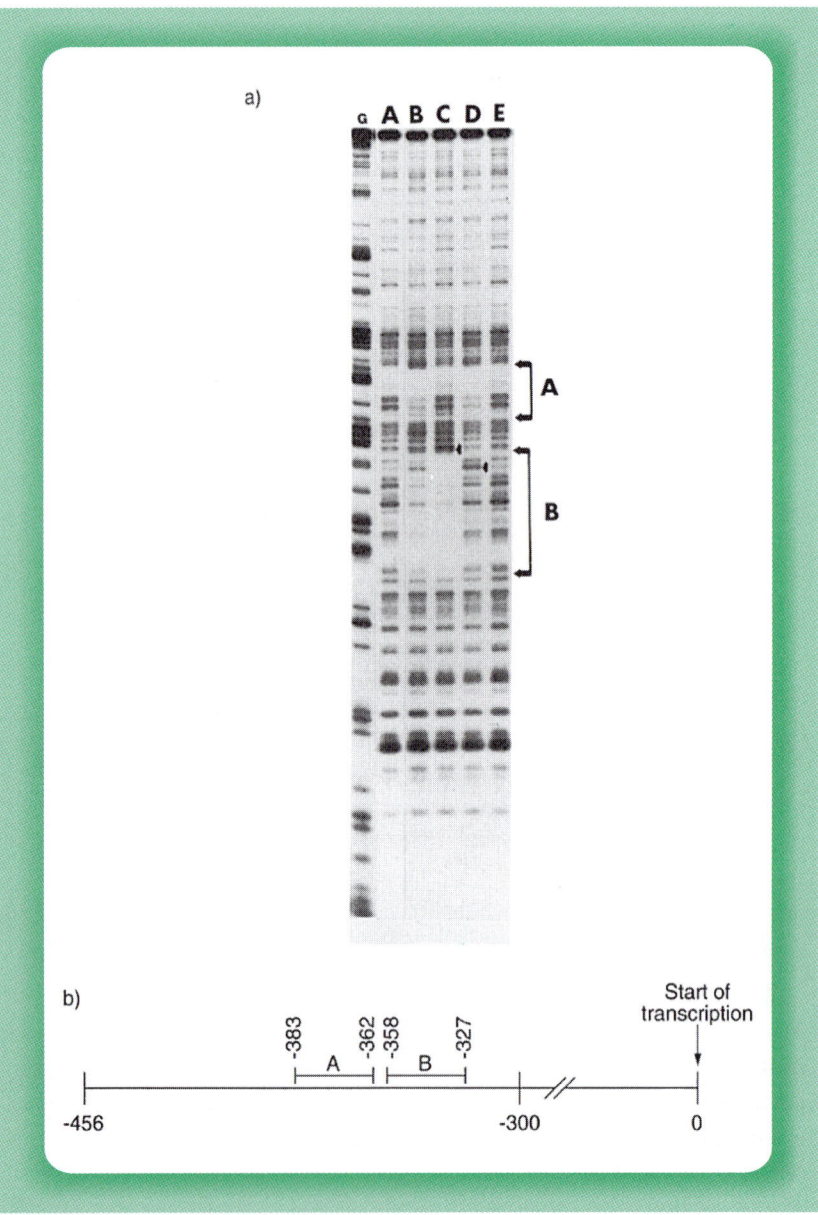

Figure 2.5

Panel (a): DNAseI footprinting assay carried out on a region of the human immunodeficiency virus (HIV) control element. The two footprints (A and B) are not observed when no cell extract is added to the reaction (track A) but are observed when cellular extract is added in the absence of competitor (track B). Addition of unlabelled oligonucleotide competitor containing the DNA sequence of site A removes the site A footprint without affecting site B (track C) whilst an unlabelled oligonucleotide containing the site B DNA sequence has the opposite effect (track D).

exact bases within the binding site which are contacted by the protein. Thus, this method relies on the ability of DMS to specifically methylate guanine residues in the DNA. These methylated G residues can then be cleaved by exposure to piperidine, whereas no cleavage occurs at unmethylated G residues (Maxam and Gilbert, 1980). A protein bound to the DNA will protect the guanine residues which it contacts from methylation and hence they will not be cleaved upon subsequent piperidine treatment. As in the other footprinting techniques therefore, specific bands produced by such treatment of naked DNA are absent in the protein–DNA sample. Unlike the other methods, however, because cleavage occurs at specific guanine residues, this method identifies specific bases within the DNA which are contacted by the transcription factor protein.

These footprinting techniques therefore offer an advance on the mobility shift assay, allowing a more precise visualization of the DNA–protein interaction. (For methodological details see Spiro and McMurray, 1999.)

2.2.3 METHYLATION INTERFERENCE ASSAY

The pattern of DNA–protein interaction can also be studied in more detail using the methylation interference assay (Siebenlist and Gilbert, 1980). Like methylation protection, this method relies on the ability of DMS to methylate G residues which can then be cleaved with piperidine. However, methylation interference is based on assessing whether the prior methylation of specific G residues in the target DNA affects subsequent protein binding. Thus, the target DNA is first partially methylated using DMS so that on average only one G residue per DNA molecule is methylated (Maxam and Gilbert, 1980). Each individual DNA molecule will therefore contain some methylated G residues, with the particular residues which are methylated being different in each molecule. These partially methylated DNAs are then used in a DNA mobility shift experiment with an appropriate cell extract containing the DNA binding protein. Following electrophoresis the band produced by the DNA which has bound protein and that produced by the DNA which has not, are excised

---

Figure 2.5 (Continued)

Both footprints are removed by a mixture of unlabelled site A and B oligonucleotides (track E). Arrows indicate the position of sites at which cleavage with DNAseI is enhanced in the presence of protein bound to an adjacent site, indicating the existence of conformational changes induced by protein binding. The track labelled G represents a marker track consisting of the same DNA fragment chemically cleaved at every guanine residue. Panel (b): Position of sites A and B within the HIV control element. The arrow indicates the start site of transcription.

from the gel and treated with piperidine to cleave the DNA at the methylated G residues and not at unmethylated Gs. Clearly, if methylation of a particular G prevents protein binding then cleavage at this particular methylated G will be observed only in the DNA which failed to bind the protein. Conversely, if a particular G residue plays no role in binding, then cleavage at this G residue will be observed equally in both the DNA which bound the protein and that which failed to do so (Fig. 2.6).

Figure 2.7 shows this type of analysis applied to the protein binding to site B within the negatively acting element in the human immunodeficiency virus promoter (for the footprint produced by the binding of this protein see Fig. 2.5). In this case the footprinted sequence was palindromic (Fig. 2.7), suggesting that the DNA–protein interaction may involve similar binding to the two halves of the palindrome. The methylation interference analysis of site B confirms this by showing that methylation of equivalent G residues in each half of the palindrome interferes with binding of the protein indicating that these residues are critical for binding.

Although the DMS method only studies contacts of the protein with G residues, interference analysis can also be used to study the interaction of DNA binding proteins with A residues in the binding site. This can be done either by methylating all purines to allow study of interference at A and G residues simultaneously (see for example Ares *et al.*, 1987) or by using diethylpyrocarbonate to specifically modify A residues (probably by carboxyethylation), rendering them susceptible to piperidine cleavage (see for example Sturm *et al.*, 1988). These techniques are of particular value when studying sequences such as the octamer motif in which there are relatively few G residues, hence limiting the information which can be obtained by studying interference at G residues alone (Sturm *et al.*, 1987; Baumruker *et al.*, 1988). Chemical interference techniques can therefore be used to supplement footprinting methodologies and identify the precise DNA–protein interactions within the footprinted region. (For methodological details see Spiro and McMurray, 1999.)

2.2.4 IN VIVO *FOOTPRINTING ASSAY*

Although the methods described so far can provide considerable information about DNA–protein contacts, they all suffer from the deficiency that the DNA–protein interaction occurs *in vitro* when cell extract and the DNA are mixed. Hence they indicate what factors can bind to the DNA, rather than whether such factors actually do bind to the DNA in the intact cell where a particular factor may be sequestered in the cytoplasm or where its binding may be impeded by the association of DNA with other proteins such as histones.

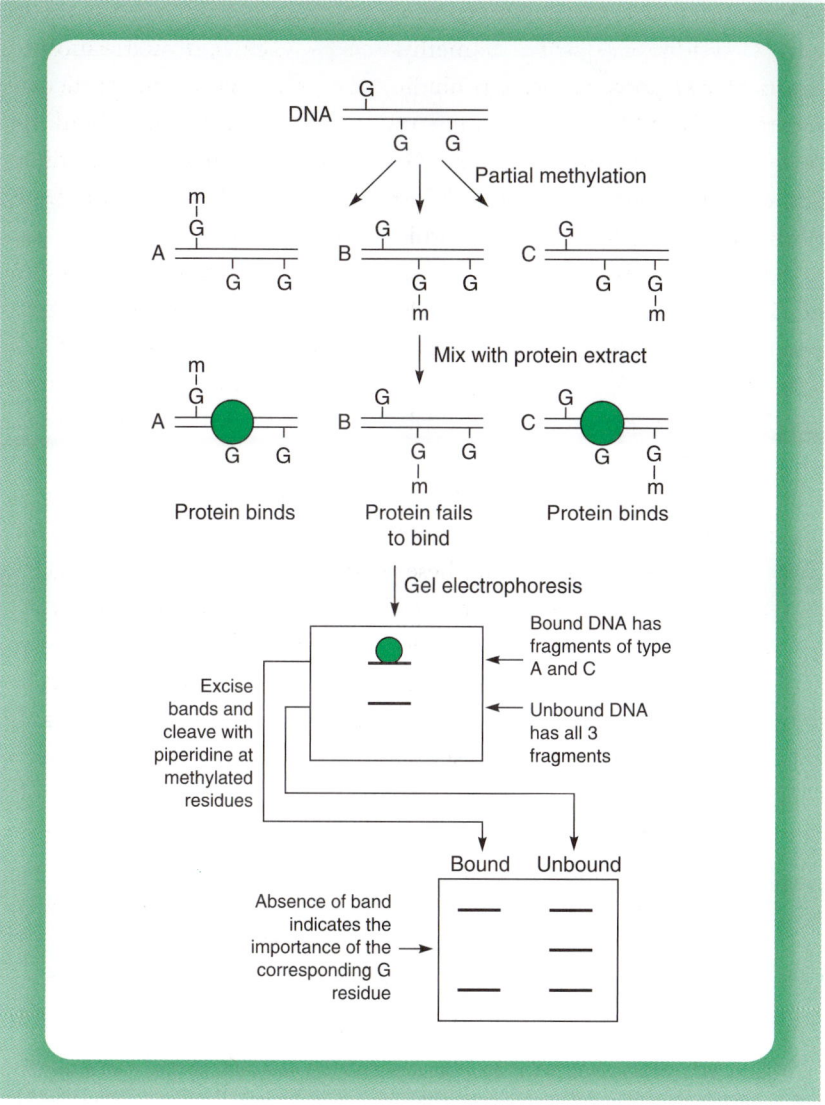

Figure 2.6

Methylation interference assay. Partially methylated DNA is used in a DNA mobility shift assay and both the DNA which has failed to bind protein and that which has bound protein and formed a retarded band are subsequently cleaved at methylated G residues with piperidine. If methylation at a specific G residue has no effect on protein binding (types A and C), the bound and unbound DNA will contain equal amounts of methylated G at this position. In contrast, if methylation at a particular G prevents binding of the protein (type B), only the unbound DNA will contain methylated G at this position.

These problems are overcome by the technique of *in vivo* footprinting, which is an extension of the *in vitro* DMS protection footprinting technique described in section 2.2.2. Thus intact cells are freely permeable to

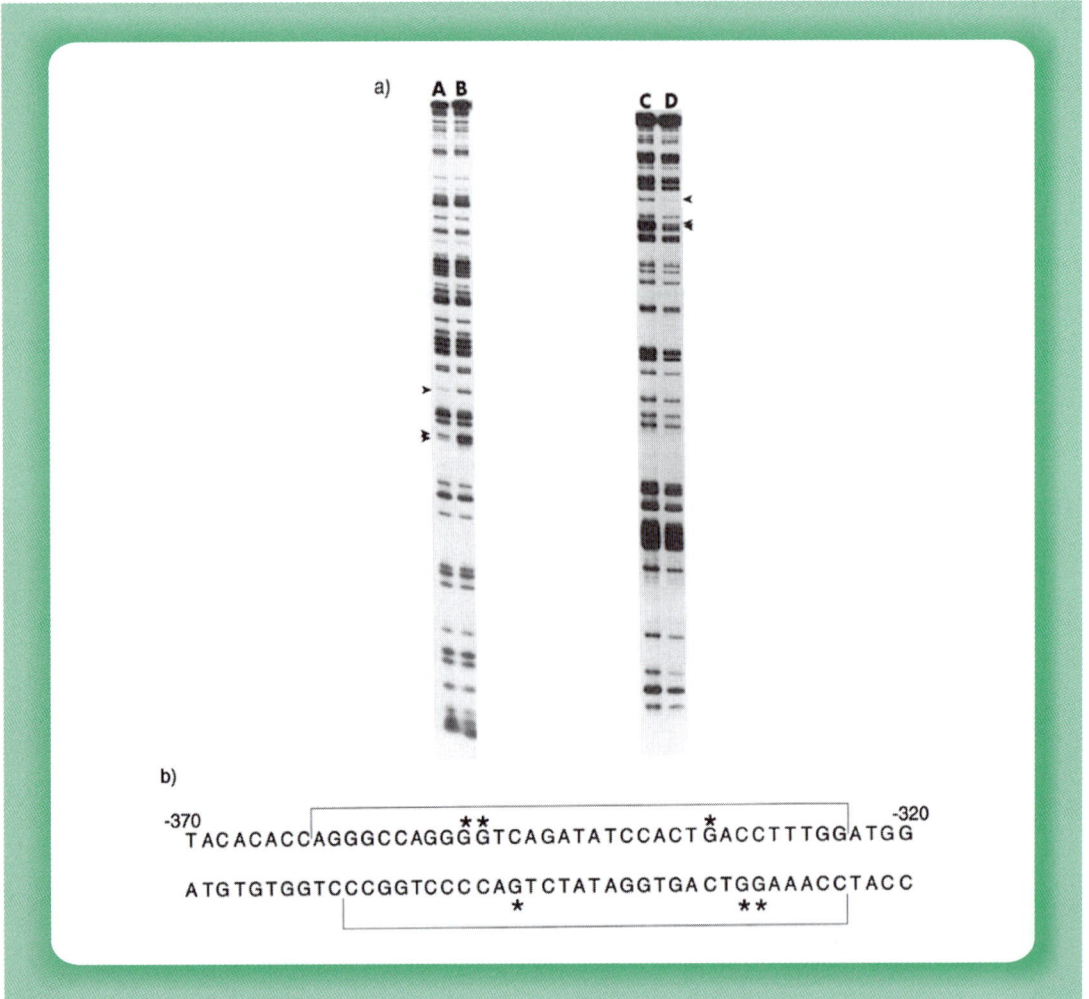

Figure 2.7

Panel (a): Methylation interference assay applied to the DNA of site B in the HIV control element as defined in the footprinting experiment shown in Figure 2.5. Both the upper (tracks A and B) and lower (tracks C and D) strands of the double stranded DNA sequence were analysed. Tracks B and C show the methylation pattern of the unbound DNA which failed to bind protein whereas tracks A and D show the methylation pattern of DNA which has bound protein. The arrows show G residues whose methylation is considerably lower in the bound compared to the unbound DNA and which are therefore critical for binding the specific cellular protein which interacts with this DNA sequence. Panel (b): DNA sequence of site B. The extent of the footprint region is indicated by the square brackets and the critical G residues defined by the methylation interference assay in panel (a) are asterisked. Note the symmetrical pattern of critical G residues within the palindromic DNA sequence.

DMS which can therefore be used to methylate the DNA within its native chromatin structure in such cells. Exactly as in the *in vitro* technique, G residues to which a protein has bound will be protected from such methylation and will therefore not be cleaved when the DNA is subsequently isolated and treated with piperidine. Hence the bands produced by cleavage at these residues will be absent when the pattern produced by intact chromatin is compared to that produced by naked DNA (Fig. 2.8).

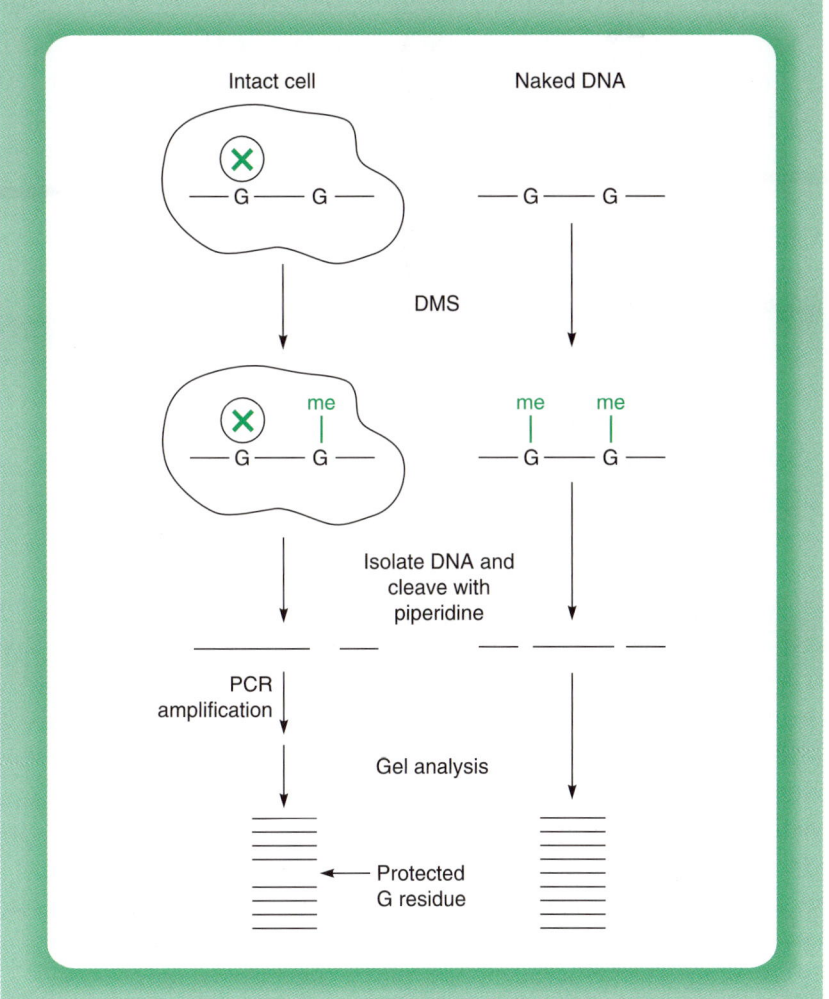

Figure 2.8

In vivo footprinting using the methylation protection assay in which specific G residues are protected by bound protein (X) from methylation by DMS treatment of intact cells. Hence following DNA isolation, cleavage of methylated G residues with piperidine and subsequent amplification by the polymerase chain reaction (PCR), the band corresponding to cleavage at this protected residue will be absent. In contrast, cleavage at this position will be observed in naked DNA where no protein protects this residue from methylation.

Obviously the amounts of any specific DNA sequence obtained from total chromatin in this procedure are vanishingly small compared to when a cloned DNA fragment is used in the *in vitro* procedure. It is hence necessary to amplify the DNA of interest from within total chromatin by the polymerase chain reaction in order to obtain sufficient material for analysis by this method. When this is done, however, *in vivo* footprinting provides an excellent means for analysing DNA–protein contacts within intact cells *in vivo* as well as determining the changes in such contacts which occur in response to specific treatments (see Herrera *et al.*, 1989; Mueller and Wold, 1989 for examples of this approach and Spiro and McMurray, 1999 for a full description of the methodologies involved).

Taken together, therefore, the three methods of DNA mobility shift, footprinting and methylation interference can provide considerable information on the nature of the interaction between a particular DNA sequence and a transcription factor. They serve as an essential prelude to a detailed study of the transcription factor itself.

2.3 METHODS FOR PURIFYING AND/OR CLONING TRANSCRIPTION FACTORS

2.3.1 PROTEIN PURIFICATION

As discussed above, once a particular DNA sequence has been shown to be involved in transcriptional regulation, a number of techniques are available for characterizing the binding of transcription factors to this sequence. Although such studies can be carried out on crude cellular extracts containing the protein, ultimately they need to be supplemented by studies on the protein itself. This can be achieved by purifying the transcription factor from extracts of cells containing it. Unfortunately, however, conventional protein purification techniques such as conventional chromatography and high pressure liquid chromatography (HPLC) result in the isolation of transcription factors at only 1–2% purity (Kadonaga and Tjian, 1986).

To overcome this problem and purify the transcription factor Sp1, Kadonaga and Tjian (1986) devised a method involving DNA affinity chromatography. In this method (Fig. 2.9), a DNA sequence containing a high affinity binding site for the transcription factor is synthesized and the individual molecules joined to form a multimeric molecule. This very high affinity binding site is then coupled to an activated sepharose support on a column and total cellular protein passed down the column. The Sp1 protein binds specifically to its corresponding DNA sequence whilst all other cellular proteins do not bind. The bound Sp1 can be eluted simply by raising the salt concentration. Two successive affinity chromatography steps of this type successfully resulted in the isolation of Sp1 at 90% purity, 30% of the Sp1 in the original extract being recovered, representing a 500- to 1000-fold purification (Kadonaga and Tjian, 1986).

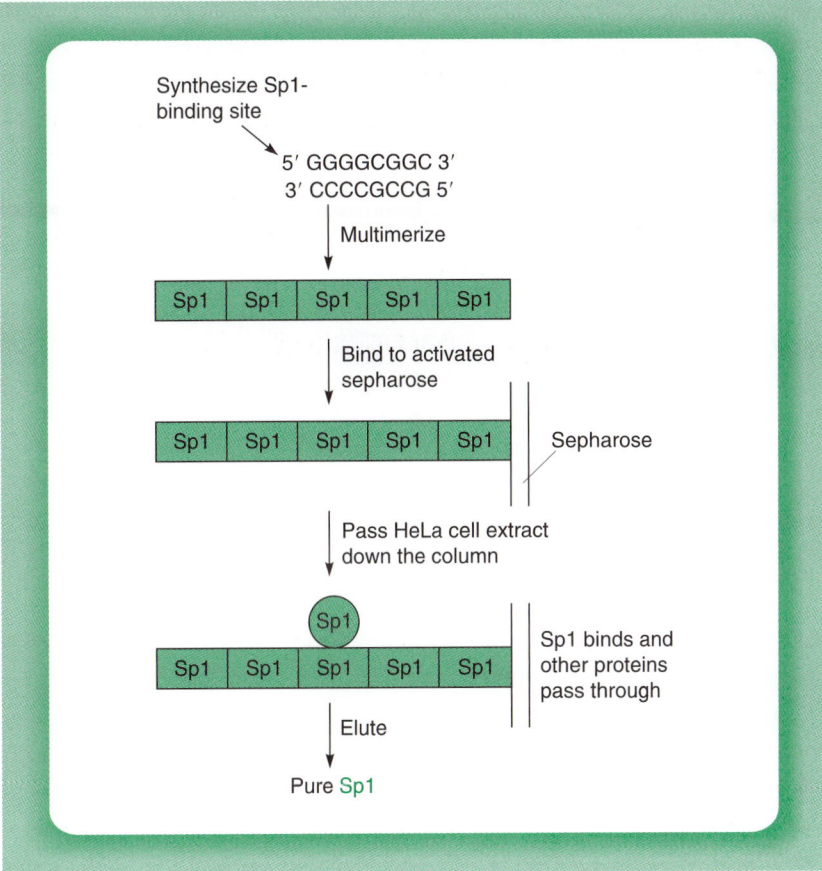

Figure 2.9
Purification of transcription factor Sp1 on an affinity column in which multiple copies of the DNA sequence binding Sp1 have been coupled to a sepharose support (Kadonaga and Tjian, 1986).

Although this simple one step method was successful in this case, it relies critically on the addition of exactly the right amount of non-specific DNA carrier to the cell extract. Thus this added carrier acts to remove proteins which bind to DNA in a non-sequence specific manner and which would hence bind non-specifically to the Sp1 affinity column and contaminate the resulting Sp1 preparation. This contamination will occur if too little carrier is added. If too much carrier is added, however, it will bind out the Sp1 since, like all sequence specific proteins, Sp1 can bind with low affinity to any DNA sequence. Hence in this case no Sp1 will bind to the column itself (Fig. 2.10).

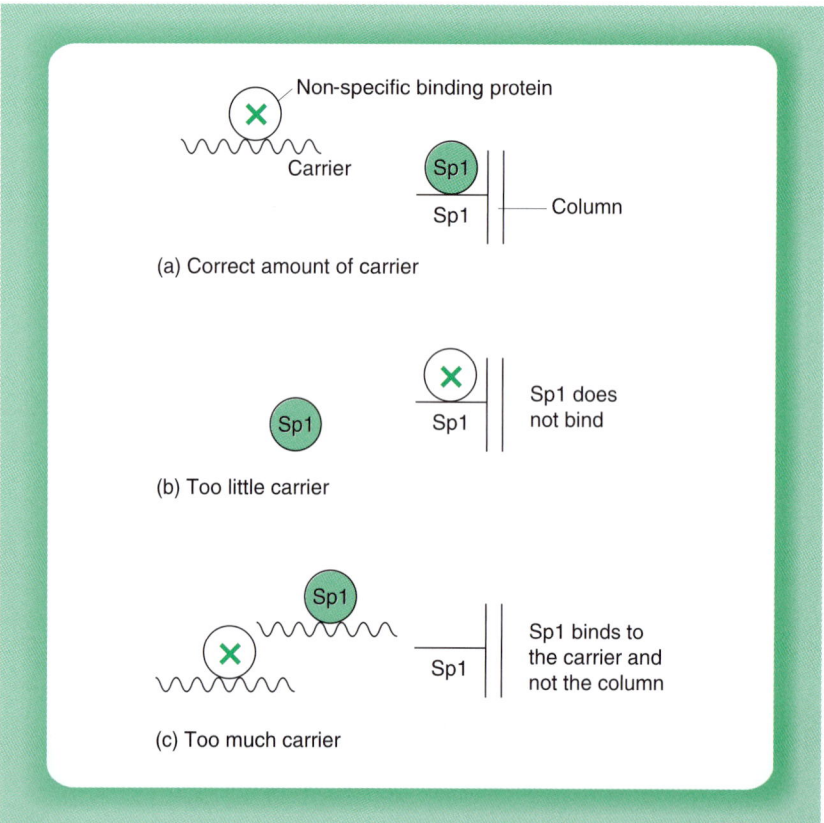

Figure 2.10

Consequences of adding different amounts of non-specific carrier DNA to the protein passing through the Sp1 affinity column. If the correct amount of non-specific carrier is added it will bind proteins which interact with DNA in a non-sequence specific manner allowing Sp1 to bind to the column (a). However, addition of too little carrier will result in non-sequence specific proteins binding to the column, thereby preventing the binding of Sp1 (b) whereas in the presence of too much carrier both the non-specific proteins and Sp1 will bind to the carrier (c).

To overcome this problem Rosenfeld and Kelley (1986) devised a method in which proteins capable of binding to DNA with high affinity in a non-sequence specific manner are removed prior to the affinity column. To do this the bulk of cellular protein was removed on a Biorex 70 high capacity ion exchange column, and proteins which can bind to any DNA with high affinity were then removed on a cellulose column to which total bacterial DNA had been bound. Subsequently, the remaining proteins which had bound to non-sequence specific DNA with only low affinity were applied to a column containing a high affinity binding site for transcription factor NF-1 (Fig. 2.11). NF-1 bound to this site with high affinity and could be eluted in essentially pure form by raising the salt concentration (Table 2.1). It should be noted that in this and other purification procedures the fractions containing the transcription factor can readily be identified by carrying out a DNA mobility shift or footprinting assay with each fraction, using the specific DNA binding site of the transcription factor.

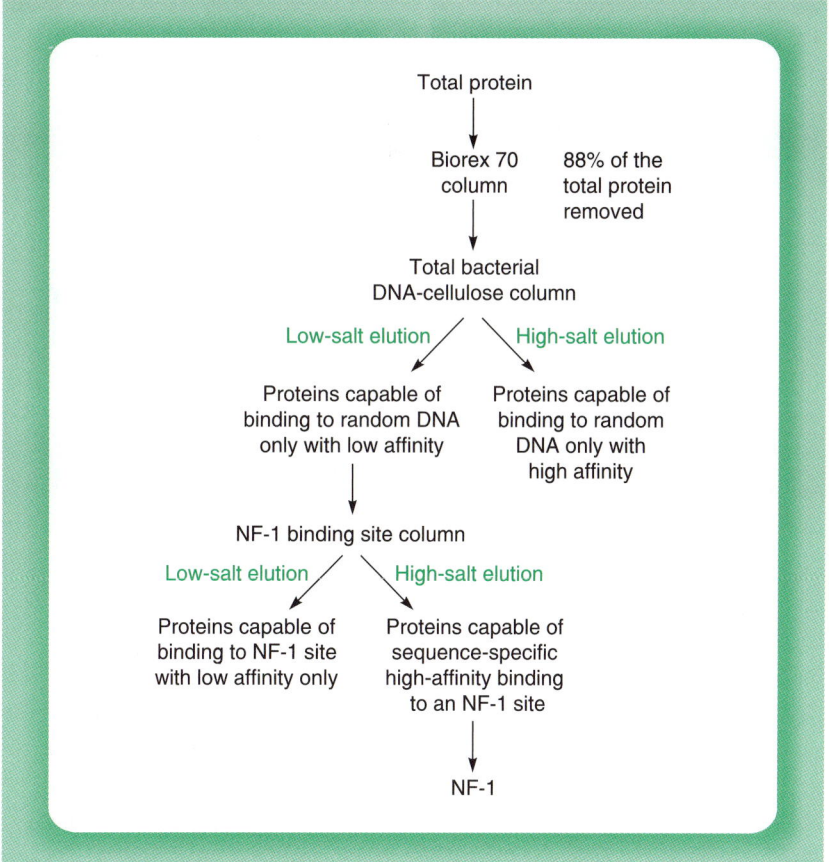

Figure 2.11

Purification of transcription factor NF-1 (Rosenfeld and Kelley, 1986). Following removal of most cellular proteins on a Biorex 70 ion exchange column, proteins which bind to all DNA sequences with high affinity were removed on a bacterial DNA-cellulose column. Subsequent application of the remaining proteins to a column containing the NF-1 binding site results in the purification of NF-1 since it is the only protein which binds with low affinity to random DNA but with high affinity to an NF-1 site.

Table 2.1

Purification of transcription factor NF-1 from HeLa cells

	Total protein (mg)	Specific binding of ^{32}P DNA (fmol/ mg protein) $\times 10^{-3}$	Purification (fold)	Yield (%)
HeLa cell extract*	4590	3.1	1.0	100
Biorex 70 column	550	27.1	8.7	104
E. coli DNA cellulose	65.2	181	58.4	8.3
NF-1 affinity matrix				
1st passage	2.1	4510	1455	67
2nd passage	1.1	7517	2425	57

*Prepared from 6×10^{10} cells or 120 g cells.

The purified protein obtained in this way can obviously be used to characterize the protein, for example by determining its molecular weight or by raising an antibody to it to characterize its expression pattern in different cell types. Similarly, the activity of the protein can be assessed by adding it to cellular extracts and assessing its effect on their ability to transcribe an exogenously added DNA in an *in vitro* transcription assay. Unfortunately, however, because of the very low abundance of transcription factors in the cell, these purification procedures yield very small amounts of protein. For example, Treisman (1987) succeeded in purifying only 1.6 micro-grams of the serum response factor starting with 2×10^{10} cells or 40 grams of cells. Such difficulties clearly limit the experiments which can be done with purified material. Indeed the primary use of purified factor in most cases has simply been to provide material to isolate the gene encoding the protein. This gene can then be expressed either *in vitro* or in bacteria to provide a far more abundant source of the corresponding protein than could be obtained from cells which naturally express it.

2.3.2 GENE CLONING

Several methods are available for cloning the gene encoding a particular transcription factor and these will be discussed in turn.

(a) Use of oligonucleotide probes predicted from the protein sequence of the factor

If a particular transcription factor has been purified it is possible to obtain portions of its amino acid sequence. In turn, such sequences can be used to predict oligonucleotides containing DNA sequence capable of encoding these protein fragments. Due to the redundancy of the genetic code whereby several different DNA codons can encode a particular amino acid, there will be multiple different oligonucleotides capable of encoding a particular amino acid sequence. All these possible oligonucleotides are synthesized chemically, made radioactive and used to screen a cDNA library prepared from mRNA isolated from a cell type expressing the factor. The oligonucleotide in the mixture which does correspond to the transcription factor amino acid sequence will hybridize to the corresponding sequence in a cDNA clone derived from mRNA encoding the factor. Hence such a clone can be readily identified in the cDNA library (Fig. 2.12).

In cases where purified protein is available, as in those discussed in the previous section, this approach represents a relatively simple method for isolating cDNA clones. It has therefore been widely used to isolate cDNA clones corresponding to purified factors such as Sp1 (Kadonaga *et al.*, 1987; see Fig. 2.12), NF1 (Santoro *et al.*, 1988) and the serum response factor (Norman *et al.*, 1988) (for methodological details see Nicolas *et al.*, 1999).

(b) Use of oligonucleotide probes derived from the DNA binding site of the factor

Although relatively simple, the use of oligonucleotides derived from protein sequence does require purified protein. As we have seen, purification of a transcription factor requires a vast quantity of cells and is technically difficult. Moreover, eventual determination of the partial amino acid sequence of the protein requires access to expensive protein sequencing apparatus.

To bypass these problems Singh *et al.* (1988) devised a procedure which is based on the fact that information is usually available about the specific DNA sequence to which a particular transcription factor binds. Hence a cDNA clone expressing the factor can be identified in a library by its ability to bind the appropriate DNA sequence. This method relies therefore on DNA–protein binding rather than DNA–DNA binding. Hence the library must be prepared in such a way that the cloned cDNA inserts are translated by the bacteria into their corresponding proteins. This is normally achieved by inserting the cDNA into the coding region of the bacteriophage lambda beta-galactosidase gene resulting in

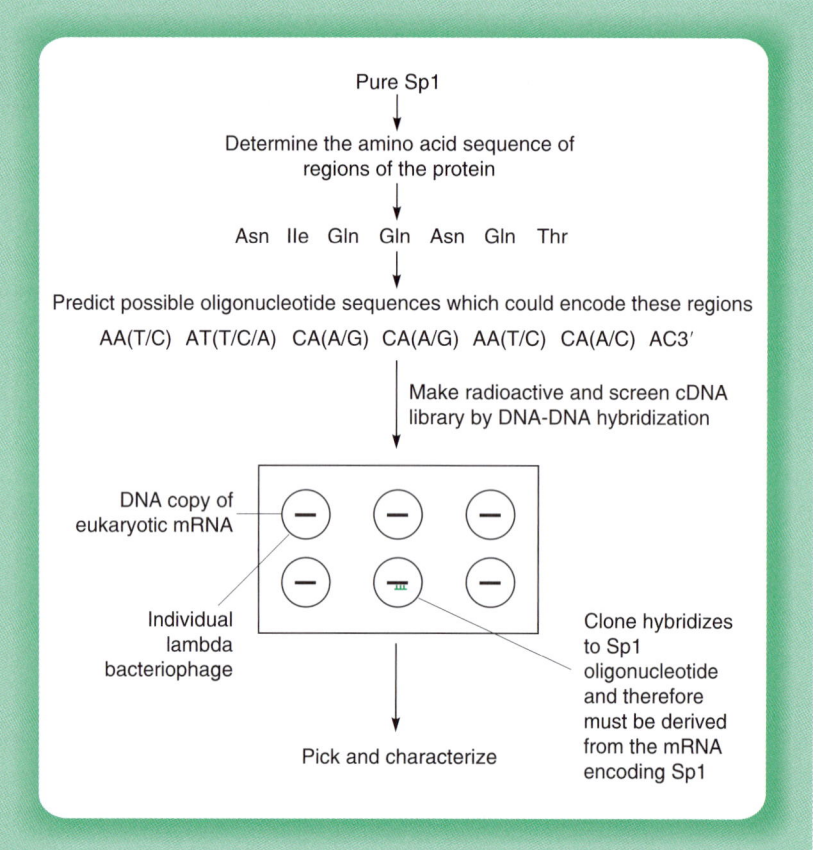

Figure 2.12

Isolation of cDNA clones for the Sp1 transcription factor by screening with short oligonucleotides predicted from the protein sequence of Sp1. Because several different triplets of bases can code for any given amino acid, multiple oligonucleotides that contain every possible coding sequence are made. Positions at which these oligonucleotides differ from one another are indicated by the brackets containing more than one base.

its translation as part of the bacteriophage protein. The resulting fusion protein binds DNA with the same sequence specificity as the original factor. Hence a cDNA clone encoding a particular factor can be identified in the library by screening with a radioactive oligonucleotide containing the binding site (Fig. 2.13).

This technique has been used to isolate cDNA clones encoding several transcription factors such as the CCAAT box binding factor C/EBP (Vinson *et al.*, 1988) and the octamer binding proteins Oct-1 (Sturm *et al.*, 1988) and Oct-2 (Staudt *et al.*, 1988) (for methodological details see Cowell and Hurst, 1999).

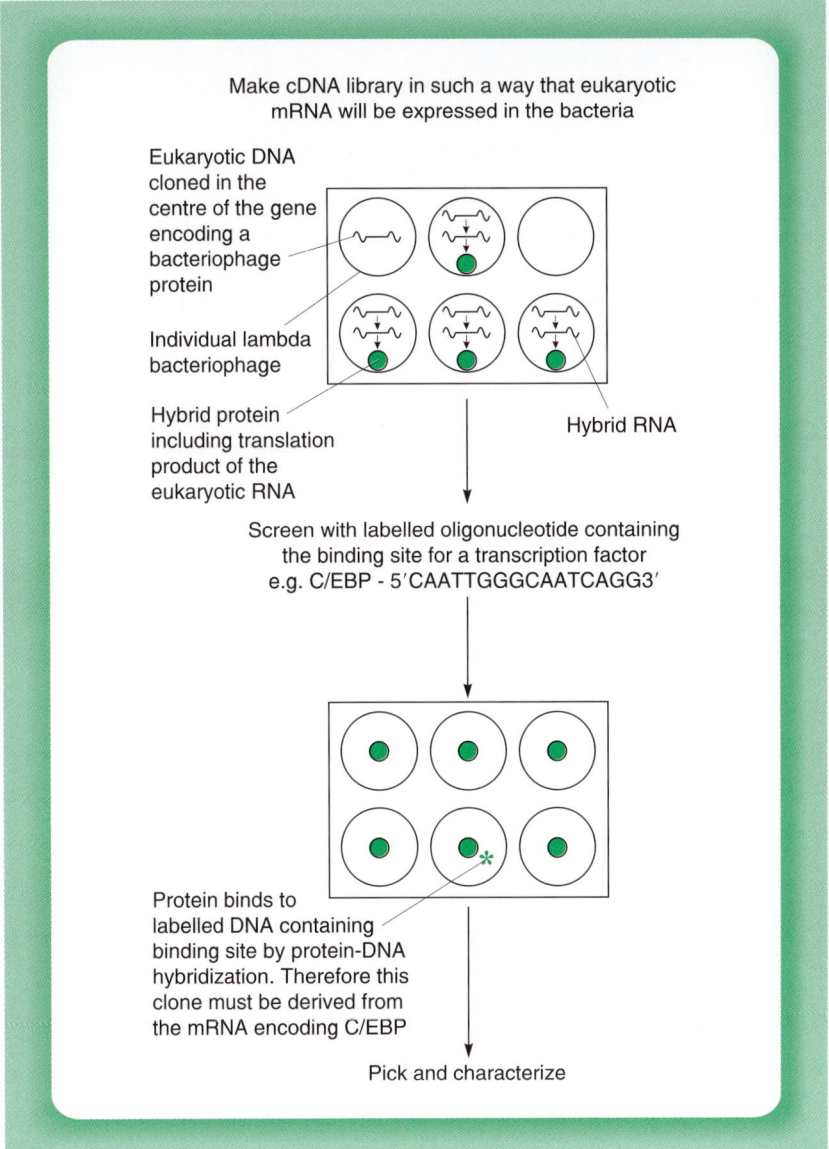

Figure 2.13
Isolation of cDNA clones for the C/EBP transcription factor by screening an expression library with a DNA probe containing the binding site for the factor.

(c) Cloning of novel transcription factors by homology to known factors

The development of the two methods described above involving screening with oligonucleotides derived from the protein sequence or oligonucleotides derived from the binding site has therefore resulted in the isolation of cDNA clones corresponding to very many transcription factors.

More recently, however, novel transcription factors are increasingly being cloned on the basis of their relationship to previously character-ized factors. In an early example of this approach, He *et al.* (1989) iden-tified short amino acid sequences which were highly conserved in the known members of the POU family of transcription factors (Fig. 2.14) (see Chapter 4, section 4.2.6 for a description of this family of proteins). They then prepared degenerate oligonucleotides which contained all the possible DNA sequences able to encode these sequences. Two of these degenerate oligonucleotides were then used in a polymerase chain reaction (PCR) to amplify cDNA prepared from the mRNA of different tissues. Evidently, cDNAs derived from mRNAs encoding novel POU pro-teins which contain these sequences will be amplified in the PCR pro-cedure and can be isolated and characterized. Indeed, He *et al.* (1989) cloned several novel POU factors by this means and this approach has been applied by a number of others to both the POU family and other transcription factor families (for review and full description of the meth-ods involved see Ashworth, 1999).

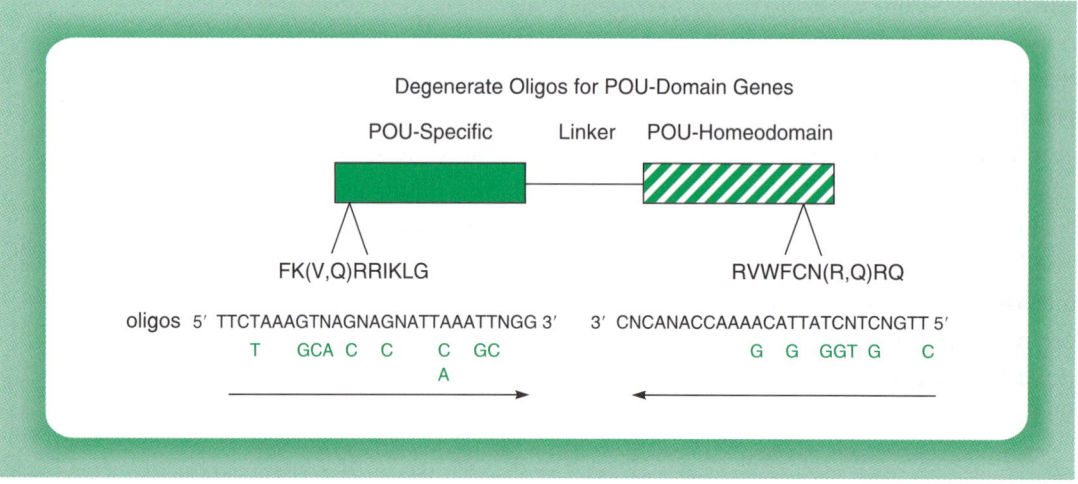

Figure 2.14

Cloning of novel members of the POU family of transcription factors on the basis of all family members having two conserved amino acid sequences, one in the POU-specific domain and one in the POU-homeodomain. Degenerate oligonucleotides containing all possible sequences able to encode these conserved sequences are used in a polymerase chain reaction with cDNA prepared from mRNA of a particular tissue. Novel POU factors expressed in this tissue will be amplified on the basis that they contain the conserved sequences and can then be characterized.

Of course, as more and more genomes including the human genome are fully sequenced, this approach can now be conducted *'in silico'* by using the DNA sequences of known transcription factors to search for related sequences in computer databases and this is now perhaps the most common means by which DNA sequences able to encode novel transcription factors are identified.

2.4 USE OF CLONED GENES

2.4.1 DOMAIN MAPPING OF TRANSCRIPTION FACTORS

The cloning of transcription factors by the means described above has in turn resulted in an explosion of information on these factors. Thus once a clone has been isolated, its DNA sequence can be obtained allowing prediction of the corresponding protein sequence and comparison with other factors. Similarly, the clone can be used to identify the mRNA encoding the protein and examine its expression in various tissues by Northern blotting, to study the structure of the gene itself within genomic DNA by Southern blotting and as a probe to search for related genes expressed in other tissues or other organisms.

Most importantly, however, the isolation of cDNA clones provides a means of obtaining large amounts of the corresponding protein for functional study. This can be achieved either by coupled *in vitro* transcription and translation (Fig. 2.15a: see for example Sturm *et al.*, 1988) or by expressing the gene in bacteria either in the original expression vector used in the screening procedure (see above section 2.3.2b) or more commonly by sub-cloning the cDNA into a plasmid expression vector (Fig. 2.15b; see for example Kadonaga *et al.*, 1987).

The protein produced in this way has similar activity to the natural protein, being capable of binding to DNA in footprinting or mobility shift assays (see for example Kadonaga *et al.*, 1987) and of stimulating the transcription of appropriate DNAs containing its binding site when added to a cell free transcription system (see for example Mueller *et al.*, 1990).

Moreover, once a particular activity has been identified in a protein produced in this way, it is possible to analyse the features of the protein which produce this activity in a way that would not be possible using the factor purified from cells which normally express it. Thus because the cDNA clone of the factor can be readily cut into fragments and each fragment expressed as a protein in isolation, particular features exhibited by the intact protein can readily be mapped to a particular region. Using the approach outlined in Figure 2.16 for example, it has proved

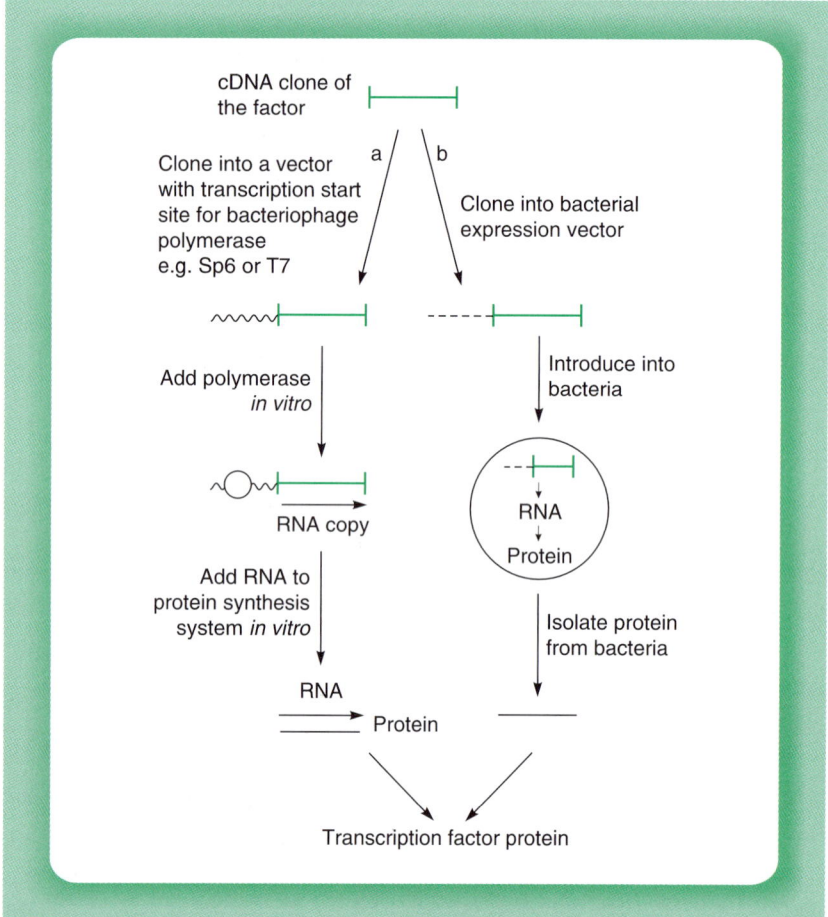

Figure 2.15

Methods of producing transcription factor protein from a cloned transcription factor cDNA. In the coupled *in vitro* transcription and translation method (a) the cDNA is cloned downstream of a promoter recognized by a bacteriophage polymerase and transcribed *in vitro* by addition of the appropriate polymerase. The resulting RNA is translated in an *in vitro* protein synthesis system to produce transcription factor protein. Alternatively the cDNA can be cloned downstream of a prokaryotic promoter in a bacterial expression vector (b). Following introduction of this vector into bacteria, the bacteria will transcribe the cDNA into RNA and translate the RNA into protein which can be isolated from the bacteria.

possible to map the DNA binding abilities of specific transcription factors such as the octamer binding proteins Oct-1 (Sturm *et al.*, 1987) and Oct-2 (Clerc *et al.*, 1988) to a specific short region of the protein. Once this has been done, particular bases in the DNA encoding the DNA binding domain of the factor can then be mutated so as to alter its amino acid sequence. The effect of these mutations on DNA binding can then be assessed as before by expressing the mutant protein and measuring its ability to bind to DNA.

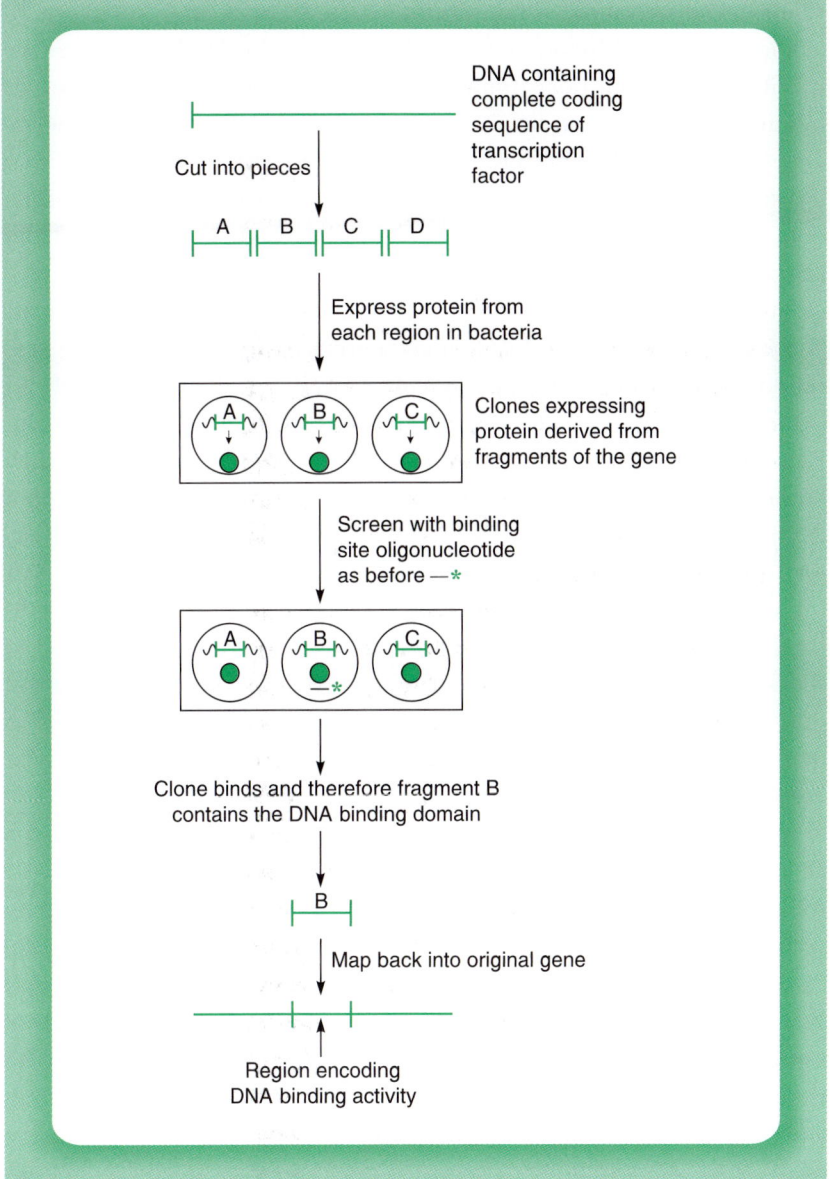

Figure 2.16

Mapping of the DNA binding region of a transcription factor by testing the ability of different regions to bind to the appropriate DNA sequence when expressed in bacteria.

Approaches of this type have proved particularly valuable in defining DNA binding motifs present in many factors and in analysing how differences in the protein sequence of related factors define which DNA sequence they bind. This is discussed in Chapter 4.

One other piece of information to emerge from these studies is that the binding to DNA of a small fragment of the factor does not normally result in the activation of transcription. Thus, a sixty amino acid region of the yeast transcription factor GCN4 can bind to DNA in a sequence specific manner but does not activate transcription of genes bearing its binding site (Hope and Struhl, 1986). Although DNA binding is necessary for transcription therefore, it is not sufficient. This indicates that transcription factors have a modular structure in which the DNA binding domain is distinct from another domain of the protein which mediates transcriptional activation.

The identification of the activation domain in a particular factor is complicated by the fact that DNA binding is necessary prior to activation. Hence the activation domain cannot be identified simply by expressing fragments of the protein and monitoring their activity. Rather, the various regions of the cDNA encoding the factor must each be linked to the region encoding the DNA binding domain of another factor and the hybrid proteins produced. The ability of the hybrid factor to activate a target gene bearing the DNA binding site of the factor supplying the DNA binding domain is then assessed (Fig. 2.17). In these so-called 'domain swap' experiments, binding of the factor to the appropriate DNA binding site will be followed by gene activation only if the hybrid factor contains the region encoding the activation domain of the factor under test, allowing the activation domain to be identified.

Thus, if another sixty amino acid region of GCN4 distinct from the DNA binding domain is linked to the DNA binding domain of the bacterial LexA protein, it can activate transcription in yeast from a gene containing a binding site for LexA. This cannot be achieved by the LexA DNA binding domain or this region of GCN4 alone, indicating that this region of GCN4 contains the activation domain of the protein which can activate transcription following DNA binding and is distinct from the GCN4 protein DNA binding domain (Hope and Struhl, 1986).

As with DNA binding domains, the identification of activation domains and comparisons between the domains in different factors has provided considerable information on the nature of activation domains and the manner in which they function. This is discussed in Chapter 5.

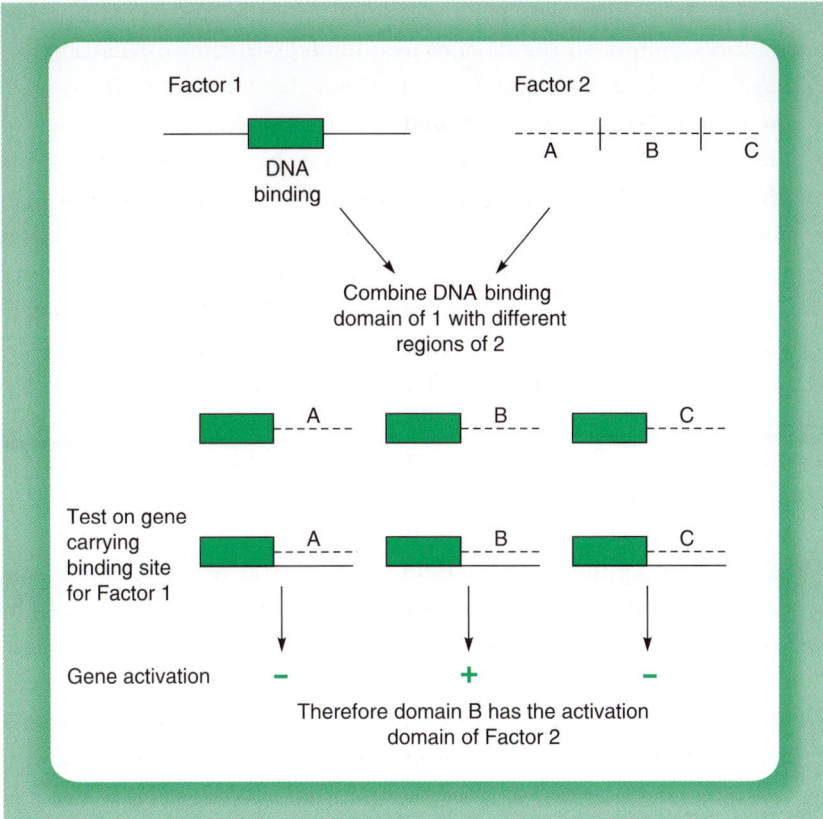

Figure 2.17

Domain swapping experiment in which the activation domain of Factor 2 is mapped by combining different regions of Factor 2 with the DNA binding domain of Factor 1 and assaying the hybrid proteins for the ability to activate transcription of a gene containing the DNA binding site of Factor 1.

2.4.2 DETERMINING THE DNA BINDING SPECIFICITY OF AN UNCHARACTERIZED FACTOR

As indicated above it is common for a transcription factor to be identified on the basis of its binding to a known DNA sequence and the gene encoding the factor then cloned. It is also possible, however, for a novel gene to be cloned on the basis, for example, that its expression changes in response to a particular stimulus (see Chapter 7) or that it is mutated in a specific disease (see Chapter 9). On inspection of the DNA sequence and predicted protein sequence, it then appears that this gene encodes a transcription factor either because it is homologous to known transcription factors or because it contains regions with structures similar to those known to mediate DNA binding (see Chapter 4) or transcriptional activation (see Chapter 5). Alternatively, as described above (section 2.3.2c), the novel factor may have been identified by experimental or computer methods simply on the basis of its homology to known transcription factors.

Obviously all the techniques for analysing a cloned factor described in section 2.4.1 above can be applied to analysing this factor, examining, for example, its expression pattern or determining whether regions within it mediate transcriptional activation when linked to the DNA binding domain of another factor. Unlike the situation for transcription factors which were identified on the basis of their DNA binding specificity however, no information will be available on the DNA sequences to which this novel factor binds. It is evidently essential for the further study of this novel factor that such sequences are identified, so allowing, for example, an analysis of the effect of the factor on artificial promoters carrying its binding site and the identification of its target genes.

To do this, Pollock and Treisman (1990) used a method in which oligonucleotides containing a randomized central twenty-six base pair sequence flanked by two defined twenty-five nucleotide sequences were prepared (Fig. 2.18). These sequences were then mixed with transcription factor protein. An antibody to the transcription factor was then used to immunoprecipitate the factor together with the oligonucleotides to which it had bound. This procedure should select from the pool of random oligonucleotides those which contain the binding site for the factor within their central twenty-six base pair sequence whilst removing those which contain all other sequences. However, after a single round of immunoprecipitation these oligonucleotides will be present in insufficient amounts and purity for further analysis. The immunoprecipitated sequences are therefore amplified by the polymerase chain reaction (PCR) using primers corresponding to the defined twenty-five base pair sequences at the ends of each oligonucleotide. Further cycles of transcription factor binding, immunoprecipitation and PCR are then carried out to further purify the binding sequences. Ultimately, the oligonucleotides which bind the factor are cloned and subjected to sequence analysis to identify the common sequence which they contain and which is therefore the binding site for the factor.

This method thus allows the identification of specific binding sites for the transcription factor and has been used, for example, to identify the DNA binding site for the Brn-3 POU family transcription factors (Gruber et al., 1997) which were originally isolated on the basis of homology to other members of the POU family as described in section 2.3.2c (He et al., 1989) (see Chapter 4, section 4.2.6 for further discussion of POU family transcription factors). Binding sites identified in this way can then, for example, be linked to a gene promoter and introduced

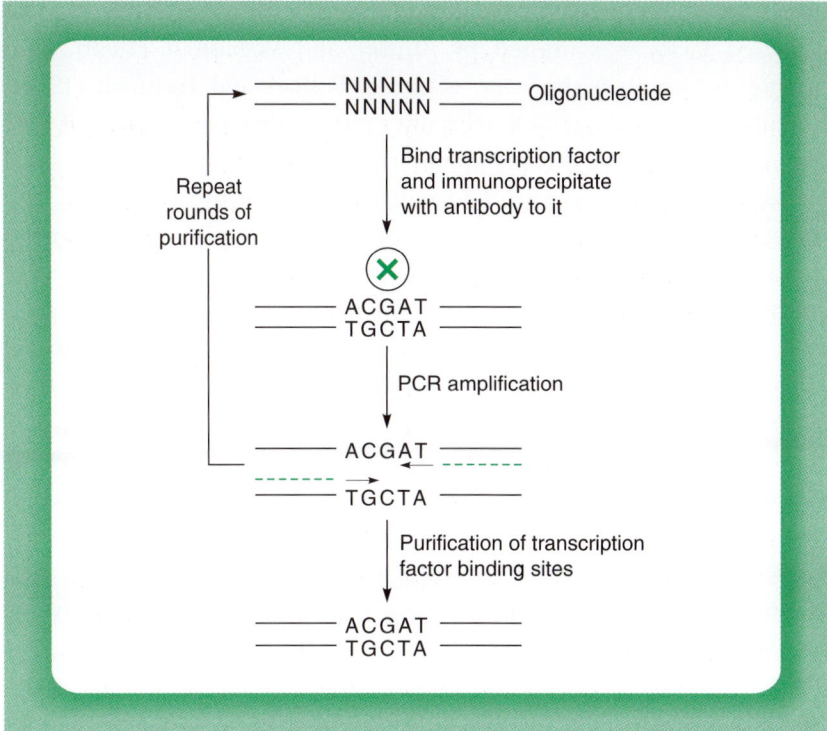

Figure 2.18

Transcription factor binding sites can be cloned using oligonucleotides containing a random central sequence (NNNNN) flanked by defined sequences (solid lines). Repeated cycles of trancription factor binding (X), immunoprecipitation and PCR amplification with primers complimentary to the defined end sequences (dotted lines) will eventually result in the purification of oligonucleotides containing the binding site for the factor (ACGAT in this case).

into cells with an expression vector encoding the transcription factor itself to determine whether the factor acts as an activator or repressor of gene expression. Similarly, by inspecting the sequences of promoter or enhancer elements of known genes, it may be possible to identify putative target genes for the factor.

2.4.3 IDENTIFICATION OF TARGET GENES FOR TRANSCRIPTION FACTORS

(a) In vitro analysis of transcription factor binding to genomic DNA fragments

Although the approach described above can identify binding sites for transcription factors, it does not directly identify their target genes. A direct approach to identify such target genes for a previously uncharacterized factor was devised by Kinzler and Vogelstein (1989). This method is essentially the same as that of Pollock and Treisman (1990), except that the starting material is not random oligonucleotides but total genomic DNA. This DNA is digested with a restriction enzyme and small defined DNA sequences are added to the ends of the fragments. The transcription factor binding and immunoprecipitation steps are carried out as before, resulting in the purification of pieces of genomic DNA containing the binding site for the transcription factor. These are then PCR amplified as before using primers corresponding to the defined DNA sequences which were added at the fragment ends and are then cloned.

Although this method is more technically difficult than the use of oligonucleotides due to the complexity of genomic DNA, it has the great advantage that the DNA binding sites are obtained linked to the sequences to which they are normally joined in the genome rather than in isolation (Fig. 2.19). Hence these linked sequences can immediately be characterized and used to identify a target gene for the factor. This method has thus been used, for example, to identify novel target genes for members of the nuclear receptor transcription factor family discussed in Chapter 4 (section 4.4), such as the estrogen receptor (Inoue *et al.*, 1993) and the thyroid hormone receptor (Caubin *et al.*, 1994).

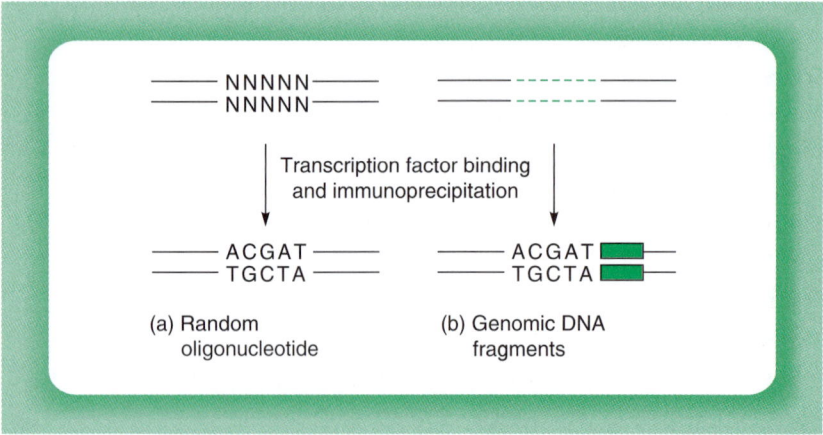

Figure 2.19

Whilst the purification of transcription factor binding sites using random oligonucleotides (as in Fig. 2.18) simply isolates the binding site (a), the use of genomic DNA sequences (dotted lines) in the purification results in the isolation of the binding site linked to a fragment of its target gene (boxed) which can then be characterized (b).

(b) Chromatin immunoprecipitation (ChIP)

The above method using genomic fragments thus represents an advance over the oligonucleotide method in the identification of potential target genes for a specific factor. However, since the genome DNA fragments and the transcription factor are mixed in the test tube, it indicates which genomic fragments can bind the factor of interest *in vitro* rather than identifying those genes to which it actually binds in the cell.

In a further advance, the chromatin immunoprecipitation method (ChIP) actually involves the direct identification of target genes for known or unknown factors in the intact cell. In this method (for review see Orlando, 2000), living cells are first fixed with formaldehyde. This has the effect of stably cross-linking transcription factors to the DNA sequences to which they are bound in the cell (Fig. 2.20). The chromatin in the cell is then broken up into small pieces and isolated. An antibody to the transcription factor is then added to immunoprecipitate it, together with the target DNA to which it is cross-linked.

Following breakage of the cross-links and release of the immunoprecipitated DNA from the transcription factor protein, the DNA can be analysed in a number of ways. In the simplest method (Fig. 2.20a) one can test whether a particular gene has been immunoprecipitated by carrying out a PCR amplification with primers for that gene. This will test whether a particular transcription factor binds to a specific gene in intact cells as well as *in vitro*. Similarly, by carrying out the ChIP assay in cells incubated under different conditions or in different cell types, one can detect the changes in such binding which occur in these situations.

In addition, however, methods exist to identify all the genes immunoprecipitated by the ChIP assay rather than testing for the presence of individual genes (for review see Weinmann and Farnham, 2002). Thus, the immunoprecipitated DNA can be cloned and subjected to DNA sequence analysis to identify all the different DNA fragments (Fig. 2.20b). Alternatively, it is now possible to prepare microarrays containing thousands of DNA sequences representing the entire genome of an organism. The immunoprecipitated DNA can be labelled and hybridized to such an array, allowing all the genes to which the protein is bound in the cell to be identified and characterized (Fig. 2.20c). This method, known as genome-wide location analysis, was initially used to define all the binding sites for specific transcription factors in yeast (Ren *et al.*, 2000) but has now been extended to more complex organisms including mammals. Thus, for example, by carrying out the immunoprecipitation with antibodies to a component of TFIID, which is part of the basal transcriptional complex (see Chapter 3, section 3.5), it has been possible to map the active promoters in the human genome (Kim *et al.*, 2005).

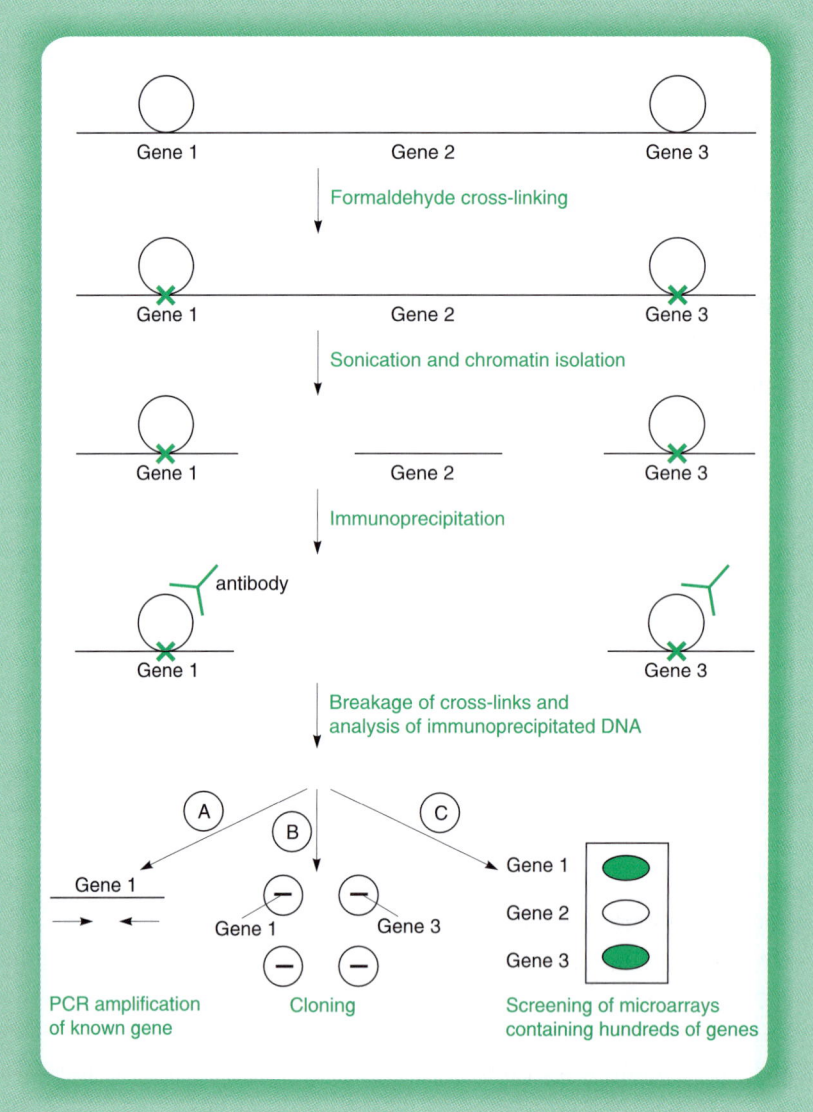

Figure 2.20

In the chromatin immunoprecipitation (ChIP) assay, transcription factors (circle) are cross-linked to their target DNA in the intact cell. The chromatin is then fragmented by sonication and immunoprecipitated with an antibody to the transcription factor of interest. The resulting immunoprecipitated DNA is then analysed by either (a) PCR to detect known potential target genes or (b) cloning and DNA sequencing or (c) hybridization to a microarray which contains a very large number of genes. Note that in this case, gene 1 and gene 3 bind the factor and gene 2 does not. Methods (b) and (c) detect both genes 1 and 3, whereas method (a) with primers for gene 1 evidently detects only that gene.

This approach is becoming more and more powerful, as the entire genomes of more and more organisms are sequenced. Thus, for example, in yeast, a number of putative regulatory sequences have been identified on the basis of their DNA sequence and their conservation between different strains of yeast (Lee *et al.*, 2002; Cliften *et al.*, 2003; Kellis *et al.*, 2003) and similar approaches are now being extended to mammalian cells (Encode Project Consortium, 2004; Xie *et al.*, 2005).

By combining such DNA sequence data with genome-wide location analysis it is possible to build a global picture of the transcription factors which are bound to all the different genes in the genome, in a specific cell type or under specific conditions (Fig. 2.21a). This can then be combined with expression data identifying the genes which are expressed in that cell type and functional data on the effects of inactivating individual transcription factors (Fig. 2.21b). This will indicate, for example, whether a particular factor acts as an activator or repressor of gene expression (see Chapters 5 and 6). Together the two sets of data can thus be used to define a regulatory network

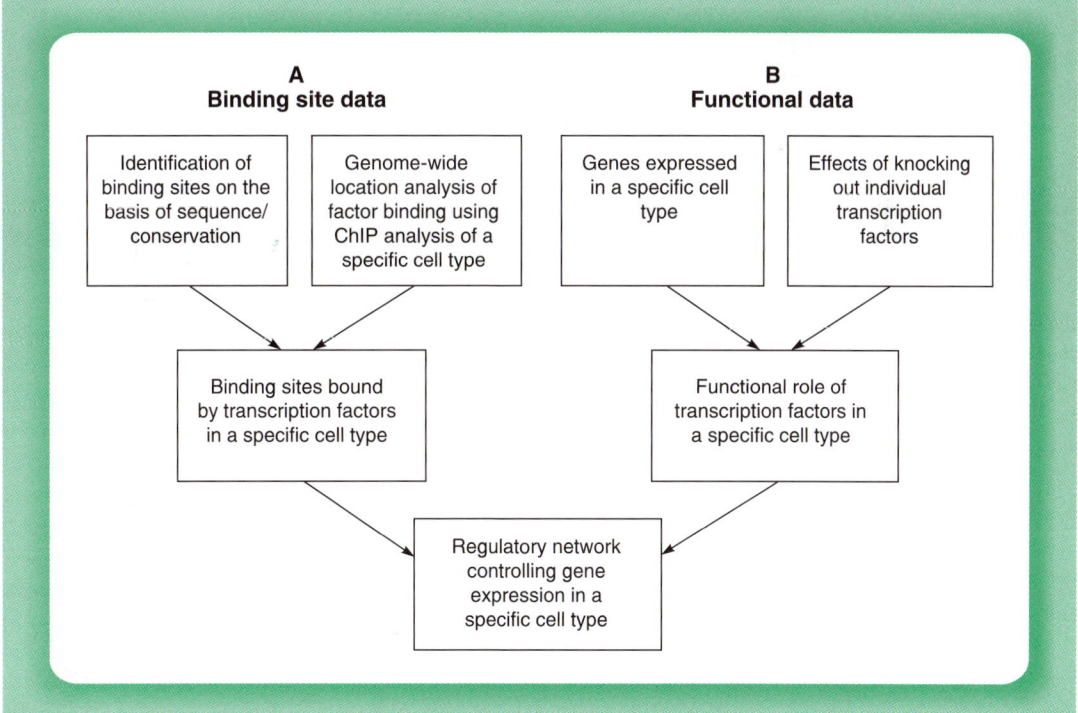

Figure 2.21

Regulatory networks can be defined by combining data on the sequence/conservation of DNA binding sites and on the factors binding to these sites (panel a) with data on the expression of specific genes and the functional consequences of inactivating particular transcription factors (panel b).

which controls gene expression in that cell type (Fig. 2.21) (for reviews see Blais and Dynlacht, 2005; Imai *et al.*, 2006).

Genome-wide location/ChIP analysis can thus be used to define regulatory networks in a single cell type. Clearly, however, its use can also be extended by carrying out parallel analysis using cellular material from different cell types or the same cell type under different conditions. Thus, it has been used to compare the set of genes in the yeast genome which is bound by a specific transcription factor under different conditions (Zeitlinger *et al.*, 2003; Harbison *et al.*, 2004). For example, the yeast transcription factor Ste 12 has been shown to bind to distinct but overlapping sets of genes under basal conditions and following induction of either mating or filamentous phenotypes (Fig. 2.22) (Zeitlinger *et al.*, 2003) (for further discussion see Chapter 4, section 4.2.4).

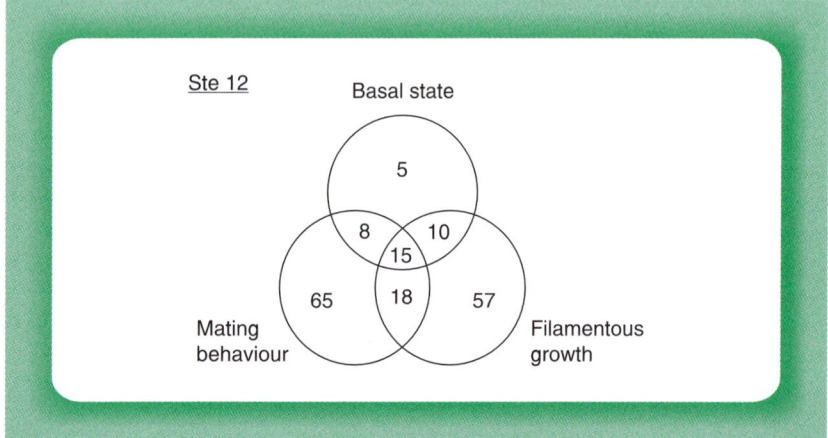

Figure 2.22
Venn diagram showing the number of genes bound by the yeast transcription factor Ste 12, under three different growth conditions. Note that only fifteen genes are bound by the factor under all three conditions, whereas all the others are bound in only one or two different conditions.

Hence, these powerful methods allow the experimenter to move from the level of the individual gene to define regulatory networks operating under different conditions. Although initially applied in the relatively simple yeast system, the progressive application of these methods to mammalian and other complex eukaryotes offers considerable potential for the future.

2.5 CONCLUSIONS

This chapter has described a number of methods which allow the investigation of the interaction of a transcription factor with DNA, its purification,

gene cloning and dissection of its functional domains, as well as the identification of its DNA binding site and its target genes and the definition
of regulatory networks in which it is involved. The information obtained
by the application of these procedures to particular factors is discussed in
subsequent chapters.

REFERENCES

Ares, M. Jr., Chung, J-S., Giglio, L. and Weiner, A.M. (1987) Distinct factors with
Sp1 and NF-A specificities bind to adjacent functional elements of the
human U2 snRNA gene enhancer. Genes and Development 1, 808–817.

Ashworth, A. (1999) Cloning transcription factors by sequence homology. In:
Transcription factors: a practical approach. Second edition (Latchman, D.S.,
ed.). Oxford University Press, Oxford and New York, pp. 145–164.

Baumruker, T., Sturm, R. and Herr, W. (1988) OBP 100 binds remarkably degenerate octamer motifs through specific interactions with flanking sequences.
Genes and Development 2, 1400–1413.

Blais, A. and Dynlacht, B.D. (2005) Constructing transcriptional regulatory networks. Genes and Development 19, 1499–1511.

Caubin, J., Iglesias, T., Bernal, J., Munoz, A., Marquez, G., Barbero, J.L. and
Zaballos, A. (1994) Isolation of genomic DNA fragments corresponding to
genes modulated *in vivo* by a transcription factor. Nucleic Acids Research
22, 4132–4138.

Clerc, R.G., Corcoran, L.M., LeBowitz, J.H., Baltimore, D. and Sharp, P.A. (1988)
The B-cell specific Oct-2 protein contains POU box and homeo box type
domains. Genes and Development 2, 1570–1581.

Cliften, P., Sudarsanam, P., Desikan, A., Fulton, L., Fulton, B., Majors, J., Waterston,
R., Cohen, B.A. and Johnston, M. (2003) Finding functional features in
Saccharomyces genomes by phylogenetic footprinting. Science 301, 71–76.

Cowell, I.G. and Hurst, H.C. (1999) Cloning transcription factors from a cDNA
expression library. In: *Transcription factors: a practical approach*. Second
edition (Latchman, D.S., ed.). Oxford University Press, Oxford and New York,
pp. 123–143.

Dignam, J.D., Lebovitz, R.M. and Roeder, R.G. (1983) Accurate transcription initiation by RNA polymerase II in a soluble extract from isolated mammalian
nuclei. Nucleic Acids Research 11, 1575–1589.

Dynan, W.S. and Tjian, R. (1983) The promoter specific transcription factor Sp1
binds to upstream sequences in the SV40 promoter. Cell 35, 79–87.

ENCODE project consortium (2004) The ENCODE (ENCyclopedia Of DNA
Elements) Project. Science 306, 636–640.

Fried, M. and Crothers, D.M. (1981) Equilibria and kinetics of lac repressor–operator interactions by polyacrylamide gel electrophoresis. Nucleic Acids Research 9, 6505–6525.

Galas, D. and Schmitz, A. (1978) DNAse footprinting: a simple method for the detection of protein-DNA binding specificity. Nucleic Acids Research 5, 3157–3170.

Garner, M.M. and Revzin, A. (1981) A gel electrophoresis method for quantifying the binding of proteins to specific DNA regions: application to components of the *Escherichia coli* lactose operon regulatory system. Nucleic Acids Research 9, 3047–3060.

Gruber, C.A., Rhee, J.M., Gleiberman, A. and Turner, E.E. (1997) POU domain factors of the Brn-3 class recognize functional DNA elements which are distinctive, symmetrical and highly conserved in evolution. Molecular and Cellular Biology 17, 2391–2400.

Harbison, C.T., Gordon, D.B., Lee, T.I., Rinaldi, N.J., Macisaac, K.D., Danford, T.W., Hannett, N.M., Tagne, J.B., Reynolds, D.B., Yoo, J., Jennings, E.G., Zeitlinger, J., Pokholok, D.K., Kellis, M., Rolfe, P.A., Takusagawa, K.T., Lander, E.S., Gifford, D.K., Fraenkel, E. and Young, R.A. (2004) Transcriptional regulatory code of a eukaryotic genome. Nature 431, 99–104.

He, X., Treacey, M.N., Simmonds, D.M., Ingraham, H.A., Swanson, L.W. and Rosenfeld, M.G. (1989) Expression of a large family of POU-domain genes in mammalian brain development. Nature 340, 35–42.

Herrera, R.E., Shaw, P.E. and Nordheim, A. (1989) Occupation of the c-*fos* serum response element *in vivo* by a multi-protein complex is unaltered by growth factor induction. Nature 340, 68–71.

Hope, I.A. and Struhl, K. (1986) Functional dissection of a eukaryotic transcriptional activator GCN4 of yeast. Cell 46, 885–894.

Imai, K.S., Levine, M., Satoh, N. and Satou, Y. (2006) Regulatory blueprint for a chordate embryo. Science 312, 1183–1187.

Inoue, S., Orimo, A., Hosoi, T., Kondo, S., Toyoshima, H., Kondo, T., Ikegami, A., Ouchi, Y., Orimoto, H. and Muramatsu, M. (1993) Genomic binding site cloning reveals an estrogen responsive gene that encodes a RING finger protein. Proceedings of the National Academy of Sciences, USA 90, 11117–11121.

Kadonaga, J.T. and Tjian, R. (1986) Affinity purification of sequence-specific DNA binding proteins. Proceedings of the National Academy of Sciences, USA, 83, 5889–5893.

Kadonaga, J.T., Carner, K.R., Masiarz, F.R. and Tjian, R. (1987) Isolation of cDNA encoding the transcription factor Sp1 and functional analysis of the DNA binding domain. Cell 51, 1079–1090.

Kellis, M., Patterson, N., Endrizzi, M., Birren, B. and Lander, E.S. (2003) Sequencing and comparison of yeast species to identify genes and regulatory elements. Nature 423, 241–254.

Kim, T.H., Barrera, L.O., Zheng, M., Qu, C., Singer, M.A., Richmond, T.A., Wu, Y., Green, R.D. and Ren, B. (2005) A high-resolution map of active promoters in the human genome. Nature 436, 876–880.

Kinzler, K.W. and Vogelstein, B. (1989) Whole genome PCR: application to the identification of sequences bound by gene regulatory proteins. Nucleic Acids Research 17, 3645–3653.

Kreale, G.G. (ed.) (1994) *DNA–protein interactions*. Humana Press, Urbana, NJ.

Latchman, D.S. (ed.) (1999) *Transcription factors: a practical approach*. Second edition. Oxford University Press, Oxford and New York.

Lee, T.I., Rinaldi, N.J., Robert, F., Odom, D.T., Bar-Joseph, Z., Gerber, G.K., Hannett, N.M., Harbison, T., Thompson, C.M., Simon, I., Zeitlinger, J., Jennings, E.G., Murray, H.L., Gordon, D.B., Ren, B., Wyrick, J.J., Tagne, J-B., Volkert, T.L., Fraenkel, E., Gifford, D.K. and Young, R.A. (2002) Transcriptional regulatory networks in *Saccharomyces cerevisiae*. Science 298, 799–804.

Manley, J.L., Fire, A., Cano, A., Sharp, P.A. and Gefter, M.L. (1980) DNA-dependent transcription of adenovirus genes in a soluble whole-cell extract. Proceedings of the National Academy of Sciences, USA, 77, 3855–3859.

Maxam, A.M. and Gilbert, W. (1980) Sequencing end labelled DNA with base-specific chemical cleavages. Methods in Enzymology 65, 499–560.

Mueller, C.R., Macre, P. and Schibler, U. (1990) DBP a liver-enriched transcriptional activator is expressed late in ontogeny and its tissue specificity is determined post-transcriptionally. Cell 61, 279–291.

Mueller, P.R. and Wold, B. (1989) *In vivo* footprinting of a muscle specific enhancer by ligation mediated PCR. Science 246, 780–786.

Nicolas, R.H., Hynes, G. and Goodwin, G.H. (1999) Purification and cloning of DNA binding transcription factors. In: *Transcription factors: a practical approach*. Second edition (Latchman, D.S., ed.). Oxford University Press, Oxford and New York, pp. 97–122.

Norman, C., Runswick, M., Pollock, R. and Treisman, R. (1988) Isolation and properties of cDNA clones encoding SRF, a transcription factor that binds to the c-*fos* serum response element. Cell 55, 989–1003.

Orchard, K., Perkins, N.D., Chapman, C., Harris, J., Emery, V., Goodwin, G., Latchman, D.S. and Collins, M.K.L. (1990) A novel T cell protein recognizes a palindromic element in the negative regulatory element of the HIV-1 LTR. Journal of Virology 64, 3234–3239.

Orlando, V. (2000) Mapping chromosomal proteins *in vivo* by formaldehyde-crosslinked-chromatin immunoprecipitation. Trends in Biochemical Sciences 25, 99–104.

Papavassilou, A.G. (1995) Chemical nucleases as probes for studying DNA–protein interactions. Biochemical Journal 305, 345–357.

Pollock, R. and Treisman, R. (1990) A sensitive method for the determination of protein DNA binding specificities. Nucleic Acids Research 18, 6197–6204.

Ren, B., Robert, F., Wyrick, J.J., Aparicio, O., Jennings, E.G., Simon, I., Zeitlinger, J., Schreiber, J., Hannett, N., Kanin, E., Volkert, T.L., Wilson, C J., Bell, S.P. and Young, R.A. (2000) Genome-wide location and function of DNA binding proteins. Science 290, 2306–2309.

Rosenfeld, P.J. and Kelley, T.J. (1986) Purification of nuclear factor 1 by DNA recognition site affinity chromatography. Journal of Biological Chemistry 261, 1398–1408.

Santoro, C., Mermod, N., Andrews, P.C. and Tjian, R. (1988) A family of human CCAAT box binding proteins active in transcription and DNA replication: cloning and expression of multiple cDNAs. Nature 334, 218–224.

Siebenlist, U. and Gilbert, W. (1980) Contacts between the RNA polymerase and an early promoter of phage T7. Proceedings of the National Academy of Sciences, USA, 77, 122–126.

Singh, H., Le Bowitz, J.H., Baldwin, A.S. and Sharp, P.A. (1988) Molecular cloning of an enhancer binding protein: isolation by screening of an expression library with a recognition site DNA. Cell 52, 415–429.

Smith, M.D., Dent, C.L. and Latchman, D.S. (1999) The DNA mobility shift assay. In: *Transcription factors: a practical approach*. Second edition (Latchman, D.S., ed.). Oxford University Press, Oxford and New York, pp. 1–25.

Spiro, C. and McMurray, C.T. (1999) Footprint analysis of DNA–protein complexes *in vitro* and *in vivo*. In: *Transcription factors: a practical approach*. Second edition (Latchman, D.S., ed.). Oxford University Press, Oxford and New York, pp. 27–62.

Staudt, L.M., Clerc, R.G., Singh, H., Le Bowitz, J.H., Sharp, P.A. and Baltimore, D. (1988) Cloning of a lymphoid-specific cDNA encoding a protein binding the regulatory octamer DNA motif. Science 241, 577–580.

Sturm, R., Baumruker, T., Franza, R. Jr. and Herr, W. (1987) A 100-kD HeLa cell octamer binding protein (OBP 100) interacts differently with two separate octamer-related sequences within the SV40 enhancer. Genes and Development 1, 1147–1160.

Sturm, R.A., Das, G. and Herr, W. (1983) The ubiquitous octamer-binding protein Oct-1 contains a POU domain with a homeobox subdomain. Genes and Development 2, 1582–1599.

Treisman, R. (1987) Identification and purification of a polypeptide that binds to the c-fos serum response element. EMBO Journal 6, 2711–2717.

Vinson, C.R., La Marco, K.L., Johnson, P.F., Landschulz, W.H. and McKnight, S.Z. (1988) In situ detection of sequence-specific DNA binding activity specified by a recombinant bacteriophage. Genes and Development 2, 801–806.

Weinmann, A.S. and Farnham, P.J. (2002) Identification of unknown target genes of human transcription factors using chromatin immunoprecipitation. Methods 26, 37–47.

Xie, X., Lu, J., Kulbokas, E.J., Golub, T.R., Mootha, V., Lindblad-Toh, K., Lander, E.S. and Kellis, M. (2005) Systematic discovery of regulatory motifs in human promoters and 3' UTRs by comparison of several mammals. Nature 434, 338–345.

Zeitlinger, J., Simon, I., Harbison, C.T., Hannett, N.M., Volkert, T.L., Fink, G.R. and Young, R.A. (2003) Program-specific distribution of a transcription factor dependent on partner transcription factor and MAPK signalling. Cell 113, 395–404.

CHAPTER 3

RNA POLYMERASES AND THE BASAL TRANSCRIPTIONAL COMPLEX

3.1 RNA POLYMERASES

Transcription involves the polymerization of ribonucleotide precursors into an RNA molecule using a DNA template. The enzymes which carry out this reaction are known as RNA polymerases. In all eukaryotes, three different enzymes of this type exist in the nucleus and they are discussed in this chapter. A fourth RNA polymerase, which is involved in transcriptional repression, has been identified only in plants and is discussed in Chapter 6 (section 6.4.2).

Each of the three conserved RNA polymerases is active on a different set of genes and the polymerases can be distinguished on the basis of their different sensitivities to the fungal toxin alpha-amanitin (Table 3.1; for review see Sentenac, 1985). All the genes which code for proteins as well as those encoding some of the small nuclear RNAs involved in RNA splicing are transcribed by RNA polymerase II. Because of the very wide variety of regulatory processes which these genes exhibit, much of this book is concerned with the interaction of different transcription factors with RNA polymerase II. Information is also available, however, on the interaction of such factors with RNA polymerase I which transcribes the genes encoding the 28S, 18S and 5.8S ribosomal RNAs (Somerville, 1984) and with RNA polymerase III which transcribes the transfer RNA and 5S ribosomal RNA genes (Cilberto *et al.*, 1983). These interactions are therefore discussed where appropriate.

All three RNA polymerases are large multi-subunit enzymes, RNA polymerase II for example having ten to fourteen subunits with sizes ranging from 220 to 10 kilo-daltons (Sentenac, 1985; Saltzman and Weinmann, 1989) which interact with one another to form a highly complex multimeric molecule that has been crystallized allowing structural analysis (Cramer *et al.*, 2001; Klug, 2001; Landick, 2001; Asturias and

Table 3.1

Eukaryotic RNA polymerases

Genes transcribed	Sensitivity to α-amanitin
I Ribosomal RNA (45S precursor of 28S, 18S and 5.8S rRNA)	Insensitive
II All protein-coding genes, small nuclear RNAs U1, U2, U3, etc.	Very sensitive (inhibited 1 μg/ml)
III Transfer RNA, 5S ribosomal RNA, small nuclear RNA U6, repeated DNA sequences: Alu, B1, B2 etc., 7SK, 7SL, RNA	Moderately sensitive (inhibited 10 μg/ml)

Craighead, 2003). Interestingly, the cloning of the genes encoding the largest subunits of each of the three polymerases has revealed that they show homology to one another (Memet *et al.*, 1988). Similarly, chemical labelling experiments have indicated that the second largest subunit of each polymerase contains the active site of the enzyme (Riva *et al.*, 1987) whilst at least three smaller, non-catalytic subunits are shared by the three yeast polymerases (Woychik *et al.*, 1990). Such relationships evidently indicate a basic functional similarity between the three eukaryotic RNA polymerases and may also be indicative of a common evolutionary origin.

In addition to the conservation of function between the three eukaryotic enzymes, each individual enzyme exhibits a strong conservation between different organisms. Thus the largest subunit of the mammalian RNA polymerase II enzyme is 75% homologous to that of the fruit fly *Drosophila* (Saltzman and Weinmann, 1989) and also shows homology to the equivalent enzymes in yeast (Memet *et al.*, 1988) and even *E. coli* (Ahearn *et al.*, 1987). All the eukaryotic RNA polymerase II enzymes contain a repeated region at the carboxyl end of the largest subunit which contains multiple copies of the sequence Tyr-Ser-Pro-Thr-Ser-Pro-Ser. This sequence is unique to the largest subunit of RNA polymerase II and is present in multiple copies, being repeated fifty-two times in the mouse protein and twenty-six times in the yeast protein. This repeated region is highly evolutionarily conserved (for review see Stiller and Hall, 2002) and, as expected from this, is essential for the proper functioning of the enzyme and hence for cell viability, although its size can be reduced to some extent without affecting the activity of the enzyme (for review see Meinhart *et al.*, 2005).

Interestingly, this repeated region serves as a site for phosphorylation and it is likely that such phosphorylation is critical for functioning of the polymerase. Thus it appears that the dephosphorylated form of RNA polymerase II is the form which enters the basal transcriptional complex (see section 3.5.1) whilst its phosphorylation triggers the start of transcription to produce the RNA product (see sections 3.5.1 and 3.7 for further discussion).

In addition, as will be discussed in Chapters 5 (section 5.6) and 6 (section 6.5), this region is a target for transcriptional activators and repressors regulating transcriptional elongation. Moreover, recent studies have indicated that factors involved in post-transcriptional processes such as RNA splicing, associate with this region of the polymerase so that the nascent RNA transcript produced by the polymerase can actually be spliced by factors which are bound to the polymerase itself (see section 3.7 for further discussion). Hence this region appears to represent a critical target for cellular transcriptional and post-transcriptional regulatory processes.

Although the RNA polymerases therefore posses the enzymatic activity necessary for transcription, they cannot function independently. Rather, transcription involves numerous transcription factors which must interact with the polymerase and with each other if transcription is to occur. The role of these factors is to organize a stable transcriptional complex containing the RNA polymerase and which is capable of repeated rounds of transcription.

3.2 THE STABLE TRANSCRIPTIONAL COMPLEX

For all three eukaryotic polymerases, the initiation of transcription requires a multi-component complex containing the RNA polymerase and transcription factors. This complex has several characteristics which have led to it being referred to as a stable transcriptional complex (Brown, 1984).

These are:

1. The assembled complex is stable to treatment with low concentrations of specific detergents or to the presence of a competing DNA template, both of which would prevent its assembly.
2. The complex contains factors which are necessary for its assembly but not for transcription itself. These factors can therefore be dissociated once the complex has formed without affecting transcription.
3. The complex of RNA polymerase and other factors necessary for transcription is stable through many rounds of transcription, resulting in the production of many RNA copies from the gene.

These characteristics are illustrated in Figure 3.1.

Much of the information on these complexes has been obtained by studying the relatively simple systems of RNA polymerases I and III and applying the information obtained to the RNA polymerase II situation. The stable complex formed by each of these enzymes will therefore be discussed in turn.

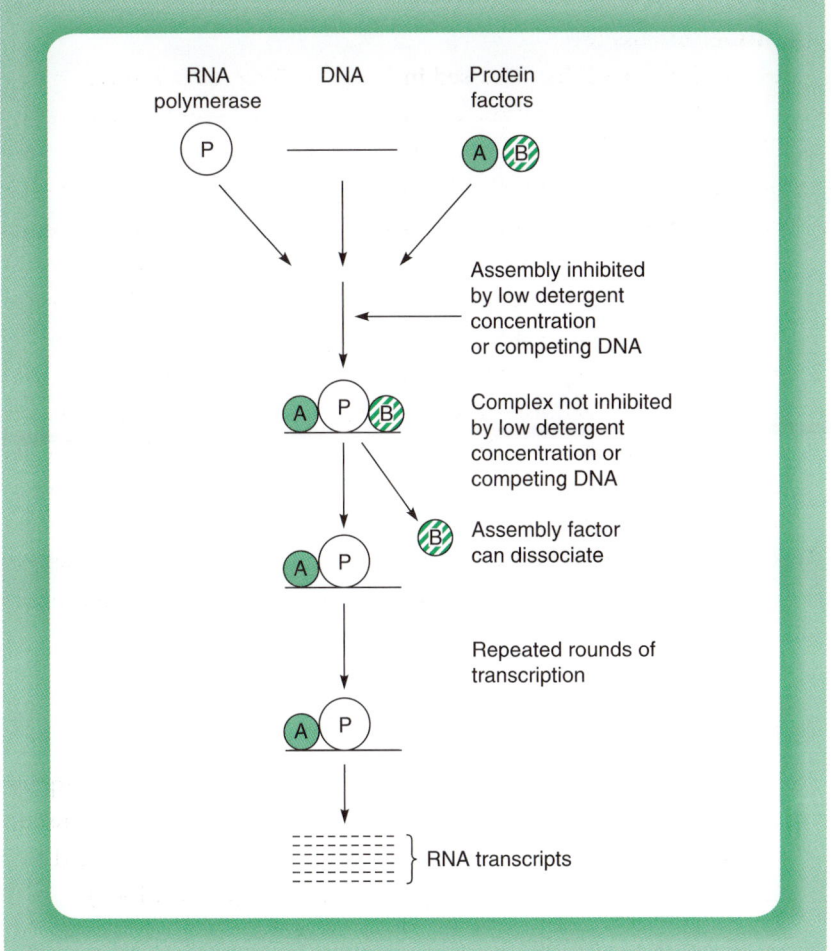

Figure 3.1

Stages in the formation of the stable transcriptional complex. The initial binding of the transcription factor (A) and the assembly factor (B) results in a metastable complex which can be dissociated by low levels of detergent or competing DNA. Following RNA polymerase binding, however, a stable complex is formed. This complex cannot be dissociated by low levels of detergent or competing DNA, is stable through multiple rounds of transcription and retains activity if the assembly factor (B) is removed.

3.3 RNA POLYMERASE I

The simplest complex known is found for the transcription of the ribosomal RNA genes by RNA polymerase I in *Acanthamoeba* (for review see Paule and White, 2000). In this organism only one transcription factor, known as TIF-1, is required for transcription by the polymerase. This factor binds to the ribosomal RNA promoter protecting a region from

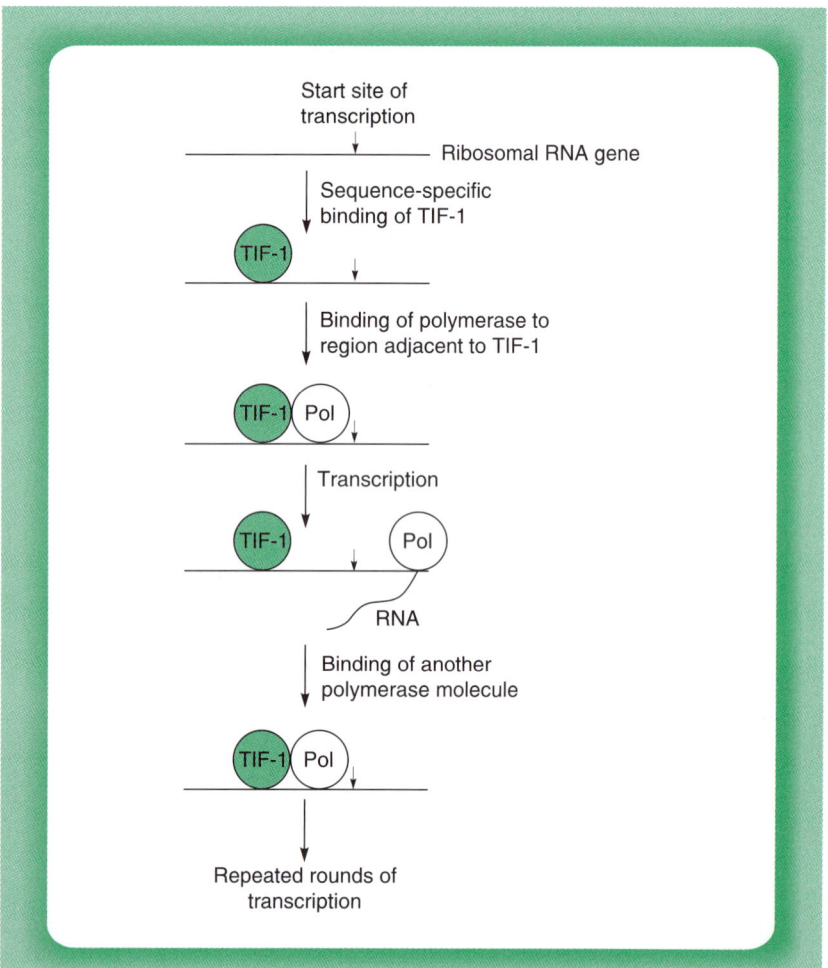

Figure 3.2

Transcription by *Acanthamoeba* RNA polymerase I involves the binding of transcription factor TIF-1 to a specific DNA sequence followed by binding of the polymerase in a non-sequence specific manner to the DNA region adjacent to TIF-1. When the RNA polymerase moves away as it transcribes the gene, TIF-1 remains bound at the promoter allowing another RNA polymerase molecule to bind and initiate a new round of transcription.

twelve to seventy bases upstream of the transcriptional start site from DNAseI digestion. Subsequently, the polymerase itself binds to the DNA just downstream of TIF-1 protecting a region between eighteen and fifty-two bases upstream of the start site. Interestingly, binding of the polymerase is not dependent on the specific DNA sequence within this region since it can be replaced with a completely random sequence without affecting binding of the polymerase. Hence RNA polymerase is positioned on the promoter by protein–protein interaction with TIF-1 which has previously bound in a sequence specific manner (Fig. 3.2). When the RNA polymerase moves along the DNA transcribing the gene, TIF-1 remains bound at the promoter allowing subsequent rounds of transcription to occur following binding of another polymerase molecule.

This system therefore represents a simple one in which one single factor is necessary for transcription and is active through multiple rounds of transcription. In vertebrate rRNA gene transcription, the situation is more complex, however, with an additional factor UBF (upstream binding factor) also being involved (for review see Paule and White, 2000; Grummt, 2003; Russell and Zomerdijk, 2005). UBF binds specifically to the promoter and upstream elements of the ribosomal RNA genes and stimulates transcription. This is achieved, however, by interaction with the vertebrate TIF-1 homologue, known as SL1. Thus, although a low basal rate of transcription is observed in the absence of UBF, no transcription is detectable unless SL1 is present. Unlike TIF-1, SL1 does not exhibit sequence specific binding to the ribosomal RNA promoter. Hence UBF acts by binding to the DNA in a sequence specific manner and facilitating the binding of SL1. Thus, whilst both SL1 and its homologue TIF-1 act as transcription factors necessary for polymerase I binding, UBF is an additional assembly factor required for binding of SL1 in vertebrates but not of TIF-1 in *Acanthamoeba*. This example therefore illustrates the distinction between factors required only for assembly of the complex or for binding of the polymerase and transcription itself (Fig. 3.3).

3.4 RNA POLYMERASE III

The different role of transcription factors and assembly factors is also well illustrated by the RNA polymerase III system (for reviews see Paule and White, 2000; Geiduschek and Kassavetis, 2001; Schramm and Hernandez, 2002). Thus three different classes (I–III) of RNA polymerase III transcription unit exist, all of which require the essential factor TFIIIB for transcription (for review see Hernandez, 1993).

In the case of class I transcription units encoding the 5S ribosomal RNAs, transcription by RNA polymerase III requires the binding of three additional factors: TFIIIA, TFIIIB and TFIIIC. Although both TFIIIA and TFIIIC exhibit the ability to bind to 5S DNA in a sequence specific manner, TFIIIB, like SL1, cannot do so unless TFIIIC has already bound. Once the complex of all these factors has formed and the RNA polymerase has bound, TFIIIA and TFIIIC can be removed and transcription continues with only TFIIIB and the polymerase bound to the DNA. Hence, like UBF, TFIIIA and TFIIIC are assembly factors which are required for the binding of the transcription factor TFIIIB. In turn, bound TFIIIB is

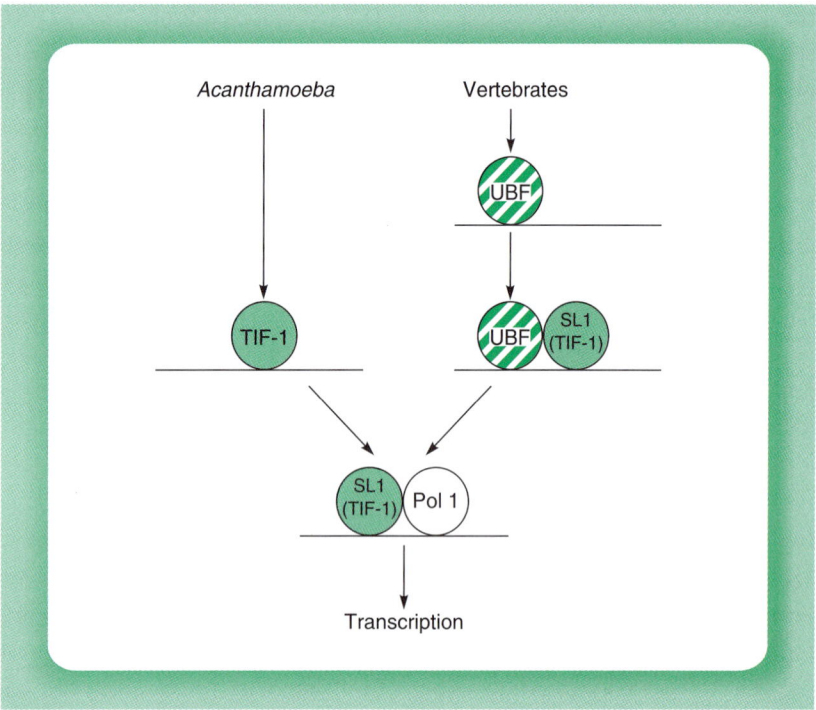

Figure 3.3

Comparison of ribosomal RNA gene transcription in *Acanthamoeba* and vertebrates. In vertebrates, transcription requires both the TIF-1 homologue SL1 and an additional assembly factor UBF whose prior binding is necessary for subsequent binding of SL1.

recognized by the polymerase itself and transcription begins (Fig. 3.4). As with RNA polymerase I, RNA polymerase III binds to the region of DNA adjacent to that which has bound the transcription factor, binding of the polymerase being independent of the DNA sequence in this region.

Although the transcription of the class II RNA polymerase III transcription units such as those encoding the tRNAs is similar to that described for the 5S RNA genes, TFIIIA is not required. Rather transcription is dependent only upon TFIIIB and TFIIIC, with binding of TFIIIC being sufficient for subsequent binding of TFIIIB and the polymerase. Similarly, the class III RNA polymerase III transcription units, which have a TATA box in the promoter (for review see Sollner-Webb, 1988) that resembles that found in RNA polymerase II promoters (see Chapter 1, section 1.3.2) also require TFIIIB for transcription together with other accessory factors (for discussion see Hernandez, 1993).

The process of transcription by RNA polymerases I and III therefore involves the binding of a single transcription factor to the promoter

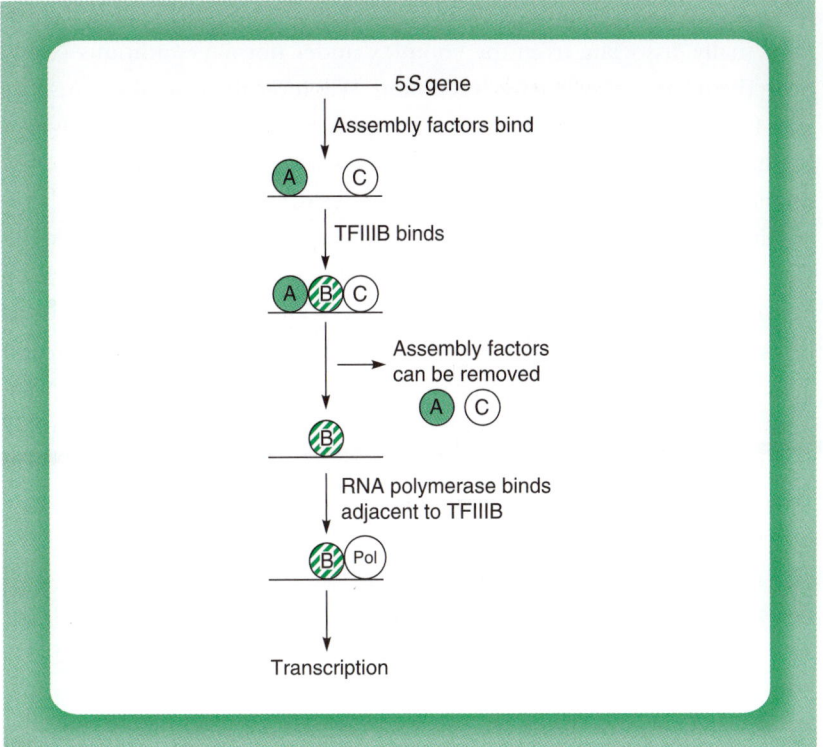

Figure 3.4
Binding of factors to the 5S RNA gene. Transcription requires the initial binding of the assembly factors TFIIIA and TFIIIC with subsequent binding of the transcription factor TFIIIB and of RNA polymerase III itself.

allowing subsequent binding of the RNA polymerase to an adjacent region of DNA. The transcription factor remains bound at the promoter as the polymerase moves down the DNA allowing repeated binding of polymerase molecules and hence repeated rounds of transcription. Binding of the polymerase to the promoter requires prior binding of the transcription factor since the polymerase does not recognize a specific sequence in the promoter but rather makes protein–protein contact with the transcription factor and binds to the adjacent region of the DNA.

In different systems, however, different requirements exist for the binding of the transcription factor itself. Thus in the *Acanthamoeba* system, TIF-1 can bind to DNA in a sequence specific manner and hence is the only factor required. In most other systems, this is not the case and the transcription factors do not bind to the DNA unless other assembly factors which exhibit sequence specific DNA binding are present. Once the transcription factor has bound, these assembly factors can be removed, for example, by detergent treatment without affecting subsequent transcription. It is unclear, however, whether these factors do actually dissociate from the complex under normal conditions *in vivo* once the transcription factor has bound. Whatever the case, the transcription factor itself remains bound at the promoter even after the polymerase has moved down the gene, allowing repeated binding of polymerase molecules and hence repeated rounds of transcription.

Although assembly factors play only an accessory role in transcription itself, they are essential if the complex is to assemble. Hence both assembly factors and transcription factors can be the target for processes which regulate the rate of transcription (for review see Brown *et al.*, 2000). Thus whilst the high rates of polymerase III transcription observed in embryonal carcinoma cells and in hypertrophic growth of cardiac cells are dependent on a high level of transcription factor TFIIIB (Goodfellow *et al.*, 2006), the increase in transcription by this polymerase following adenovirus infection is due to an increase in the activity of the assembly factor TFIIIC. Similarly, alterations in the level of TFIIIA during *Xenopus* development control the nature of the 5S RNA genes which are transcribed at different developmental stages. In addition, as will be discussed in Chapter 9 (section 9.4.3), the retinoblastoma anti-oncoprotein inhibits cellular growth by interacting with UBF to inhibit RNA polymerase I activity and with TFIIIB to inhibit RNA polymerase III activity.

3.5 RNA POLYMERASE II

3.5.1 STEPWISE ASSEMBLY OF THE RNA POLYMERASE II BASAL TRANSCRIPTIONAL COMPLEX

Although some regulation of RNA polymerase I and III activity does occur therefore, this is much less extensive compared to the very wide variety of regulatory events affecting the activity of genes transcribed by RNA polymerase II. As discussed above, this results in a bewildering array of transcription factors interacting with this enzyme and conferring particular patterns of regulation. Interestingly, however, even the basal transcriptional complex which is essential for any transcription by this enzyme contains far more components than is the case for the other RNA polymerases (for reviews see Orphanides *et al.*, 1996; Woychick and Hampsey, 2002; Roeder, 2003; Hahn, 2004).

One component of this complex which has been intensively studied and plays an essential role in RNA polymerase II mediated transcription is TFIID (for review see Burley and Roeder, 1996). In promoters containing a TATA box (see Chapter 1, section 1.3.2), TFIID binds to this element, protecting a region from thirty-five bases to nineteen bases upstream of the start site of transcription in the human hsp70 promoter, for example. The binding of TFIID to the TATA box or equivalent region is the earliest step in the formation of the stable transcriptional complex, such binding being facilitated by another factor TFIIA (Fig. 3.5a).

Interestingly, as TFIID is progressively purified, its requirement for TFIIA to aid its activity decreases. This is because in less purified preparations and in the intact cell, TFIID is associated with a number of inhibitory factors such as DR1 and DR2 which act by preventing its binding to the DNA and/or its interaction with other components of the basal complex such as TFIIB (see below) (for further discussion of the role of DR1, see Chapter 6, section 6.3.3). One role of TFIIA appears to be to bind to TFIID and overcome this inhibition, thereby stimulating the activity of TFIID. Hence the need for TFIIA decreases as TFIID is purified away from these inhibitory factors, although it is likely to play a critical role in the intact cell. In addition, TFIIA may also play a role in the response to transcriptional activators acting as a co-activator molecule linking DNA-bound activators and the basal transcriptional complex.

Hence rather than acting as a basal transcription factor essential for all transcription, TFIIA appears to play a key role in the response of the complex to activating and inhibiting molecules. Such a role is of particular importance since the antagonism between positively and negatively acting factors in the assembly of the basal transcriptional complex may play a critical role in regulating the rate of transcription, representing a major target for activators and repressors of transcription (see Chapters 5 and 6 for a further discussion of the mechanisms by which specific factors activate or inhibit transcription).

Once TFIID has bound to the DNA, another transcription factor, TFIIB, joins the complex by binding to TFIID (Fig. 3.5b). This binding of TFIIB is an essential step in initiation complex formation since, as well as binding to TFIID, TFIIB can also bind to the RNA polymerase itself. Hence it acts as a bridging factor allowing the recruitment of RNA polymerase to the complex in association with another factor TFIIF (Fig. 3.5c). Following polymerase binding, three other transcription factors, TFIIE, TFIIH and TFIIJ, rapidly associate with the complex (Fig. 3.5d). At this point, TFIIH, which has a DNA helicase activity, unwinds the double-stranded DNA so allowing it to be copied into RNA. Subsequently, the kinase activity of TFIIH which allows it to phosphorylate other proteins, phosphorylates the C-terminal domain of RNA polymerase on the serine amino acid at position five in the conserved sequence Tyr-Ser-Pro-Thr-Ser-Pro-Ser (see section 3.1) (for review see Orphanides et al., 1996). This converts it from the non-phosphorylated form which joins the complex to the phosphorylated form which is capable of beginning transcription (Fig. 3.6).

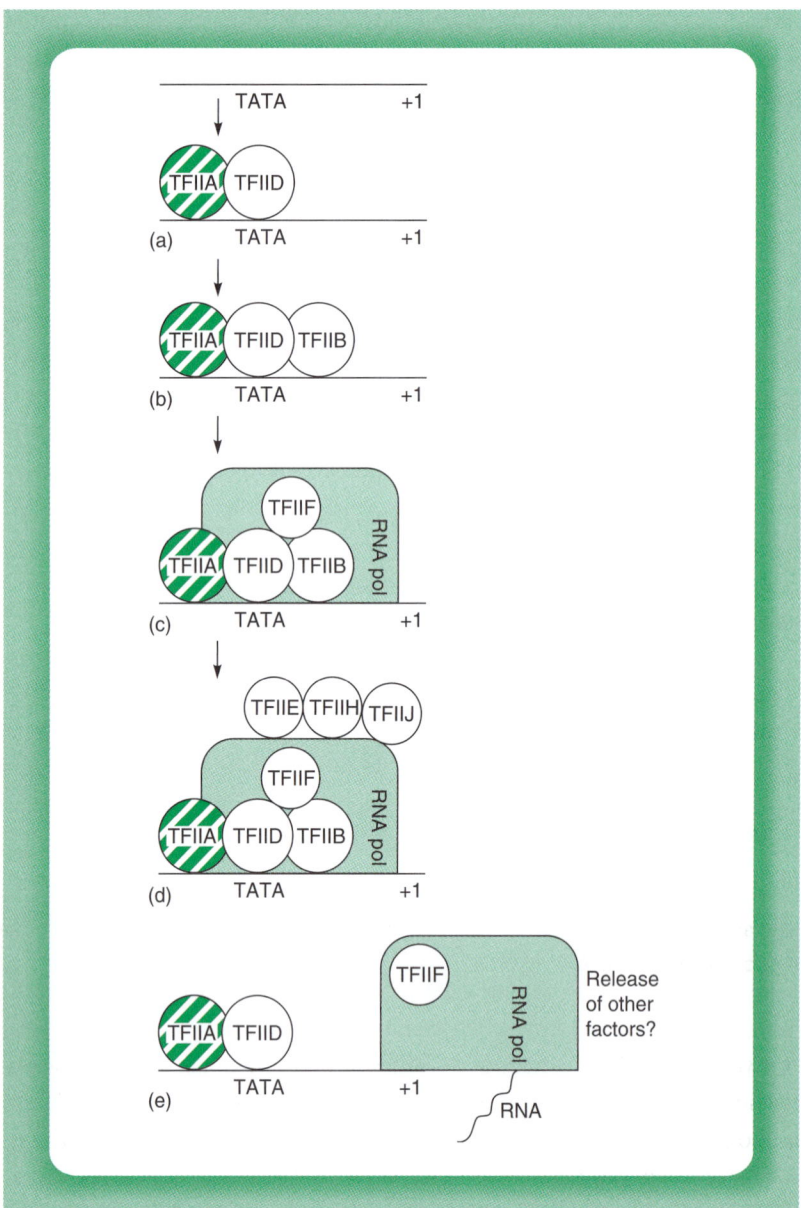

Figure 3.5

Stages in the assembly of the stable transcriptional complex for RNA polymerase II transcription. As the polymerase moves away from the promoter to transcribe the gene, TFIIF remains associated with it whilst TFIIA and TFIID remain bound at the TATA box allowing the formation of a new stable complex and further rounds of transcription.

Hence TFIIH via its kinase and helicase activities plays a critical role in allowing the basal transcriptional complex to initiate transcription. Moreover, TFIIH also plays a critical role in the repair of damaged DNA providing a possible link between the processes of DNA repair and transcription

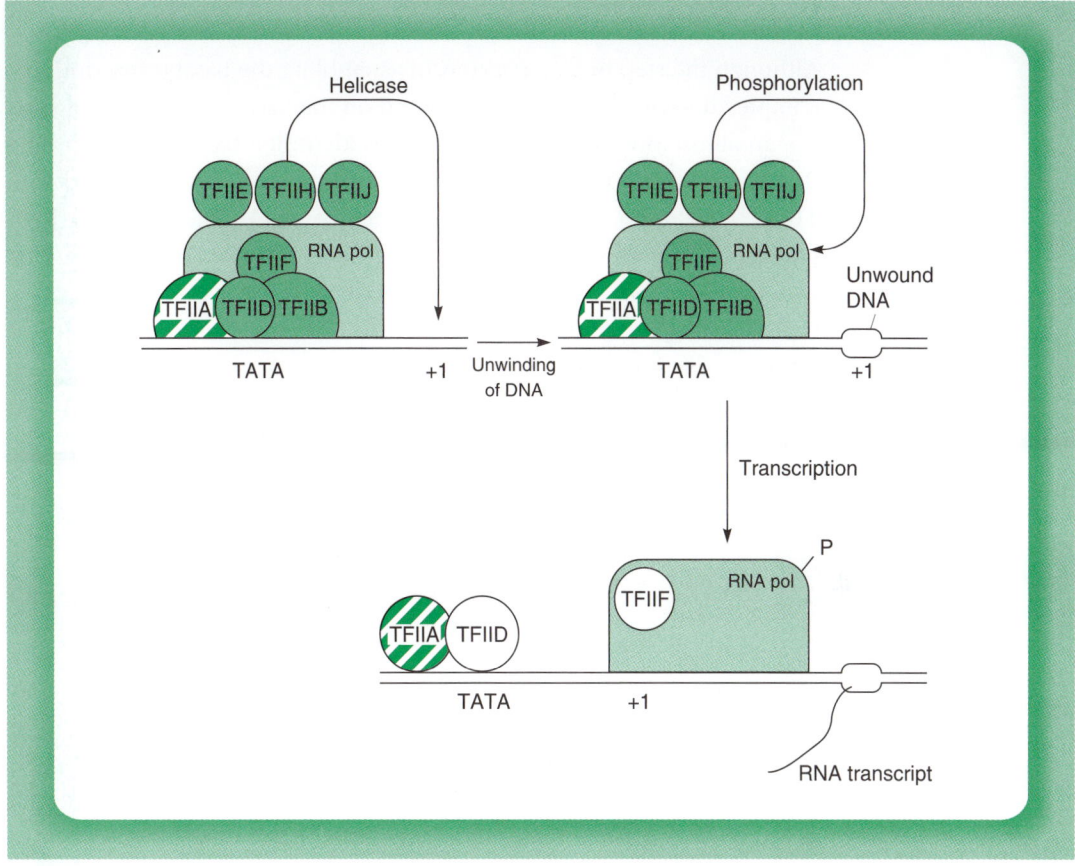

Figure 3.6

TFIIH has a helicase activity which unwinds the DNA allowing its transcription into RNA and a kinase activity that phosphorylates the C-terminal region of RNA polymerase which allows it to begin transcription.

(for reviews of TFIIH see Hoeijmakers *et al.*, 1996; Zurita and Merino, 2003). Interestingly, it has been shown that the kinase activity associated with TFIIH can also phosphorylate the retinoic acid receptor, which is a member of the nuclear receptor transcription factor family discussed in Chapter 4 (section 4.4). This phosphorylation stimulates the ability of the retinoic acid receptor to activate transcription (Rochette-Egly *et al.*, 1997), indicating that TFIIH may play a role in the regulation of transcription factor activity by phosphorylation (see Chapter 8, section 8.4.2).

The complex of the seven factors (TFIIA, B, D, E, F, H and J) and the polymerase is thus sufficient for transcription to occur. As the polymerase moves down the gene during this process, TFIIF remains associated with it, whilst TFIIA and TFIID remain bound at the promoter and are capable of binding another molecule of polymerase allowing repeated rounds of transcription as with the other polymerases (see Fig. 3.5e).

3.5.2 THE RNA POLYMERASE HOLOENZYME

Although the step by step pathway of assembling the basal transcriptional complex described above was proposed on the basis of a number of studies, an alternative pathway has also been identified based on the finding that some RNA polymerase is found in solution already associated with TFIIB, TFIIF and TFIIH in the absence of DNA. This so-called RNA polymerase holoenzyme has now been observed in a wide range of organisms ranging from yeast to man. It is clear therefore that, in some cases, following binding of TFIIA and TFIID to the promoter, this complex of RNA polymerase and associated factors may bind, resulting in a reduced number of steps being required for complex formation (Fig. 3.7) (for discussion see Greenblatt, 1997; Myer and Young, 1998).

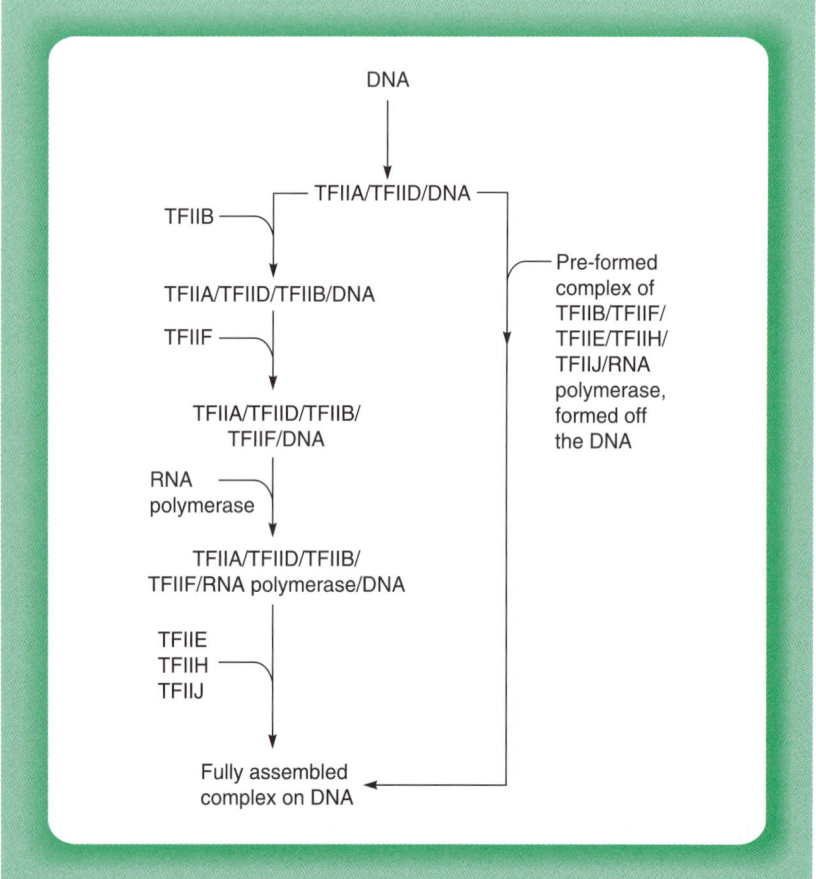

Figure 3.7

Alternative pathways in the assembly of the stable transcriptional complex for RNA polymerase II involving either the step by step pathway (see Fig. 3.5) or the binding of a pre-formed complex of RNA polymerase and its associated factors to DNA which has already bound TFIIA and TFIID.

Interestingly, the RNA polymerase holoenzyme also contains a number of other components apart from RNA polymerase itself and the basal transcription factors. Thus, it includes a complex of proteins known as the mediator complex which appears to be required, at least in yeast, for the response to transcriptional activators (see Chapter 5, section 5.4.1). Hence the mediator may serve as a link between these activators and the components of the basal transcriptional complex whose activity they stimulate. In addition, the holoenzyme can also associate with the SWI/SNF complex discussed in Chapter 1 (section 1.2.2), whose role is to remodel the chromatin into a form which allows the binding of transcriptional activators and transcription itself. Hence, at least in some cases, this remodelling complex can be recruited to DNA together with the RNA polymerase and its associated proteins (see Chapter 5, section 5.5.1).

The RNA polymerase holoenzyme is thus a highly complex structure which, as well as RNA polymerase itself and basal transcription factors, also contains factors involved in the response to transcriptional activators and others which remodel chromatin structure. Although this holoenzyme represents only one of the two possible methods by which the basal transcription complex assembles on the DNA, it is clear that regardless of its method of assembly, the basic stable transcriptional complex for RNA polymerase II requires a number of factors in addition to the polymerase itself and is therefore much more complex than that of RNA polymerase I or III.

3.6 TBP: THE UNIVERSAL TRANSCRIPTION FACTOR?

Most of the transcription factors described in the previous sections were isolated by the biochemical fractionation of cellular extracts and were then shown to have a particular functional activity in modulating the rate of transcription when mixed with RNA polymerase and other sub-cellular fractions. When these factors were characterized in more detail by further fractionation, and subsequent cloning, however, many of them were shown to consist of several different proteins which together are responsible for the properties ascribed to the original factor. Thus although these factors have been dealt with for simplicity in the previous sections as single factors, most of them are in fact complexes of several different proteins; for example, TFIIE and TFIIF both contain two distinct proteins. Similarly, TFIIH is a multi-protein complex whose structure has been determined (Chang and Kornbeg, 2000; Schultz et al., 2000) with one of the component proteins having the kinase activity which results in phosphorylation of the RNA polymerase whilst another has the helicase activity which unwinds the DNA (see section 3.5.1) (for review see Hoejmakers *et al.*, 1996; Zurita and Merino, 2003).

This responsibility of one component of the complex for an activity formerly ascribed to the whole complex is seen most clearly in TFIID. Thus, TFIID is a multi-protein complex in which only one protein, known as TBP (TATA-binding protein), directs the binding to the TATA box whilst the other components of the complex, known as TAFs (TBP-associated factors), do not bind directly to the TATA box and appear to allow TFIID to respond to stimulation by transcriptional activators (see Chapter 5, section 5.4.2) (for review see Hahn, 1998; Green, 2000). They thus represent co-activator molecules, linking transcriptional activators and the basal transcriptional complex.

Hence TBP plays a critical role in the transcription of TATA box-containing RNA polymerase II promoters by binding to the TATA box as the first step in assembly of the basal transcriptional complex. In view of this critical role, it is not surprising that TBP is one of the most highly conserved eukaryotic proteins. The structure of this protein has been defined by X-ray crystallography and shown to have a saddle structure in which the concave underside binds to DNA and the convex outer surface is accessible for interactions with other factors. Most interestingly, binding of TBP to the DNA deforms the DNA so that it follows the concave curve of the saddle (Fig. 3.8). Moreover, structural studies of the TFIID complex (consisting of TBP and the TAFs) bound to DNA have indicated that it resembles the complex of the eight histone molecules around which DNA is wound in the nucleosome to form the normal chromatin structure (see Chapter 1, section 1.2.1). Hence the DNA may bend around TFIID at the promoter in a similar manner to the folding of the rest of DNA in the basic nucleosome structure of chromatin (for reviews see Hoffmann *et al.*, 1997; Gangloff *et al.*, 2001). This role for TFIID in altering nucleosome structure at the promoter is also supported by the finding that $TAF_{II}250$, one of the subunits of TFIID, has histone acetyltransferase activity (Mizzen *et al.*, 1996), since acetylation of histones appears to play a key role in modulating chromatin structure (see Chapter 1, section 1.2.3).

The bent DNA with TFIID bound to it serves as the central platform on which the basal transcriptional complex assembles. Thus, structural studies have shown that TFIIA binds to the amino terminal stirrup of the TBP saddle and interacts only with the DNA upstream of the TATA box. This allows it to fulfil its role of protecting TFIID from inhibition by transcriptional repressors and allowing it to respond to activators bound to upstream DNA sequences (see section 3.5.1). In contrast, TFIIB binds to the carboxyl-terminal stirrup of the TBP saddle and binds to the DNA downstream (as well as upstream) of the TATA box (Andel *et al.*, 1999).

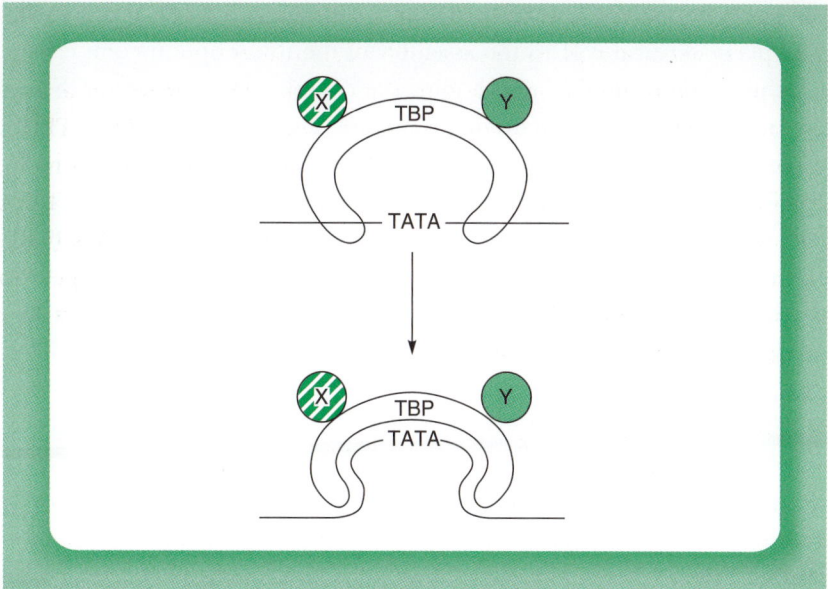

Figure 3.8

The saddle structure of TBP as determined by X-ray crystallography allows the concave surface to interact with the TATA box whilst the convex surface associates with other accessory transcription factors (X and Y). The initial binding induces the bending of the DNA so that it follows the concave under surface of the saddle. See also Plate 1.

This allows it to fulfil its role of acting as a bridge between TBP and RNA polymerase II so positioning the start site of transcription by the polymerase relative to the TATA box (see Plate 1; Geiger *et al.*, 1996) (for reviews see Woychick and Hampsey, 2002; Hahn, 2004; Svejstrup *et al.*, 2006). Interestingly, however, it appears that TFIIB does more than simply promote recruitment of factors to the complex. Thus, it has been shown that binding of TFIIB promotes bending of the DNA by TBP (Zhao and Herr, 2002). Moreover, structural studies indicate that, by interacting with both TFIID and the polymerase, TFIIB ensures that the DNA is correctly positioned so that it can enter the interior of the polymerase molecule, so allowing transcription to occur (Bushnell *et al.*, 2004; Chen and Hahn, 2004; for review see Hahn, 2004).

Paradoxically in view of its TATA box binding ability, TBP also plays a critical role in the transcription of the subset of RNA polymerase II genes which do not contain a TATA box (see Chapter 1, section 1.3.2). In this case, however, TBP does not bind to the DNA but is recruited to the promoter by another DNA binding protein which binds to the initiator element overlapping the transcriptional start site. TBP then binds to this initiator binding protein allowing the recruitment of TFIIB and the

RNA polymerase itself, as for promoters containing a TATA box. Hence, TBP plays a critical role in the assembly of the transcription complex for RNA polymerase II, although it joins the complex by binding to DNA in the case of TATA box-containing promoters (Fig. 3.9a) and is recruited by protein–protein interactions in the case of promoters which lack a TATA box (Fig. 3.9b). Interestingly, a recent study has suggested that in promoters lacking a TATA box, the basal transcriptional complex can actually assemble on an enhancer element (see Chapter 1, section 1.3.4) and then move to the promoter by sliding along the DNA or by looping out of the intervening DNA (George *et al.*, 2006; for review see Szutorisz *et al.*, 2005).

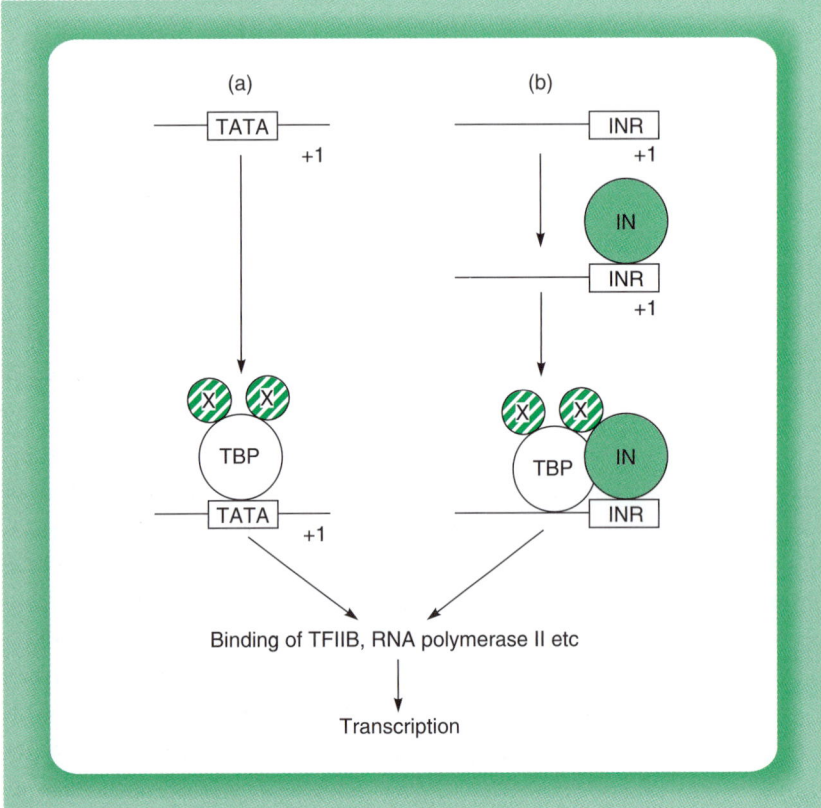

Figure 3.9

Transcription of promoters by RNA polymerase II involves the recruitment of TBP (and associated factors (X) forming the TFIID complex) to the promoter. This may be achieved by direct DNA binding to the TATA box where this is present (panel a) or by protein–protein interaction with a factor (IN) bound to the initiator element where the TATA box is absent (panel b).

The key role of TBP in different types of RNA polymerase II-transcribed genes has led to the suggestion that TBP represents the basic transcription factor for RNA polymerase II, paralleling the role of SL1 for RNA polymerase I and TFIIIB for RNA polymerase III. This idea was supported by the amazing finding that TBP is actually also a component of both SL1 and TFIIIB (for review see White and Jackson, 1992). Thus the SL1 factor is actually a complex of four factors, one of which is TBP. Hence when SL1 is recruited to the promoter by UBF (see section 3.3), TBP is delivered to the DNA exactly as in the non-TATA box-containing RNA polymerase II promoters where TBP is recruited by the prior binding of another protein to the initiator element.

Similarly, in the case of RNA polymerase III transcription where TBP is part of the multi-component TFIIIB complex (for review see Rigby, 1993), TBP is delivered to class I polymerase III promoters by protein–protein interaction following the prior binding of TFIIIA and TFIIIC and is delivered to class II polymerase III promoters by the prior binding of TFIIIC (section 3.5) (Fig. 3.10a). Interestingly however, as noted in

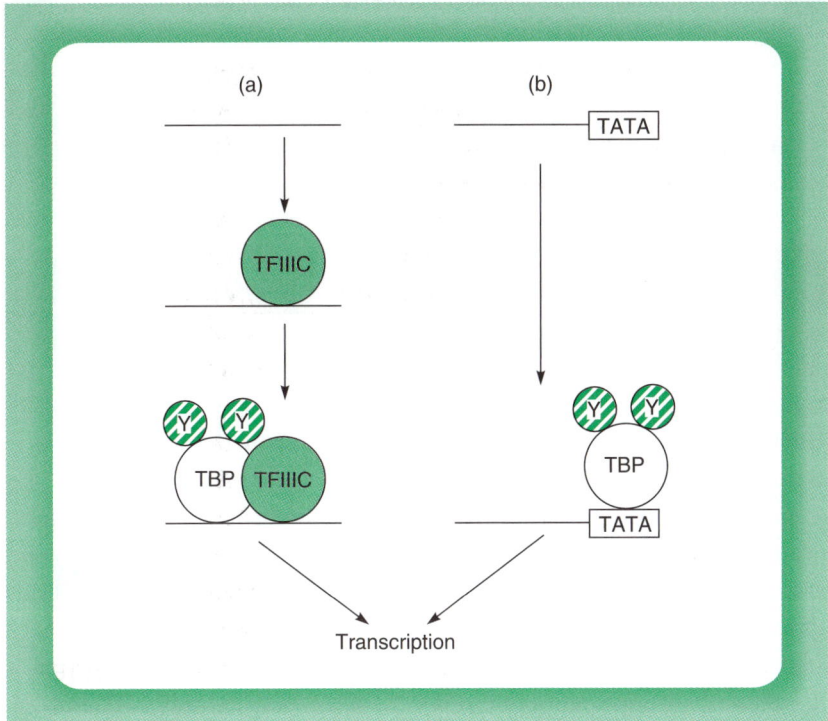

Figure 3.10

Transcription of promoters by RNA polymerase III involves the recruitment of TBP (and associated factors (Y) forming the TFIIIB complex) to the promoter. This may be achieved by protein–protein interactions with TFIIIA and TFIIIC in the case of class I promoters, with TFIIIC alone in the case of class II promoters (panel a) or by direct DNA binding to the TATA box in class III promoters where the TATA box is present (panel b).

section 3.5, the class III group of RNA polymerase III promoters contain a TATA box and hence in this case TBP can bind directly (Fig. 3.10b). As in RNA polymerase II promoters, distinct mechanisms therefore ensure the recruitment of TBP to all RNA polymerase III promoters.

The similarities between the three RNA polymerases discussed in section 3.1 are therefore paralleled by the involvement of a common factor, TBP, in transcription by all three RNA polymerases (Fig. 3.11). Interestingly, this relationship has been extended further by the finding that TFIIH also plays an essential role in transcription by RNA polymerase I as well as that mediated by RNA polymerase II (Iben *et al.*, 2002). Similarly, the structural proteins, β-actin and nuclear myosin appear to play an important role in transcription by all three RNA polymerases (see for example Hu *et al.*, 2004; Philimonenko *et al.*, 2004; Vreugde *et al.*, 2006).

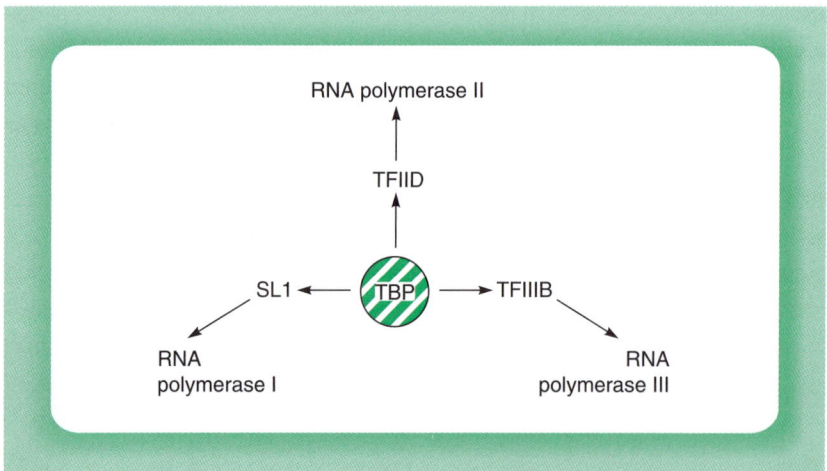

Figure 3.11
TBP is involved in transcription by all three RNA polymerases.

In all three RNA polymerase complexes, TBP forms a part of the multi-protein complexes which have been shown to be essential for transcription binding directly via the TATA box or by protein–protein interactions with assembly factors (for review see Struhl, 1994). In turn, structural analysis has shown that TBP then makes contact with other factors within the basal transcriptional complexes for RNA polymerases I, II and III, enhancing the assembly and/or activity of the complexes (Schröder *et al.*, 2003; Bric *et al.*, 2004).

This role of TBP in transcription by all three polymerases has suggested that it represents an evolutionarily ancient transcription factor preceding the division of the three RNA polymerases and having a universal and essential role in eukaryotic transcription (for review see Hernandez,

1993). Indeed, a TBP homologue is also found in the archaebacteria, which constitutes a separate kingdom distinct from the eukaryotes and the eubacteria. Hence the existence of TBP appears to predate not only the divergence of the three RNA polymerases but also the divergence of the eukaryotic and archaebacterial kingdoms (for review of archaebacterial transcription see Reeve *et al.*, 1997).

It was initially believed that each organism would have only one form of TBP encoded by a single gene. However, studies of the genomes of multi-cellular organisms have identified genes encoding other TBP-related proteins with humans, for example, having one such TBP-like factor (TLF) whereas *Drosophila* has two (for reviews see Berk, 2000; Veenstra and Wolffe, 2001).

It has been shown that in some cases the basal transcriptional complex contains a TLF rather than a TBP. For example, specific stages of development in the amphibian *Xenopus* require a TLF activity which cannot be substituted by TBP (Veenstra *et al.*, 2000). Similarly, the *Drosophila* PCNA gene has two promoters, one of which is recognized by a basal transcriptional complex containing TBP and the other by a complex containing TRF2 which is a TLF (Hocheimer *et al.*, 2002), and specific genes regulated by TRF2 rather than TBP have also been identified in mammalian cells (Chong *et al.*, 2005). Hence, it is clear that some specific transcription complexes contain a TLF rather than TBP and that this is required for their proper functioning (Fig. 3.12b). The existence of the TLFs thus offers a further means of regulating gene transcription in specific situations.

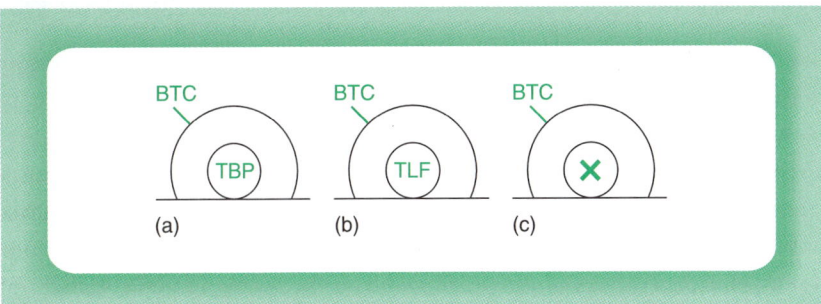

Figure 3.12

On different target promoters, the basal transcriptional complex (BTC) for RNA polymerase II can contain TBP (panel a), a TBP-like factor (TLF) (panel b) or lack either TBP or TLF which are presumably replaced by an unrelated factor (X) (panel c).

Interestingly, as well as basal transcriptional complexes containing either TBP or a TLF, it has also been shown that RNA polymerase II transcription can be driven by a complex which does not contain TBP or a TLF (Wieczorek *et al.*, 1998). This suggests that under some circumstances neither TBP nor TLF is required for transcription. In agreement with this, some RNA polymerase II transcription occurs in early stage embryos from knock out mice lacking functional TBP, although transcription by RNA polymerases I and III does not occur. As early stage mouse embryos do not express a TLF, this indicates the existence of TBP/TLF independent transcription, at least for RNA polymerase II (Martianov *et al.*, 2002) (Fig. 3.12c).

Hence, TBP is a highly ancient transcription factor which is involved in transcription by all three RNA polymerases (see Fig. 3.11). However, in some situations transcription can occur in a TBP-independent manner involving either a TBP-like factor or a complex which lacks TBP or a TLF (Fig. 3.12).

3.7 TRANSCRIPTIONAL ELONGATION

So far in this chapter we have discussed the mechanisms by which transcription is initiated and this is clearly a key target for regulatory processes. It has become increasingly clear, however, that it is also necessary to consider the process by which this initial transcript is elongated, since this can also be a target for regulation (for reviews see Arndt and Kane, 2003; Sims *et al.*, 2004).

Thus, as described in section 3.5.1, following the formation of the basal transcriptional complex, TFIIH phosphorylates the polymerase on the serine at position five of the conserved sequence Tyr-Ser-Pro-Thr-Ser-Pro-Ser (YSPTSPS in the one letter code for amino acids) and transcription is initiated (Fig. 3.13a). However, transcription proceeds for only twenty to thirty bases and the polymerase then pauses and does not proceed further (Fig. 3.13b). At this stage, the short nascent RNA transcript becomes capped by the addition of a modified G nucleotide at its 5' end. This is achieved by an enzyme complex whose recruitment is dependent on the prior phosphorylation of the polymerase on the serine at position five.

In turn, capping of the short nascent RNA transcript serves as a signal to recruit the P-TEFb kinase (for review see Peterlin and Price, 2006; Phatnani and Greenleaf, 2006). This phosphorylates the RNA polymerase on serine two of the conserved YSPTSPS sequence and this allows transcriptional elongation to occur (Fig. 3.13c) (for review see Orphanides and Reinberg, 2002; Sims *et al.*, 2004). Hence, the process of transcriptional elongation is linked to the post-transcriptional modification of the RNA transcript by capping. This close linkage of transcriptional and

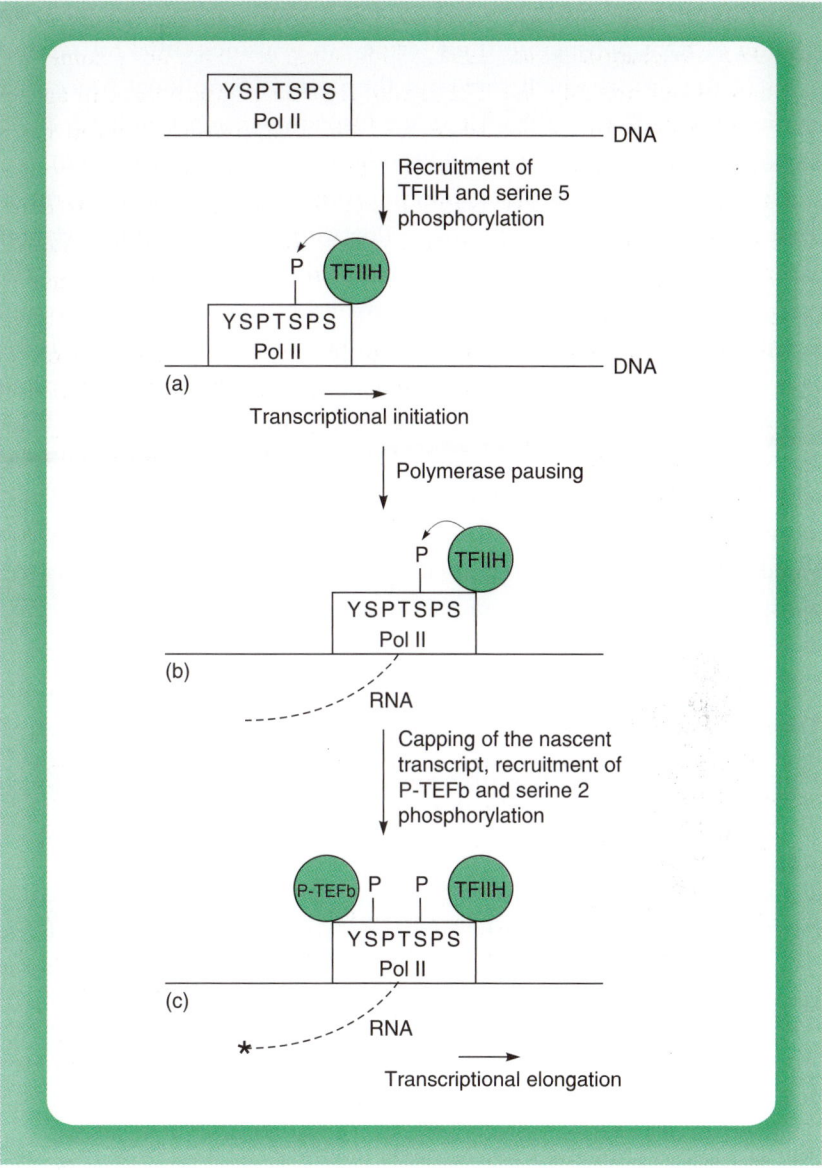

Figure 3.13

Recruitment of the TFIIH factor results in the phosphorylation of RNA polymerase
II (Pol II) on serine 5 of the conserved C-terminal domain sequence Tyr-Ser-
Pro-Thr-Ser-Pro-Ser (YSPTSPS) in the one letter code for amino acids and this
allows transcriptional initiation to occur (panel a). However, the polymerase then
pauses after transcribing only twenty to thirty nucleotides (panel b). Transcriptional
elongation requires capping of the RNA transcript at its 5′ end (indicated by
the star). This allows recruitment of the P-TEFb kinase, which phosphorylates
the polymerase on serine 2 of the conserved sequence, allowing transcriptional
elongation to occur (panel c).

post-transcriptional processes can be extended further since the phosphorylation of RNA polymerase II on serine two is also essential for recruitment of the factors which carry out the post-transcriptional processing of the RNA transcript (for reviews see Orphanides and Reinberg, 2002; Calvo and Manley, 2003; Ares and Proudfoot, 2005; Latchman, 2005).

Hence, transcriptional elongation to produce the full RNA transcript is distinct from the initiation of transcription with different phosphorylation events being required, although the conserved C-terminal domain of RNA polymerase II plays a key role in these processes and in coupling transcriptional and post-transcriptional events (for review see Meinhart *et al.*, 2005). Interestingly, these effects are not unique to RNA polymerase II. For example, it has been suggested that the UBF basal transcription factor for RNA polymerase I (see section 3.3) plays a key role in the transition from transcriptional initiation to transcriptional elongation (Panov *et al.*, 2006). Moreover, it has been shown that growth factors enhance RNA polymerase I-mediated transcription by stimulating transcriptional elongation via the phosphorylation of UBF (Stefanovsky *et al.*, 2006).

3.8 CONCLUSIONS

The binding of each of the three eukaryotic RNA polymerases to appropriate gene promoters and subsequent transcription is dependent on the prior binding of a specific transcription factor to the promoter. Binding of the polymerase to the DNA adjacent to this factor occurs by recognition of the bound protein rather than by recognition of the specific DNA sequence in this region. In most cases, the binding of the transcription factor itself requires the prior binding of other factors to the DNA. These assembly factors therefore play a critical role in the formation of the stable transcriptional complex but can be dissociated once the complex has formed without affecting its activity. In the case of RNA polymerase II transcription, either the stability of the complex or its activity is greatly affected by the binding of other proteins to sequences upstream of the promoter. The roles of these transcription factors and the mechanisms by which they function are described in the remainder of this book.

REFERENCES

Ahearn, J.M. Jr., Bartolomei, M.S., West, M.L., Cisek, L.J. and Corden, J.L. (1987) Cloning and sequence analysis of the mouse genomic locus encoding the large subunit of RNA polymerase II. Journal of Biological Chemistry 262, 10695–10705.

Andel, F., Ladurner, A.G., Inouye, C., Tjian, R. and Nogales, E. (1999) Three-dimensional structure of the human TFIID-IIa-IIB complex. Science 286, 2153–2156.

Ares, M. and Proudfoot, N.J. (2005) The Spanish connection: transcription and mRNA processing get even closer. Cell 120, 163–166.

Arndt, K.M. and Kane, C.M. (2003) Running with RNA polymerase: eukaryotic transcript elongation. Trends in Genetics 19, 543–550.

Asturias, F.J. and Craighead, J.L. (2003) RNA polymerase II at initiation. Proceedings of the National Academy of Sciences, USA 100, 6893–6895.

Berk, A.J. (2000) TBP-like factors come into focus. Cell 103, 5–8.

Bric, A., Radebaugh, C.A. and Paule, M.R. (2004) Photocross-linking of the RNA polymerase I preinitiation and immediate postinitiation complexes. Journal of Biological Chemistry 279, 31259–31267.

Brown, D.D. (1984) The role of stable complexes that repress and activate eukaryotic genes. Cell 37, 359–365.

Brown, T.R., Scott, PH., Stein, T., Winter, AG. and White, R.J. (2000) RNA polymerase III transcription: its control by tumour suppressors and its deregulation by transforming agents. Gene Expression 9, 15–28.

Burley, S.K. and Roeder, R.G. (1996) Biochemistry and structural biology of transcription factor IID (TFIID). Annual Review of Biochemistry 65, 769–799.

Bushnell, D.A., Westover, K.D., Davis, R.E. and Kornberg, R.D. (2004) An RNA polymerase II-TFIIB cocrystal at 4.5 angstroms. Science 303, 983–988.

Calvo, O. and Manley, J.L. (2003) Strange bedfellows: polyadenylation factors at the promoter. Genes and Development 17, 1321–1327.

Chang, W-H. and Kornberg, R.D. (2000) Electron crystal structure of the transcription factor and DNA repair complex, core TFIIH. Cell 102, 609–613.

Chen, H.T. and Hahn, S. (2004) Mapping the location of TFIIB within the RNA polymerase II transcription preinitiation complex: a model for the structure of the PIC. Cell 119, 169–180.

Chong, J.A., Moran, M.M., Teichmann, M., Kaczmarek, J.S., Roeder, R. and Clapham, D.E. (2005) TATA-binding protein (TBP)-like factor (TLF) is a functional regulator of transcription: reciprocal regulation of the neurofibromatosis type 1 and c-fos genes by TLF/TRF2 and TBP. Molecular and Cellular Biology 25, 2632–2643.

Cilberto, G., Castagnoli, L. and Cortese, R. (1983) Transcription by RNA polymerase III. Current Topics in Developmental Biology 18, 59–88.

Cramer, P., Bushnell, D.A. and Kornberg, R.D. (2001) Structural basis of transcription: RNA polymerase II at 2.8 angstrom resolution. Science 292, 1863–1876.

Gangloff, Y-G., Romier, C., Thuault, S., Werten, S. and Davidson, I. (2001) The histone fold is a key structural motif of transcription factor TFIID. Trends in Biochemical Sciences 26, 250–256.

Geiduschek, E.P. and Kassavetis, G.A. (2001) The RNA polymerase III transcription apparatus. Journal of Molecular Biology 310, 1–26.

Geiger, J.H., Hahn, S., Lee, S. and Sigler, P.B. (1996) Crystal structure of the yeast TFIIA/TBP/DNA complex. Science 272, 830–836.

George, A.A., Sharma, M., Singh, B.N., Sahoo, N.C. and Rao, K.V. (2006) Transcription regulation from a TATA and INR-less promoter: spatial segregation of promoter function. EMBO Journal 25, 811–821.

Goodfellow, S.J., Innes, F., Derblay, L.E., Maclellan, W.R., Scott, P.H. and White, R.J. (2006) Regulation of RNA polymerase III transcription during hypertrophic growth. EMBO Journal 25, 1522–1533.

Green, M.R. (2000) TBP-associated factors (TAF$_{II}$s): multiple, selective transcriptional mediators in common complexes. Trends in Biochemical Sciences 25, 59–63.

Greenblatt, J. (1997) RNA polymerase II holoenzyme and transcriptional regulation. Current Opinion in Cell Biology 9, 310–319.

Grummt, I. (2003) Life on a planet of its own: regulation of RNA polymerase I transcription in the nucleolus. Genes and Development 17, 1691–1702.

Hahn, S. (1998) The role of TAFs in RNA polymerase II transcription. Cell 95, 579–582.

Hahn, S. (2004) Structure and mechanism of the RNA polymerase II transcription machinery. Nature Structural and Molecular Biology 11, 394–403.

Hernandez, N. (1993) TBP, a universal eukaryotic transcription factor. Genes and Development 7, 1291–1308.

Hochheimer, A., Zhou, S., Zheng, S., Holmes, M.C. and Tjian, R. (2002) TRF2 associates with DREF and directs promoter-selective gene expression in *Drosophila*. Nature 420, 439–445.

Hoeijmakers, J.H.J., Egly, J.M. and Vermueler, W. (1996) TFIIH: a key component in multiple DNA transactions. Current Opinion in Genetics and Development 6, 26–33.

Hoffman, A., Oelgeschlage, T. and Roeder, R. (1997) Considerations of transcriptional control mechanisms: do TFIID-core promoter complexes recapitulate nucleosome-like functions. Proceedings of the National Academy of Sciences, USA, 94, 8923–8925.

Hu, P., Wu, S. and Hernandez, N. (2004) A role for β-actin in RNA polymerase III transcription. Genes and Development 18, 3010–3015.

Iben, S., Tschochner, H., Bier, M., Hoogstraten, D., Hozák, P., Egly, J-M. and Grummt, I. (2002) TFIIH plays an essential role in RNA polymerase I transcription. Cell 109, 297–306.

Klug, A. (2001) A marvellous machine for making messages. Science 292, 1844–1846.

Landick, R. (2001) RNA polymerase clamps down. Cell 105, 567–570.

Latchman, D.S. (2005) *Gene regulation: a eukaryotic perspective.* Fifth edition. Taylor and Francis, Oxford and New York.

Martianov, I., Viville, S. and Davidson, I. (2002) RNA polymerase II transcription in murine cells lacking the TATA-binding protein. Science 298, 1036–1039.

Meinhart, A., Kamenski, T., Hoeppner, S., Baumli, S. and Cramer, P. (2005) A structural perspective of CTD function. Genes and Development 19, 1401–1415.

Memet, S., Saurn, W. and Sentenac, A. (1988) RNA polymerases B and C are more closely related to each other than to RNA polymerase A. Journal of Biological Chemistry 263, 10048–10051.

Mizzen, C.A., Yang, X-J., Kokubo, T., Brownwell, J.E., Bannister, A.J., Owen-Hughes, T., Workman, J., Wang, L., Berger, S.L., Kouzarides, T., Nakatani, Y. and Allis, C.D. (1996) The TAFII 250 subunit of TFIID has histone acetyltransferase activity. Cell 87, 1261–1270.

Myer, V.E. and Young, R.A. (1998) RNA polymerase II holoenzymes and subcomplexes. Journal of Biological Chemistry 273, 27757–27760.

Orphanides, G. and Reinberg, D. (2002) A unified theory of gene expression. Cell 108, 439–451.

Orphanides, G., Lagrange, T. and Reinberg, D. (1996) The general transcription factors of RNA polymerase II. Genes and Development 10, 2657–2683.

Panov, K.I., Friedrich, J.K., Russell, J. and Zomerdijk, J.C. (2006) UBF activates RNA polymerase I transcription by stimulating promoter escape. EMBO Journal 25, 3310–3322.

Paule, M.R. and White, R.J. (2000) Transcription by RNA polymerases I and III. Nucleic Acids Research 28, 1283–1298.

Peterlin, B.M. and Price, D.H. (2006) Controlling the elongation phase of transcription with P-TEFb. Molecular Cell 23, 297–305.

Phatnani, H.P. and Greenleaf, A.L. (2006) Phosphorylation and functions of the RNA polymerase II CTD. Genes and Development 20, 2922–2936.

Philimonenko, V.V., Zhao, J., Iben, S., Dingova, H., Kysela, K., Kahle, M., Zentgraf, H., Hofmann, W.A., de Lanerolle, P., Hozak, P. and Grummt, I. (2004) Nuclear actin and myosin I are required for RNA polymerase I transcription. Nature Cell Biology 6, 1165–1172.

Reeve, J.N., Sandman, K. and Daniels, C.J. (1997) Archaeal histones nucleosomes and transcriptional initiation. Cell 89, 999–1002.

Rigby, P.W.J. (1993) Three in one and one in three: it all depends on TBP. Cell 72, 7–10.

Riva, M., Schaffner, A.R., Sentenac, A., Hartmann, G.R., Mustner, A.A., Zaychikov, F. and Grachev, M.A. (1987) Active site labelling of the RNA polymerases A, B and C from yeast. Journal of Biological Chemistry 262, 14377–14380.

Rochette-Egly, C., Adam, S., Rossignol, M., Egly, J-M. and Chambon, P. (1997) Stimulation of RARα activation function AF-1 through binding to the general transcription factor TFIIH and phosphorylation by CDK7. Cell 90, 97–107.

Roeder, R.G. (2003) The eukaryotic transcriptional machinery: complexities and mechanisms unforeseen. Nature Medicine 9, 1239–1244.

Russell, J. and Zomerdijk, J.C.B.M. (2005) RNA-polymerase-I-directed rDNA transcription, life and works. Trends in Biochemical Sciences 30, 87–96.

Saltzman, A.G. and Weinmann, R. (1989) Promoter specificity and modulation of RNA polymerase II transcription. FASEB Journal 3, 1723–1733.

Schramm, L. and Hernandez, N. (2002) Recruitment of RNA polymerase III to its target promoters. Genes and Development 16, 2593–2620.

Schröder, O., Bryant, G.O., Geiduschek, E.P., Berk, A.J. and Kassavetis, G.A. (2003) A common site on TBP for transcription by RNA polymerases II and III. EMBO Journal 22, 5115–5124.

Schultz, P., Fribourg, S., Poterszman, A., Mallouh, V., Moras, D. and Egly, J.M. (2000) Molecular structure of human TFIIH. Cell 102, 599–607.

Sentenac, A. (1985) Eukaryotic RNA polymerases. CRC Critical Reviews in Biochemistry 1, 31–90.

Sims, R.J., III, Belotserkovskaya, R. and Reinberg, D. (2004) Elongation by RNA polymerase II: the short and long of it. Genes and Development 18, 2437–2468.

Sollner-Webb, B. (1988) Surprises in RNA polymerase III transcription Cell 52,153–154.

Somerville, J. (1984) RNA polymerase I promoters and cellular transcription factors. Nature 310, 189–190.

Stefanovsky, V., Langlois, F., Gagnon-Kugler, T., Rothblum, L.I. and Moss, T. (2006) Growth factor signalling regulates elongation of RNA polymerase I transcription in mammals via UBF posphorylation and r-chromatin remodeling. Molecular Cell 21, 629–639.

Stiller, J.W. and Hall, B.D. (2002) Evolution of the RNA polymerase II C-terminal domain. Proceedings of the National Academy of Sciences, USA 99, 6091–6096.

Struhl, K. (1994) Duality of TBP: the universal transcription factor. Science 263, 1103–1104.

Svejstrup, J.Q., Conaway, R.C. and Conaway, J.W. (2006) RNA polymerase II: a 'Nobel' enzyme demystified. Molecular Cell 24, 637–642.

Szutorisz, H., Dillon, N. and Tora, L. (2005) The role of enhancers as centres for general transcription factor recruitment. Trends in Biochemical Sciences 30, 593–599.

Veenstra, G.J.C. and Wolffe, A.P. (2001) Gene-selective developmental roles of general transcription factors. Trends in Biochemical Sciences 26, 665–671.

Veenstra, G.J.C., Weeks, D.L. and Wolffe, A.P. (2000) Distinct roles for TBP and TBP-like factor in early embryonic gene transcription in *Xenopus*. Science 290, 2312–2315.

Vreugde, S., Ferrai, C., Miluzio, A., Hauben, E., Marchisio, P.C., Crippa, M.P., Bussi, M. and Biffo, S. (2006) Nuclear myosin VI enhances RNA polymerase II-dependent transcription. Molecular Cell 23, 749–755.

White, R.J. and Jackson, S.P. (1992) The TATA-binding protein: a central role in transcription by RNA polymerases I and III. Trends in Genetics 8, 284–288.

Wieczorek, E., Brand, M., Jacq, X. and Tora, L. (1998) Function of TAF_{II}-containing complex without TBP in transcription by RNA polymerase II. Nature 393, 187–191.

Woychik, N.A. and Hampsey, M. (2002) The RNA polymerase II machinery: structure illuminates function. Cell 108, 454–463.

Woychik, N.A., Liao, S-M., Koldrieg, P.A. and Young, R.A. (1990) Subunits shared by eukaryotic RNA polymerases. Genes and Development 4, 313–323.

Zhao, X. and Herr, W. (2002) A regulated two-step mechanism of TBP binding to DNA: a solvent-exposed surface of TBP inhibits TATA box recognition. Cell 108, 615–627.

Zurita, M. and Merino, C. (2003) The transcriptional complexity of the TFIIH complex. Trends in Genetics 19, 578–584.

CHAPTER 4

FAMILIES OF DNA BINDING TRANSCRIPTION FACTORS

4.1 INTRODUCTION

In previous chapters we have considered the role of chromatin structure, DNA sequences and RNA polymerases in the process of transcription and its regulation. The remaining aspect of this process, and the major subject of this book, are the transcription factors which bind to specific DNA sequences that have been exposed by changes in chromatin structure and then alter transcription by interacting directly or indirectly with RNA polymerase. To fulfil this role, transcription factors must possess certain features allowing them to modulate gene expression.

Clearly the first feature that many of these factors require is the ability to bind to DNA in a sequence specific manner, and this is discussed in this chapter. Following binding, the factor must interact with other factors or with the RNA polymerase itself in order to influence transcription either positively or negatively and these aspects are discussed in Chapters 5 and 6, respectively. Finally, in the case of factors modulating inducible, tissue specific or developmentally regulated gene expression, some means must exist to regulate the synthesis or activity of the factor so that it is active only in a particular situation. This regulation of factor synthesis or activity is discussed in Chapters 7 and 8, respectively.

Following the cloning of many different eukaryotic transcription factors, the domain mapping experiments described in Chapter 2 (section 2.4.1) have led to the identification of several distinct structural elements in different factors which can mediate DNA binding. These motifs have been used to classify transcription factors into families. These families and the DNA motifs that define them will be discussed in turn using transcription factors which contain them to illustrate their properties (for reviews see Harrison, 1991; Pabo and Sauer, 1992; Travers, 1993; Garvie and Wolberger, 2001).

4.2 THE HOMEODOMAIN

4.2.1 TRANSCRIPTION FACTORS IN DROSOPHILA DEVELOPMENT

Detailed genetic studies in the fruit fly *Drosophila melanogaster* have led to the identification of a very large number of mutations which affect the development of this organism, and their corresponding genes have been named on the basis of the observed phenotype of the mutant fly (for reviews see Ingham, 1988; Lawrence and Morata, 1994). Thus mutations in the so-called homeotic genes result in the transformation of one particular segment of the body into another; mutations in the Antennapedia gene, for example, cause the transformation of the segment which normally produces the antenna into one which produces a middle leg (Fig. 4.1). Similarly, mutations in genes

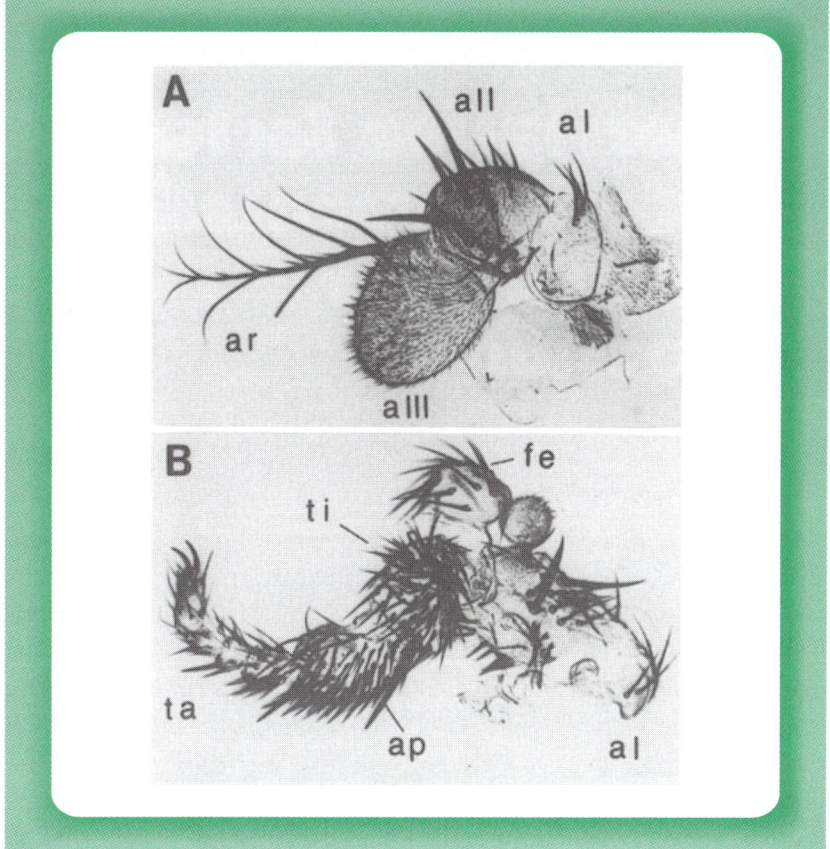

Figure 4.1

Effect of the homeotic mutation Antennapedia, which produces a middle leg (B) in the region that would contain the antenna of a normal fly (A), a1, aII, aIII: 1st, 2nd, and 3rd antennal segments; ar: arista; ta: tarsus; ti: tibia; fe: femur; ap: apical bristle.

of the gap class result in the total absence of particular segments; mutations in the Knirps gene, for example, result in the absence of most of the abdominal segments, although the head and thorax develop normally.

The products of genes of this type therefore play critical roles in *Drosophila* development. The products of the gap genes, for example, are necessary for the production of particular segments whilst the homeotic gene products specify the identify of these segments. Given that these processes are likely to require the activation of genes whose protein products are required in the particular segment, it is not surprising that many of these genes have been shown to encode transcription factors. Thus the Knirps gene product and that of another gap gene, Kruppel, contain multiple zinc finger motifs characteristic of DNA binding transcription factors and can bind to DNA in a sequence specific manner (see section 4.3). Similarly, the tailless gene whose product plays a key role in defining the anterior and posterior regions of the *Drosophila* embryo has been shown to be a member of the nuclear receptor family (see section 4.4).

It is clear therefore that the genes identified by mutation as playing a role in *Drosophila* development can encode several different types of transcription factors. However, of the first twenty-five such genes which were cloned, allowing a study of their protein products, well over half (fifteen) contain a motif known as the homeobox or homeodomain (Gehring *et al.*, 1994), which was originally identified in the homeotic genes of *Drosophila*. The features of these homeodomain proteins and the manner in which they mediate DNA binding and transcriptional regulation will be extensively discussed since they serve as a paradigm for the manner in which transcription factors function and can control highly complex processes such as development.

4.2.2 THE HOMEOBOX

When the first homeotic genes were cloned, it was found that they shared a region of homology approximately one hundred and eighty base pairs long, and therefore capable of encoding sixty amino acids, which was flanked on either side by regions that differed dramatically between the different genes. This region was named the homeobox or homeodomain (for review see Gehring *et al.*, 1994). Subsequently, the homeobox was shown to be present in many other *Drosophila* regulatory genes. These include the Fushi-tarazu gene (Ftz), which is a member of the pair rule class of regulatory loci whose mutation causes alternate segments to be absent, and the engrailed gene (eng), which is a member of the class of genes whose products regulate segment polarity. The close similarity of the homeoboxes encoded by the homeotic genes Antennapedia and Ultrabithorax and that encoded by the Ftz gene is shown in Figure 4.2.

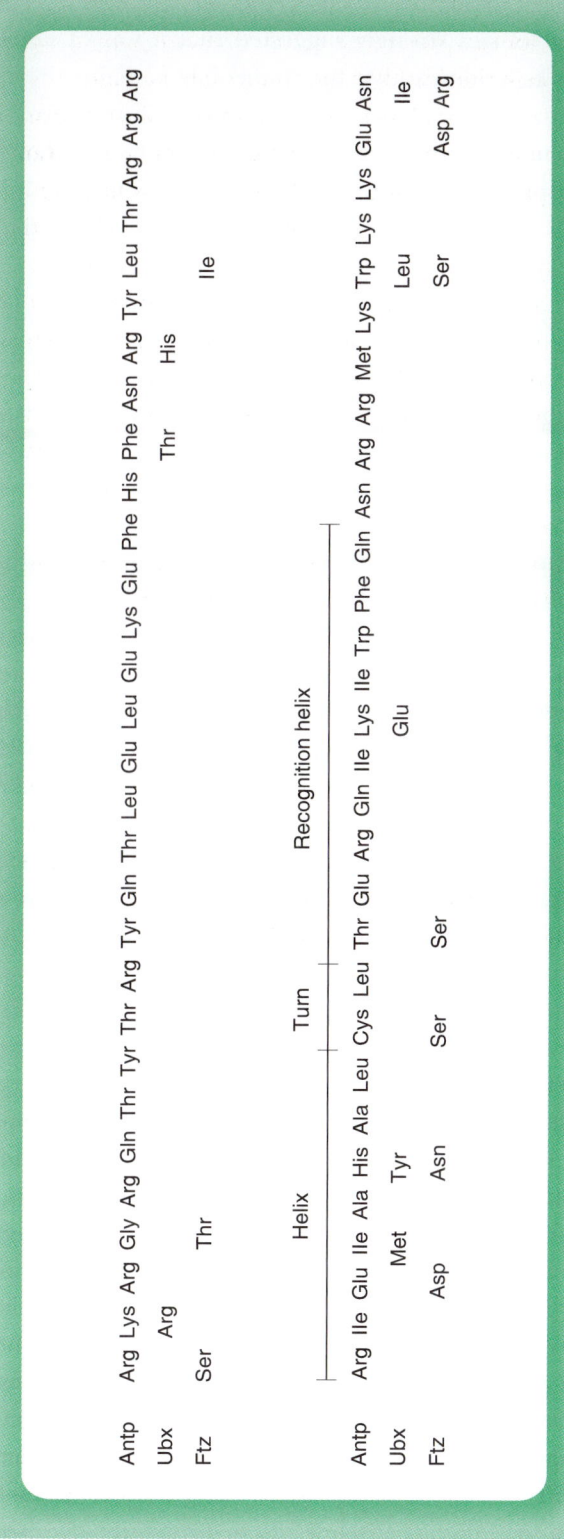

Figure 4.2

Amino-acid sequences of several *Drosophila* homeodomains, showing the conserved helical motifs. Differences between the sequences of the Ubx and Ftz homeodomains from that of Antp are indicated, a blank denotes identity in the sequence. The helix–turn–helix region is indicated.

The presence of this motif in a large number of different regulatory genes of different classes strongly suggested that it was of importance in their activity. The evidence that the homeobox contains proteins are transcription factors whose DNA binding activity is mediated by the homeobox is discussed in the next section (for reviews see Hayashi and Scott, 1990; Gehring *et al.*, 1994).

4.2.3 DNA BINDING BY THE HELIX-TURN-HELIX MOTIF IN THE HOMEOBOX

The first indication that the homeobox proteins were indeed transcription factors came from the finding that the homeobox was also present in the yeast mating type a and α gene products which are known to be transcription factors that regulate the activity of a and α-specific genes (for review see Dolan and Fields, 1991), hence suggesting, by analogy, that the *Drosophila* proteins also fulfilled such a role.

Direct evidence that this is the case is available from a number of different approaches. Thus it has been shown that many of these proteins bind to DNA in a sequence specific manner as expected for transcription factors (Hoey and Levine, 1988). Moreover, binding of a specific homeobox protein to the promoter of a particular gene correlates with the genetic evidence that the protein regulates expression of that particular gene. For example, the Ultrabithorax (Ubx) protein has been shown to bind to specific DNA sequences within its own promoter and in the promoter of the Antennapedia gene, in agreement with the genetic evidence that Ubx represses Antennapedia expression (Fig. 4.3).

The ability of the homeobox-containing proteins to bind to DNA is directly mediated by the homeobox itself. Thus if the homeobox of the Antennapedia protein is synthesized in isolation either in bacteria or by chemical synthesis, it is capable of binding to DNA in the identical sequence specific manner characteristic of the intact protein.

This ability to define the sixty amino acid homeodomain as the region binding to DNA has led to intensive study of its structure in the hope of elucidating how the protein binds to DNA in a sequence specific manner (for reviews see Kornberg, 1993; Gehring *et al.*, 1994). In particular the crystal structure of the Antennapedia (Antp) homeodomain bound to DNA has been determined by nuclear magnetic resonance spectroscopy (NMR) whilst similar structural studies of the engrailed (eng) and the yeast MATα2 homeodomains bound to DNA have been carried out by X-ray crystallography.

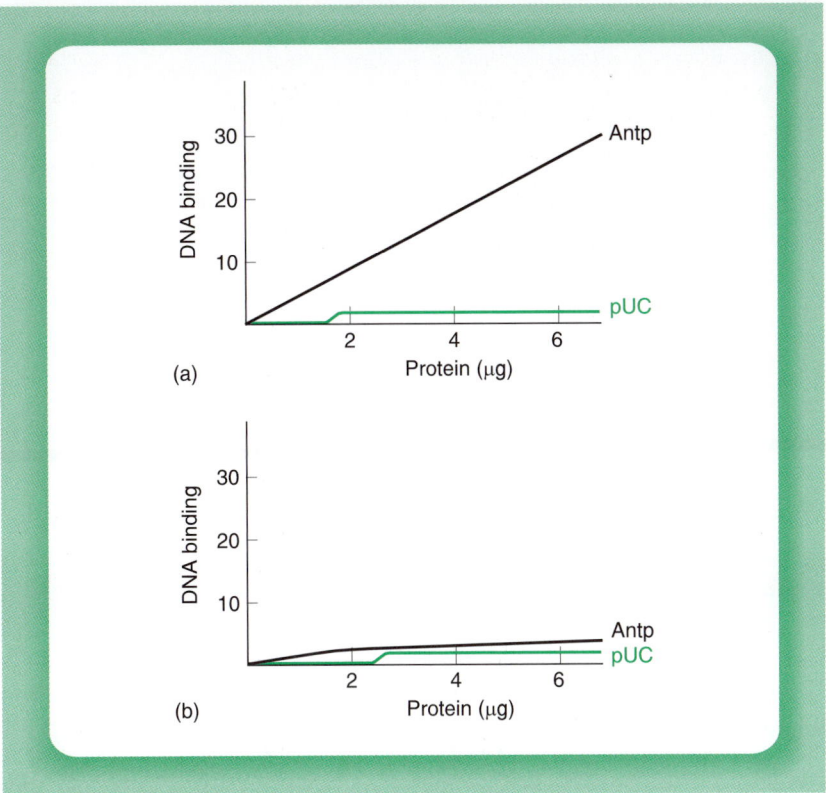

Figure 4.3

Assay of protein binding to a DNA fragment from the Antennapedia gene promoter (Antp) or a control fragment of plasmid DNA (pUC) using protein extracts from *E. coli* which have been genetically engineered to express the *Drosophila* Ubx protein (top panel) or protein extracts from control *E. coli* not expressing Ubx (bottom panel). Note the specific binding of Ubx protein to the Antennapedia DNA fragment.

By this means the Antp homeodomain was shown to contain a short N-terminal arm of six residues followed by four alpha-helical regions (Fig. 4.4). The first two helices are virtually anti-parallel to each other with the other two helices arranged at right-angles to the first. Most interestingly, helices II and III are separated by a beta turn forming a helix-turn-helix motif (Fig. 4.5). The eng and MATα2 homeodomains also have a similar structure with an N-terminal arm and a subsequent helix-turn-helix motif. In this case, however, the third and fourth helices observed in Antp form a single helical region. Interestingly, the helix-turn-helix structure typical of the homeodomain is very similar to the DNA binding motif of several bacteriophage regulatory proteins such as the lambda cro protein or the phage 434 repressor which have also been crystallized and subjected to intensive structural study.

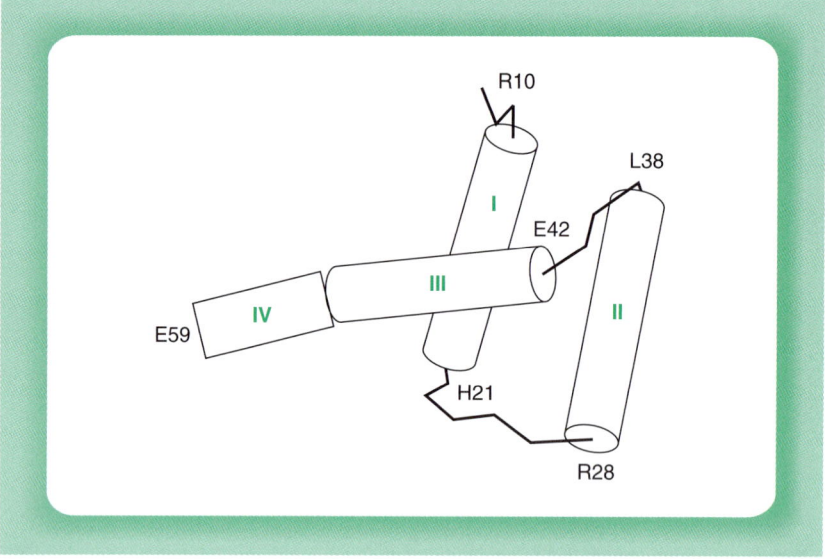

Figure 4.4

Structure of the Antennapedia homeodomain as determined by nuclear magnetic resonance spectroscopy. Note the four alpha-helical regions (I–IV) represented as cylinders with the amino acids at their ends indicated by numbers and the one letter amino acid code.

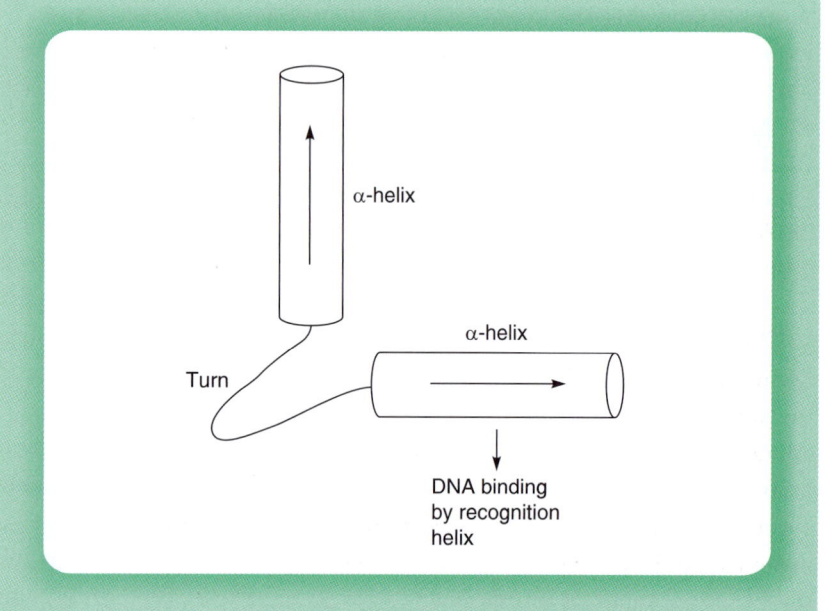

Figure 4.5

The helix-turn-helix motif.

In these bacteriophage proteins, X-ray crystallographic studies have shown that the helix-turn-helix motif does indeed contact DNA. One of the two helices lies across the major groove of the DNA whilst the other lies partly within the major groove where it can make sequence specific contacts with the bases of DNA. It is this second helix (known as the recognition helix) that therefore controls the sequence specific DNA binding activity of these proteins (Fig. 4.6).

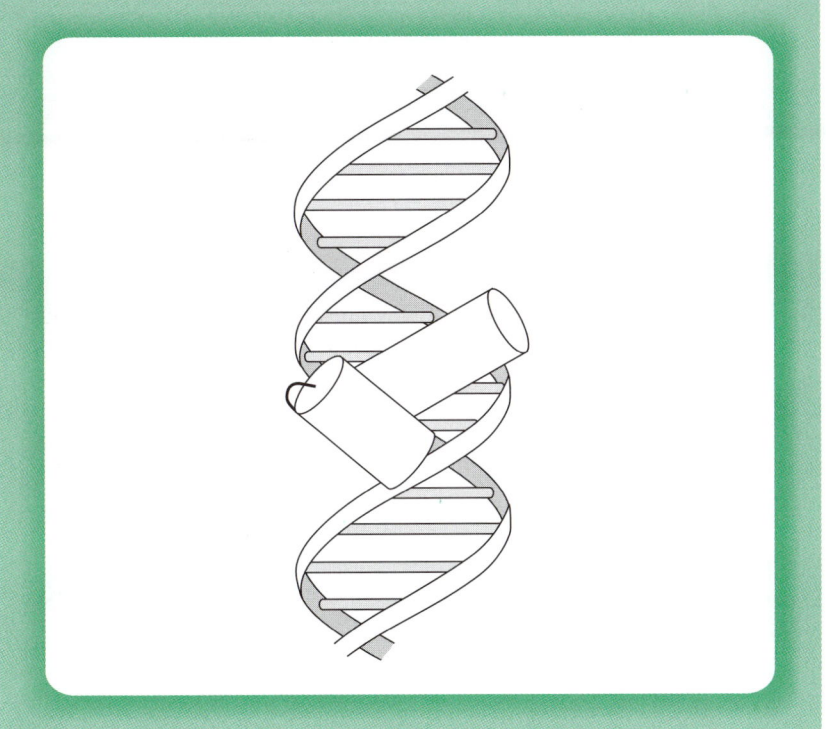

Figure 4.6
Binding of the helix-turn-helix motif to DNA with the recognition helix in the major groove of the DNA.

The similarity in structure of helices II and III in the eukaryotic homeodomains to the two helices of the bacteriophage proteins led to the suggestion that these two helices in the homeodomain are similarly aligned relative to the DNA with helix III constituting the recognition helix responsible for sequence specific DNA binding. Hence the precise amino acid sequence in the recognition helix in different homeodomain proteins would determine the DNA sequence which they bound (for review see Treisman *et al.*, 1992).

In agreement with this idea, exchanging the recognition helix in the Bicoid (Bcd) homeodomain for that of Antp resulted in a protein with the DNA binding specificity of Antp and not that of Bicoid. Most interestingly a Bcd protein with the DNA binding specificity of Antp could also be obtained by exchanging only the ninth amino acid in the recognition helix, replacing the lysine residue in Bcd with the glutamine residue found in the Antp protein (Fig. 4.7), whereas the exchange of other residues which differ between the two proteins has no effect on the DNA binding specificity. Hence the ninth amino acid within the recognition helix of the homeodomain plays a critical role in determining DNA binding specificity.

	Recognition helix										Binding to Bicoid site TCTAATCCC	Binding to Antp site TCAATTAAAT
Bicoid	T	A	Q	V	K	I	W	F	K	N	+	–
	A	–	–	–	–	–	–	–	–	–	+	–
	–	–	–	–	A	–	–	–	–	–	+	–
	–	–	–	–	–	–	–	–	A	–	–	–
	A	–	–	–	A	A	–	–	A	–	–	–
	E	R	–	–	–	–	–	–	Q	–	–	+
	E	R	–	–	–	–	–	–	–	–	+	–
	–	–	–	–	–	–	–	–	Q	–	–	+
Antp	E	R	Q	I	K	I	W	F	Q	N	–	+

Figure 4.7

Effect of changing the amino acid sequence in the recognition helix of the Bicoid protein on its binding to its normal recognition site and that of the Antennapedia (Antp) protein. Note the critical effect of changing the ninth amino acid in the helix which completely changes the specificity of the Bicoid protein.

It is likely that the amino group of lysine found at the ninth position in the Bcd protein makes hydrogen bonds with the O6 and N7 positions of a guanine residue in the Bcd-specific DNA binding site whereas the amide group of glutamine found at the corresponding position in the Antp recognition helix forms hydrogen bonds with the N6 and N7 positions of an adenine residue at the equivalent position within the Antp-specific DNA binding site. Hence the replacement of lysine with glutamine results in the loss of two potential hydrogen bonds to a Bcd site and the gain of two potential hydrogen bonds to an Antp site, explaining the observed change in DNA binding specificity (Fig. 4.8).

A similar critical role for the ninth amino acid in determining the precise DNA sequence which is recognized is also seen in other homeobox-containing proteins, replacement of the serine found at this position

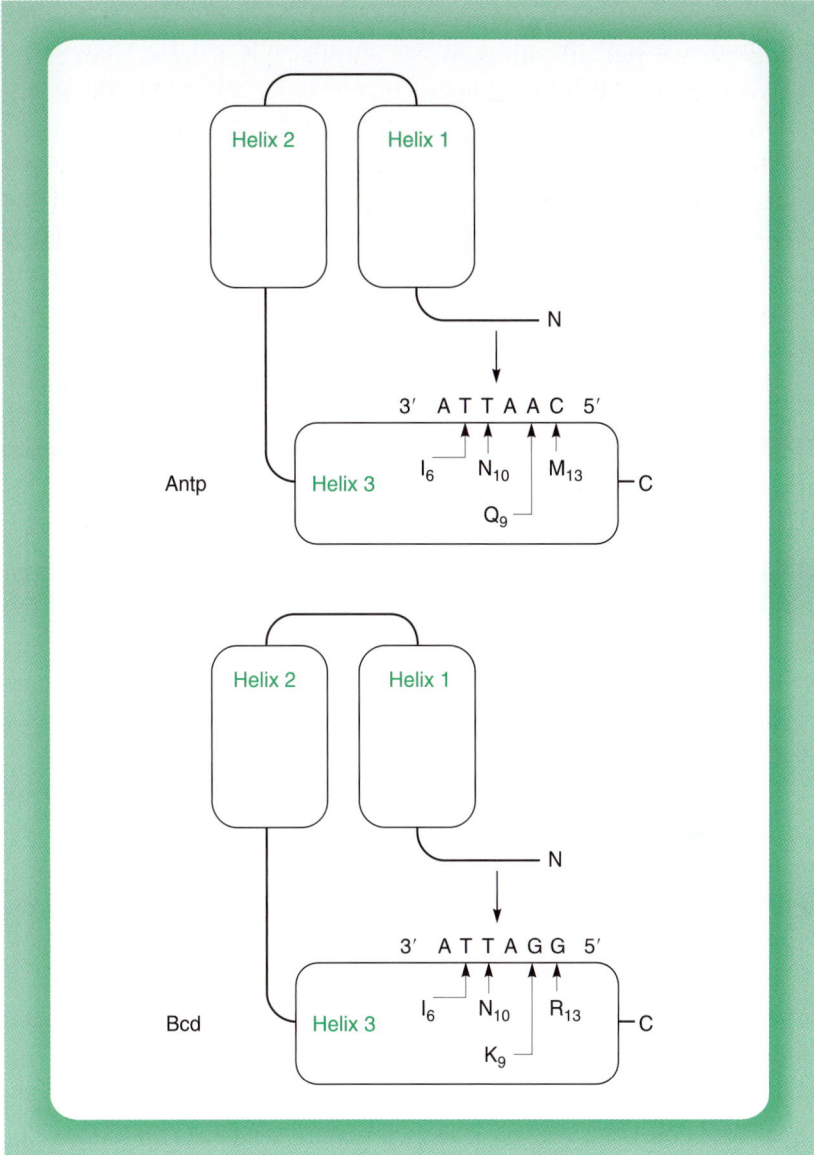

Figure 4.8

Contacts between DNA and the Antp or Bcd homeodomains. Note that the change in the ninth amino acid of the recognition helix (helix 3) alters the base which is preferentially bound from a G for Bcd to an A for Antp as discussed in the text whilst the N-terminal arm of the homeodomain contacts the ATTA sequence common to the recognition site of both proteins.

in the paired protein with the lysine found in Bicoid or the glutamine found in Antp allowing the paired protein to recognize respectively Bcd or Antp-specific DNA sequences. Hence the DNA sequence recognized by a homeobox-containing protein appears to be primarily determined by the ninth amino acid in the recognition helix, proteins with different amino acids at this position recognizing different DNA sequences whereas proteins such as Antp and Ftz, which have the same amino acid at this position, recognize the same DNA sequence.

This critical role of the ninth amino acid is in contrast to the situation in the bacteriophage proteins in which the helix-turn-helix motif was originally defined. In these proteins, the most N-terminal residues (1–3) in the recognition helix play a critical role in determining DNA binding specificity (for review see Pabo and Sauer, 1992). As shown in Figure 4.7, however, these amino acids appear to play little or no role in determining the DNA binding specificity of eukaryotic helix-turn-helix proteins, suggesting therefore that the recognition helix of these proteins is oriented differently in the major groove of the DNA.

This idea is in agreement with the structural studies of the eukaryotic homeodomains bound to DNA which have identified the actual protein–DNA contacts. These studies have shown that, as in the bacteriophage proteins, the recognition helix directly contacts the bases of DNA in the major groove. However, in the eukaryotic homeobox proteins this helix is oriented within the major groove somewhat differently, such that the critical base-specific contacts are, as predicted, made by the C terminal end of the helix which contains residue nine (Fig. 4.8).

It is clear therefore that the helix-turn-helix motif in the homeobox mediates both the DNA binding of the protein and also, via the recognition helix, controls the precise DNA sequence that is recognized. Interestingly, however, the short N-terminal arm of the homeodomain also contacts the bases of the DNA, although it makes contact in the minor groove rather than the major groove. Removal of this short N-terminal arm dramatically reduces the DNA binding affinity of the homeodomain, indicating that this region contributes significantly to the DNA binding ability of the homeodomain probably by contacting the ATTA bases common to the DNA binding sites of several homeodomain proteins (Fig. 4.8).

Although DNA binding is important for the modulation of transcription, it is necessary to demonstrate that the homeobox proteins do actually affect transcription following such binding. In the case of the Ubx protein, this was achieved by showing that co-transfection of a plasmid expressing Ubx with a plasmid in which the Antennapedia promoter drives a marker gene resulted in the repression of gene expression driven by the Antennapedia promoter (Fig. 4.9). Hence the observed binding of Ubx to the Antp promoter (see above) results in down regulation of its activity in agreement with the results of genetic experiments.

Most interestingly, the Ubx expression plasmid was able to up regulate activity of its own promoter in co-transfection experiments, this ability being dependent on the previously defined binding sites for Ubx within its own promoter. Similarly, although Ubx normally has no effect on expression of the alcohol dehydrogenase (Adh) gene it can stimulate the Adh promoter following linkage of the promoter to a DNA sequence containing multiple binding sites for Ubx. Hence a homeobox protein can produce distinct effects following binding, Ubx activating its own promoter and a hybrid promoter containing Ubx binding sites but repressing the activity of the Antp promoter (Fig. 4.9).

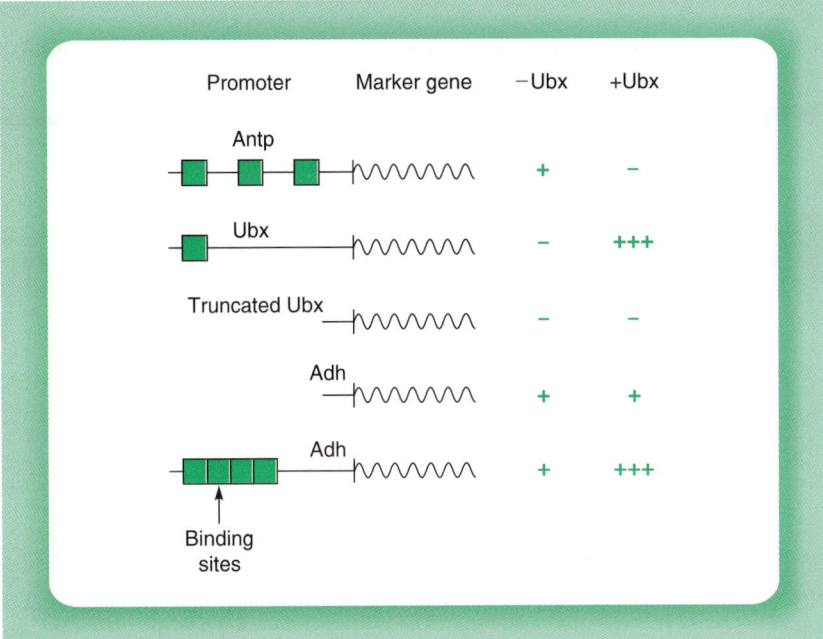

Figure 4.9
Effect of Ubx on various marker genes with or without binding sites (hatched boxes) for the Ubx protein. Note that Ubx can stimulate its own promoter which contains a Ubx binding site and this effect is abolished by deleting the Ubx binding site. Similarly the alcohol dehydrogenase (Adh) gene which is normally unaffected by Ubx, is rendered responsive to Ubx stimulation by addition of Ubx binding sites. In contrast, the Antennapedia promoter which also contains Ubx binding sites, is repressed by Ubx. Hence binding of Ubx can activate or repress different promoters.

A similar transcriptional activation effect of DNA binding has been demonstrated for the Fushi-tarazu (Ftz) protein. This protein binds specifically to the sequence TCAATTAAATGA. As with Ubx, linkage of this sequence to a marker gene confers responsivity to activation by Ftz, such activation being dependent upon binding of Ftz to its target sequence, a one base pair change which abolishes binding, also abolishing the induction of transcription (Fig. 4.10).

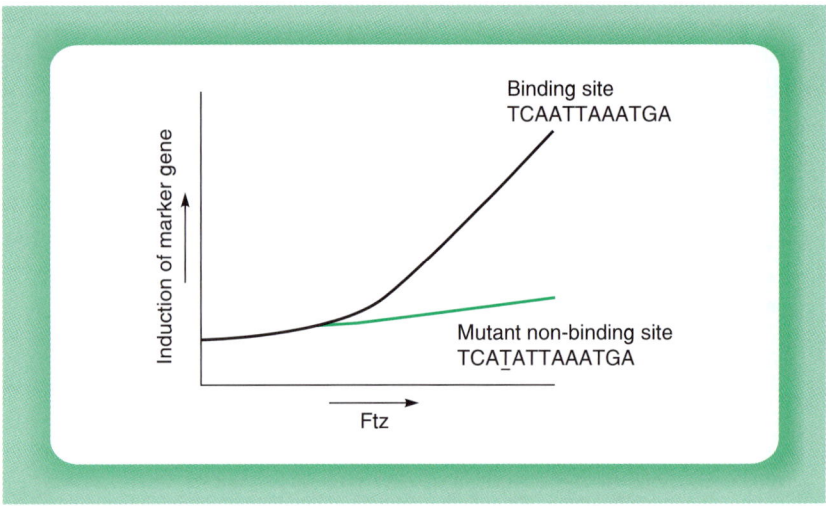

Figure 4.10

Effect of expression of the Ftz protein on the expression of a gene containing its binding site, or a mutated binding site containing a single base pair change which abolishes binding of Ftz.

Interestingly, the ability of Ubx to induce its own transcription provides a mechanism for the long term maintenance of Ubx gene expression during development since once expression has been switched on and some Ubx protein made, it will induce further transcription of the gene via a simple positive feedback loop even if the factors which originally stimulated its expression are no longer present (Fig. 4.11). This long term maintenance of Ubx expression is essential since if the Ubx gene is mutated within the larval imaginal disc cells which eventually produce the adult fly, the cells which would normally produce the haltere (balancer) will produce a wing instead. Thus although these cells are known to already be committed to form the adult haltere at the larval stage, the continued expression of the Ubx gene is essential to maintain this commitment and allow eventual overt differentiation (see Hadorn, 1968 for a review of imaginal discs and their role in *Drosophila* development).

4.2.4 REGULATION OF DNA BINDING SPECIFICITY BY INTERACTIONS BETWEEN DIFFERENT HOMEOBOX PROTEINS

Although we have previously described the DNA binding specificity of individual homeobox proteins, it is possible for the DNA binding specificity of one factor to be altered in the presence of another factor. Thus

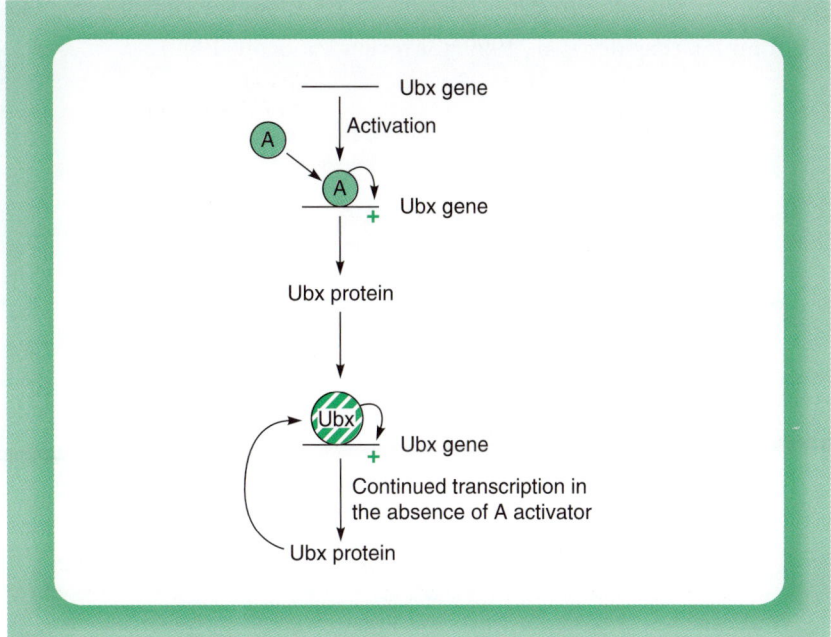

Figure 4.11

The stimulatory effect of the Ubx protein on the transcription of its own gene ensures that once Ubx gene transcription is initially switched on by an activator protein (A), transcription will continue even if the activator protein is removed.

several homeobox proteins such as Ubx and Antp bind to the same DNA sequences when tested in isolation *in vitro* (Hoey and Levine, 1988). Paradoxically, however, the effects of mutations which inactivate the genes encoding each of these proteins are different, indicating that they cannot substitute for one another. Similarly, *in vivo* Ubx can bind to a site in the promoter of the decapentaplegic (dpp) gene and regulate its expression whereas Antp cannot do so.

This paradox is explained by the presence in the dpp promoter of a binding site for another homeobox protein extradenticle (Exd) which lies adjacent to the site to which Ubx binds. The Exd protein interacts with the Ubx protein and both enhances its DNA binding affinity and modifies its DNA binding specificity so it can bind strongly to the dpp gene promoter and regulate its expression (Fig. 4.12) (for review see Mann and Chan, 1996). As Antp does not interact with Exd, its specificity is not modified in this way. Hence, it does not bind to the dpp gene promoter and therefore cannot regulate this promoter. Interestingly, structural studies have shown that Ubx and Exd bind to opposite sides of the DNA and that a short region of Ubx N-terminal to the homeodomain extends round the DNA and inserts into a cleft in the Exd homeodomain resulting in interaction of the proteins and enhanced DNA binding by the complex (Passner *et al.*, 1999; for review see Scott, 1999).

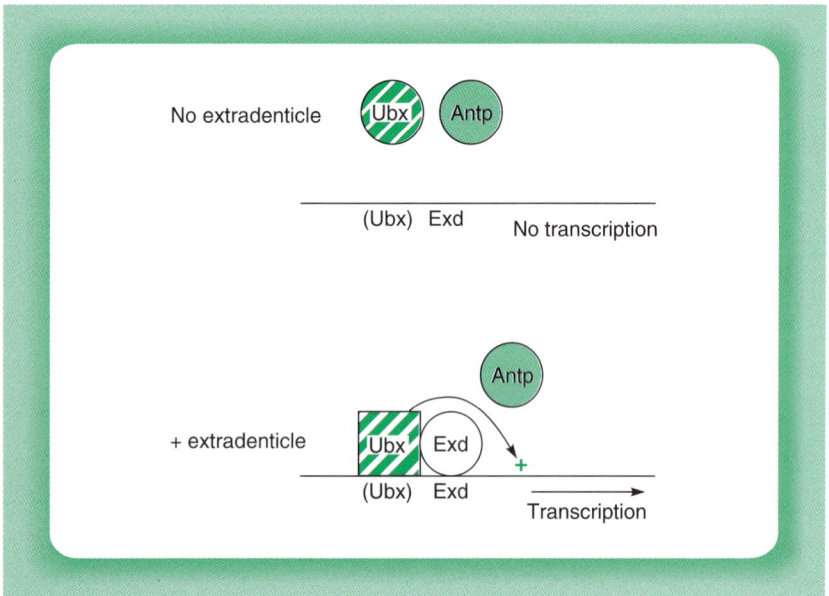

Figure 4.12
The Exd protein interacts with the Ubx protein to allow it to bind with high affinity to its potential binding site in the dpp promoter (indicated as (UBX)) and activate its expression. In contrast the Antp protein cannot interact with Exd and so does not bind to the dpp promoter.

A similar interaction is observed in the case of the yeast homeodomain proteins a1 and α2, which control the mating type in this organism (for review see Dolan and Fields, 1991). Thus, in the absence of a1, the α2 protein has a weak DNA binding ability. However, in the presence of a1, an a1/α2 heterodimer forms and binds to specific gene promoters. As the α2 protein is a strong transcriptional repressor, this results in the repression of the genes which bind the a1/α2 heterodimer. In this case, however, unlike the Ubx/Exd case, the interaction is mediated by the C-terminal region of the α2 homeodomain which forms an additional α-helix and interacts with the homeodomain of the a1 protein (Andrews and Donoviel, 1995; Li *et al.*, 1995) (Plate 2).

Interestingly, α2 can also interact with the non-homeodomain protein MCM1 to form a heterodimer which has a different DNA binding specificity to that of the a1/α2 heterodimer and which therefore binds to and represses a different set of genes (Fig. 4.13). Hence, α2 is a repressor protein with a weak DNA binding specificity which is guided to different sets of target genes depending on whether it interacts with a1 or MCM1 to form heterodimers with different DNA binding specificities.

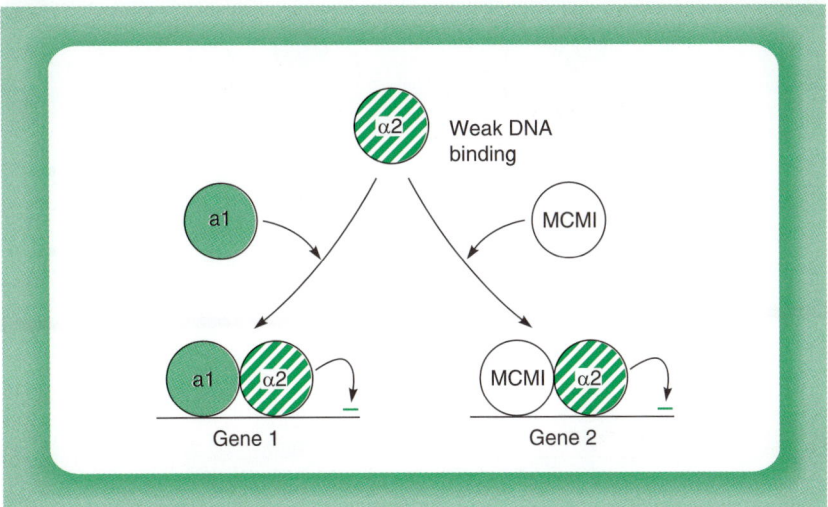

Figure 4.13

The yeast α2 repressor protein is a homeodomain protein with weak DNA binding activity. However, it can form DNA binding heterodimers with either the a1 homeodomain protein or the MCM1 protein. As these heterodimers have different DNA binding specificities, they bind different genes which are then repressed by the α2 protein.

The DNA binding specificity of homeodomain proteins can therefore be altered by interactions with other homeodomain and non-homeodomain-containing proteins with different regions within or adjacent to the homeodomain mediating this interaction in different cases.

This effect is also responsible, at least in part, for the phenomenon described in Chapter 2 (section 2.4.2) in which the yeast Ste 12 transcription factor (whose DNA binding domain is related to the homeodomain) binds to different target genes under different conditions. Thus, in filamentous cells, Ste 12 binds to specific target genes together with the Tec 1 transcription factor and Tec 1 is required for binding of Ste 12 to these genes. In other situations, such as mating cells, Tec 1 is not active and so Ste 12 cannot bind to these target genes and binds only to target genes which have a different binding sequence to which Ste 12 can bind without Tec 1 (Zeitlinger *et al.*, 2003).

It is clear therefore that the genes targeted by a specific transcription factor can be modulated by other factors, which alter the precise DNA binding specificity of the factor (Fig. 4.14).

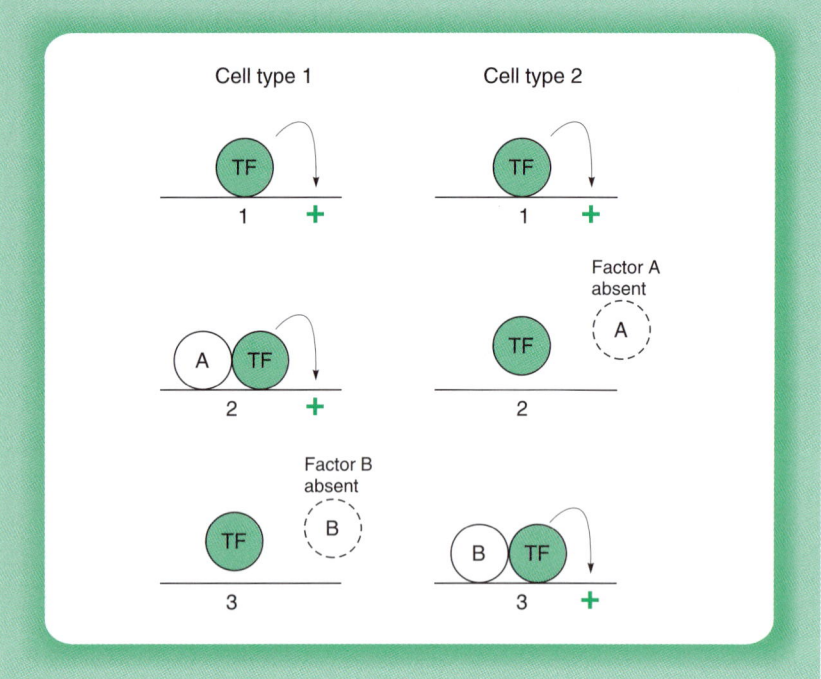

Figure 4.14

The genes bound by a particular transcription factor in a specific cell type can be regulated by the presence or absence of other factors which are required for binding to a specific binding site. Thus, in the sample shown, a transcription factor (TF) binds to binding site one without the need for any other factor and so the gene carrying this binding site is activated in both cell types. In contrast, binding to binding sites two or three needs another factor (A or B respectively) and hence the genes carrying those binding sites bind the factor and are activated only when the appropriate additional factor is present.

4.2.5 HOMEODOMAIN TRANSCRIPTION FACTORS IN OTHER ORGANISMS

The critical role played by the homeobox genes in the regulation of *Drosophila* development suggests that they may also play a similar role in other organisms. Thus, in the nematode *C. elegans*, homeoboxes have been identified in several genes whose mutation affects development, such as the *mec-3* gene which controls the terminal differentiation of specific sensory cells.

As in *Drosophila*, studies in the nematode have been facilitated by the availability of well-characterized mutations affecting development, allowing the corresponding genes to be isolated and the homeobox identified. In higher organisms where such genetic evidence was unavailable, numerous investigators have used Southern blot hybridization with labelled probes derived from *Drosophila* homeoboxes in an attempt to identify home-obox-containing genes in these species. Thus, for example, Holland and Hogan (1986) used a probe from the Antennapedia homeobox to identify homeobox genes in a wide range of species including not only other inver-tebrates such as the molluscs but also chordates such as the sea urchin and vertebrates including the mouse (Fig. 4.15). Subsequent studies have resulted in the identification of a large number of different homeobox-containing genes from a wide variety of organisms including both mouse and human and many of these genes have been isolated and their DNA sequence obtained (for reviews see Duboule, 2000; Briscoe and Wilkinson, 2004; Morgan, 2006).

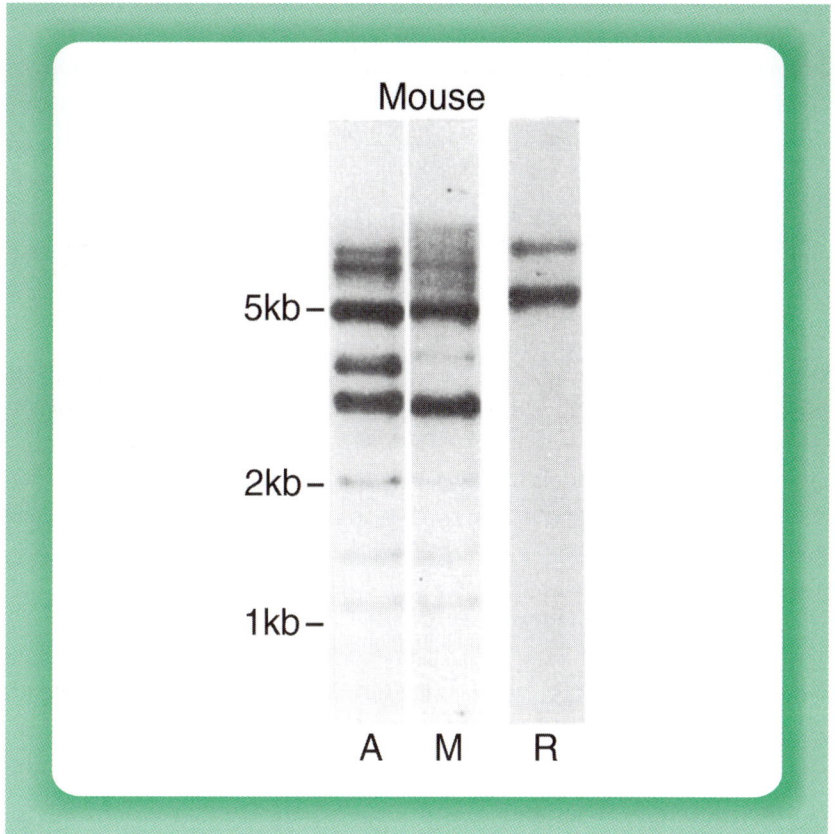

Figure 4.15

Southern blot of mouse DNA hybridized with a probe from the *Drosophila* Antennapedia gene (A), a mouse Antennapedia-like gene (M) and mouse ribosomal DNA (R). Note the presence of DNA fragments which hybridize to both Antennapedia-like DNAs but not to ribosomal DNA and which represent Antennapedia-like sequences in the mouse genome.

It is clear from these studies that homeobox-containing genes are not confined to invertebrates such as *Drosophila* or yeast but are found also in vertebrates, including mammals such as mouse and human. Interestingly, this evolutionary conservation is not confined to the homeobox portion of these genes. Thus, homologues of individual homeobox genes of *Drosophila*, such as engrailed and deformed have been identified in mouse and human, the fly and mammalian proteins showing extensive sequence homology which extends beyond the homeobox to include other regions of the proteins.

Moreover, the similarity between the *Drosophila* and mammalian systems extends also to the manner in which the homeobox-containing genes are organized in the genome. Thus, in both *Drosophila* and mammals, these genes are organized into clusters containing several homeobox-containing genes with homologous genes in the different organisms occupying equivalent positions in the clusters. For example, in a detailed comparison of the genes in the *Drosophila* Bithorax and Antennapedia complexes with those of one mouse homeobox gene complex Hoxb (Hox2), Graham *et al.* (1989) showed that the first gene in the mouse complex, Hoxb -9 (2.5), was most homologous to the first gene in the *Drosophila* Bithorax complex, Abd-B and so on across the complex (Fig. 4.16). Hence both the homeobox genes and their arrangement are highly conserved in evolution, the common ancestor of mammals and insects having presumably possessed a similar cluster of homeobox-containing genes. Interestingly, the DNA sequences and arrangement in the genome of different homeobox genes has been used as a means of determining evolutionary relationships amongst multicellular organisms (for review see Martindale and Kourakis, 1999).

As well as the simple homeobox/homeodomain proteins we have discussed so far, other families of transcription factor exist which contain the homeodomain as part of a larger, more complex, DNA binding structure. Two such families are discussed in the next two sections.

4.2.6 POU PROTEINS

As discussed above, the homeobox-containing genes were first identified in *Drosophila* and only subsequently in other organisms. The reverse is true, however, for another set of transcription factors which possess a homeobox as part of a much larger motif and which were first identified in mammalian cells. Thus, the transcription factors Oct-1 and Oct-2, which bind to the octamer motif ATGCAAAT play an important role in regulating the expression of specific genes such as those encoding histone H2B, the SnRNA molecules and the immunoglobulins. Similarly, the transcription factor Pit-1, which binds to a sequence two bases different

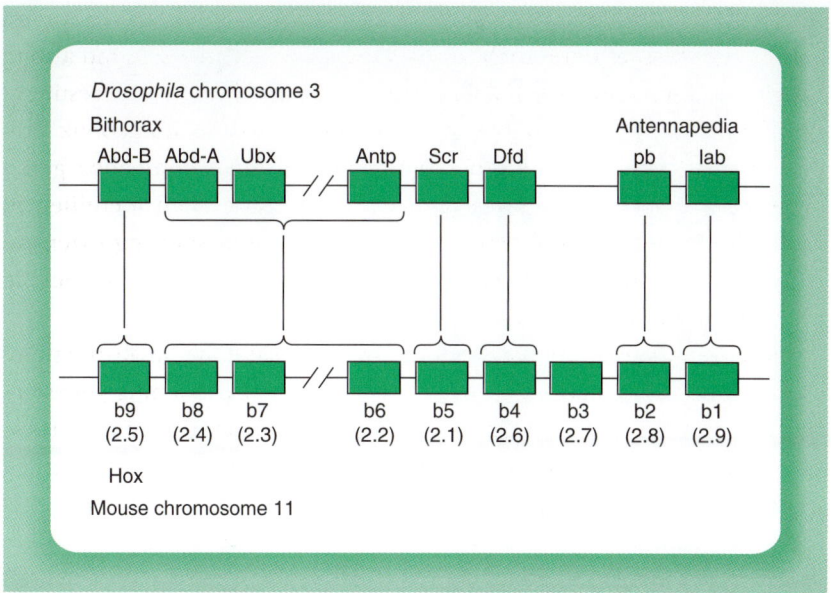

Figure 4.16

Comparison of the Bithorax/Antennapedia complex on *Drosophila* chromosome 3 with the Hox b complex on mouse chromosome 11. Individual genes are indicated by open boxes. Note that each gene in the *Drosophila* complex is most homologous to the equivalent gene in the mouse complex as indicated by the vertical lines. The *Drosophila* Abd-A, Ubx and Antp genes are too closely related to each other to be individually related to a particular mouse gene but are most closely related to the Hox b-6, b-7 and b-8 genes which occupy the equivalent positions in the Hox b cluster as indicated by the brackets. The two alternative nomenclatures for mouse Hox genes are indicated.

from the octamer sequence, plays a critical role in pituitary-specific gene expression (Chapter 1, section 1.3.3).

When the genes encoding these factors were cloned, they were found to share a 150–160 amino acid sequence which was also found in the protein encoded by the nematode gene *unc*-86 whose mutation affects sensory neuron development. This common POU (Pit-Oct-Unc) domain contains both a homeobox sequence and a second conserved domain, the POU specific domain (Fig. 4.17; for reviews see Verrijzer and Van der Vliet, 1993; Ryan and Rosenfeld, 1997).

Interestingly, whilst the homeoboxes of the different POU proteins are closely related to one another (53 out of 60 homeobox residues are the same in Oct-1 and Oct-2 and 34 out of 60 in Oct-1 and Pit-1), they show less similarity to the homeoboxes of other mammalian genes lacking the POU specific domain, sharing at best only 21 out of 60 homeobox residues. Hence they represent a distinct class of homeobox proteins containing both a POU specific domain and a diverged homeodomain.

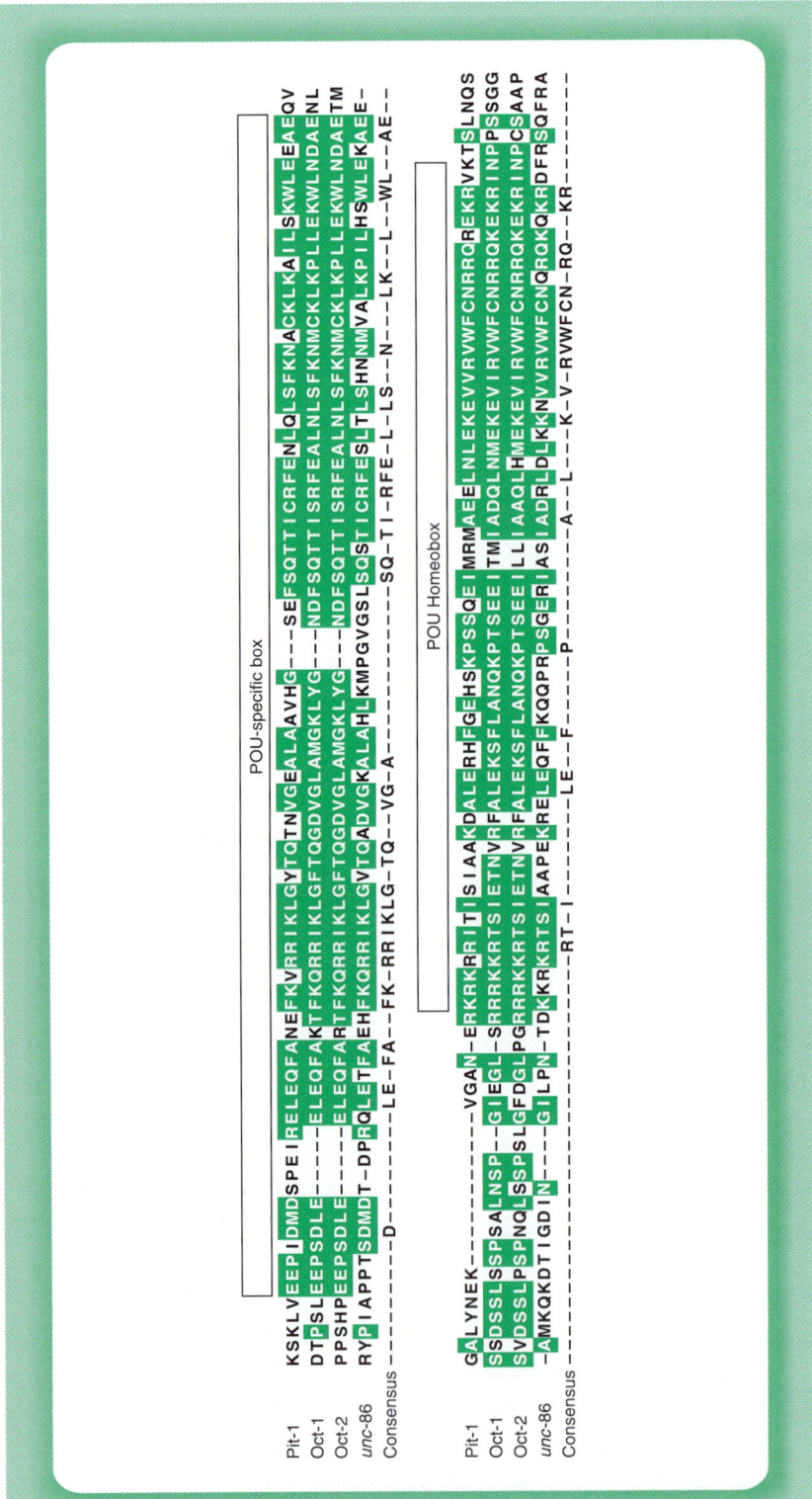

Figure 4.17

Amino-acid sequences of the POU proteins. The homeodomain and the POU specific domain are indicated. Solid boxes indicate regions of identity between the different POU proteins. The final line shows a consensus sequence obtained from the four proteins. Note the highly conserved sequences near each end of the POU domain which have been used as a method of isolating novel POU proteins (see Chapter 2, section 2.3.2c and Fig. 2.14).

As with the *Drosophila* homeobox proteins, however, the isolated homeo-domains of the Pit-1 and Oct-1 proteins are capable of mediating sequence specific DNA binding in the absence of the POU specific domain. The affinity and specificity of binding by such an isolated homeodomain is much lower, however, than that exhibited by the intact POU domain, indicating that the POU specific domain plays a critical role in producing high affinity binding to specific DNA sequences. Hence the POU homeo-domain and the POU specific domain form two parts of a DNA binding element which are held together by a flexible linker sequence.

The crystal structure of the Oct-1 POU domain bound to DNA (Klemm *et al.*, 1994) has shown that the Oct-1 homeodomain binds in a similar manner to the classical homeobox proteins, with the recogni-tion helix lying in the major groove and the N-terminal arm in the minor groove. Like the homeodomain, the POU specific domain forms a helix-turn-helix motif, which allows it to bind to the adjacent bases within the DNA to those contacted by the homeodomain with binding of the two regions occurring on opposite sides of the DNA double helix (Fig. 4.18).

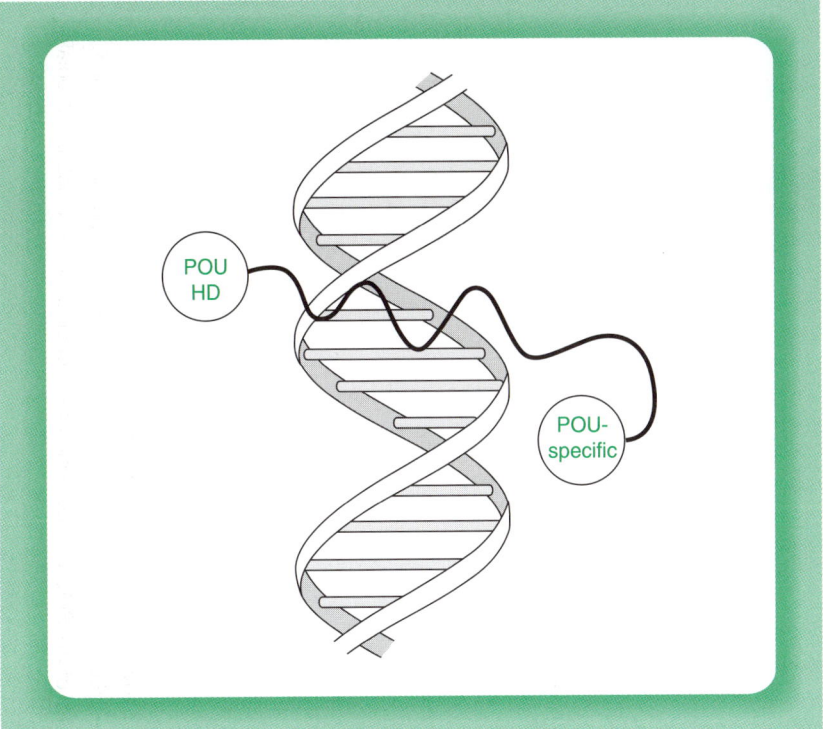

Figure 4.18

Binding of the POU specific domain and POU homeodomain to opposite sides of the DNA double helix. Note the flexible linker region joining the two DNA binding motifs.

The POU domain appears to allow factors that contain it to bind to highly divergent DNA sequences. Thus, Oct-1 binds to a sequence in the SV40 enhancer, which shares less than thirty per cent homology (four out of fourteen bases) or little more than a random match with another Oct-1 binding sequence in the herpes simplex virus (HSV) immediate–early (IE) gene promoters (Fig. 4.19). By analysing a series of other Oct-1 binding elements, however, Baumruker *et al.* (1988) were able to show that the two apparently unrelated Oct-1 binding sites could be linked by a smooth progression via a series of other binding sites which were related to one another (Fig. 4.19). This suggests therefore that Oct-1 can bind to very dissimilar sequences because there are few, if any, obligatory contacts with specific bases in potential binding sites. Rather, specific binding to a particular sequence can occur via many possible independent interactions with DNA, only some of which will occur with any particular binding site. Hence, the binding to apparently unrelated sequences does not reflect two distinct binding specificities but indicates that the protein can make many different contacts with DNA, the sequences which can specifically bind the protein being those with which it can make a certain proportion of these possible contacts.

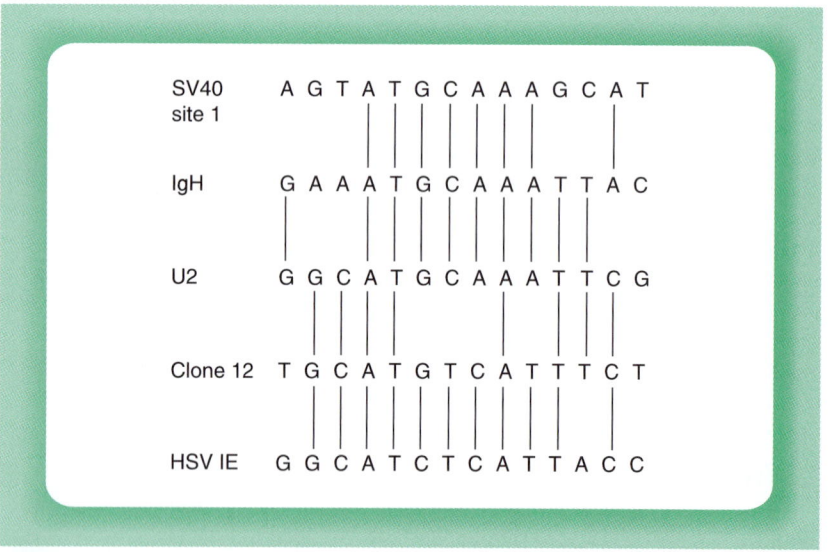

Figure 4.19

Relationship between the various diverse sequences bound by the Oct-1 transcription factor in the simian virus 40 enhancer, the immunoglobulin IgH chain gene enhancer (IgH), the U2 snRNA gene, clone 12 (a mutated version of a site in the SV40 enhancer which binds Oct-1) and the herpes simplex virus immediate-early genes (HSV IE).

Interestingly, it has been shown that the secondary structure of Oct-1 bound to these sites differs so that its configuration when bound to the HSV IE sequence is different to that observed when it is bound to the other sequences (Walker *et al.*, 1994). Moreover, this configurational change allows the Oct-1 bound to the HSV promoter to be recognized by the HSV VP16 (Vmw 65) protein whereas this does not occur with Oct-1 bound to other sequences. As VP16 is a much stronger trans-activator than Oct-1 alone, this therefore results in the strong activation of the HSV IE promoters by the Oct-1/VP16 complex whereas other promoters in which Oct-1 has bound to different sequences are insensitive to such trans-activation by VP16. Hence, this provides a novel example of gene regulation in which the nature of the sequence bound by a factor controls its recognition by another factor, resulting in strong trans-activation only from a subset of sequences bound by Oct-1 (Fig. 4.20) (for review see Wysocka and Herr, 2003).

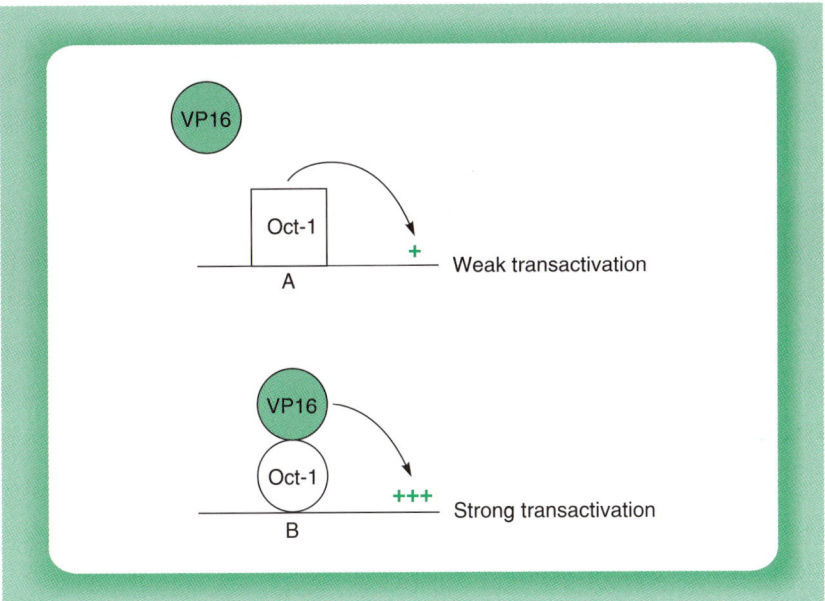

Figure 4.20

The octamer binding protein Oct-1 binds to most binding sites (A) in a configuration which is not recognized by VP16. This results in only the weak transactivation characteristic of Oct-1 alone. In contrast, when it binds to its binding sites in the HSV IE promoters (B), Oct-1 undergoes a conformational change allowing it to be recognized by the strong trans-activator VP16 leading to strong trans-activation.

As well as the different configuration Oct-1 adopts when binding to viral sequences, it has been shown that it can also adopt different configurations when binding to different cellular DNA targets and this also has consequences for its effect on gene transcription. Thus, when Oct-1 binds as a dimer to a DNA element known as the PORE sequence, it exposes a region of the POU domain which can recruit a cellular co-activator, OBF-1, resulting in strong activation of transcription. In contrast, when it binds to a distinct DNA sequence, known as the MORE sequence, this region of the POU domain is masked at the interface between the two Oct-1 molecules. Hence, in this case OBF-1 cannot be recruited and only weak transactivation results (Fig. 4.21) (Reményi *et al.*, 2001; Tomilin *et al.*, 2001; for review see Latchman, 2001).

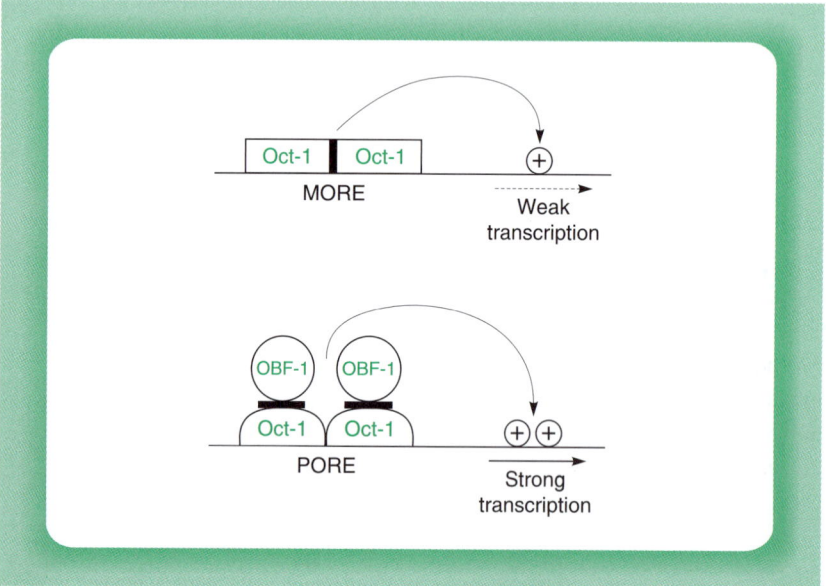

Figure 4.21

Binding of the Oct-1 dimer to the PORE DNA target sequence exposes a region of Oct-1 (heavy line) which can recruit the cellular co-activator OBF-1 resulting in strong activation of transcription. In contrast, binding of the Oct-1 dimer to the MORE DNA sequence produces a configuration in which this region is hidden in the interface between the two Oct-1 molecules. Hence, OBF-1 cannot be recruited and only weak transactivation occurs.

A more extreme example of this effect of DNA binding sequence is seen in the case of the Pit-1 member of the POU family. When Pit-1 binds as a dimer to its binding site in the prolactin promoter, it activates transcription. However, its binding site in the growth hormone promoter contains an extra two T bases. This results in a different binding configuration of the Pit-1 dimer, which allows it to recruit a co-repressor molecule and thereby inhibit rather than activate the growth hormone gene (Fig. 4.22) (Scully *et al.*, 2000; for review see Marx, 2000; Latchman, 2001).

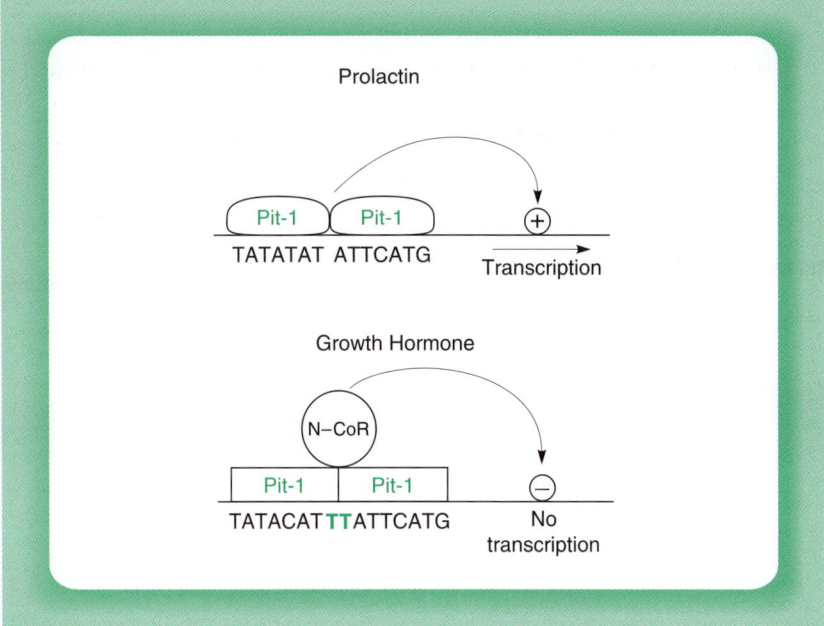

Figure 4.22
Binding of the Pit-1 dimer to its DNA binding site in the prolactin promoter allows it to activate transcription. In contrast, the extra two T bases in the binding site in the growth hormone promoter results in a different configuration of the Pit-1 dimer, leading to recruitment of the N-CoR co-repressor molecule and transcriptional repression.

Hence, the DNA binding sequence which is bound by a particular factor can have profound effects. Indeed, in the case of Pit-1 this is critical to its role in specifying the production of lactotrope cells in the pituitary gland, where expression of prolactin and not of growth hormone must occur.

This effect of the DNA binding site on the configuration of a transcription factor is not confined to the POU factors. Thus, for example, it has also been observed in the case of the NFκB transcription factor which has a DNA binding domain unrelated to the homeodomain or the POU domain (see section 4.6). In this case, NFκB binds to binding sites with different DNA sequences, in a different configuration that affects the co-activator molecules that bind to NFκB at each site (Leung *et al.*, 2004; for review see Natoli, 2004).

Hence, the different configuration of a factor bound to different binding sites can affect its ability to recruit other molecules which induce activation (co-activators) or inhibition (co-repressors) (see Chapter 5, section 5.4.3 and Chapter 6, section 6.3.2 for further discussion of co-activators and co-repressors, respectively) and hence produce different effects on transcription (Fig. 4.23).

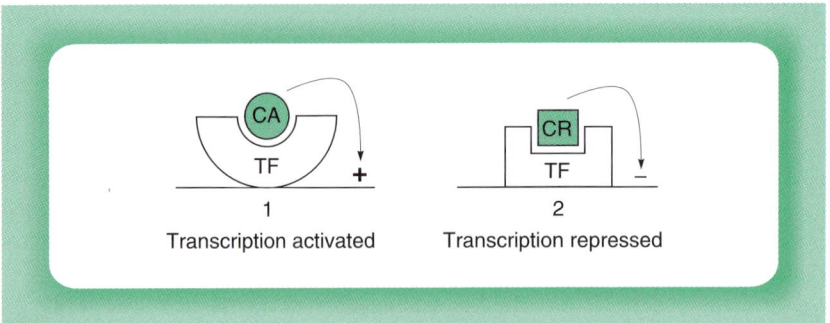

Figure 4.23
A transcription factor can bind to different DNA binding sites (1 and 2) in different configurations which affect its ability to recruit other proteins such as co-activators (CA) and co-repressors (CR) and hence its ability to activate or repress transcription.

As well as control of recruitment of such proteins at the level of a single factor, another level of control can operate by different POU proteins differing in their ability to recruit these factors. Thus, for example, the ability of Oct-1 and not Oct-2 to interact with the herpes simplex virus trans-activator protein VP16 is controlled by a single difference in the homeodomain region of the POU domains in the two proteins. Thus the replacement of a single amino acid residue at position 22 in the homeodomain of Oct-2 with the equivalent amino acid of Oct-1 allows Oct-2 to interact with VP16 which is normally a property only of Oct-1 (Lai *et al.*, 1992) (Fig. 4.24).

Interestingly, the key role of position 22 in the homeodomain is not confined to the interaction of Oct-1/Oct-2 with VP16. Thus, the closely related mammalian POU factors Brn-3a and Brn-3b differ in that Brn-3a activates the promoter of several genes expressed in neuronal cells whereas Brn-3b represses them. Alteration of the isoleucine residue found at position 22 in Brn-3b to the valine found in Brn-3a converts Brn-3b from a repressor into an activator whereas the reciprocal mutation in Brn-3a converts it into a repressor (for review see Latchman, 1999). This effect suggests that the activating/repressing effects of Brn-3a/Brn-3b are mediated by their binding of cellular co-activator or co-repressor molecules whose binding to Brn-3a/Brn-3b is affected by the nature of the amino acid at position 22. More generally, this finding provides the first example of a single amino acid change which can

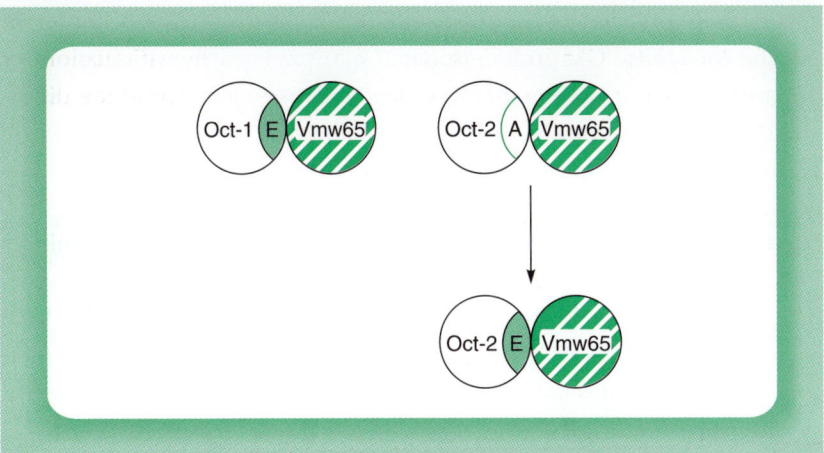

Figure 4.24

Alteration of an alanine residue (A) in the homeodomain of Oct-2 to the glutamic acid residue (E) found at the equivalent position in the homeodomain of Oct-1 allows Oct-2 to interact with the herpes simplex virus transactivator Vmw65 which is normally a property of Oct-1 only.

reverse the functional activity of a transcription factor, from activator to repressor and vice versa.

As in the case of the homeobox-containing proteins, the POU proteins appear to play a critical role in the regulation of developmental gene expression and in the development of specific cell types. Thus, the *unc-86* mutation in the nematode results, for example, in the lack of touch receptor neurons or male specific cephalic companion neurons indicating that this POU protein is required for the development of these specific neuronal cell types. Similarly, inactivation of the gene encoding Pit-1 leads to a failure of pituitary gland development resulting in dwarfism in both mice and humans (for review see Andersen and Rosenfeld, 1994). Interestingly, however, one type of dwarfism in mice (the Ames dwarf) is produced not by a mutation in Pit-1 but by a mutation in a gene encoding a homeobox-containing factor, which was named Prophet of Pit-1 (Sornson *et al.*, 1996). This factor appears to control the activation of the Pit-1 gene in pituitary cells so that Pit-1 is not expressed when this factor is inactivated. This example illustrates how hierarchies of regulatory transcription factors are required in order to control the highly complex process of development.

Following the initial identification of the original four POU factors, a number of other members of this family have been described both in mammals and other organisms such as *Drosophila, Xenopus* and zebra fish. Like the original factors, these novel POU proteins also play a critical role in the regulation of developmental gene expression. Thus, for example, the *Drosophila* POU protein drifter (CFla) has been shown to be of vital importance in the development of the nervous system (Anderson *et al.*, 1995), whilst mutations in the gene encoding the Brn-4 factor appear to be the cause of the most common form of deafness in humans (de Kok *et al.*, 1995). Moreover, all the novel POU domain-containing genes isolated by He *et al.* (1989) from the rat, on the basis of their containing a POU domain (see Chapter 2, section 2.3.2c), are expressed in the embryonic and adult brain, suggesting a similar role for these proteins in the regulation of neuronal-specific gene expression. Such a close connection of POU proteins and the central nervous system is also supported by studies using the original POU domain genes which revealed expression in the embryonic brain, even in the case of Oct-2 which had previously been thought to be expressed only in B lymphocytes (He *et al.*, 1989).

It is clear therefore that like the homeobox proteins, POU proteins occur in a wide variety of organisms and play an important role in the regulation of gene expression in development. Moreover, these proteins may be of particular importance in the development of the central nervous system.

4.2.7 PAX PROTEINS

As well as being found as part of the POU domain which gives the POU factors their name, a homeodomain is also found in some members of another family of transcription factors, the Pax factors (for reviews see Mansouri *et al.*, 1996; Chi and Epstein, 2002). These factors are defined on the basis that they contain a common DNA binding domain, known as the paired domain because it was originally identified in the *Drosophila* paired gene. In addition, however, some Pax proteins also contain a full size or truncated homeodomain whilst some, but not all, members of the family contain an eight amino acid element known as the octapeptide, which is of unknown function. All combinations of the paired domain with or without a homeodomain and/or the octapeptide are found in the various mammalian PAX factors (Fig. 4.25).

Obviously, in the Pax factors which lack the homeodomain, the paired domain is necessary and sufficient for DNA binding. Hence, this case is distinct from that of the POU factors where the POU specific and POU homeodomains are both necessary for high affinity DNA binding. Nonetheless, in factors such as Pax3, which have both a paired domain and a full length homeodomain, both domains participate in DNA binding. This produces very high affinity binding to a DNA binding site which contains the recognition sequence for both the DNA binding domains and the affinity of binding to such sites is greatly reduced when either

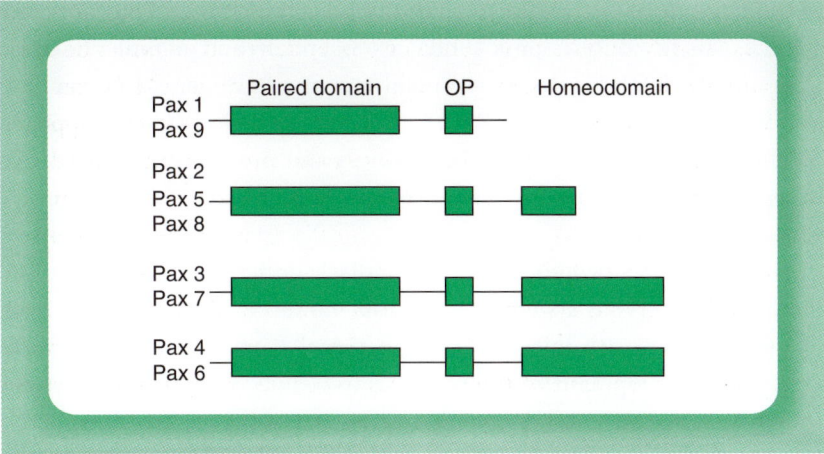

Figure 4.25
Structure of the mammalian Pax factors which contain an N-terminal paired domain linked in some cases to an octapeptide (OP) of unknown function and/or a full length or truncated homeodomain.

the paired domain or the homeodomain is deleted. Interestingly, the paired domain itself is distantly related to the homeodomain in terms of its structure and mechanism of DNA binding.

Thus, like the homeodomain, the paired domain also binds to DNA via a helix-turn-helix motif. Structural analysis of this motif, however, reveals that it is more similar to that in the bacteriophage proteins (see section 4.2.3) than in the eukaryotic homeodomain proteins with the residues at the N-terminus of the recognition helix being critical for DNA binding (Xu *et al.*, 1995). Indeed, one form of Waardenburg's syndrome which results from inactivation of PAX3 (see Chapter 9, section 9.1) is due to mutation in a glycine residue at the N-terminus of the PAX3 recognition helix resulting in a failure of the factor to bind to DNA. Hence, the helix-turn-helix motif is a widely used DNA binding domain which exists in at least two different forms that differ in the manner in which the recognition helix contacts the DNA.

As with the POU proteins, Pax factors play a critical role in gene regulation during development, particularly in the developing nervous system (for review see Robson *et al.*, 2006). Thus, for example, Pax6 has been shown to be of critical importance in specifying which cells will develop into different types of motor neurons during development (Ericson *et al.*, 1997) and also appears to play a critical role in eye development in a wide range of organisms (Gehring and Ikeo, 1999). In agreement with the critical role of these genes in development, knock out mice in which specific Pax genes have been inactivated show defects in the development of the nervous system, whilst the naturally occurring mutant mouse strain splotch which exhibits spina bifida, exencephaly and neural crest and limb muscle defects is due to a mutation in the Pax3 gene. Interestingly, as noted above, mutations in Pax3 in humans result in Waardenburg syndrome, which is characterized by deafness and eye defects, whilst mutations in Pax6 also result in severe eye defects such as aniridia (for review see Latchman, 1996).

Hence, the Pax proteins play a particularly critical role in the development of the nervous system. In addition, however, they also play a role in other tissues, with mice lacking functional Pax6 showing abnormalities in the development of the pancreas as well as of the nervous system (Sander *et al.*, 1997) whilst, as discussed in Chapter 7 (section 7.2.1), Pax3 is involved in activating the expression of the muscle determining factor, MyoD.

4.3 THE TWO CYSTEINE TWO HISTIDINE ZINC FINGER

4.3.1 TRANSCRIPTION FACTORS WITH THE TWO CYSTEINE TWO HISTIDINE FINGER

Transcription factor TFIIIA plays a critical role in regulating the transcription of the 5S ribosomal RNA genes by RNA polymerase III (see Chapter 3, section 3.4). When this transcription factor was purified, it was found to have a repeated structure and to be associated with between seven and eleven atoms of zinc per molecule of purified protein (Miller *et al.*, 1985). When the gene encoding TFIIIA was cloned, it was shown that this repeated structure consisted of the unit Tyr/Phe-X-Cys-X-Cys-$X_{2,4}$-Cys-X_3-Phe-X_5-Leu-X_2-His-$X_{3,4}$-His-X_5, which is repeated nine times within the TFIIIA molecule. This repeated structure therefore contains two invariant cysteine and two invariant histidine residues which were predicted to bind a single zinc atom accounting for the multiple zinc atoms bound by the intact molecule.

This motif is referred to as a zinc finger on the basis of its proposed structure in which a loop of twelve amino acids containing the conserved leucine and phenylalanine residues as well as several basic amino acids projects from the surface of the molecule, being anchored at its base by the cysteine and histidine residues which tetrahedrally co-ordinate an atom of zinc (Fig. 4.26). The proposed interaction of zinc with the conserved cysteine and histidine residues in this structure was subsequently confirmed by X-ray adsorption spectroscopy of the purified TFIIIA protein.

Following its identification in the RNA polymerase III transcription factor TFIIIA, similar cys_2 his_2-containing zinc finger motifs were identified in

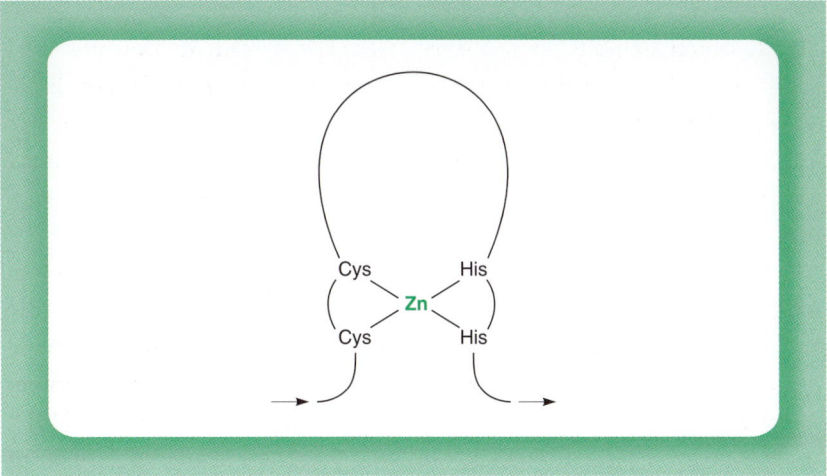

Figure 4.26

Schematic representation of the zinc finger motif. The finger is anchored at its base by the conserved cysteine and histidine residues which tetrahedrally co-ordinate an atom of zinc.

a number of RNA polymerase II transcription factors such as Sp1, which contains three contiguous zinc fingers, and the *Drosophila* Kruppel protein, which contains four finger motifs (see section 4.2.1). A list of some zinc finger containing transcription factors is given in Table 4.1 (for reviews see Evans and Hollenberg, 1988; Klug and Schwabe, 1995; Bieker, 2001; Klug, 2005; Gamsjaeger *et al.*, 2007).

In all cases studied the zinc finger motifs have been shown to constitute the DNA binding domain of the protein, with DNA binding being dependent upon their activity. Thus, in the case of TFIIIA, DNA binding is dependent on the presence of zinc, allowing the finger structures to form whilst progressive deletion of more and more zinc finger repeats in the molecule results in a parallel loss of DNA binding activity. Similarly, in the case of Sp1, DNA binding is dependent on the presence of zinc and most importantly the sequence specific binding activity of the intact protein can be reproduced by a protein fragment containing only the zinc finger region.

A similar dependence of DNA binding on the zinc finger motif is also seen in the *Drosophila* Kruppel protein, which is essential for correct thoracic and abdominal development. In this case a single mutation in one of the conserved cysteine residues in the finger, replacing it with a serine that cannot bind zinc, results in the production of a mutant fly indistinguishable from that produced by a complete deletion of the gene (Redemann *et al.*, 1988) indicating the vital importance of the zinc finger (Fig. 4.27).

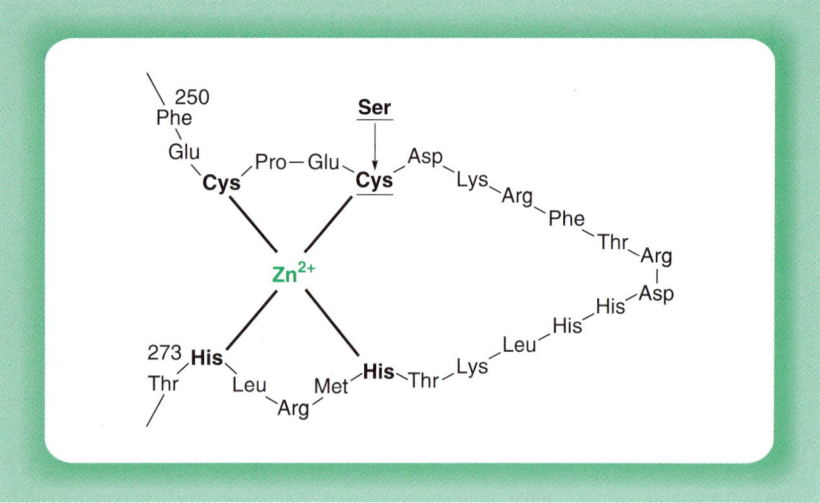

Figure 4.27
Zinc finger in the *Drosophila* Kruppel protein indicating the cysteine to serine change which abolishes the ability to bind zinc and results in a mutant fly indistinguishable from that obtained when the entire gene is deleted.

As with the helix-turn-helix motif of the homeobox, therefore, the zinc finger motif forms the DNA binding element of the transcription factors which contain it. Interestingly, however, a single zinc finger taken from the yeast ADRI protein is unable to mediate sequence specific DNA binding in isolation whereas a protein fragment containing both the two fingers present in the intact protein can do so. This suggests therefore that DNA binding by the zinc finger is dependent upon interactions with adjacent fingers and explains why zinc finger-containing transcription factors always contain multiple copies of the zinc finger motif (see Table 4.1).

4.3.2 DNA BINDING BY THE TWO CYSTEINE TWO HISTIDINE FINGER

In the zinc finger structure the zinc co-ordination via cysteine and histidine serves as a scaffold for the intervening region which makes direct contact with the DNA. Detailed structural analysis has shown that these intervening amino acids do not form a simple loop structure as proposed in the original model (for review see Rhodes and Klug, 1993; Klug and Schwabe, 1995). Rather, the finger region forms a motif consisting of two anti-parallel beta-sheets with an adjacent alpha-helix packed against one face of the beta-sheet (Fig. 4.28; see Plate 3; Lee *et al.*, 1989). Upon contact with DNA, the alpha-helix lies in the major groove of the DNA and

Table 4.1

Transcriptional regulatory proteins containing Cys_2-His_2 zinc fingers

Organism	Gene	Number of fingers
Drosophila	Kruppel	4
	Hunchback	6
	Snail	4
	Glass	5
Yeast	ADR1	2
	SW15	3
Xenopus	TFIIIA	9
	Xfin	37
Rat	NGF-1A	3
Mouse	MK1	7
	MK2	9
	Egr 1	3
	Evi 1	10
Human	Sp1	3
	TDF	13

makes sequence specific contacts with the bases of DNA whilst the beta-sheets lie further away from the helical axis of the DNA and contact the DNA backbone (for review see Klug, 2005).

Most interestingly, this structure indicates that a critical role in sequence specific DNA binding will be played by amino acids at the amino terminus of the alpha-helix, most notably the amino acids immediately preceding the first histidine residue. In agreement with this idea, two amino acids in this region play a critical role in determining the DNA binding specificity of the *Drosophila* Krox-20 transcription factor (Nardelli *et al.*, 1991). Thus this factor contains three zinc fingers and interacts with the DNA sequence 5' GCGGGGGCG 3'. If each finger contacts three bases within this sequence, then the central finger must recognize the sequence GGG whereas the two outer fingers will each recognize the sequence GCG (Fig. 4.29).

When the amino acid sequence of each of the Krox-20 fingers was compared, it was found that the two outer fingers contain a glutamine residue at position 18 of the finger and an arginine at position 21 whereas the central finger differs in that it has histidine and threonine residues at these positions. As expected, if these two amino acid differences are critical in determining the DNA sequence which is recognized,

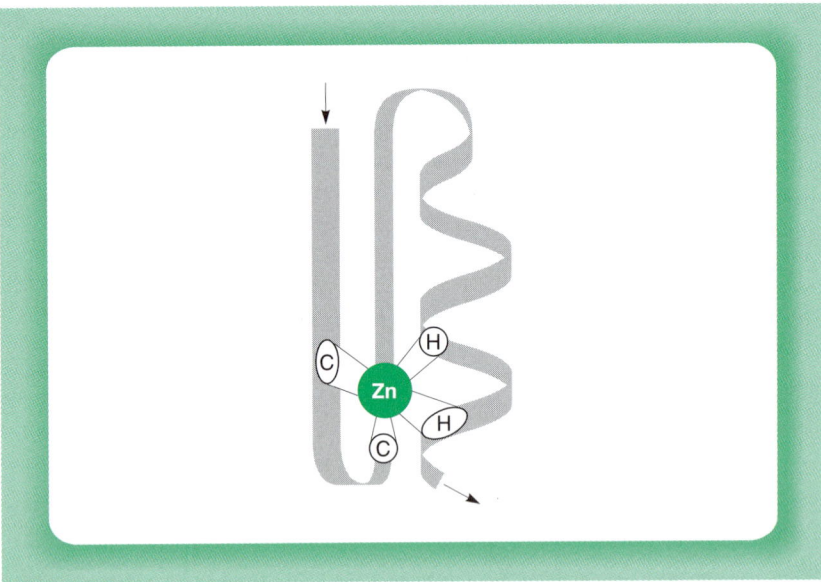

Figure 4.28

Structure of the zinc finger in which two anti-parallel beta-sheets (straight lines) are packed against an adjacent alpha-helix (wavy line).

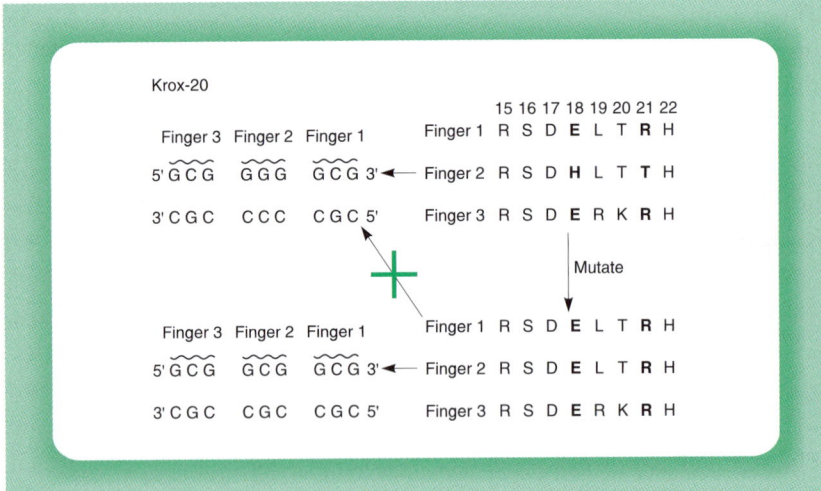

Figure 4.29

DNA binding specificity and amino acid sequence of the three cysteine–histidine zinc fingers in the *Drosophila* Krox 20 protein. Note that each finger binds to three specific bases in the recognition sequence and that finger 2 which differs from fingers 1 and 3 in the DNA sequence it recognizes also differs in the amino acids at positions 18 and 21 in the finger (bold letters). Mutating these amino acids to their equivalents in fingers 1 and 3 changes the DNA binding specificity of finger 2 to that of fingers 1 and 3 indicating that these amino acids play a critical role in determining the DNA sequence which is recognized.

altering these two residues in the central finger to their equivalents in the outer two fingers resulted in a factor which failed to bind to the normal Krox-20 binding site. Instead it bound to the sequence 5′ GCGGCGGCG 3′ in which each finger binds the sequence GCG. This experiment therefore indicates the critical role of two amino acids at the amino terminus of the alpha-helix in producing the DNA binding specificity of zinc fingers of this type and also shows that at least in the case of Krox-20, each successive finger interacts with three bases of DNA within the recognition sequence.

The importance of these amino acids has also been confirmed in experiments in which the amino acids at different positions in the zinc finger were randomly altered and their interaction with a wide range of DNA sequences assessed (Choo and Klug, 1994; Rebar and Pabo, 1994; for review see Klug, 2005). Clearly such an important role for the amino acids at the amino terminus of an alpha helix parallels the similar critical role for the equivalent amino acids in the recognition helix of the bacteriophage DNA binding proteins and in the paired domain (see section 4.2).

Interestingly, using this type of information on the DNA binding properties of individual fingers, it has recently proved possible to create novel zinc finger transcription factors with a defined DNA binding specificity. The potential use of such engineered zinc fingers to control gene expression to produce a therapeutic benefit in human patients is discussed in Chapter 9 (section 9.5).

Hence, like the helix-turn-helix motif, the cysteine–histidine zinc finger plays a critical role in mediating the DNA binding abilities of transcription factors which contain it, with sequence specific recognition of DNA being determined in both cases by amino acids within an alpha-helix.

4.4 THE MULTI-CYSTEINE ZINC FINGER

4.4.1 NUCLEAR RECEPTORS

The steroid hormones are a group of substances derived from cholesterol which exert a very wide range of effects on biological processes such as growth, metabolism and sexual differentiation (for review see King and Mainwaring, 1974). Early studies using radioactively labelled hormones showed that they act by interacting with specific receptor proteins. This binding of hormone to its receptor activates the receptor and allows it to bind to a limited number of specific sites in chromatin. In turn, this DNA binding activates transcription of genes carrying the receptor binding site. Hence, these receptor proteins are transcription factors, becoming activated in response to a specific signal and in turn activating specific genes (for reviews see Weatherman *et al.*, 1999; Khorasanizadeh and Rastinejad, 2001; Olefsky, 2001; McKenna and O'Malley, 2002; Rochette-Egly, 2005). These receptor proteins were therefore amongst the earliest transcription factors to be identified, well before the techniques described in Chapter 2 were in routine use, simply on the basis of their ability to bind radioactively labelled steroid ligand.

Genes which are induced by a particular steroid hormone contain a specific binding site for the receptor–hormone complex. The responses to different steroid hormones such as glucocorticoids and estrogen are mediated by distinct palindromic sequences which are related to one another. In turn, such sequences are related to one of the sequences which mediates induction by other substances which are related to steroids such as thyroid hormone and retinoic acid. Similarly, repeated elements with different spacings between the repeats also mediate responses to these different substances (Table 4.2).

Table 4.2

Relationship of various hormone response elements

(a) Palindromic repeats	
Glucocorticoid	RGRACANNNTGTYCY
Estrogen	RGGTCANNNTGACCY
Thyroid	RGGTCA - - - TGACCY
(b) Direct repeats	
9-cis retinoic acid	AGGTCAN$_1$AGGTCA
All-trans retinoic acid	AGGTCAN$_2$AGGTCA
	AGGTCAN$_5$AGGTCA
Vitamin D$_3$	AGGTCAN$_3$AGGTCA
Thyroid hormone	AGGTCAN$_4$AGGTCA

N indicates that any base can be present at that position; R indicates a purine, i.e. A or G; Y indicates a pyrimidine, i.e. C or T; W indicates A or T. A dash indicates that no base is present, the gap having been introduced to align the sequence with the other sequences.

The basis of this binding site relationship was revealed when the genes encoding the receptor proteins were cloned. Thus, they were found to constitute a family of genes encoding closely related proteins of similar structure with particular regions being involved in DNA binding, hormone binding and transcriptional activation (Fig. 4.30). This has led to the idea that these receptors are encoded by an evolutionarily related gene family which is known as the steroid-thyroid hormone receptor or nuclear receptor gene super family (for reviews see Weatherman *et al.*, 1999; Khorasanizadeh and Rastinejad, 2001; Olefsky, 2001; McKenna and O'Malley, 2002; Rochette-Egly, 2005).

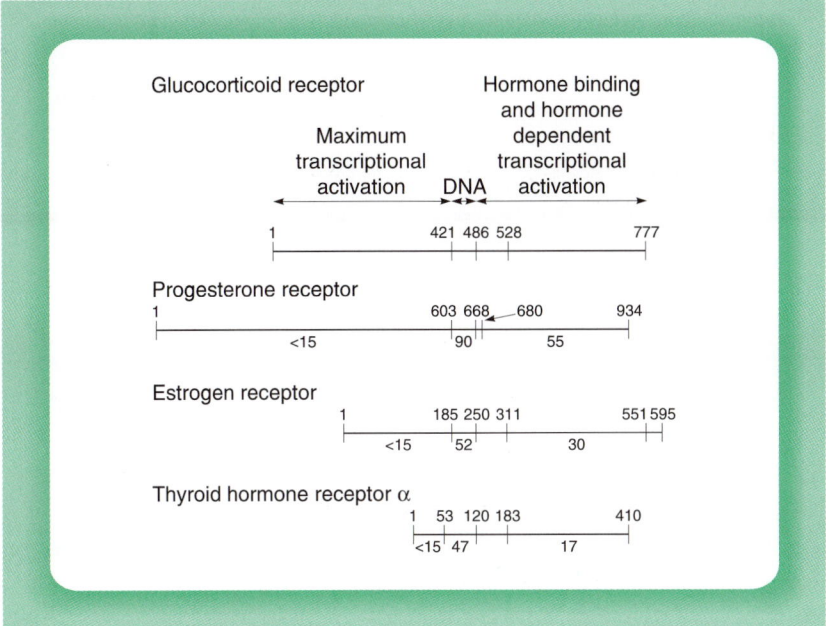

Figure 4.30
Domain structure of individual members of the nuclear receptor super family. The proteins are aligned on the DNA binding domain, which shows the most conservation between different receptors. The percentage homologies in each domain of the receptors to that of the glucocorticoid receptor are indicated.

As shown in Fig.4.30, the most conserved region between the different receptors is the DNA binding domain explaining the ability of the receptors to bind to similar DNA sequences. Interestingly, both DNAseI protection and methylation studies support the idea that the receptor binds to DNA as a dimer, each receptor molecule binding to one half of the recognition sequence.

4.4.2 DNA BINDING BY THE MULTI-CYSTEINE ZINC FINGER

Analysis of the nuclear receptor DNA binding domains identified a similar zinc binding motif to that discussed in section 4.3. As with the cysteine–histidine fingers, this motif has been shown by X-ray adsorption spectroscopy to bind zinc in a tetrahedral configuration. However, in this case, co-ordination is achieved by four cysteine residues rather than the two cysteine–two histidine structure discussed above. Similar multi-cysteine motifs have also been identified in several other DNA binding transcription factors such as the yeast proteins GAL4, PRRI and LAC9 as well as in the adenovirus transcription factor E1A (see Table 4.3, for review see Evans and Hollenberg, 1988; Klug and Schwabe, 1995), indicating that this type of motif is not confined to the nuclear receptors.

Table 4.3

Transcriptional regulatory proteins with multiple cysteine fingers

Finger type	Factor	Species
Cys_4-Cys_5	Nuclear receptors	Mammals
Cys_4	E1A	Adenovirus
Cys_6	Gal4, PPRI, LAC9	Yeast

In the case of the nuclear receptors, the DNA binding domain has the consensus sequence Cys-X_2-Cys-X_{13}-Cys-X_2-Cys-$X_{15,17}$-Cys-X_5-Cys-X_9-Cys-X_2-Cys-X_4-Cys. This motif is therefore capable of forming a pair of fingers each with four cysteines co-ordinating a single zinc atom (Fig. 4.31), and as with the cysteine–histidine finger proteins, DNA binding of the receptors is dependent on the presence of zinc.

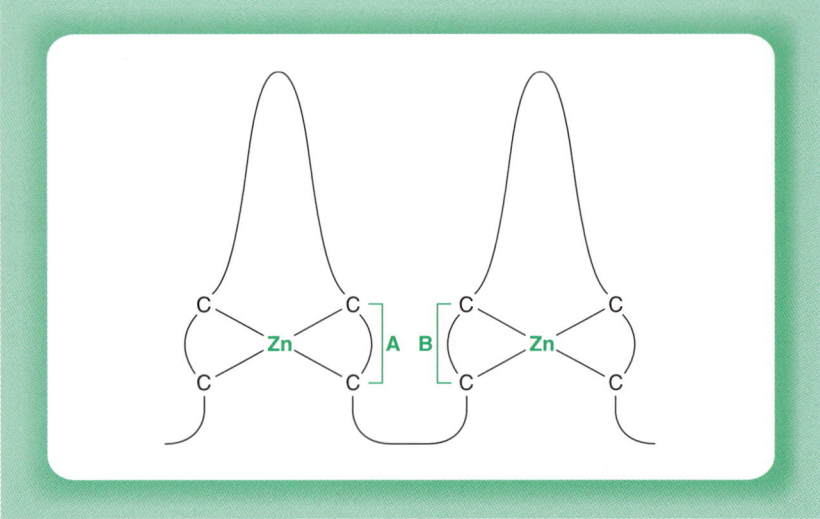

Figure 4.31

Schematic representation of the four cysteine zinc finger. Regions labelled A and B are of critical importance in determining respectively the DNA sequence which is bound by the finger and the optimal spacing between the two halves of the palindromic sequence which is recognized.

However, the multi-cysteine finger cannot be converted into a functional cysteine–histidine finger by substituting two of its cysteine residues with histidines, indicating that the two types of finger are functionally distinct. Moreover, unlike the cysteine–histidine zinc finger which is present in multiple copies within the proteins that contain it, the unit of two multi-cysteine fingers present in the steroid receptors is found only once in each receptor. Interestingly, structural studies of the two multi-cysteine fingers in the glucocorticoid and estrogen receptors (for review see Schwabe and Rhodes, 1991; Klug and Schwabe, 1995) have indicated that the two fingers form one single structural motif consisting of two alpha-helices perpendicular to one another with the cysteine–zinc linkage holding the base of a loop at the N-terminus of each helix (Fig. 4.32; see Plate 4; Hard *et al.*, 1990). This is quite distinct from the modular structure of the two cysteine–two histidine finger where each finger constitutes an independent structural element whose configuration is unaffected by the presence or absence of adjacent fingers.

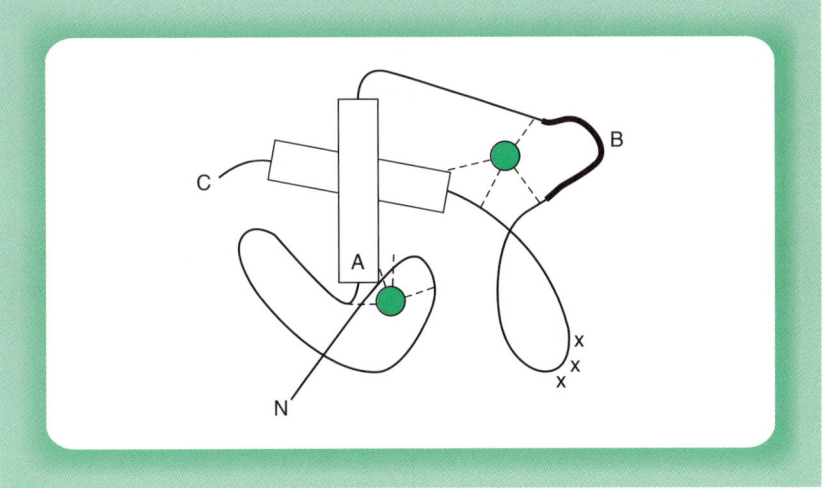

Figure 4.32

Schematic model of a pair of zinc fingers in a single molecule of the estrogen receptor. Note the helical regions (indicated as cylinders) with the critical residues for determining the DNA sequence which is bound located at the terminus of the recognition helix (indicated as A), the zinc atoms (shaded), conserved basic residues (+++) and the region which interacts with another receptor molecule and determines the optimal spacing between the two halves of the palindromic sequence which is recognized (indicated as B). Note that A and B indicate the same regions as in Figure 4.31.

Thus, although these two DNA binding motifs are similar in their co-ordination of zinc, they differ in the lack of histidines and of the conserved phenylalanine and leucine residues in the multi-cysteine finger, as well as structurally. It is clear therefore that they represent distinct functional elements and are unlikely to be evolutionarily related (for review see Schwabe and Rhodes, 1991; Rhodes and Klug, 1993; Klug and Schwabe, 1995).

Whatever the precise relationship between these motifs, it is clear that the multi-cysteine finger mediates the DNA binding of the nuclear receptors. Thus, mutations which eliminate or alter critical amino acids in this motif interfere with DNA binding by the receptor (Fig. 4.33).

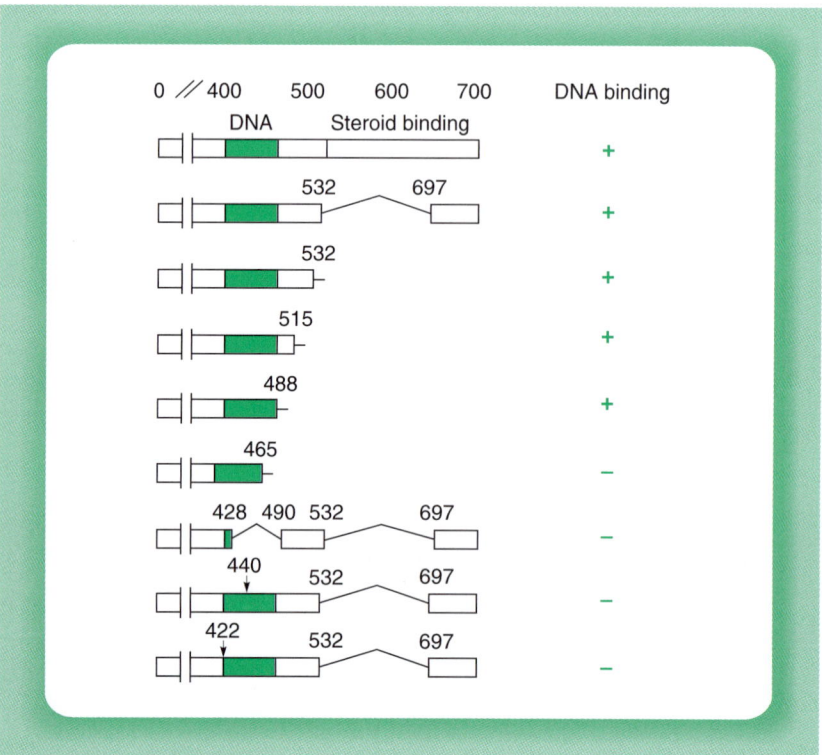

Figure 4.33

Effect of various deletions or mutations on the DNA binding of the glucocorticoid receptor. Note that DNA binding is only prevented by deletions which include part of the DNA binding domain (shaded) or by mutations within it (arrows) but not by deletions in other regions such as the steroid binding domain. Numbers indicate amino acid residues.

The role of the cysteine fingers in mediating DNA binding by the nuclear receptors can also be demonstrated by taking advantage of the observation that the different nuclear receptors bind to distinct but related palindromic sequences in the DNA of hormone responsive genes (see Khorasanizadeh and Rastinejad, 2001 for review and Table 4.2 for a comparison of these binding sites). Thus if the cysteine-rich region of the estrogen receptor is replaced by that of the glucocorticoid receptor, the resulting chimaeric receptor has the DNA binding specificity of the glucocorticoid receptor but continues to bind estrogen since all the other regions of the molecule are derived from the estrogen receptor (Fig. 4.34). Hence, the DNA binding specificity of the hybrid receptor is determined by its cysteine-rich region, resulting in the hybrid receptor inducing the expression of glucocorticoid responsive genes (which carry its DNA binding site) in response to estrogen (to which it binds).

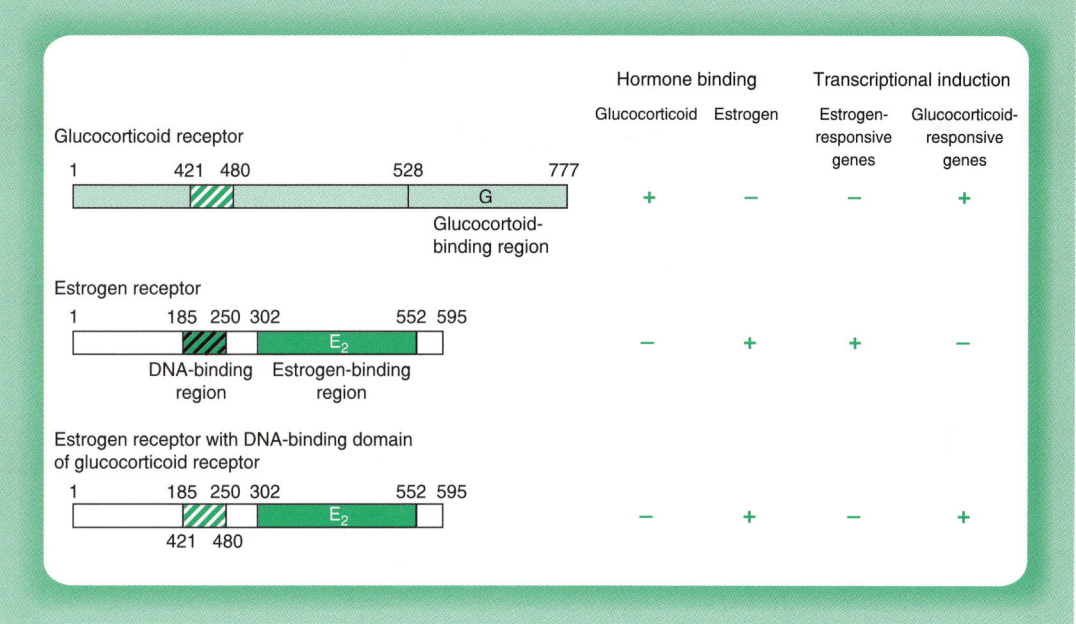

Figure 4.34

Effect of exchanging the DNA binding domain of the estrogen receptor with that of the glucocorticoid receptor on the binding of hormone and gene induction by the hybrid receptor.

These so-called 'finger swap' experiments therefore provide further evidence in favour of the critical role for the multi-cysteine fingers in DNA binding, exchanging the fingers of two receptors exchanging the DNA binding specificity. In addition, however, because of the existence of short distinct DNA binding regions of this type in receptors which bind to distinct but related DNA sequences, they provide a unique opportunity to dissect the elements in a DNA binding structure which mediate binding to specific sequences.

Thus, by exchanging one or more amino acids between two different receptors it is possible to investigate the effects of these changes on DNA binding specificity and hence elucidate the role of individual amino acid differences in producing the different patterns of sequence specific binding. For example, the alteration of the two amino acids between the third and fourth cysteines of the N-terminal finger in the glucocorticoid receptor for their equivalents in the estrogen receptor changes the DNA binding specificity of the chimaeric receptor to that of the estrogen receptor (Fig. 4.35). Hence, the exchange of two amino acids in a critical region of a protein of 777 amino acids (indicated as A in Fig. 4.31) can completely change the DNA binding specificity of the glucocorticoid receptor resulting in it binding to and activating genes which are normally estrogen responsive. The specificity of this hybrid receptor for such estrogen responsive genes can be further enhanced by exchanging another amino acid located between the two fingers (Fig. 4.35), indicating that this region also plays a role in controlling the specificity of DNA binding.

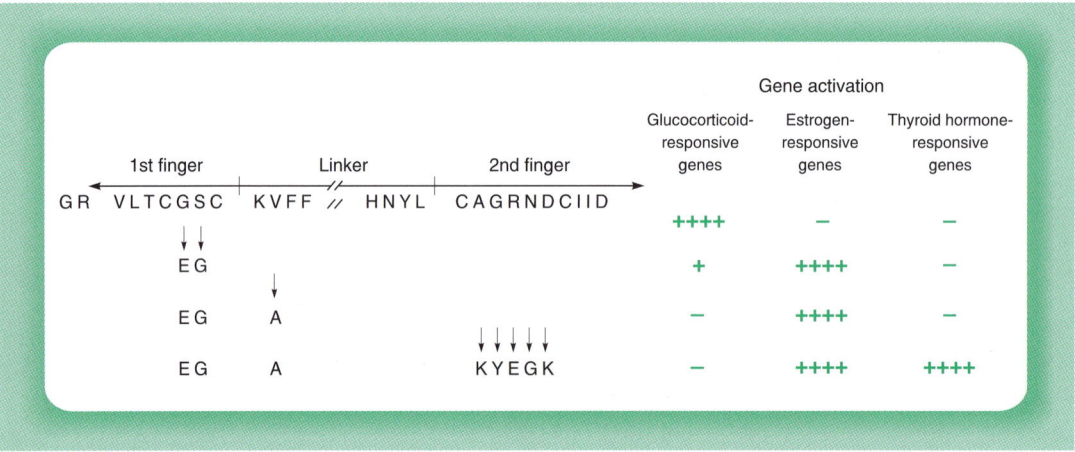

Figure 4.35
Effect of amino acid substitutions in the zinc finger region of the glucocorticoid receptor on the ability to bind to and activate genes which are normally responsive to different steroid hormones.

As noted above (section 4.4.1), the nuclear receptors bind to palindromic recognition sequences within DNA, with the receptor binding to DNA as a homodimer in which each receptor molecule interacts with one half of the palindrome. In addition to differences in the actual sequence recognized, nuclear receptors can also differ in the optimal spacing between the two separate halves of the palindromic DNA sequence which is recognized (Table 4.2a). Thus the estrogen receptor and the thyroid hormone receptor both recognize the identical palindromic sequence in the DNA but differ in that in the thyroid receptor binding sites the two halves of the palindrome are adjacent whereas in the estrogen receptor binding sites they are separated by three extra bases. The further alteration of the chimaeric receptor illustrated in Figure 4.35 by changing five amino acids in the second finger to their thyroid hormone receptor equivalents is sufficient to allow the receptor to recognize thyroid hormone receptor binding sites (Umesono and Evans, 1989; Fig. 4.35). These amino acids in the second finger (indicated as B in Fig. 4.31) appear to play a critical role therefore in determining the optimal spacing of the palindromic sequence which is recognized.

As discussed above, structural studies of the two zinc fingers in the estrogen and glucocorticoid receptors suggest that they form a single structural motif with two perpendicular alpha-helices (Fig. 4.32). In this structure, the critical amino acids for determining the spacing in the palindromic sequence recognized are located on the surface of the molecule allowing them to interact with equivalent residues on another receptor monomer during dimerization (indicated as B in Fig. 4.35; see Plate 5; Schwabe et al., 1993). Hence, differences in the interaction of these regions in the different receptors determine the spacing of the two monomers within the receptor dimer and thus the optimal spacing in the palindromic DNA sequence which is recognized.

Interestingly, within this structure, the critical residues for determining the precise DNA sequence which is recognized are located at the N-terminus of the first alpha-helix (indicated as A in Fig. 4.32 and Fig. 4.36), further supporting the critical role of such helices in DNA binding. Moreover, in the proposed structure of the estrogen receptor dimer the DNA binding helices in each monomer will be separated by 34Å allowing each of these recognition helices to make sequence specific contacts in adjacent major grooves of the DNA molecule.

Differences in the DNA binding domain also regulate the binding of members of the nuclear receptor family to directly repeated sequences with different spacings between the two halves of the repeat (see Table 4.2b). Thus, when the direct repeats are separated by only one base, they can bind a homodimer of the retinoid X-receptor (RXR) and hence confer a response to 9-cis retinoic acid, which binds to this receptor (Fig. 4.37). In contrast, the RXR homodimer cannot bind to the direct repeats when they are separated by between 2 and 5 base pairs. Rather, on these elements RXR forms a heterodimer with other members of the nuclear receptor family (Fig. 4.37).

Moreover, the nature of the heterodimers which form on a particular response element controls the response it mediates, with the nature of

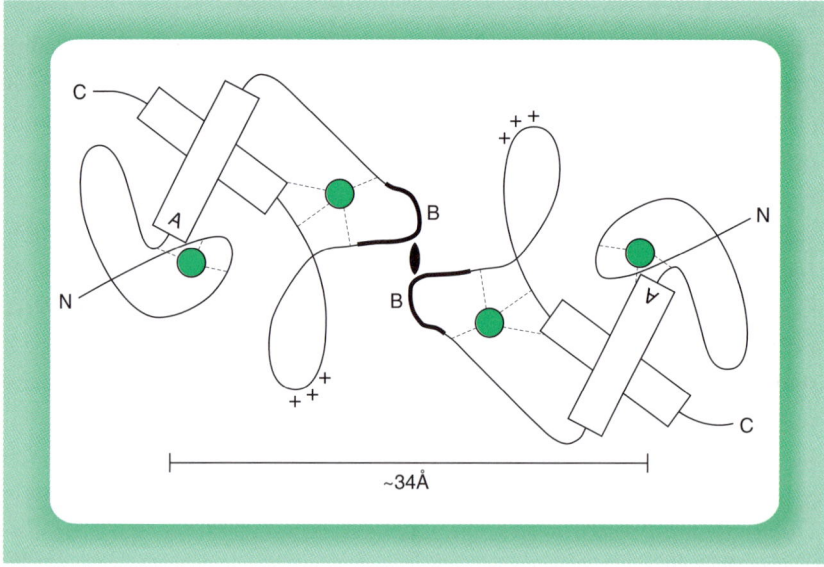

Figure 4.36
Interaction of two estrogen receptor molecules to form a DNA binding dimer. Compare with Figure 4.32 and note the interaction of the B regions on each molecule. The resulting dimer has a spacing of 34 Angstroms between the two DNA binding regions allowing binding in successive major grooves of the DNA molecule.

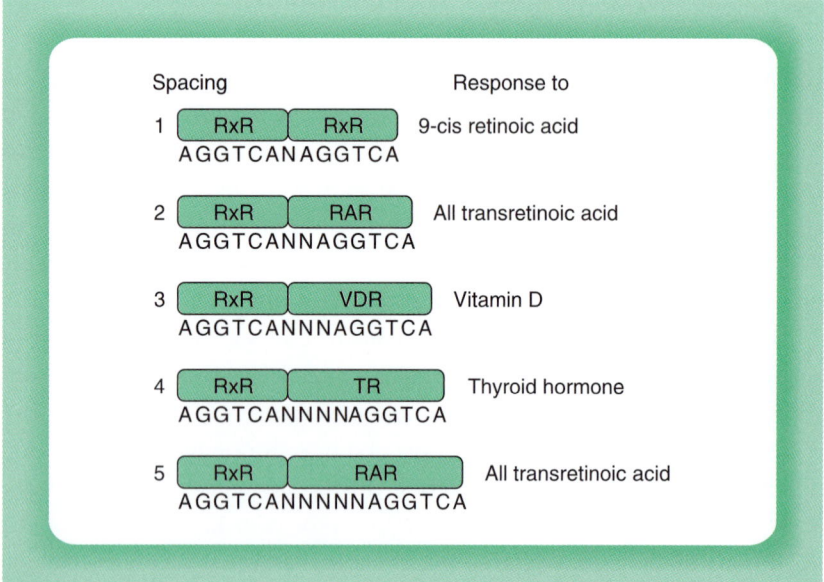

Figure 4.37
Binding of different nuclear receptor heterodimers to directly repeated elements with different spacings between the repeats determines the response mediated by each element.

the non-RXR component determining the response. Thus a spacing of two or five base pairs binds a heterodimer of RXR and the retinoic acid receptor (RAR) and therefore mediates responses to all transretinoic acid which binds to RAR. In contrast, a spacing of four base pairs binds a heterodimer of RXR and the thyroid hormone receptor (TR) and therefore can mediate responses to thyroid hormone.

As on the palindromic repeats, it is the DNA binding domain of the receptors which controls which heterodimers can form on particular spacings of the direct repeat. Interestingly, the crystal structure of the RXR-TR heterodimer bound to a direct repeat with a four base spacing indicates that the dimerization interface involves amino acids in the first finger of the thyroid hormone receptor and the second finger of RXR rather than only residues in the second finger, as occurs for homodimerization of receptors on palindromic repeats (Rastinejad *et al.*, 1995) (Fig. 4.38).

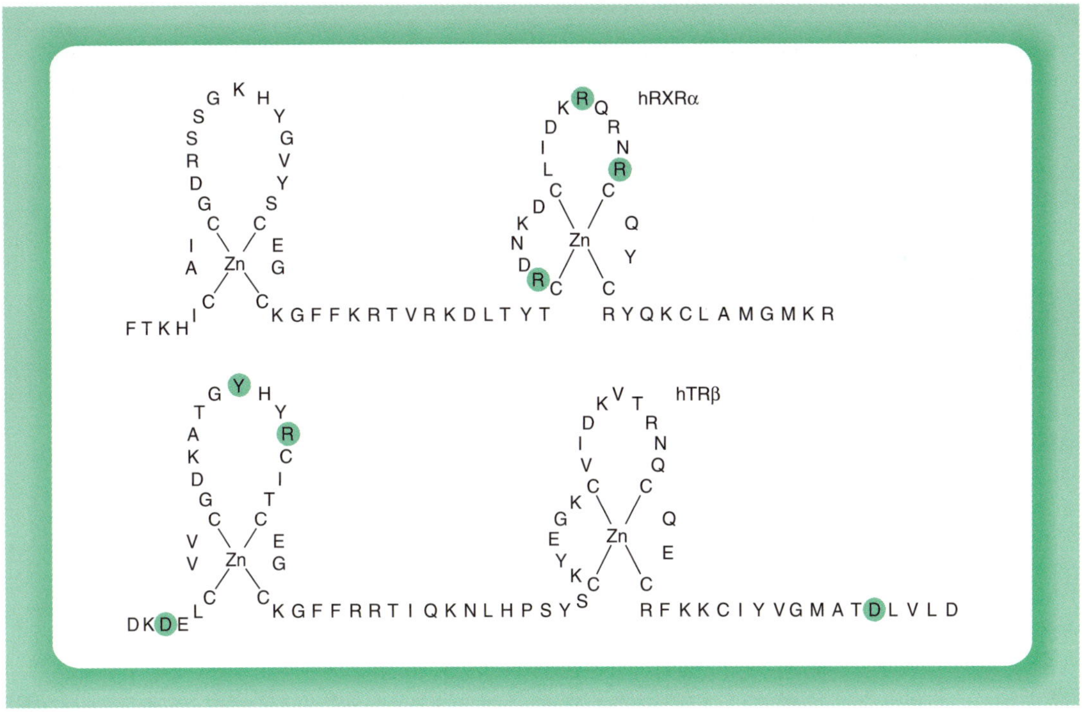

Figure 4.38
Zinc fingers in the retinoid X-receptor α and the thyroid hormone receptor β. The residues in each receptor which are involved in heterodimer formation with the other receptor are indicated.

The definition of the DNA binding domain of the nuclear receptors as a short sequence containing two multi-cysteine fingers has therefore allowed the elucidation of the features in this motif which mediate the different sequence specificities of the different receptors and their relationship to the structure of the motif. In particular, a helical region of the first finger plays a critical role in determining the precise DNA sequence which is recognized by binding in the major groove of the DNA. Similarly, other regions in either the first or second fingers control the spacing of adjacent palindromic or directly repeated sequences which is optimal for the binding of receptor homo or heterodimers, by interacting with another receptor monomer and hence affecting the structure of the receptor dimer which forms.

4.5 THE BASIC DNA BINDING DOMAIN

4.5.1 THE LEUCINE ZIPPER AND THE BASIC DNA BINDING DOMAIN

As discussed in the preceding sections of this chapter, the study of motifs common to several different transcription factors has led to the identification of the role of these motifs in DNA binding. A similar approach led to the identification of the leucine zipper motif (for reviews see Lamb and McKnight, 1991; Kerppola and Curran, 1995; Hurst, 1996). Thus, this structure has been detected in several different transcription factors such as the CAAT box binding protein C/EBP, the yeast factor GCN4 and the oncogene products Myc, Fos and Jun (see Chapter 9, sections 9.3.1 and 9.3.3). It consists of a leucine-rich region in which successive leucine residues occur every seventh amino acid (Fig. 4.39).

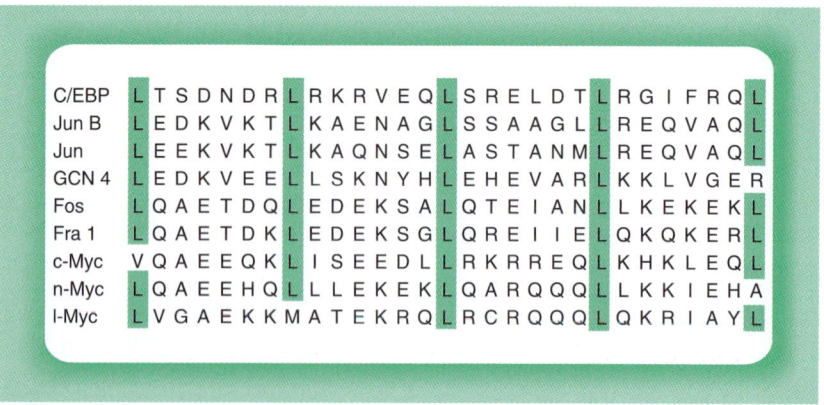

C/EBP	L TSDNDR	L RKRVEQ	L SRELDT	L RGIFRQ L
Jun B	L EDKVKT	L KAENAG	L SSAAGL	L REQVAQ L
Jun	L EEKVKT	L KAQNSE	L ASTANM	L REQVAQ L
GCN 4	L EDKVEE	L LSKNYH	L EHEVAR	L KKLVGE R
Fos	L QAETDQ	L EDEKSA	L QTEIAN	L LKEKEK L
Fra 1	L QAETDK	L EDEKSG	L QREIIE	L QKQKER L
c-Myc	V QAEEQK	L ISEEDL	L RKRREQ	L KHKLEQ L
n-Myc	L QAEEHQ	L LLEKEK	L QARQQQ	L LKKIEHA
I-Myc	L VGAEKKMATEKRQ	L RCRQQQ	L QKRIAY L

Figure 4.39

Alignment of the leucine-rich region in several cellular transcription factors. Note the conserved leucine residues (L) which occur every seven amino acids.

In all these cases, the leucine-rich region can be drawn as an alpha-helical structure in which adjacent leucine residues occur every two turns on the same side of the helix. Moreover, these leucine residues appear to play a critical role in the functioning of the protein. Thus with one exception (a single methionine in the Myc protein), the central leucine residues of the motif are conserved in all the factors which contain it (Fig. 4.39). It was therefore proposed (Landshultz *et al.*, 1988) that the long side chains of the leucine residues extending from one polypeptide would interdigitate with those of the analogous helix of a second polypeptide, forming a motif known as the leucine zipper which would result in the dimerization of the factor (Fig. 4.40). This effect could also be achieved by a methionine residue (which like leucine has a long side chain with no lateral methyl groups) but not by other hydrophobic amino acids such as valine or isoleucine (which have methyl groups extending laterally from the beta carbon atom).

Figure 4.40

Model of the leucine zipper and its role in the dimerization of two molecules of a transcription factor.

In agreement with this idea, substitutions of individual leucine residues in C/EBP or other leucine zipper-containing proteins such as Myc, Fos and Jun with isoleucine or valine, abolish the ability of the intact protein to form a dimer indicating the critical role of this region in dimerization. A comparison of the effects of various mutations of this type on the ability of the mutant protein to dimerize, suggested that the two leucine-rich regions associate in a parallel manner with both helices oriented in the same direction (as illustrated in Fig. 4.40) rather than in an anti-parallel configuration as originally suggested (Landshultz *et al.*, 1989). This idea was confirmed by structural studies of the leucine zipper regions in GCN4 and in the Fos/Jun dimer bound to DNA (Glover and Harrison, 1995). These studies indicated that each zipper motif forms a right-handed alpha-helix with dimerization occurring via the association of two parallel helices that coil around each other to form a coiled coil motif similar to that found in fibrous proteins such as the keratins and myosins (Fig. 4.41).

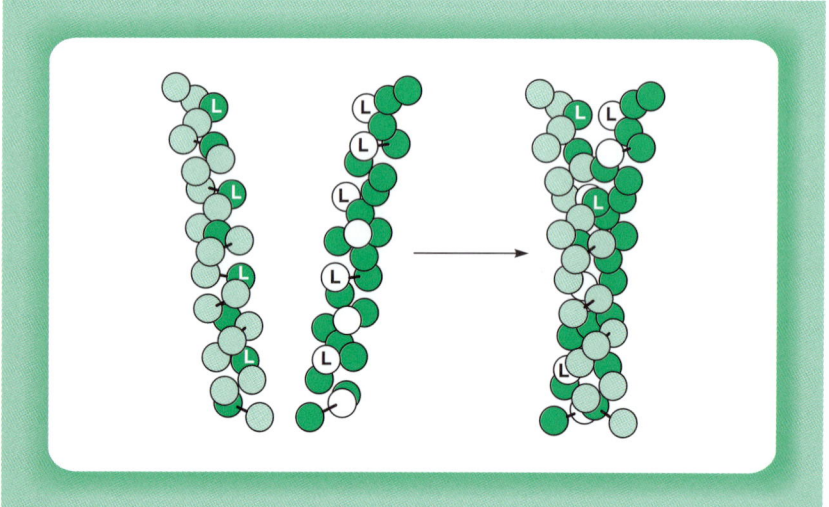

Figure 4.41

Coiled coil structure of the leucine zipper formed by two helical coils wrapping around each other. L indicates a leucine residue.

In addition to its role in dimerization, the leucine zipper is also essential for DNA binding by the intact molecule. Thus mutations in the zipper which prevent dimerization also prevent DNA binding from occurring (Landshultz *et al.*, 1989). Unlike the zinc finger or helix-turn-helix motifs, however, the zipper is not itself the DNA binding domain of the molecule and does not directly contact the DNA. Rather it facilitates DNA binding by an adjacent region of the molecule which in C/EBP, Fos and Jun is rich in basic amino acids and can therefore interact directly with the acidic DNA. The leucine zipper is believed therefore to serve an indirect structural role in DNA binding, facilitating dimerization which in turn results in the correct positioning of the two basic DNA binding domains in the dimeric molecule for DNA binding to occur (Fig. 4.42).

In agreement with this idea, mutations in the basic domain abolish the ability to bind to DNA without affecting the ability of the protein to

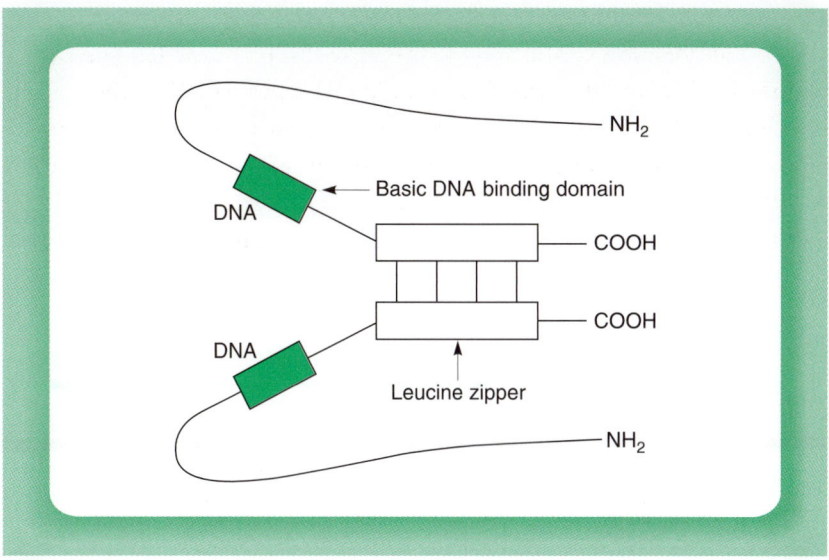

Figure 4.42

Model for the structure of the leucine zipper and the adjacent DNA binding domain following dimerization of the transcription factor C/EBP.

dimerize, as expected for mutations which directly affect the DNA binding domain (Landshultz *et al.*, 1989). Similarly, exchange of the basic region of GCN4 for that of C/EBP results in a hybrid protein with the DNA binding specificity of C/EBP whilst exchange of the leucine zipper region has no effect on the DNA binding specificity of the hybrid molecule (Fig. 4.43).

Hence, the DNA binding specificity of leucine zipper-containing transcription factors is determined by the sequence of their basic domain with the leucine zipper allowing dimerization to occur and hence facilitating DNA binding by the basic domain. As expected from this idea, the basic DNA binding domain can interact with DNA in a sequence specific manner in the absence of the leucine zipper if it is first dimerized via an inter-molecular disulphide bond (Fig. 4.44). Interestingly, the basic DNA binding domain can bind to DNA as a monomer in the case of the Skn-1 factor, which lacks a leucine zipper. In this factor, however, the basic domain is part of a composite DNA binding domain which also contains a region homologous to the N-terminal arm of the homeobox (see section 4.2).

In factors having a simple basic DNA binding domain, following dimerization via the leucine zipper, the intact transcription factor will form a rotationally symmetric dimer that contacts the DNA via the bifurcating basic regions (see Fig. 4.42), which form alpha-helical structures.

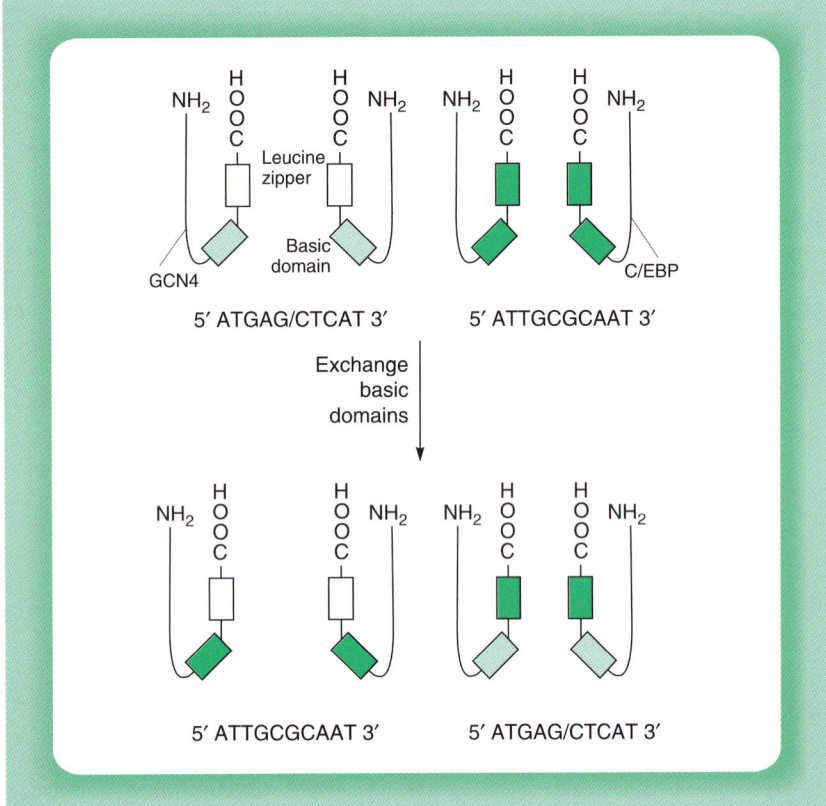

Figure 4.43

Effect of exchanging the basic domains of GCN4 and C/EBP on the DNA binding
specificity. Note that the DNA binding specificity is determined by the origin of
the basic domain and not that of the leucine zipper.

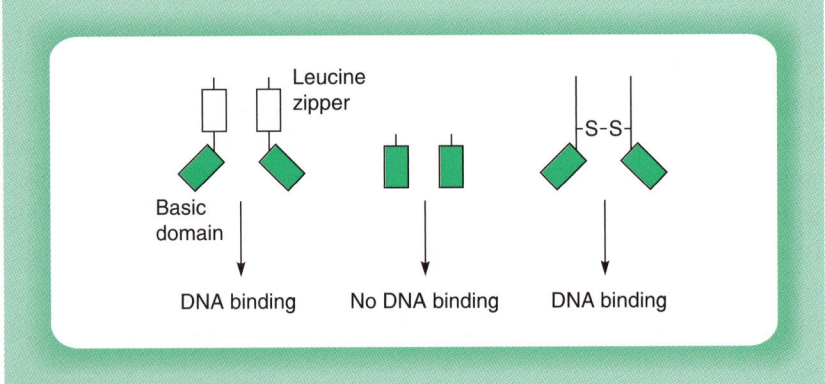

Figure 4.44

DNA binding of molecules containing basic DNA binding domains can occur
following dimerization mediated by leucine zippers or by a disulphide bridge (S–S)
but cannot be achieved by unlinked monomeric molecules.

These two helices then track along the DNA in opposite directions corresponding to the dyad symmetric structure of the DNA recognition
site and form a clamp or scissors grip around the DNA, similar to the
grip of a wrestler on his opponent, resulting in very tight binding of the
protein to DNA (Glover and Harrison, 1995). Most interestingly, structural studies have suggested that the basic region does not assume a
fully alpha-helical structure until it contacts the DNA when it undergoes
a configurational change to a fully alpha-helical form. Hence, the association of the transcription factor with the appropriate DNA sequence
results in a conformational change in the factor leading to a tight association with that sequence (for discussion see Sauer, 1990).

4.5.2 THE HELIX-LOOP-HELIX MOTIF AND THE BASIC DNA BINDING DOMAIN

Although originally identified in the leucine zipper-containing proteins,
the basic DNA binding domain has also been identified by a homology
comparison in a number of other transcription factors which do not contain a leucine zipper (Prendergast and Ziff, 1989). These factors include
the MyoD transcription factor, which plays a key role in activating specific
genes in skeletal muscle (see Chapter 7, section 7.2.1) and the E12 and
E47 factors, which play a key role in the development of immunoglobulin
producing B lymphocytes and in the development of the nervous system.

In these cases, the basic DNA binding domain is juxtaposed to a region
which can form a helix-loop-helix motif (for reviews see Littlewood and
Evan, 1995; Massari and Murre, 2000; Kewley *et al.*, 2004). This helix-loop-
helix motif is distinct from the helix-turn-helix motif in the homeobox
(section 4.2) in that it can form two amphipathic helices, containing all
the charged amino acids on one side of the helix, which are separated by
a non-helical loop (Murre *et al.*, 1989a). This helix-loop-helix motif plays
a similar role to the leucine zipper, allowing dimerization of the transcription factor molecule and thereby facilitating DNA binding by the basic
motif (Murre *et al.*, 1989b; for discussion see Jones, 1990).

In agreement with this, deletion or mutations in the basic domain
of the MyoD protein do not abolish dimerization but do prevent DNA
binding, paralleling the effect of similar mutations in C/EBP (Fig. 4.45).
Similarly, mutations or deletions in the helix-loop-helix region abolish
both dimerization and DNA binding, paralleling the effects of similar
mutations in leucine zipper-containing proteins. Moreover, the DNA
binding ability of MyoD from which the basic DNA binding domain has
been deleted can be restored by substituting the basic domain of the E12
protein (Davis *et al.*, 1990). However, such substitution does not allow the

hybrid protein to activate muscle-specific gene expression, suggesting that, in addition to mediating DNA binding, the basic region of MyoD also contains elements involved in the activation of muscle-specific genes (Davis *et al.*, 1990; Fig. 4.45).

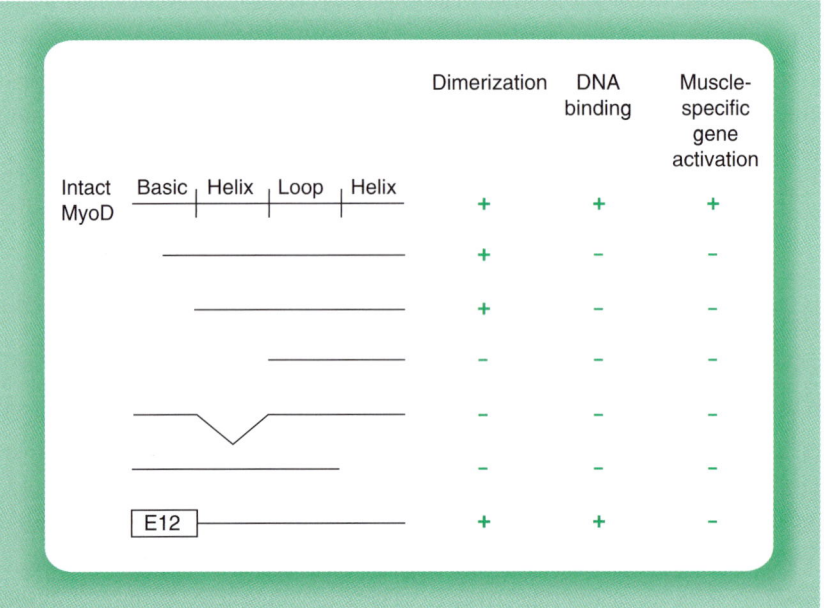

Figure 4.45

Effect of deleting the basic domain or the adjacent helix-loop-helix motif on dimerization, DNA binding and activation of muscle-specific gene expression by the MyoD transcription factor. Note that deletion of any part of the helix-loop-helix motif abolishes dimerization and consequent DNA binding and gene activation, whilst deletion of the basic domain directly abolishes DNA binding and consequent gene activation. Substitution of the basic domain of the constitutive factor E12 for that of MyoD restores DNA binding but not the ability to activate muscle-specific gene expression.

Interestingly, it has been shown that the conversion of three amino acids within the E12 basic region to their MyoD equivalents allows the E12 basic region to activate muscle-specific gene expression following DNA binding (Fig. 4.46) (Davis and Weintraub, 1992). The crystal structure of MyoD bound to DNA (Ma *et al.*, 1994) suggests that these amino acids may play a critical role in allowing the MyoD basic region to assume a particular structural configuration in which it can interact with other activating transcription factors. In agreement with this idea, the substitution of these same three amino acids in E12 for their MyoD equivalents allows the mutant E12 protein to bind to another muscle-specific transcription factor MEF2A which is normally a property of MyoD alone (Fig. 4.46; Kanshal *et al.*, 1994). Hence, like the POU domain (see section 4.2.6), the basic domain appears to function both as a DNA binding domain and as a site for protein–protein interactions critical for transcriptional activation.

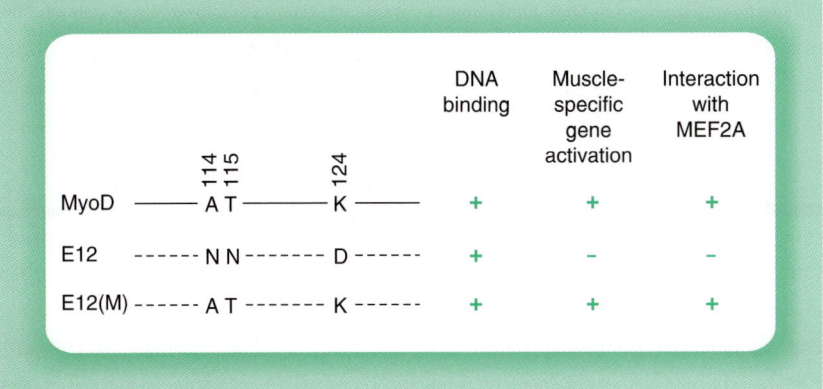

Figure 4.46
Alterations of three amino acids (positions 114, 115 and 124) in the E12 basic domain to their MyoD equivalents confers on the resulting protein (E12(M)) the ability to interact with the muscle-specific transcription factor MEF2A and activate muscle-specific genes following DNA binding which are normally properties of MyoD alone.

4.5.3 DIMERIZATION OF BASIC DNA BINDING DOMAIN-CONTAINING FACTORS

Both the leucine zipper and the helix-loop-helix motif therefore act by causing dimerization, allowing DNA binding by the adjacent basic motif. Interestingly, the Myc oncoproteins contain both a helix-loop-helix motif and a leucine zipper region adjacent to the basic DNA binding region (Landshultz *et al.*, 1988; Murre *et al.*, 1989a). Moreover, the leucine zipper can also be found as a dimerization motif in proteins that use DNA binding motifs other than the basic region. For example, in the *Arabidopsis* Athb-1 and 2 proteins, the leucine zipper facilitates dimerization with DNA binding being produced by the adjacent homeobox (Sessa *et al.*, 1993). Thus, individual DNA binding and dimerization motifs can be combined in different combinations to produce molecules capable of dimerizing and binding to DNA.

The essential role of dimerization (mediated by the leucine zipper or the helix-loop-helix motifs) in allowing DNA binding by basic DNA binding domain proteins provides an additional aspect to the regulation of these factors (for discussion see Jones, 1990; Lamb and McKnight, 1991). Thus in addition to the formation of homodimers, it is possible to hypothesize that heterodimers will also form between two different leucine zipper or two different helix-loop-helix-containing factors allowing the production of dimeric factors with novel DNA binding specificities or affinities for different sites.

One example of this process is seen in the oncogene products Fos and Jun. Thus, as discussed in Chapter 9 (section 9.3.1), the Fos protein cannot bind to AP1 sites in DNA when present alone but can form a heterodimer with the Jun protein which is capable of binding to such sites with thirty-fold greater affinity than a Jun homodimer (Fig. 4.47). The formation of Jun homodimers and Jun/Fos heterodimers is dependent upon the leucine zipper regions of the proteins. Moreover, the failure of Fos to form homodimers is similarly dependent on its leucine zipper region. Thus, if the leucine zipper domain of Fos is replaced by that of Jun, the resulting protein can dimerize and the chimaeric protein can bind to DNA through the basic DNA binding region of Fos which is therefore a fully functional DNA binding domain.

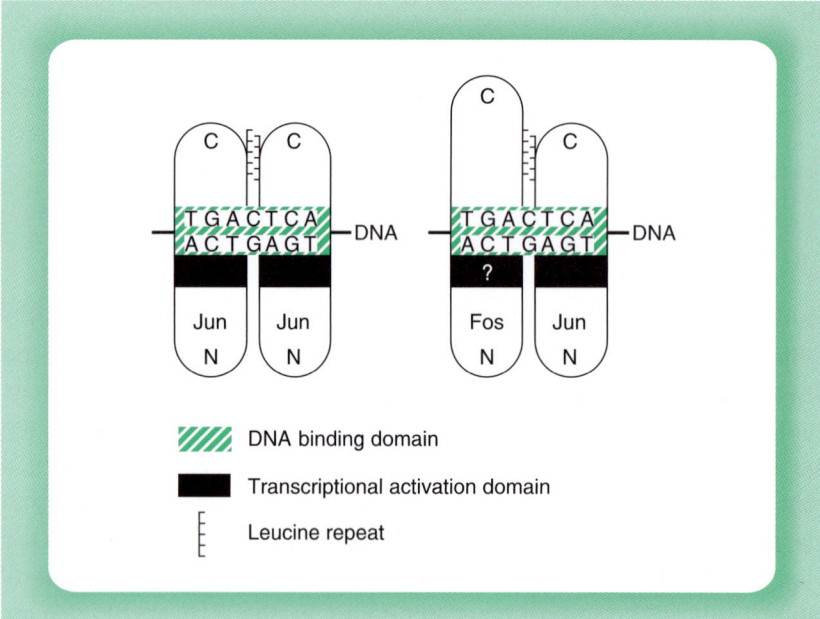

Figure 4.47

Model for DNA binding by the Jun homodimer and the Fos–Jun heterodimer.

Hence, the ability of leucine zipper proteins to bind to DNA is determined both by the nature of the leucine zipper which facilitates homodimerization and/or heterodimerization as well as by the basic

DNA binding motif which allows DNA binding following dimerization (for discussion see Kerppola and Curran, 1995). Moreover, this heterodimerization is a very specific process. Thus, a study of forty-nine human leucine zipper-containing proteins, showed that they form heterodimers only in very specific combinations even though they all have very similar leucine zipper motifs (Newman and Keating, 2003).

In addition to its positive role in allowing DNA binding by factors which cannot do so as homodimers, heterodimerization between two related factors can also have an inhibitory role. Thus, the DNA binding ability of functional helix-loop-helix proteins which contain a basic DNA binding domain can be inhibited by association with the Id protein. This protein contains a helix-loop-helix motif allowing it to associate with other members of this family but lacks the basic DNA binding domain. The heterodimer of Id and a functional protein therefore lacks the dimeric basic regions necessary for DNA binding and the activity of the functional transcription factor is thereby inhibited by Id (see Chapter 6, section 6.2.2 for further discussion of transcriptional repression by Id).

Hence, the role of the leucine zipper and helix-loop-helix motifs in dimerization can be put to use in gene regulation in both positive and negative ways, either allowing DNA binding by factors which could not do so in isolation or inhibiting the binding of fully functional factors. Moreover, only specific combinations of proteins can form heterodimers with one another, indicating that this is a very specific process and is therefore likely to play an important role in gene regulation.

4.6 OTHER DNA BINDING MOTIFS

Although the majority of DNA binding domains which have been identified in known transcription factors fall into the families we have discussed in the preceding sections, not all do so. As more and more factors are characterized, other relationships between the DNA binding domains of different factors have emerged, leading to the identification of new transcription factor families. Thus, for example, the UBF ribosomal RNA transcription factor (see Chapter 3, section 3.3) contains a DNA binding domain that has also been identified in several other factors, including high mobility group (HMG) proteins and which is therefore known as the HMG box (Grosschedel *et al.*, 1994), whilst the DNA binding domain in the p53 protein discussed in Chapter 9 (section 9.4.2) has been shown to be related to that of the NFκB family (Muller *et al.*, 1995; for review see Baltimore and Beg, 1995).

Interestingly, however, as the structure of more and more DNA binding domains is understood, relationships have emerged between different domains which were originally thought to be entirely distinct. For example, structural analysis of the Ets DNA binding domain which is found in the Ets-1 proto-oncogene protein (see Chapter 9, section 9.3.1) and the mouse PU-1 factor has revealed it to be identical to the winged helix-turn-helix motif originally identified in the *Drosophila* fork head factor and in the mammalian liver transcription factor HNF-3 (Donaldson *et al.*, 1996).

Moreover, as its name suggests, this domain contains a helix-turn-helix motif which is similar to that found in the homeobox proteins discussed in section 4.2. However, the winged helix-turn-helix motif also contains an additional β-sheet structure with loops that appear as wings protruding from the DNA bound factor, giving this motif its name (for review see Brennan, 1993). In the majority of winged-helix-containing proteins, the helix-turn-helix motif is responsible for DNA binding. However, in the hRFX1 member of the family, it is the β-sheet structure which binds to DNA rather that the helix-turn-helix motif, indicating that members of this family can use one of two distinct structures to bind to DNA (Gajiwala *et al.*, 2000).

As discussed in section 4.2, both the POU specific domain of the POU factors and the paired box of the PAX proteins also bind to DNA via helix-turn-helix motifs, indicating that this is one of the most commonly used motifs mediating the DNA binding of factors whose DNA binding domains appear distinct at first sight.

4.7 CONCLUSIONS

In this chapter we have discussed a number of different DNA binding motifs, common to several different transcription factors, which can mediate DNA binding. These motifs are listed in Table 4.4.

Interestingly, it is also possible for the same DNA sequence to be bound by more than one factor. Although in many cases, the factors binding to a particular DNA sequence share a common DNA binding domain, this is not always the case. Thus, whilst the transcription factors CTF/NFI and C/EBP both bind to the CAAT box sequence, they do so via completely different DNA binding domains, with C/EBP having a basic DNA binding domain (section 4.5) whilst CTF/NFI has a DNA binding domain distinct from that of any other factor.

It is unlikely therefore that the existence of several distinct DNA binding domains reflects the need of the factors which contain them to bind to distinct types of DNA sequences. Rather, it seems perfectly possible that one DNA binding motif could be present in all factors with variations of it in different factors producing the observed binding to different DNA

Table 4.4

DNA binding motifs

Motif	Structure	Factors containing domain	Comments
Homeobox	Helix-turn-helix	Numerous *Drosophila* homeotic genes, related genes in other organisms	Structurally related to similar motif in bacteriophage proteins
POU	Helix-turn-helix and adjacent helical region	Mammalian Oct-1, Oct-2, Pit-1, nematode *unc86*	Related to homeodomain
Paired	Helix-turn-helix	Mammalian Pax factors, *Drosophila* paired factor	Often found in factors which also contain a homeobox
Cysteine–histidine zinc finger	Multiple fingers, each co-ordinating a zinc atom	TFIIIA, Kruppel, Sp1, etc.	May form β-sheet and adjacent α-helical structure
Cysteine–cysteine zinc finger	Single pair of fingers each co-ordinating a zinc atom	Nuclear receptor family	Related motifs in EIA, GAL4, etc.
Basic domain	α-helical	C/EBP c-*fos*, c-*jun*, c-*myc*, MyoD, etc.	Associated with leucine zipper and/or helix-loop-helix dimerization motifs
Winged HTH	Helix-turn-helix	Fork head, HNF 3A c-*ets*, c-*erg*, *Drosophila* E74, PU-1	Binds purine-rich sequences

sequences. This is particularly so in view of the fact that in diverse DNA binding motifs such as the helix-turn-helix, the basic DNA binding domain and the two types of zinc fingers, the amino acids which determine sequence specific binding to DNA are all located within similar alpha-helical structures. This idea evidently begs the question of why different DNA binding motifs exist.

It is possible that this situation has arisen simply by different motifs which could produce DNA binding having arisen in particular factors during evolution and having been retained since they efficiently fulfilled their function. Alternatively, it may be that the existence of different motifs reflects other differences in the factors containing them other than the specific DNA sequence which is recognized. For example, the highly repeated zinc finger motif may be of particular use where, as in the case of transcription factor TFIIIA, the factor must contact a large regulatory region in the DNA. Similarly, a motif such as the basic domain which can only bind to DNA following dimerization will be of particular use where the activity of the factor must be regulated whether positively or negatively via dimerization with another factor.

Whatever the case, it is clear that DNA binding by transcription factors is dependent upon specific domains of defined structure within the molecule. Following such DNA binding, the bound factor must influence the rate of transcription either positively or negatively. The manner in which this occurs and the regions of the factors which achieve this effect are discussed in the next two chapters.

REFERENCES

Andersen, B. and Rosenfeld, M.G. (1994) Pit-1 determines cell types during development of the interior pituitary gland. Journal of Biological Chemistry 269, 335–338.

Anderson, M.G., Perkins, G.L., Chittick, P., Shrigley, R.J. and Johnson, W.A. (1995) Drifter, a *Drosophila* POU-domain transcription factor, is required for correct differentiation and migration of tracheal cells and midline glia. Genes and Development 9, 123–127.

Andrews, B. and Donoviel, M.S. (1995) A heterodimeric transcriptional repressor becomes clear. Science 270, 251–253.

Baltimore, D. and Beg, A.E. (1995) DNA-binding proteins: a butterfly flutters by. Nature 373, 287–288.

Baumruker, T., Sturm, R. and Herr, W. (1988) OBP 100 binds remarkably degenerate octamer motifs through specific interaction with flanking sequences. Genes and Development 2, 1400–1413.

Bieker, J.J. (2001) Krüppel-like factors: three fingers in many pies. Journal of Biological Chemistry 276, 34355–34358.

Brennan, R.G. (1993) The winged-helix DNA-binding motif: another helix-turn-helix take off. Cell 74, 773–776.

Briscoe, J. and Wilkinson, D.G. (2004) Establishing neuronal circuitry: Hox genes make the connection. Genes and Development 18, 1643–1648.

Chi, N. and Epstein, J.A. (2002) Getting your Pax straight: Pax proteins in development and disease. Trends in Genetics 18, 41–47.

Choo, Y. and Klug, A. (1994) Toward a code for the interaction of zinc fingers with DNA: selection of randomized fingers displayed on phage. Proceedings of the National Academy of Sciences, USA 91, 11163–11167.

Cilberto, G., Castagnoli, L. and Cortese, R. (1983) Transcription by RNA polymerase III. Current Topics in Development Biology 18, 59–88.

Davis, R.L. and Weintraub, H. (1992) Acquisition of myogenic specificity by replacement of three amino acid residues from MyoD into E12. Science 256, 1027–1030.

Davis, R.L., Cheng, P-F., Lassar, A.B. and Weintraub, H. (1990) The MyoD DNA binding domain contains a recognition code for muscle-specific gene activation. Cell 60, 733–746.

Dawson, S.J., Morris, P.J. and Latchman, D.S. (1996) A single amino acid change converts a repressor into an activator. Journal of Biological Chemistry 271, 11631–11633.

de Kok, Y.J.M., Van der Maarel, S.M., Bitner-Glindzicz, M., Huber, I., Monaco, A.P., Malcolm, S., Pembrey, M.E., Ropers, H-H. and Cremers, F.P.M. (1995) Association between X-linked mixed deafness and mutations in the POU domain gene POU3F4. Science 267, 685–688.

Dolan, J.K. and Fields, S. (1991) Cell type-specific transcription in yeast. Biochimica et Biophysica Acta 1088, 155–169.

Donaldson, L.W., Peterson, J.M., Graves, B.J. and McIntosh, P. (1996) Solution structure of the ETS domain from murine Ets-1: a winged helix-turn-helix DNA binding motif. EMBO Journal 15, 125–134.

Duboule, D. (2000) A Hox by any other name. Nature 403, 607–610.

Ericson, J., Rashbass, P., Schedl, A., Brenner-Morton, S., Kawakemi, A., van Heyningen, V., Jessel, T.M. and Briscoe, J. (1997) Pax6 controls progenitor cell identity and neuronal fate in response to graded Shh signalling. Cell 90, 169–180.

Evans, R.M. and Hollenberg, S.M. (1988) Zinc fingers: gilt by association. Cell 52, 1–3.

Gajiwala, K.S., Chen, H., Cornille, F., Roques, B.P., Reith, W., Mach, B. and Burley, S.K. (2000) Structure of the winged-helix protein hRFX1 reveals a new mode of DNA binding. Nature 403, 916–921.

Gamsjaeger, R., Liew, C.K., Loughlin, F.E., Crossley, M. and Mackay, J.P. (2007) Sticky fingers: zinc-fingers as protein-recognition motifs. Trends in Biochemical Sciences 32, 63–70.

Garvie, C.W. and Wolberger, C. (2001) Recognition of specific DNA sequences. Molecular Cell 8, 937–946.

Gehring, W.J. and Ikeo, K. (1999) Pax 6 mastering eye morphogenesis and eye evolution. Trends in Genetics 15, 371–377.

Gehring, W.J., Affolter, M. and Burglin, T. (1994) Homeodomain proteins. Annual Review of Biochemistry 63, 487–526.

Gehring, W.J., Qian, Y.Q., Billeter, M., Furukubo-Tukunagu, K., Schier, A.F., Resendez-Perez, D., Affolter, M., Otting, G. and Wuthrich, K. (1994) Homeodomain-DNA recognition. Cell 78, 211–223.

Glover, J.N.M. and Harrison, S.C. (1995) Crystal structure of the heterodimeric bZip transcription factor c-Fos-c-Jun bound to DNA. Nature 373, 257–261.

Graham, A., Papalopulu, N. and Krumlauf, R. (1989) The murine and *Drosophila* homeobox gene complexes have common features of organization and expression. Cell 57, 367–378.

Grosschedel, R., Giese, K. and Pagel, J. (1994) HMG proteins: architectural elements in the assembly of nucleoprotein structures. Trends in Genetics 10, 94–100.

Hadorn, E. (1968) Transdetermination in cells. Scientific American 219 (Nov), 110–120.

Hard, T., Kellenbach, E., Boelens, R., Maler, B.A., Dahlam, K., Freedman, L.P., Carlstedt-Duke, J., Yamamoto, K.R., Gustafsson, J-A. and Kaplein, R. (1990) Solution structure of the glucocortoid receptor DNA-binding domain. Science 249, 157–160.

Harrison, S.C. (1991) A structural taxonomy of DNA binding domains. Nature 353, 715–719.

Hayashi, S. and Scott, M.P. (1990) What determines the specificity of action of *Drosophila* homeodomain proteins. Cell 63, 883–894.

He, X., Treacy, M.N., Simmons, D.M., Ingraham, H.A., Swanson, L.S. and Rosenfeld, M.G. (1989) Expression of a large family of POU-domain regulatory genes in mammalian brain development. Nature 340, 35–42.

Hoey, T. and Levine, M. (1988) Divergent homeobox proteins recognize similar DNA sequences in *Drosophila*. Nature 332, 858–861.

Holland, P.W.H. and Hogan, B.L.M. (1986) Phylogenetic distribution of Antennapedia-like homeoboxes. Nature 321, 251–253.

Hurst, H.C. (1996) bZIP proteins. Protein Profile 3, 1–72.

Ingham, P.W. (1988) The molecular genetics of embryonic pattern formation in *Drosophila*. Nature 335, 25–34.

Jones, N. (1990) Transcriptional regulation by dimerization: two sides to an incestuous relationship. Cell 61, 9–11.

Kanshal, S., Schneider, J.W., Nudal-Ginard, B. and Mahdavi, V. (1994) Activation of the myogenic lineage by MEF2A, a factor that induces and cooperates with MyoD. Science 266, 1236–1240.

Kerppola, T. and Curran, T. (1995) Zen and the art of Fos and Jun. Nature 373, 199–200.

Kewley, R.J., Whitelaw, M.L. and Chapman-Smith, A. (2004) The mammalian basic helix-loop-helix/PAS family of transcriptional regulators. International Journal of Biochemistry and Cell Biology 36, 189–204.

Khorasanizadeh, S. and Rastinejad, F. (2001) Nuclear-receptor interactions on DNA-response elements. Trends in Biochemical Sciences 26, 384–390.

King, R.J.B. and Mainwaring, W.I.P. (1974) *Steroid cell interactions*. Butterworths, London.

Klemm, J.D., Rould, M.A., Aurora, R., Herr, W. and Pabo, C.O. (1994) Crystal structure of the Oct-1 POU domain bound to an octamer site: DNA recognition with tethered DNA-binding molecules. Cell 77, 21–23.

Klug, A. (2005) The discovery of zinc fingers and their development for practical applications in gene regulation. Proceedings of the Japan Academy. Series B, Physical and Biological Sciences 81, Ser. B, 87–102.

Klug, A. and Schwabe, J.R. (1995) Zinc fingers. FASEB Journal 9, 597–604.

Kornberg, T.B. (1993) Understanding the homeodomain. Journal of Biological Chemistry 268, 26813–26816.

Lai, J.S., Cleary, M.A. and Herr, W. (1992) A single amino acid exchange transfers VP16-induced positive control from the Oct-1 to the Oct-2 homeo domain. Genes and Development 6, 2058–2065.

Lamb, P. and McKnight, S.L. (1991) Diversity and specificity in transcriptional regulation: the benefits of heterotypic dimerization. Trends in Biochemical Sciences 16, 417–422.

Landschulz, W.H., Johnson, P.F. and McKnight, S.L. (1988) The leucine zipper: a hypothetical structure common to a new class of DNA binding proteins. Science 240, 1759–1764.

Landschulz, W.H., Johnson, P.F. and McKnight, S.L. (1989) The DNA binding domain of the rat liver nuclear protein C/EBP is bipartite. Science 243, 1681–1688.

Latchman, D.S. (1996) Transcription factor mutations and human disease. New England Journal of Medicine 334, 28–33.

Latchman, D.S. (1999) POU Family transcription factors in the nervous system. Journal of Cellular Physiology 179, 126–133.

Latchman, D.S. (2001) Transcription factors: bound to activate or repress. Trends in Biochemical Sciences 26, 211–213.

Lawrence, P.A. and Morata, G. (1994) Homeobox genes: their function in *Drosophila* segmentation and pattern formation. Cell 78, 181–189.

Lee, M.S., Gippert, G.P., Soman, K.V., Case, D.A. and Wright, P.E. (1989) Three-dimensional solution structure of a single zinc finger DNA binding domain. Science 245, 635–637.

Leung, T.H., Hoffmann, A. and Baltimore, D. (2004) One nucleotide in a kappaB site can determine cofactor specificity for NF-kappaB dimers. Cell 118, 453–464.

Li, T., Stark, M.R., Johnson, A.D. and Wolberger, C. (1995) Crystal structure of the MATa1/MAT alpha 2 homeodomain heterodimer bound to DNA. Science 270, 262–269.

Littlewood, T. and Evan, G. (1995) Helix-loop-helix. Protein Profile 2, 621–702.

Ma, P.C.M., Rould, M.A., Weintraub, H. and Pabo, C.O. (1994) Crystal structure of MyoD bHLH domain – DNA complex: Perspectives on DNA recognition and implications for transcriptional activation. Cell 77, 451–459.

Mann, R.S. and Chan, S-K. (1996) Extra specificity from extradenticle: the partnership between Hox and PBX/EXD homeodomain proteins. Trends in Genetics 12, 258–262.

Mansouri, A., Hallone, T.M. and Gruss, P. (1996) Pax genes and their roles in cell differentiation and development. Current Opinion in Cell Biology 8, 851–857.

Martindale, M.Q. and Kourakis, M.J. (1999) Size doesn't matter. Nature 399, 730–731.

Marx, J. (2000) New clues to how genes are controlled. Science 290, 1066–1067.

Massari, M.E. and Murre, C. (2000) Helix-loop-helix proteins: regulators of transcription in eucaryotic organisms. Molecular and Cellular Biology 20, 429–440.

McKenna, N.J. and O'Malley, B.W. (2002) Combinational control of gene expression by nuclear receptors and coregulators. Cell 108, 465–474.

Miller, J., McLachlan, A.D. and Klug, A. (1985) Repetitive zinc-binding domains in the protein transcription factor III A from *Xenopus* oocytes. EMBO Journal 4, 1609–1614.

Morgan, R. (2006) Hox genes: a continuation of embryonic patterning? Trends in Genetics 22, 67–69.

Muller, C.W., Rey, F.A., Sodeoka, M., Verdine, G.L. and Harrison, S.C. (1995) Structure of the NF-κB p50 homodimer bound to DNA. Nature 373, 311–317.

Murre, C., McCaw, P.S. and Baltimore, D. (1989a) A new DNA binding and dimerization motif in immunoglobulin enhancer binding, daughterless, MyoD and myc proteins. Cell 56, 777–783.

Murre, C., McCaw, P.S., Vaessin, H., Caudy, M., Jan, L.Y., Jan, Y.N., Cabera, C.V., Buskin, J.N., Hauschka, S.D., Lassar, A.B., Weintraub, H. and Baltimore, D. (1989b) Interactions between heterologous helix-loop-helix proteins generate complexes that bind specifically to a common DNA sequence. Cell 58, 537–544.

Nardelli, J., Gibson, T.J., Vesque, C. and Charnay, P. (1991) Base sequence discrimination by zinc-finger DNA binding domains. Nature 349, 175–178.

Natoli, G. (2004) Little things that count in transcriptional regulation. Cell 118: 406–408.

Newman, J.R.S. and Keating, A.E. (2003) Comprehensive identification of human bZIP interactions with coiled-coil arrays. Science 300, 2097–2101.

Olefsky, J.M. (2001) Nuclear receptor minireview series. Journal of Biological Chemistry 276, 36863–36864.

Pabo, C. and Sauer, R.T. (1992) Transcription factors: structural families and principles of DNA recognition. Annual Review of Biochemistry 61, 1053–1095.

Passner, J.M., Ryoo, H.D., Shen, L., Mann, R.S. and Aggarwal, A.K. (1999) Structure of a DNA-bound ultrabithorax-extradenticle homeodomain complex. Nature 397, 714–719.

Pfeifer, K., Prezant, T. and Guarente, L. (1987) Yeast HAPI activator binds to two upstream sites of different sequence. Cell 49, 19–27.

Prendergast, G.C. and Ziff, E.B. (1989) DNA-binding motif. Nature 341, 392.

Rastinejad, F., Perlmann, T., Evans, R.M. and Sigler, P.B. (1995) Structural determinants of nuclear receptor assembly on DNA direct repeats. Nature 375, 203–211.

Rebar, E.J. and Pabo, C.O. (1994) Zinc finger phage: affinity selection of fingers with new DNA-binding specificities. Science 263, 671–673.

Redemann, N., Gaul, U. and Jackle, H. (1988) Disruption of a putative Cys-zinc interaction eliminates the biological activity of the Kruppel finger protein. Nature 332, 90–92.

Reményi, A., Tomilin, A., Pohl, E., Lins, K., Philippsen, A., Reinbold, R., Schöler, H. R. and Wilmanns, M. (2001) Differential dimer activities of the transcription factor Oct-1 by DNA-induced interface swapping. Molecular Cell 8, 569–580.

Rhodes, D. and Klug, A. (1993) Zinc finger structure. Scientific American 268 (4), 32–39.

Robson, E.J., He, S.J. and Eccles, M.R. (2006) A PANorama of PAX genes in cancer and development. Nature Reviews Cancer 6, 52–62.

Rochette-Egly, C. (2005) Dynamic combinatorial networks in nuclear receptor-mediated transcription. Journal of Biological Chemistry 280, 32565–32568.

Ryan, A.K. and Rosenfeld, M.G. (1997) POU domain family values: flexibility, partnerships and developmental codes. Genes and Development 11, 1207–1225.

Sakai, D.D., Helms, S., Carlstedt-Duke, J., Gustafsson, J-A., Rottman, F.M. and Yamamoto, K.R. (1988) Hormone-mediated repression: a negative glucocorticoid response element from the bovine prolactin gene. Genes and Development 2, 1144–1154.

Sander, M., Neubuser, A., Kalamara, J., Es, H.C., Martin, G.R. and German, M.S. (1997) Genetic analysis reveals that PAX6 is required for normal transcription of pancreatic hormone genes and islet development. Genes and Development 11, 1662–1673.

Sauer, R.T. (1990) Scissors and helical forks. Nature 347, 514–515.

Schwabe, J.W.R. and Rhodes, D. (1991) Beyond zinc fingers: steroid hormone receptors have a novel structural motif for DNA recognition. Trends in Biochemical Sciences 16, 291–296.

Schwabe, J.W.R., Chapman, L., Finch, T. and Rhodes, D. (1993) The Crystal structure of the estrogen receptor DNA binding domain bound to DNA – how receptors discriminate between their response elements. Cell 75, 567–578.

Scott, M. P. (1999) Hox proteins reach out round DNA. Nature 397, 649–651.

Scully, K.M., Jacobson, E.M., Jepsen, K., Lunyak, V., Viadiu, H., Carrière, C., Rose, D.W., Hooshmand, F., Aggarwal, A.K. and Rosenfeld, M.G. (2000) Allosteric effects of Pit-1 DNA sites on long-term repression in cell type specification. Science 290, 1127–1131.

Sessa, G., Morelli, G and Ruberti, I. (1993) The Athb-1 and -2 HD-ZIP domains homodimerise forming complexes of different DNA binding specificities. EMBO Journal 12, 3507–3517.

Sornson, M.W., Wu, W., Dasen, J.S., Flynn, S.E., Norman, D.J., O'Connell, S.M., Gukovsky, I., Carriere, C., Ryan, A.K., Miller, A.P., Zuo, L., Gleiberman, A.S., Anderson, B., Beamer, W.G. and Rosenfeld, M.G. (1996) Pituitary lineage determination by the prophet of Pit-1 homeodomain factor defective in Ames dwarfism. Nature 384, 327–333.

Tomilin, A., Reményi, A., Lins, K., Bak, H., Leidel, S., Vriend, G., Wilmanns, M. and Schöler, H.R. (2000) Synergism with the coactivator OBF-1 (OCA-B, BOB-1) is mediated by a specific POU dimer configuration. Cell 103, 853–864.

Travers, A. (1993) *DNA–protein interactions*. Chapman and Hall, London.

Treisman, J., Harris, E., Wilson, D. and Desplan, C. (1992) The homeodomain: a new face for the helix-turn-helix. Bio Essays 14, 145–150.

Verrijzer, C.R. and Van der Vliet, P.C. (1993) POU domain transcription factors. Biochimica et Biophysica Acta 1173, 1–21.

Walker, S., Hayes, S. and O'Hare, P. (1994) Site-specific conformational alteration of the Oct-1 POU domain–DNA complex as the basis for differential recognition by Vmw65 (VP16). Cell 79, 841–852.

Weatherman, R.V., Fletterick, R.J. and Scanlan, T.S. (1999) Nuclear-receptor ligands and ligand-binding domains. Annual Reviews of Biochemistry 68, 559–581.

Wysocka, J. and Herr, W. (2003) The herpes simplex virus VP16-induced complex: the makings of a regulatory switch. Trends in Biochemical Sciences 28, 294–304.

Xu, W., Rould, M.A., Jun, S., Desplan, C. and Pabo, C.O. (1995) Crystal structure of a paired domain–DNA complex at 2.5A resolution reveals structural basis for Pax developmental mutations. Cell 80, 639–650.

Zeitlinger, J., Simon, I., Harbison, C.T., Hannett, N.M., Volkert, T.L., Fink, G.R. and Young, R.A. (2003) Program-specific distribution of a transcription factor dependent on partner transcription factor and MAPK signalling. Cell 113, 395–404.

CHAPTER 5

ACTIVATION OF GENE EXPRESSION BY TRANSCRIPTION FACTORS

5.1 ACTIVATION DOMAINS

Extensive studies on a variety of transcription factors have shown that they have a modular structure in which distinct regions of the protein mediate particular functions such as DNA binding (see Chapter 4) or interaction with specific effector molecules such as steroid hormones. It is likely therefore that a specific region of each individual transcription factor will be involved in its ability to activate transcription following DNA binding. As described in Chapter 2 (section 2.4.1), such activation domains have been identified by so-called 'domain swap' experiments in which various regions of one factor are linked to the DNA binding domain of another factor and the ability to activate transcription assessed.

In general, these experiments have confirmed the modular nature of transcription factors with distinct domains mediating DNA binding and transcriptional activation. Thus, in the case of the yeast factor GCN4, two distinct regions each of sixty amino acids have been identified which mediate respectively DNA binding and transcriptional activation (Fig. 5.1a; Hope and Struhl, 1986). Similarly, domain swap experiments have identified two regions of the glucocorticoid receptor, one at the N-terminus of the molecule and the other near the C-terminus, which can independently mediate gene activation when linked to the DNA binding domain of another transcription factor (Hollenberg and Evans, 1988), and both of these are distinct from the DNA binding domain of the molecule. Interestingly, the C-terminal activation domain is located close to the hormone binding domain of the receptor (Fig. 5.1b), and can mediate the activation of transcription only following hormone addition. It therefore plays an important role in the steroid-dependent activation of transcription following hormone addition (see Chapter 8, section 8.2.2).

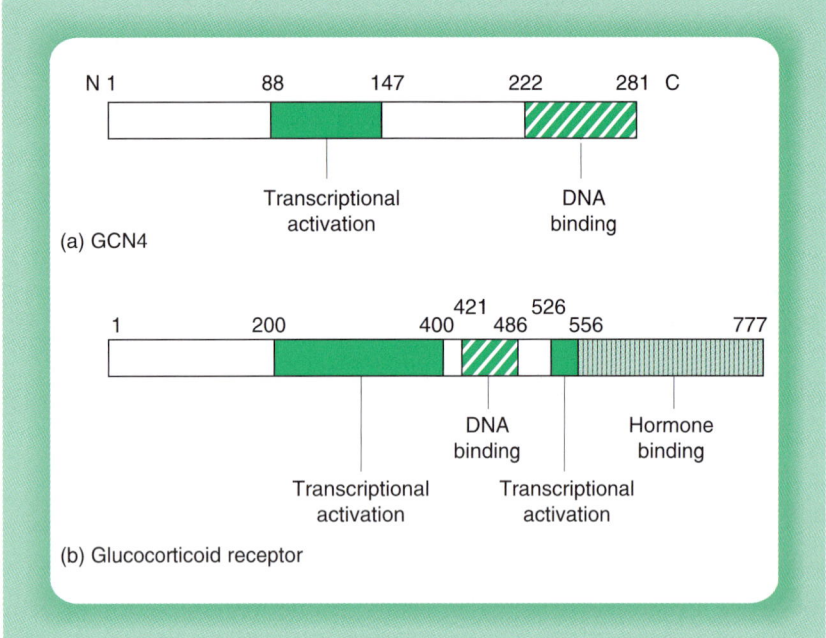

Figure 5.1

Domain structure of the yeast GCN4 transcription factor (panel a) and the mammalian glucocorticoid receptor (panel b). Note the distinct domains which are active in DNA binding or transcriptional activation.

Studies on a variety of transcription factors have therefore strongly indicated their modular nature with distinct regions of the molecule mediating DNA binding and transcriptional activation. An extreme example of this modularity is provided by the interaction of the cellular transcription factor Oct-1 (see Chapter 4, section 4.2.6) and the herpes simplex virus *trans*-activating protein VP16 (for review see Wysocka and Herr, 2003). Thus, although VP16 contains a very strong activating region which can strongly induce transcription when artificially fused to the DNA binding domain of the yeast GAL4 transcription factor, it contains no DNA binding domain and cannot therefore bind to DNA itself. Transcriptional activation by VP16 following viral infection is therefore dependent upon its ability to form a protein–protein complex with the cellular Oct-1 protein. This complex then binds to the octamer-related TAATGARAT (R = purine) motif in the viral immediate–early genes via the DNA binding domain of Oct-1 and transcription is activated by the activation domain of VP16. Hence in this case, the DNA binding and transcriptional activation domains are actually located on different proteins in the DNA binding complex (Fig. 5.2). A similar example in which the constitutively expressed Oct-1 recruits a non-DNA binding cellular co-activator molecule, OCA-B, to the promoter, resulting in its activation, was also discussed in Chapter 4 (section 4.2.6), indicating that this effect is not confined to viral trans-activating molecules.

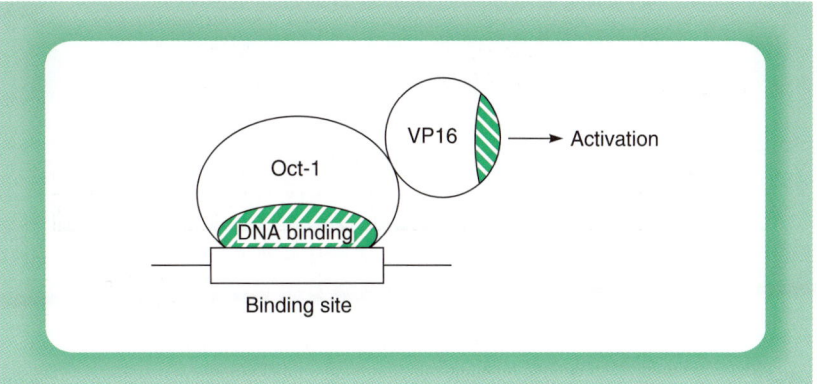

Figure 5.2
Activation of gene transcription by interaction of the cellular factor Oct-1, which contains a DNA binding domain, and the herpes simplex virus VP16 protein, which contains an activation domain but cannot bind to DNA.

5.2 NATURE OF ACTIVATION DOMAINS

Following the identification of activation domains in different transcription factors, it rapidly became clear that they fell into several distinct families with common features which will be discussed in turn (for a typical example of each of the major classes of activation domain see Fig. 5.3) (for review see Triezenberg, 1995).

5.2.1 ACIDIC DOMAINS

Comparison of several different activation domains, including those of the yeast factors GCN4 and GAL4 as well as the activation domain at the N-terminus of the glucocorticoid receptor and that of VP16 which were discussed above (section 5.1), indicated that, although they do not show any strong amino acid sequence homology to each other, they all have a large proportion of acidic amino acids producing a strong net negative charge (Fig. 5.3). Thus the eighty-two amino acid activating region of the glucocorticoid receptor contains seventeen acidic residues (Hollenberg and Evans, 1988), whilst the same number of negatively charged amino acids is found within the sixty amino acid activating region of GCN4 (Hope and Struhl, 1986). These findings indicated

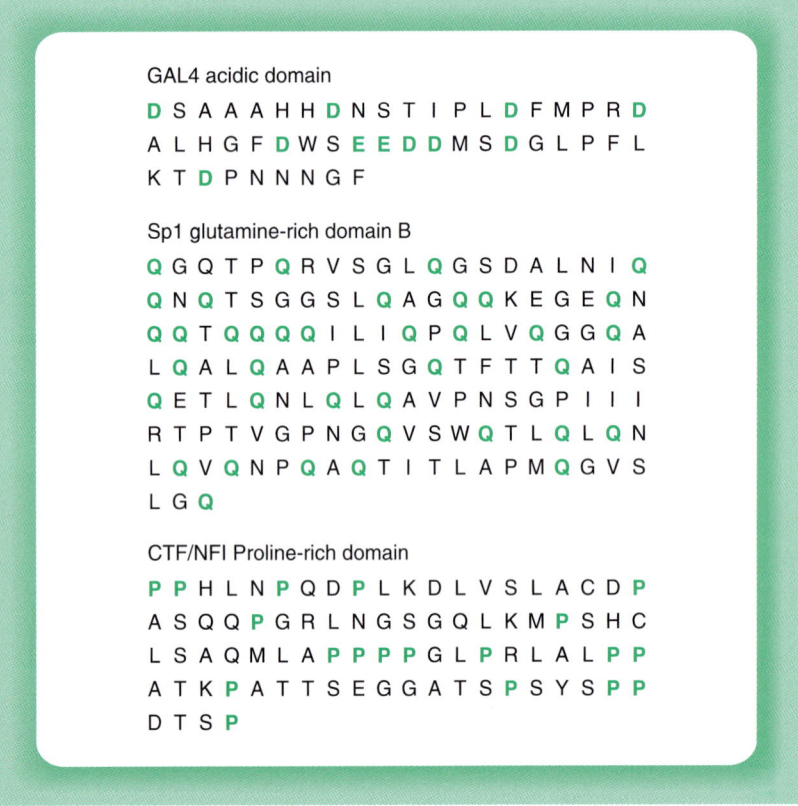

Figure 5.3
Structure of typical members of each of the three classes of activation domains. Acidic, glutamine or proline residues are highlighted in the appropriate case.

therefore that these activation regions consist of so-called 'acid blobs' or 'negative noodles' with a high proportion of negatively charged amino acids which are involved in the activation of transcription (for review see Hahn, 1993a).

In agreement with this idea, mutations in the activation domain of GAL4, which increase its net negative charge, increase its ability to activate transcription. Similarly, if recombination is used to create a GAL4 protein with several more negative charges, the effect on gene activation is additive, a mutant with four more negative charges than the parental wild-type, activating transcription nine-fold more efficiently than the wild-type. Thus the acidic nature of these domains is likely to be important in their function. It has been suggested that in the case of VP16, the negative charge of its acidic domain allows it to establish long range electrostatic interactions with the $TAF_{II}31$ component of TFIID (see section 5.4.2), with which it interacts to stimulate transcription (Uesugi *et al.*, 1997, Fig. 5.4a).

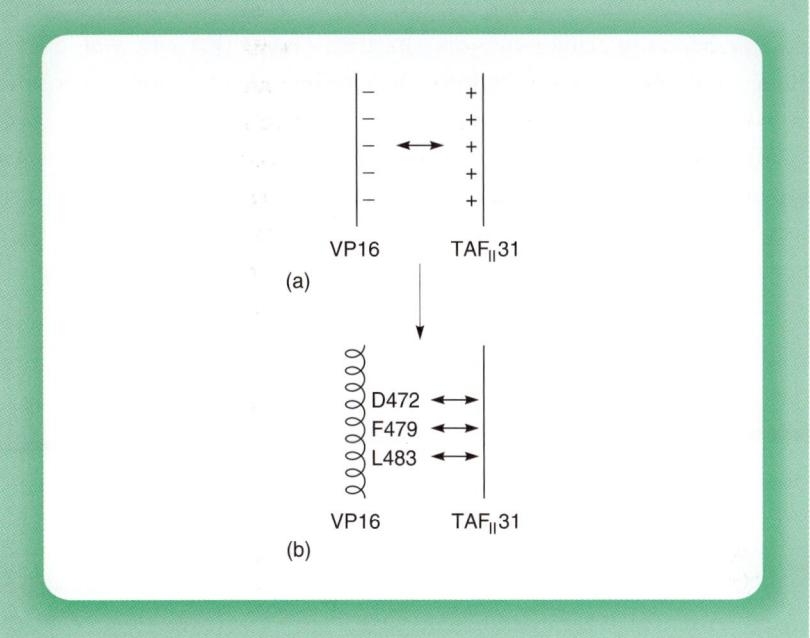

Figure 5.4

(a) The negatively charged acidic residues in the VP16 activation domain allow its initial long distance interaction with $TAF_{II}31$. (b) Interaction with $TAF_{II}31$ induces a conformational charge in the domain to an α-helical structure in which the hydrophobic residues asparagine (D) at position 472, phenylalanine (F) at position 479 and leucine (L) at position 483 are brought close to one another and bind to $TAF_{II}31$.

Although the acidic nature of the activation domain is clearly important for its function, it is not the only feature required since it is possible to decrease the activity of the GAL4 activation domain without reducing the number of negatively charged residues. Indeed, it appears that conserved hydrophobic residues in the acidic activation domains also play a key role in their ability to stimulate transcription. Thus when the VP16 activation domain interacts with the $TAF_{II}31$ component of TFIID, it undergoes a conformational change from a random coil to an α-helix which brings together three hydrophobic residues within the acidic domain and these residues then interact directly with TAF_{II} 31 (Uesugi et al., 1997; Fig. 5.4b). Hence, the acidic domain would interact with $TAF_{II}31$ via a two step process in which the initial long range attraction produced by the acidic residues allows a subsequent structural change facilitating a close interaction of the hydrophobic residues within the acidic domain with $TAF_{II}31$.

This two-stage process has recently been shown to operate also when acidic activators interact with other components of the basal transcriptional complex. It is therefore likely that this represents a general mechanism for the interaction of acidic activators with their targets (Ferreira *et al.*, 2005). Hence both the acidic and hydrophobic residues are of importance for the activity of this domain.

Although activation domains of the acidic type form the majority of the activation domains so far identified in eukaryotic transcription factors from yeast to mammals, other types of activation domains have been identified in a number of different transcription factors in higher eukaryotes and these will be discussed in turn.

5.2.2 GLUTAMINE-RICH DOMAINS

Analysis of the constitutive transcription factor Sp1, which binds to the Sp1 binding site found in many gene promoters (see Chapter 1, section 1.3.2), revealed that the two most potent activation domains contained approximately twenty-five per cent glutamine residues and very few negatively charged residues (Courey and Tjian, 1988; Fig. 5.3b). These glutamine-rich motifs are essential for the activation of transcription mediated by these domains since their deletion abolishes the ability to activate transcription. Most interestingly, however, transcriptional activation can be restored by substituting the glutamine-rich regions of Sp1 with a glutamine-rich region from the *Drosophila* homeobox transcription factor Antennapedia which has no obvious sequence homology to the Sp1 sequence. Hence, as with the acidic activation domains, the activating ability of a glutamine-rich domain is not defined by its primary sequence but rather by its overall nature in being glutamine-rich. In agreement with this a continuous run of glutamine residues with no other amino acids has been shown to act as a transcriptional activation domain (Gerber *et al.*, 1994).

Similar glutamine-rich regions have been defined in transcription factors other than Sp1 and Antennapedia, including the N-terminal activation domains of the octamer binding proteins Oct-1 and Oct-2, the *Drosophila* homeobox proteins ultra-bithorax and zeste and the yeast HAP1 and HAP2 transcription factors. This indicates that this motif is quite widespread, being found in different transcription factors in a variety of different species (for review see Mitchell and Tjian, 1989).

5.2.3 PROLINE-RICH DOMAINS

Studies on the constitutive factor CTF/NF1 which binds to the CCAAT box motif (see Chapter 1, section 1.3.2) defined a third type of activation domain distinct from those previously discussed. Thus the activation domain located at the C-terminus of CTF/NF1 is not rich in acidic or glutamine residues but instead contains numerous proline residues forming approximately one-quarter of the amino acids in this region (Mermod *et al.*, 1989; Fig. 5.3c). As with the other classes of activation domains, this region is capable of activating transcription when linked to the DNA binding domains of other transcription factors. Moreover, as with the glutamine-rich domain, a continuous run of proline residues can mediate activation, indicating that the function of this type of domain depends primarily on its richness in proline (Gerber *et al.*, 1994). Similar proline-rich domains have been identified in several other transcription factors such as the oncogene product Jun, AP2 and the C-terminal activation domain of Oct-2 (for review see Mitchell and Tjian, 1989). Thus, as with the glutamine-rich domains, proline-rich domains are not confined to a single factor, whilst a single factor such as Oct-2 can contain two activation domains of different types.

In summary, therefore, it is clear that, as with DNA binding, several distinct protein motifs can activate transcription (Fig. 5.3).

5.2.4 FUNCTIONAL RELATIONSHIP OF THE DIFFERENT ACTIVATION DOMAINS

The existence of at least three distinct classes of activation domain raises the question of whether these three domains are functionally equivalent or whether they differ in their ability to activate transcription. This question was investigated by Seipel *et al.* (1992), who linked each of the activation domains to the DNA binding domain of the GAL4 factor. They then tested the ability of these chimaeric proteins to activate transcription in mammalian cells when the GAL4 DNA binding site was placed at different positions relative to the start site of transcription (Fig. 5.5). In these experiments all three domains were able to activate transcription when the DNA binding site was placed close to the start site of transcription in the promoter region. In contrast, the glutamine-rich domain was unable to activate transcription when the binding site was located downstream of the transcription unit mimicking a position within an enhancer element (see Chapter 1, section 1.3.4). The acidic domain was strongly active from this enhancer position, whilst the proline-rich domain could also activate transcription from this position but only weakly.

These findings indicate therefore that clear differences exist in the abilities of the different activation domains to activate transcription when bound to the DNA at different positions relative to the promoter. Such differences are likely to be important in determining the functional activity of different factors. In addition, such differences in the activity of different activation domains are likely to reflect differences in the mechanisms

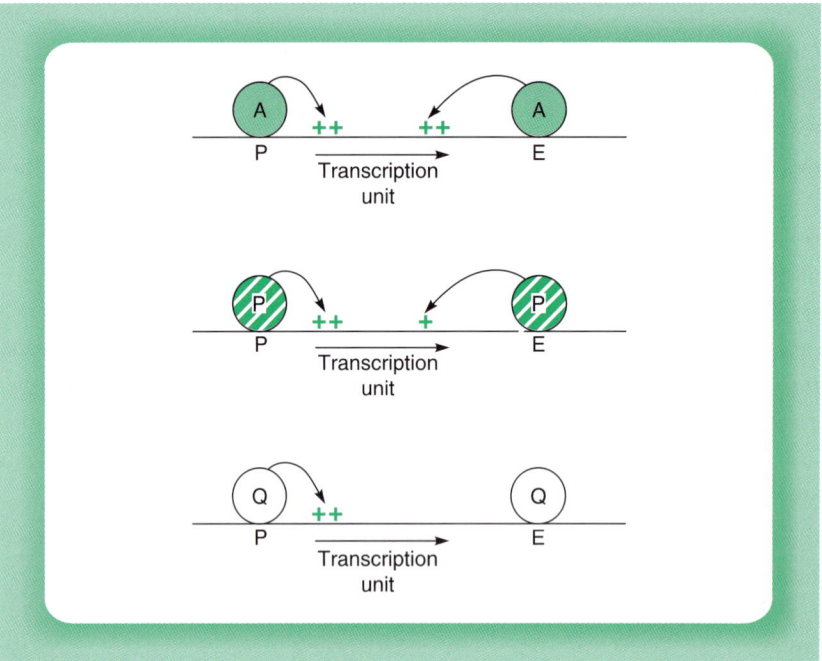

Figure 5.5

An acidic activation domain (A) can stimulate transcription when bound to DNA in the promoter (P) close to the transcriptional start site or when bound at a distant enhancer (E). In contrast, a proline-rich domain (P) stimulates only weakly from an enhancer position and a glutamine-rich domain (Q) does not stimulate at all from this position.

by which these factors act. In agreement with this idea, acidic or proline-rich activation domains derived from mammalian factors can also activate transcription when introduced into yeast cells, whereas glutamine-rich domains cannot do so (Kinzler *et al.*, 1994).

In the next sections we will consider the mechanisms by which activation domains act, focusing particularly on the acidic domains where most information is available. Similarities and differences in the mode of action of the other activation domains will be discussed where this information is available.

5.3 INTERACTION OF ACTIVATION DOMAINS WITH THE BASAL TRANSCRIPTIONAL COMPLEX

5.3.1 ACTIVATORS AND THE BASAL TRANSCRIPTIONAL COMPLEX

The widespread interchangeability of acidic activation domains from yeast, *Drosophila* and mammalian transcription factors discussed above, strongly suggests that a single common mechanism may mediate transcriptional activation by acidic activation domains in a wide range of organisms. This idea is supported by the finding noted above that mammalian transcription factors carrying such domains, such as the glucocorticoid receptor, can activate a gene carrying their appropriate DNA binding site in yeast cells whilst the yeast GAL4 factor can do so in cells of *Drosophila*, tobacco plants and mammals (reviewed by Guarente, 1988; Ptashne, 1988).

These considerations suggest that the target factor or factors with which these activators interact is likely to be highly conserved in evolution. A number of experiments have indicated that in many cases this target factor is likely to be required for the transcription of a number of different genes and not solely for that of the activated gene. Thus the over-expression of the yeast GAL4 protein, which contains a strong activation domain, results in the down regulation of genes which lack GAL4 binding sites (such as the CYC1 gene) as well as activating genes which do contain GAL4 binding sites. This phenomenon, which has been noted for a number of transcription factors with strong activation domains, is known as squelching (for review see Ptashne, 1988). Although the degree of squelching by any given factor is proportional to the strength of its activation domain, squelching differs from activation in that it does not require DNA binding and can be achieved with truncated factors containing only the activation domain and lacking the DNA binding domain. This phenomenon can therefore be explained on the basis that a transcriptional activator when present in high concentration can interact with its target factor in solution as well as on the DNA. If this target factor is present at limiting concentrations, it will therefore be sequestered away from other genes which require it for transcription, resulting in their inhibition (Fig. 5.6).

The existence of squelching indicates therefore that in many cases the target factor for activation domains is likely to be a component which is required for the transcription of a wide range of genes and which is conserved from yeast to mammals, allowing yeast activators to work in mammalian cells and vice versa. Obviously, such a common component could be part of the basal transcriptional complex required for transcription of a wide range of genes in different organisms. Clearly, an activating factor could act by stimulating the binding of such a component so that the basal complex could be assembled more efficiently. Alternatively, it could act by interacting with a factor which had already bound, so that the activity or stability of the assembled complex was stimulated. It appears that both these mechanisms are used and they will be discussed in turn.

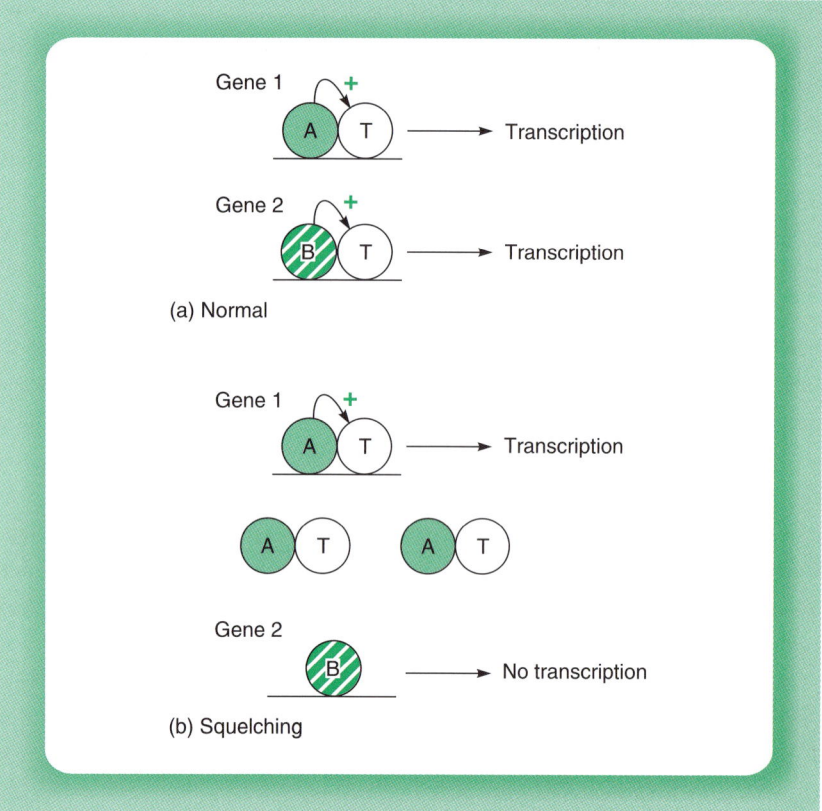

Figure 5.6

The process of squelching. In the normal case, illustrated in panel (a), two distinct activator molecules A and B involved in the activation of genes 1 and 2 respectively both act by interacting with the general transcription factor T and both genes are transcribed. In squelching, illustrated in panel (b), factor A is present at high concentration and hence interacts with T both on gene 1 and in solution. Hence factor T is not available for transcription of gene 2 and therefore only gene 1 is transcribed whilst transcription of gene 2 is squelched.

5.3.2 STIMULATION OF FACTOR BINDING

As described in Chapter 3 (section 3.5.1), the basal transcriptional complex can assemble in a stepwise manner with the binding of TFIID being followed by the binding of TFIIB and then the binding of RNA polymerase in association with TFIIF. Clearly, an activator could increase the rate of complex assembly by enhancing any one of these assembly steps. Indeed, there is evidence that activators target several of these steps in the assembly process (Fig. 5.7). Thus, for example, it appears that acidic activators interact directly with TFIID (see Chapter 3, section 3.5.1) to stimulate the binding of TFIID to the promoter (Fig. 5.7a). Interestingly, this enhanced recruitment of TFIID, induced by transcriptional activators, which was initially observed in the test tube, has been confirmed in intact cells using the ChIP assay described in Chapter 2 (section 2.4.3; Li *et al.*, 1999).

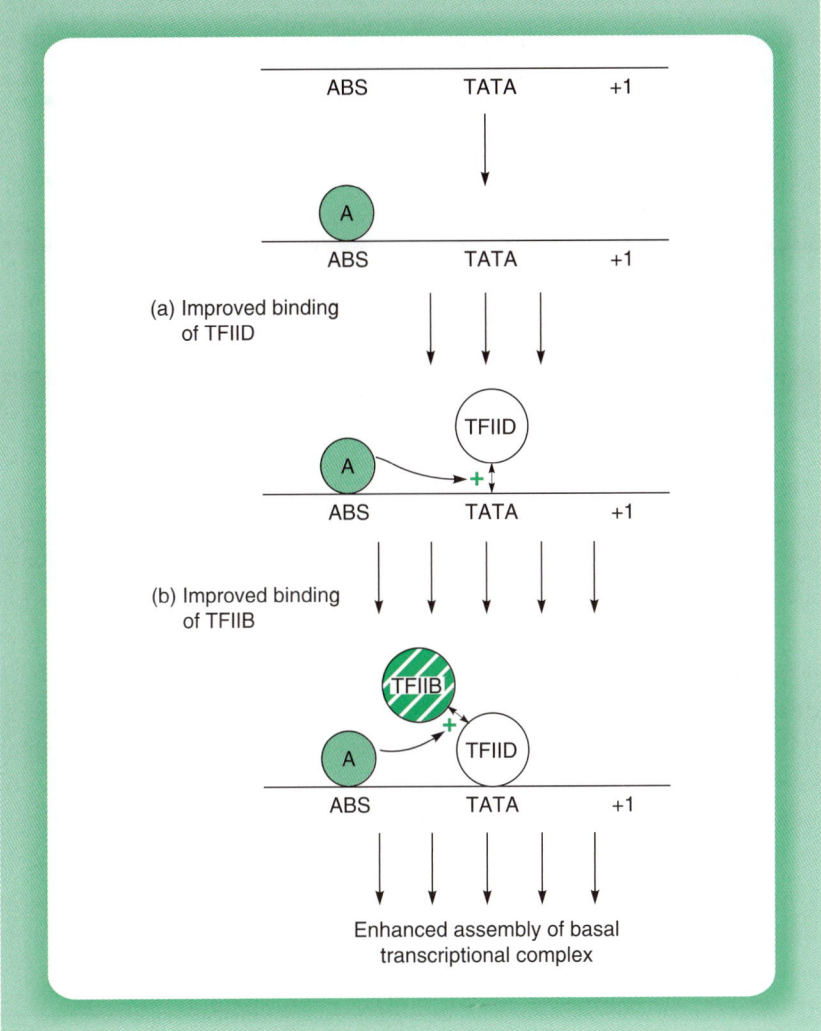

Figure 5.7

The binding of an activating molecule (A) to its binding site (ABS) can enhance both the binding of TFIID to the TATA box (a) and the recruitment of the TFIIB factor (b), so enhancing the rate of basal complex assembly and of transcription.

Although increased binding of TFIID to the promoter will directly enhance the assembly of the complex by allowing TFIIB to bind, there is evidence that activators can also act directly to improve the recruitment of TFIIB independent of their effect on TFIID (Fig. 5.7b). Thus it has been shown that both an acidic activator and glutamine or proline-rich activators can greatly stimulate the binding of TFIIB to the promoter (Choy and Green, 1993). Hence, activators can enhance the assembly of the basal transcriptional complex by independently enhancing the binding of both TFIID and TFIIB. This ability of activators to act at these two independent steps results in a strong synergistic activation of transcription in the presence of different activators targeting either TFIID or TFIIB (Gonzalez-Couto et al., 1997).

As with TFIID, it has been shown that TFIIB interacts directly with activating molecules. Thus, TFIIB can be purified on a column containing a bound acidic activator and interactions of TFIIB with non-acidic activators have also been reported. Moreover, mutations in the activator which abolish this interaction with TFIIB prevent it from activating transcription (for review see Hahn, 1993b). Thus, the effect of activators on TFIIB is mediated via a direct protein–protein interaction which is essential for their ability to stimulate transcription.

In addition to the stepwise pathway of complex assembly, it has also been shown that the basal transcriptional complex can assemble in a much simpler manner with binding of TFIID being followed by binding of the RNA polymerase holoenzyme which contains the polymerase itself, TFIIB, TFIIF and TFIIH, as well as a number of other proteins (see Chapter 3, section 3.5.2). There is evidence that activators can also act in this pathway not only by enhancing the recruitment of TFIID as described above but also by directly enhancing the binding of the RNA polymerase holoenzyme itself (Fig. 5.8a). Thus, for example, if a DNA binding domain is linked to the yeast protein Gal11, which is a component of the RNA polymerase holoenzyme, the holoenzyme is recruited to the DNA via this DNA binding domain and transcription is activated (Fig. 5.8b) (Barberis et al., 1995). Hence, the need for activators can be bypassed by recruiting the RNA polymerase holoenzyme to DNA via an artificial DNA binding domain.

Indeed, on the basis of experiments of this type, Ptashne and Gann (1997) have argued that the sole role of activators is to enhance the assembly of the basal complex by interacting with one or other of its specific components so facilitating their recruitment to the DNA. However, whilst such enhanced recruitment of specific components of the complex clearly plays a major part in the action of transcriptional activators and operates in both pathways of complex assembly, it is likely that other effects are also involved in the action of transcriptional activators. These effects are discussed in the next section.

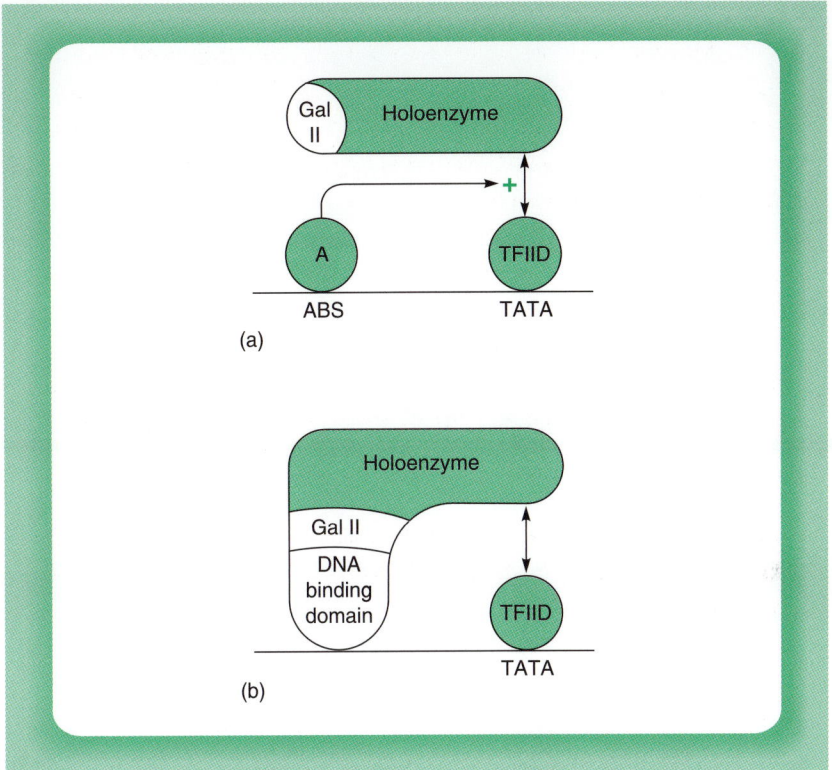

Figure 5.8

(a) Activators can act by enhancing the binding of the RNA polymerase holoenzyme complex following binding of TFIID. (b) In agreement with this idea, the need for an activator can be bypassed by artificially attaching a DNA binding domain to the Gal 11 component of the holoenzyme so enhancing holoenzyme recruitment by allowing it to bind to DNA directly.

5.3.3 STIMULATION OF FACTOR ACTIVITY

In addition to their effects on complex assembly, it is clear that activators can also stimulate transcription at a subsequent step following assembly of the complex, resulting in its enhanced stability or increased activity (Choy and Green, 1993) (Fig. 5.9).

An obvious mechanism for activation would be for activating domains to interact directly with the RNA polymerase itself to increase its activity (Fig. 5.10a). However, despite the attractiveness of a model involving direct interaction between activating factors and the polymerase itself, it is unlikely to be correct and it appears that activators interact with the polymerase indirectly via other factors (Fig. 5.10b).

TFIID is one potential candidate for the component with which activating factors interact since this factor is both required for the transcription

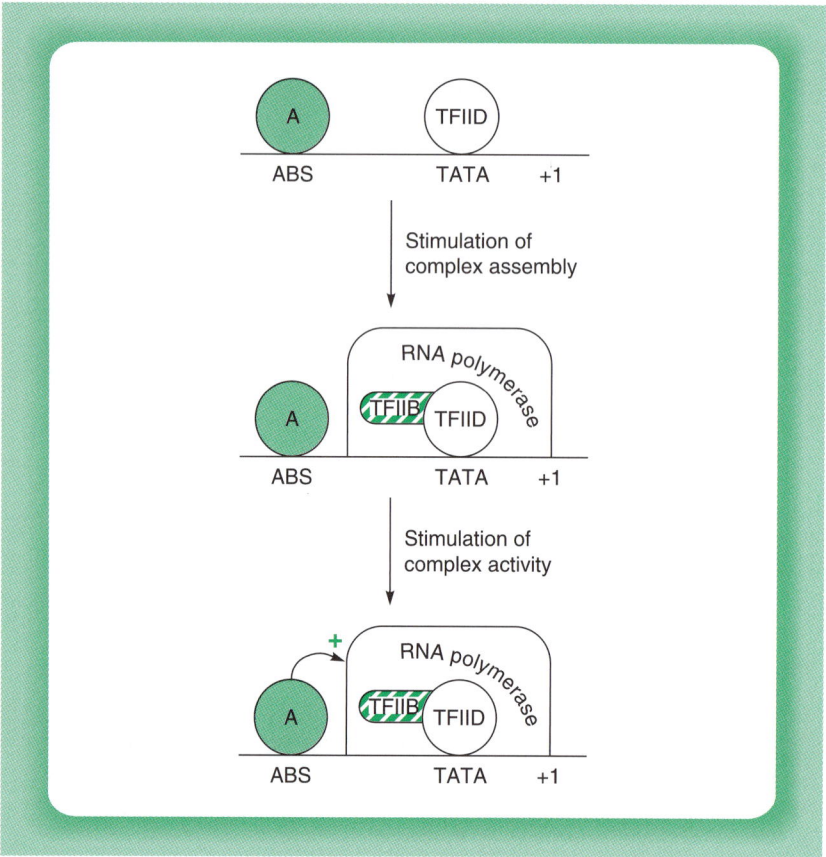

Figure 5.9

An activator can stimulate transcription both by promoting the assembly of the basal transcription complex and by stimulating its activity following assembly.

of a wide variety of genes both with and without TATA boxes (see Chapter 3, section 3.6) and is highly conserved in evolution, with the yeast factor being able to promote transcription in mammalian cell extracts and vice versa. Evidence for an effect of activating factors on TFIID has been obtained in the case of the yeast acidic activating factor GAL4 (Horikoshi *et al.*, 1988). Thus in the absence of GAL4, TFIID was shown to be bound only at the TATA box of a promoter containing both a TATA box and GAL4 binding sites. In contrast, in the presence of GAL4 bound to its upstream binding sites in the promoter, the conformation of TFIID was altered such that it now covered both the TATA box and the start site for transcription (Fig. 5.11). Moreover, no change in TFIID conformation was observed in the presence of a truncated GAL4 molecule which can bind to DNA but lacks the acidic activation domain. Hence, an acidic activator can produce a change in TFIID conformation, resulting in its binding to the start site for transcription, and this effect correlates with the

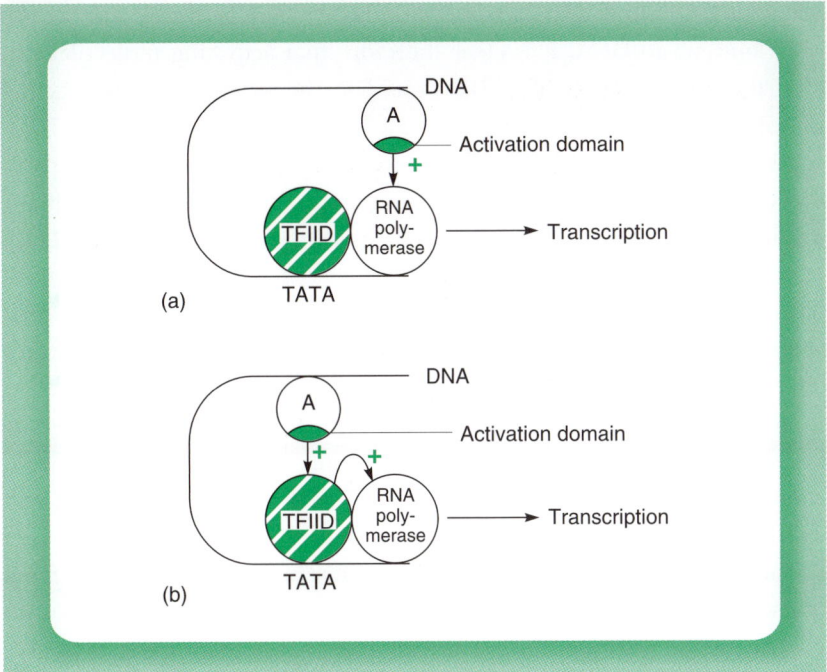

Figure 5.10

Two possible mechanisms by which an activating factor (A) could stimulate the activity of the basal transcriptional complex. This could occur via direct interaction with the RNA polymerase itself (a) or by interaction with another transcription factor such as TFIID which in turn interacts with the polymerase (b).

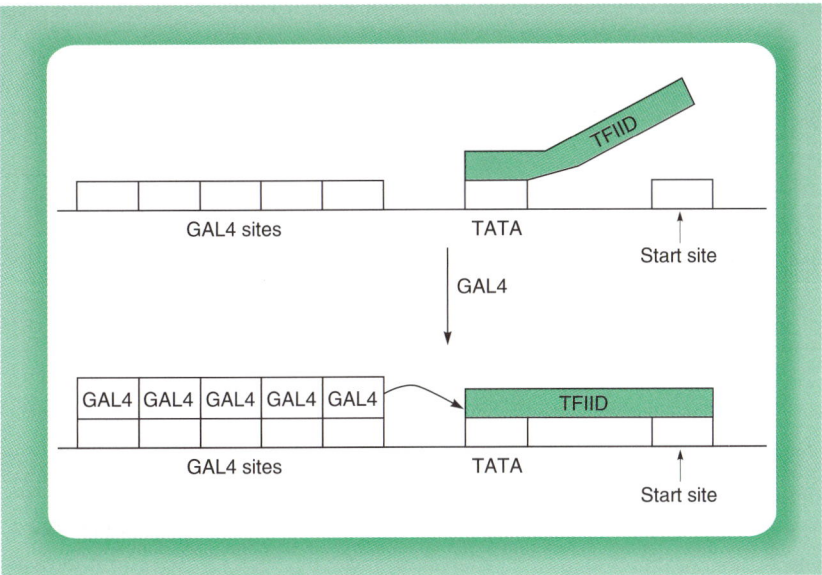

Figure 5.11

Effect of GAL4 binding on the binding of TFIID.

ability of GAL4 to activate transcription rather than being a consequence of its binding to DNA. It is clear therefore that activating molecules can alter the configuration of TFIID bound to the promoter by interacting with it.

As well as interacting with TFIID to change its configuration, activators can also interact with TFIIB, changing its conformation and enhancing its ability to recruit the complex of RNA polymerase II and TFIIF (Roberts and Green, 1994; Fig. 5.12). Hence activators appear to target both TFIID and TFIIB in two ways. First, as described in the previous section, they enhance their binding to the promoter and secondly, they alter their conformation so as to enhance their activity (Fig. 5.13).

Together with TFIIB, TFIID constitutes a major target for transcriptional activators. Interestingly, however, other components of the basal complex such as TFIIA (Ozer *et al.*, 1994), TFIIF (Joliot *et al.*, 1995) and TFIIH (Xiao *et al.*, 1994) have also been shown to interact with transcriptional activators. Hence, a number of different factors within the basal transcriptional complex can serve as targets for direct interactions with transcriptional activators. It is clear, however, that in many cases, activators

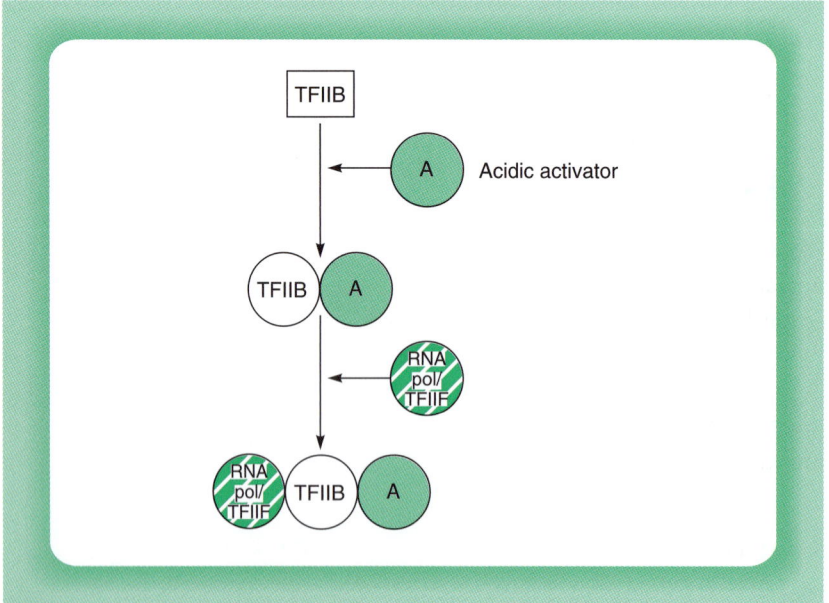

Figure 5.12

The binding of an acidic activator (A) to TFIIB produces a conformational change which enhances the ability of TFIIB to interact with the RNA polymerase/TFIIF complex, thereby enhancing its recruitment to the promoter.

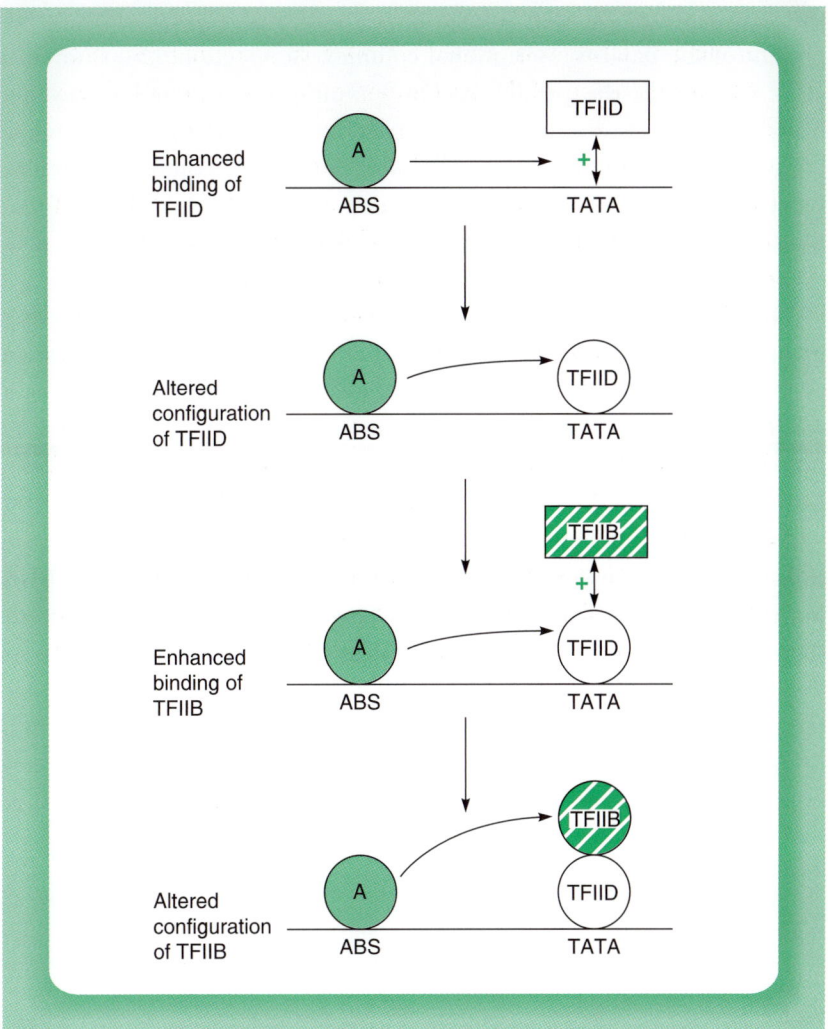

Figure 5.13
Activators can stimulate both the binding of TFIIB and TFIID and enhance their
activity by altering their conformation (square to circle).

interact with the basal complex only indirectly via other factors and such
interactions are discussed in the next section.

5.4 INTERACTION OF ACTIVATION DOMAINS WITH OTHER REGULATORY PROTEINS

5.4.1 THE MEDIATOR COMPLEX

As noted in section 5.3.1, the existence of the squelching phenomenon
indicates that activators act by contacting a factor which is involved in

the transcription of a wide range of genes. Although this could be a component of the basal transcriptional complex (see section 5.3), studies in yeast resulted in the purification of a multi-protein complex (distinct from the basal transcriptional complex) that could prevent squelching when added in excess. This so-called 'mediator' complex therefore represents a target for transcriptional activators which is present in limiting amounts so that activators compete for it. Hence, its addition in excess relieves this competition and prevents squelching.

The mediator complex consists of over twenty proteins and, following its original identification in yeast, has now been found in a wide range of multi-cellular organisms including man. It therefore appears to be a conserved component of the transcriptional machinery involved in activation of a wide range of genes (for reviews see Conaway *et al.*, 2005; Kornberg, 2005; Malik and Roeder, 2005).

As well as interacting with activators, the mediator also interacts with RNA polymerase II itself. Indeed, the mediator is part of the RNA polymerase holoenzyme discussed in Chapter 3 (section 3.5.2) which therefore consists not only of RNA polymerase II, basal factors such as TFIIB, TFIIE, TFIIF and TFIIH and a chromatin remodelling activity, but also contains the mediator complex. Hence, the mediator serves as a bridge by which activating signals are transmitted from DNA-bound transcriptional activators to RNA polymerase II (Fig. 5.14). Indeed, structural

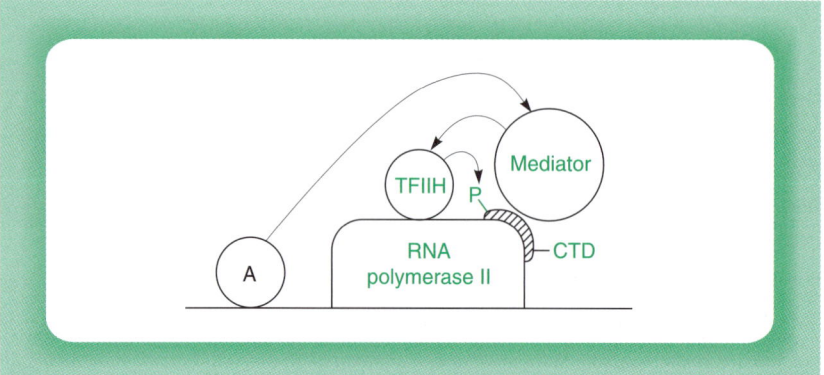

Figure 5.14

The mediator binds to the C-terminal domain (CTD) of RNA polymerase II and thereby acts as a bridge transmitting the activating signal between DNA binding activators and RNA polymerase. One mechanism for such activation involves the mediator inducing TFIIH to phosphorylate the CTD, thereby stimulating transcription.

studies suggest that the mediator partially envelops the polymerase, allowing it to receive signals from transcriptional activators and transmit them to the polymerase (for review see Chadick and Asturias, 2005) (Fig. 5.15).

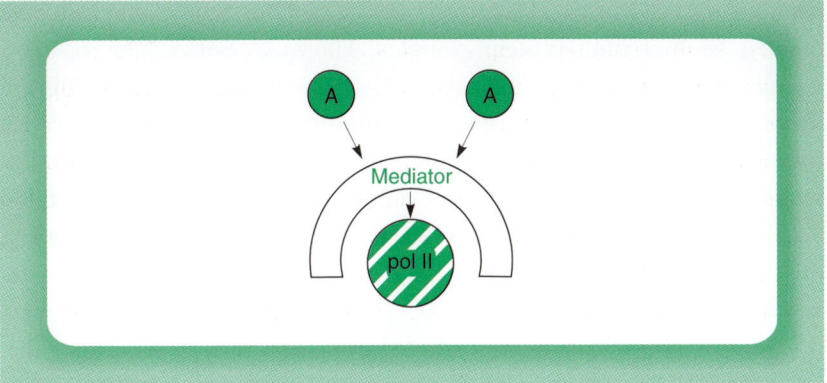

Figure 5.15

Structural studies suggest that the mediator partially envelops the polymerase allowing it to serve as a bridge between the polymerase and transcriptional activators.

Interestingly, the mediator has been shown to contact the C-terminal domain of RNA polymerase II. Hence, the involvement of this motif in activation of the polymerase, which was discussed in section 5.3.3, can be accounted for by the mediator contacting this motif and transmitting the signal from transcriptional activators. Indeed, it appears that one of the roles of the mediator is to stimulate TFIIH to phosphorylate the C-terminal domain of RNA polymerase II which, as discussed in Chapter 3 (sections 3.1 and 3.5.1), is necessary for it to begin transcribing the gene.

This association of the mediator with the RNA polymerase is reinforced by the finding that the mediator is actually a component of the RNA polymerase II holoenzyme, which was discussed in Chapter 3 (section 3.5.2). Hence, this complex contains the mediator, as well as the polymerase and basal transcription factors, such as TFIIB, TFIIE, TFIIF and TFIIH (Cantin *et al.*, 2003).

The mediator therefore plays a key role as a link between transcriptional activators and the basal transcriptional complex, and recent evidence suggests that it can act by both the methods described above (section 5.3.2 and 5.3.3) that activate the basal transcriptional complex, namely the stimulation of complex assembly and the stimulation of complex activity (for review see Struhl, 2005). Recently, however, it has been suggested that the mediator may also be important for the low basal levels of transcription which occur in the absence of transcriptional activators. Thus, yeast strains with a defect in the mediator complex are also defective for basal transcription (Takagi and Kornberg, 2006).

Although the mediator thus plays an important role in transcriptional activation and possibly in basal transcription, it is not the only multi-protein complex involved in transcriptional activation. Thus, studies in yeast have defined another multi-protein complex, known as SAGA. This contains components that between them have a variety of activities, including the ability to interact with transcriptional activators, the ability to interact with components of the basal transcriptional complex and histone acetyltransferase activity able to modify chromatin structure (see Chapter 1, section 1.2.3). Thus, the SAGA complex can link transcriptional activators both to the basal transcriptional complex and to enzymes able to alter chromatin structure (for review see Hampsey, 1997; Grant *et al.*, 1998) (Fig. 5.16).

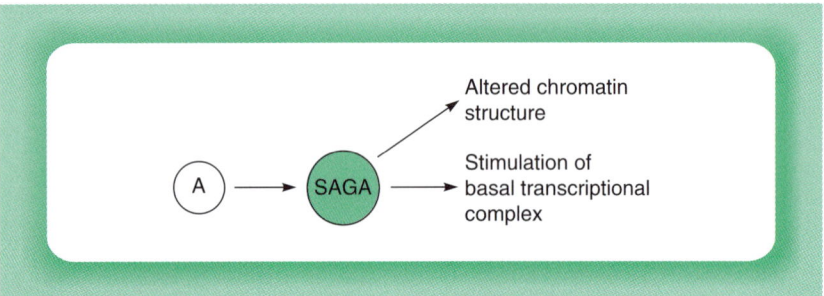

Figure 5.16
The SAGA complex contains different proteins which can respond to transcriptional activators (A), interact with the basal transcriptional complex and alter chromatin structure. It therefore links activators with both the basal transcriptional complex and the alteration of chromatin structure.

Interestingly, there is evidence that the mediator and SAGA complexes co-operate to activate transcription. Thus, Cheng *et al.* (2004) fused various components of the mediator or SAGA complexes to DNA binding domains so that they bound to DNA and then investigated whether transcription was activated. This was only observed when both a component of the mediator and a component of SAGA were bound to the target DNA and not with single components of the mediator or SAGA alone. Hence, in this case, transcriptional activation required recruitment of both the mediator and SAGA complexes (Fig. 5.17).

5.4.2 TAFs

As described in Chapter 3 (section 3.6), TFIID consists of the TBP protein which binds to the TATA box and a number of other proteins known as TAFs (TBP-associated factors). In some cases where activators interact with TFIID, such interactions can be reproduced with purified TBP. Moreover, mutations in specific acidic activators which interfere with

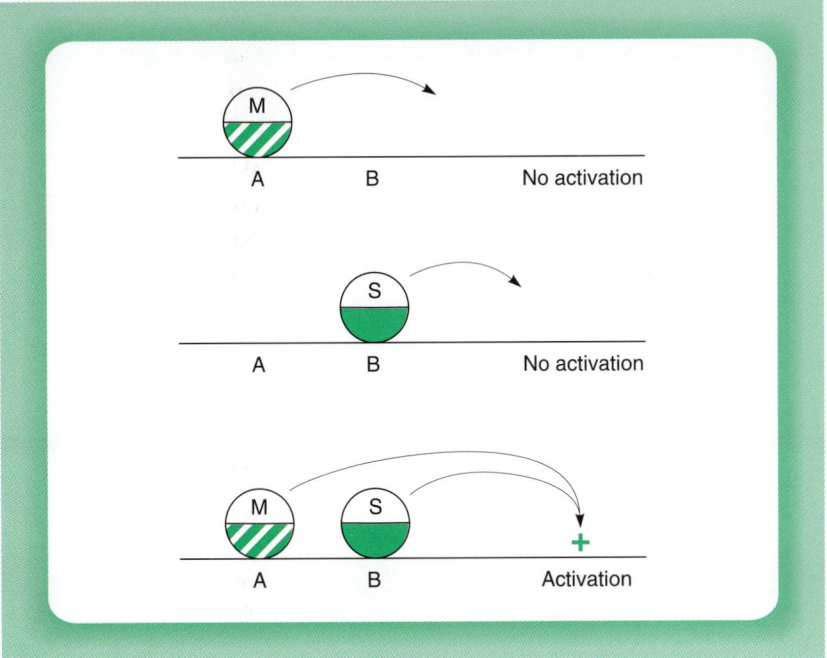

Figure 5.17

In experiments where components of the mediator (M) or SAGA (S) complexes were recruited to the DNA by linking them to different DNA binding domains (hatched or spotted), transcriptional activation required components of both the mediator and SAGA.

their ability to interact with TBP also abolish their ability to activate transcription, indicating an important functional role for these interactions.

Although there is thus evidence that the ability to interact with TBP appears to be essential for transcriptional activation in some cases (Fig. 5.18a), there is also evidence that in some circumstances such activation requires interaction of the activator with the TAFs rather than with TBP. Thus, in many cases stimulation of transcription *in vivo* by activator molecules does not occur with purified TBP but is dependent upon the presence of the TFIID complex and hence of the TAFs. This suggests a model in which the interaction of activators with TBP occurs indirectly via TAFs, with the TAFs being co-activator molecules linking the activators with the basal transcriptional complex (Fig. 5.18b) (for reviews see Hahn, 1998; Green, 2000).

Interestingly, there is evidence that different classes of activation domain may interact with different TAFs. Thus, whilst acidic activation domains have been shown to interact directly with $TAF_{II}31$ (also known as $TAF_{II}40$), the glutamine-rich domain of Sp1 interacts with $TAF_{II}110$ whilst multiple activators, including proline-rich activators, target $TAF_{II}55$.

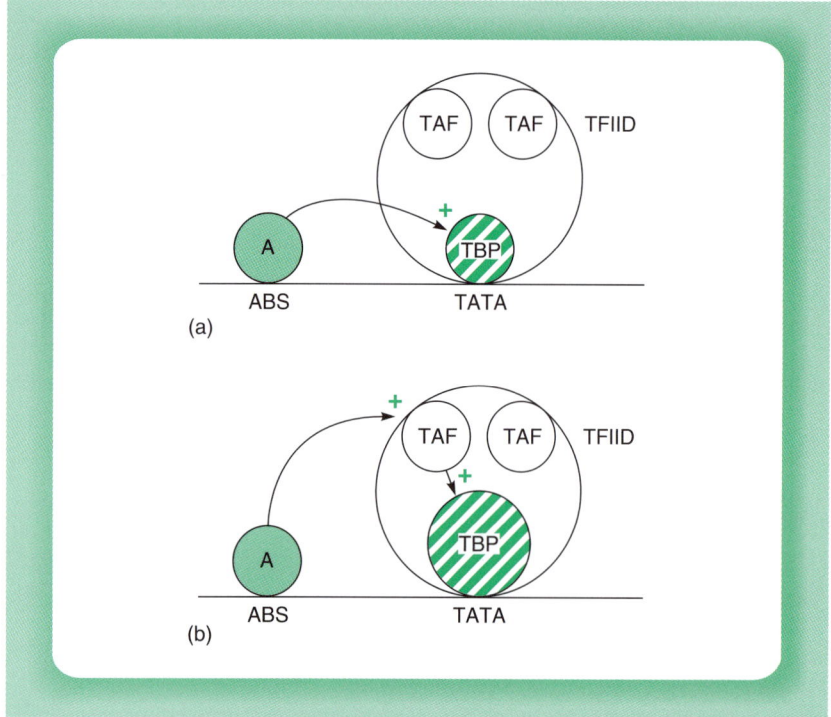

Figure 5.18

Interaction of an activator molecule with TBP can occur either directly (a) or indirectly (b) via an intermediate TBP-associated adaptor molecule (TAF).

Hence different types of activation domains may have different targets within the TFIID complex (Fig. 5.19).

In agreement with this idea, the acidic activation domain of VP16 is not capable of squelching gene activation by the non-acidic activation domain of the estrogen receptor, whereas the estrogen receptor activation domain is capable of squelching gene activation mediated both by its own activation domain and by the acidic domain of VP16, indicating that they contact different molecules. Moreover, these findings suggest that a series of TAFs within TFIID may mediate activation, with the acidic activation domain of VP16 contacting a factor that is located earlier in the series than that contacted by the non-acidic activation domain of the estrogen receptor (Fig. 5.20). Hence, the factor contacted by the activation domain of the estrogen receptor would also be essential for activation by VP16 (factor 4 in Fig. 5.20) whereas the factor contacted by the acidic activation domain of VP16 (factor 1 in Fig. 5.20) would not be required for activation by the estrogen receptor.

The functional differences which exist between different factors in their ability to activate transcription from different positions and in different

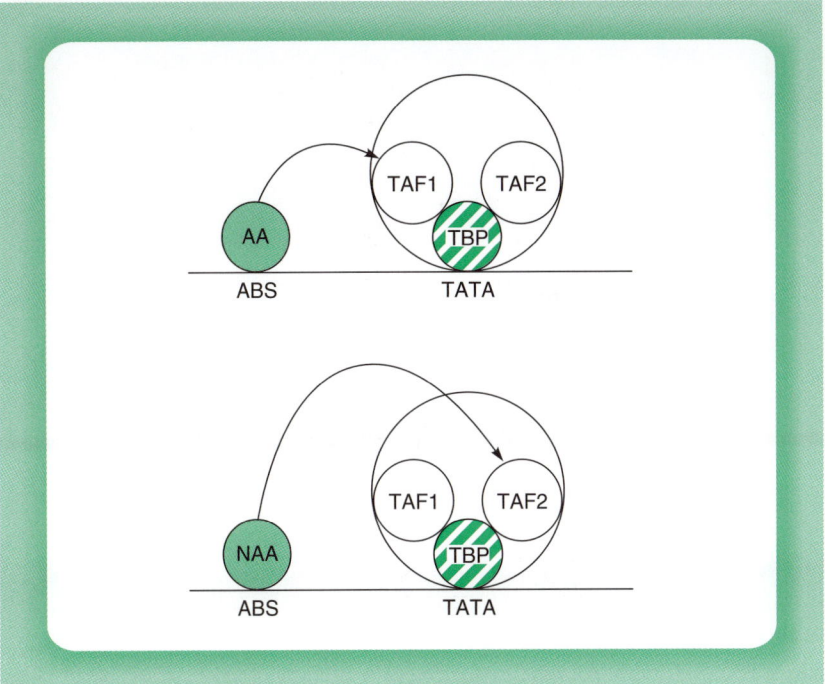

Figure 5.19
Acidic (AA) and non-acidic (NAA) activator molecules may interact with different
TBP associated factors (TAFS) within the TFIID complex.

species (see section 5.2.4) are therefore paralleled by differences in their
ability to interact with different TAFs. This ability of different activation
domains to interact with different TAFs can produce a strong synergistic activation of transcription which is far stronger than the sum of that
observed with either activation domain alone. Thus, the ability of different activators to bind to different TAFs in the TFIID complex would
result in greatly enhanced recruitment of TFIID compared to the effect
of either activator alone (Fig. 5.21) (for review see Buratowski, 1995).

These findings thus suggest that the TAFs are of importance for transcriptional activation and mediate some of the interactions between activators and TFIID which were described in section 5.3. However, it is clear
that their importance varies between different species and on different
promoters. Thus, whilst TAFs appear to be of central importance in transcriptional activation in higher eukaryotes such as man and *Drosophila*,
they are not essential for transcriptional activation at most promoters in
yeast. Similarly, even in higher eukaryotes, specific TAFs appear to be of
key importance at particular types of promoters. Thus, mutation of TAF_{II}
250 inhibits the expression of specific genes and results in cell cycle

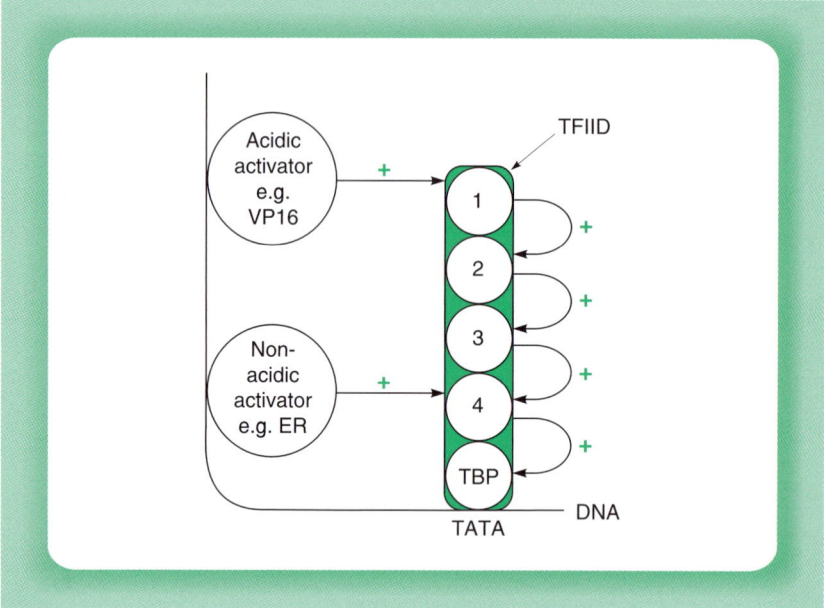

Figure 5.20

Interaction of different activator molecules with different adaptor molecules (1–4) which each activate each other and ultimately activate TBP. Note that the ability of the non-acidic activation domain of the estrogen receptor to squelch activation by the acidic activation domain of VP16 but not vice-versa can be explained if the estrogen receptor interacts with an adaptor molecule (4) closer to TBP in the series than that with which VP16 interacts (1).

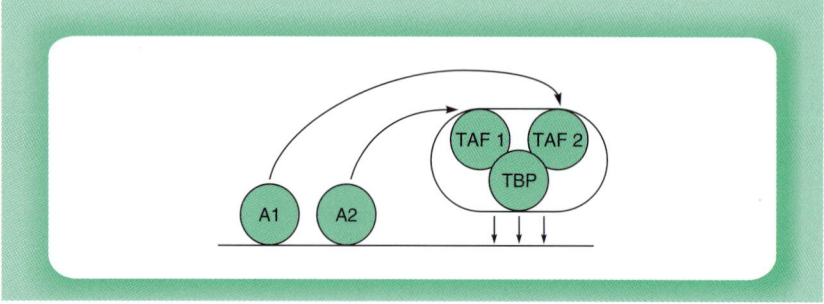

Figure 5.21

The ability of different activators (A1 and A2) to interact with different TAFs will result in a strong synergistic enhancement of TFIID recruitment and hence of transcriptional activation.

arrest in mammalian cells without affecting the transcription of other genes (Wang and Tjian, 1994).

   This idea that particular TAFs may play a critical role in mediating the response to activators at specific genes has been extended by findings suggesting that TAFs also function in promoter selectivity. Thus it appears that TFIID complexes containing particular TAFs assemble preferentially at particular promoters. This effect may be mediated by particular TAFs binding preferentially to particular core promoters (see Chapter 1, section 1.3.1) containing different sequences between the TATA box and the start site of transcription (Fig. 5.22). Thus, as noted above, most yeast genes do not require TAFs for the activation of transcription. However, a few genes involved in cell cycle progression such as the cyclin genes have been shown to be dependent upon TAF$_{II}$145 for their transcription. This dependence upon TAF$_{II}$145 is not due to the nature of the activator sequences in the promoter but is dependent upon the nature of the core promoter (Shen and Green, 1997) (Fig. 5.23). Although the yeast promoters used in this study contain a TATA box, the ability of TAFs to interact with specific core promoter sequences may be of particular importance on promoters lacking a TATA box and containing an initiator element where, as discussed in Chapter 3 (section 3.6), TBP is brought to the promoter by factors binding to the initiator element rather than by TBP binding to the TATA box.

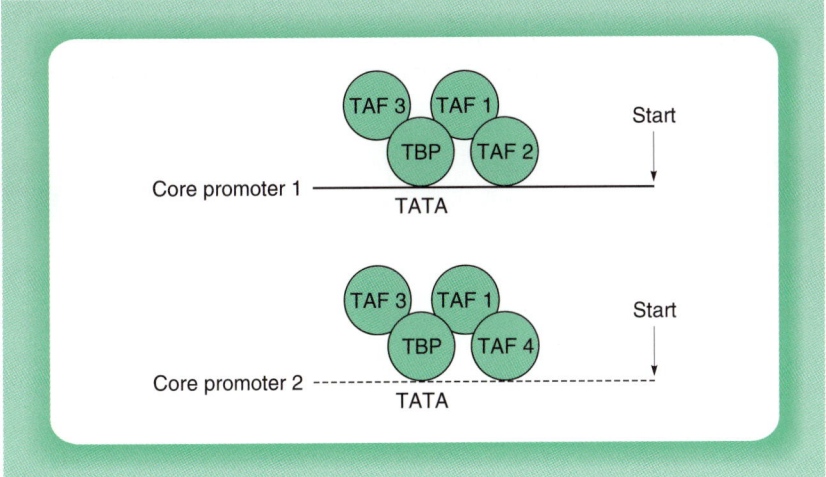

Figure 5.22

TFIID complexes containing different TAFs bind preferentially to different core promoters containing different sequences between the TATA box and the transcriptional start site.

Thus, particular TFIID complexes containing specific combinations of TAFs may bind selectively to specific promoters rather than only responding to transcriptional activators following binding. This

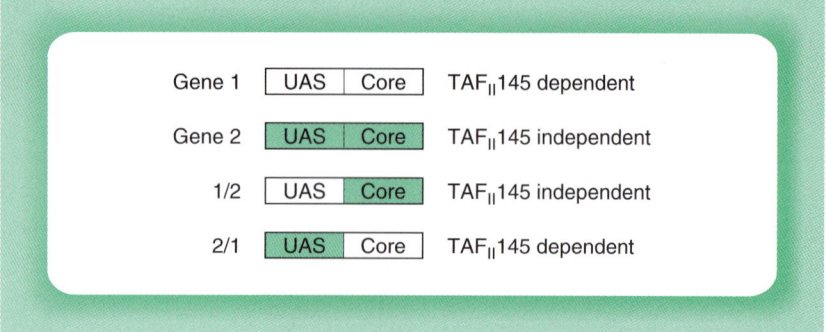

Figure 5.23

The dependence of particular yeast promoters on TAF$_{II}$145 for transcription is determined by the nature of the core promoter not by the upstream activator binding sites (UAS).

idea has been supported by the finding of a cell type specific form of TAF$_{II}$130, known as TAF$_{II}$105, which is expressed only in B lymphocytes (Dikstein *et al.*, 1996). Hence different forms of TFIID containing different TAFs may exist in different tissues and may thus play a role in the cell type specific regulation of gene expression (for review see Verrijzer, 2001) (Fig. 5.24). This is reinforced by the finding (discussed in Chapter 3, section 3.2.6) of TBP-like factors which are expressed in specific cell types.

Obviously, the different TFIID complexes formed in this manner may also differ in their responses to different transcriptional activators. Thus, for example, TAF$_{II}$30, which mediates transcriptional activation by the estrogen receptor, is found in only some TFIID complexes. In others it is replaced by TAF$_{II}$18 which does not mediate activation by the receptor (for review see Chang and Jaehning, 1997). Therefore, the ability of an activator to stimulate transcription may depend not only on its pattern of its synthesis or activation (see Chapters 7 and 8) but also on its ability to interact with different TAFs or with TBP- and TBP-like factors.

Hence, the TAF factors play a key role in transcription, by acting as co-activators mediating the response to specific activators and by regulating the binding of TFIID to specific promoters containing particular sequences adjacent to the TATA box (Fig. 5.25). This ability of the TAFs to act as an intermediate between the basal transcriptional complex and transcriptional activators evidently parallels the role of the mediator complex which acts as an intermediate between activators and the RNA polymerase itself within the RNA polymerase holoenzyme complex (see section 5.4.1).

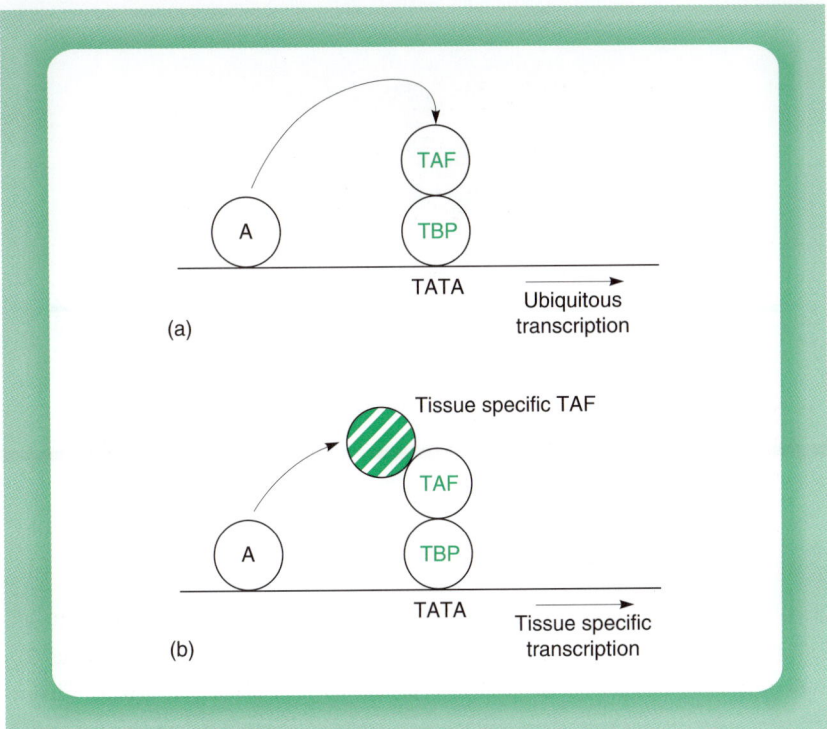

Figure 5.24

The ability of an activator (A) to stimulate transcription may be controlled by the expression pattern of the TAFs with which it interacts. Hence, an activator which interacts with a tissue specific TAF will produce tissue specific gene transcription, even if the activator itself is ubiquitously expressed.

5.4.3 CBP AND OTHER CO-ACTIVATORS

In addition to factors such as the TAFs and the mediator, which were originally defined via their association with the basal transcriptional complex, other co-activators exist which were originally defined on the basis of their essential role in transcriptional activation mediated by a specific transcriptional activator. Thus, cyclic AMP inducible genes contain a short sequence in their regulatory regions which can confer responsiveness to cyclic AMP when it is transferred to another gene that is not normally cyclic AMP inducible. This sequence, which is known as the cyclic AMP response element (CRE), consists of the eight base pair palindromic sequence TGACGTCA.

The first transcription factor shown to bind to this site was a 43 kilo-dalton protein which was named CREB (cyclic AMP response element binding protein). This factor has a basic DNA binding

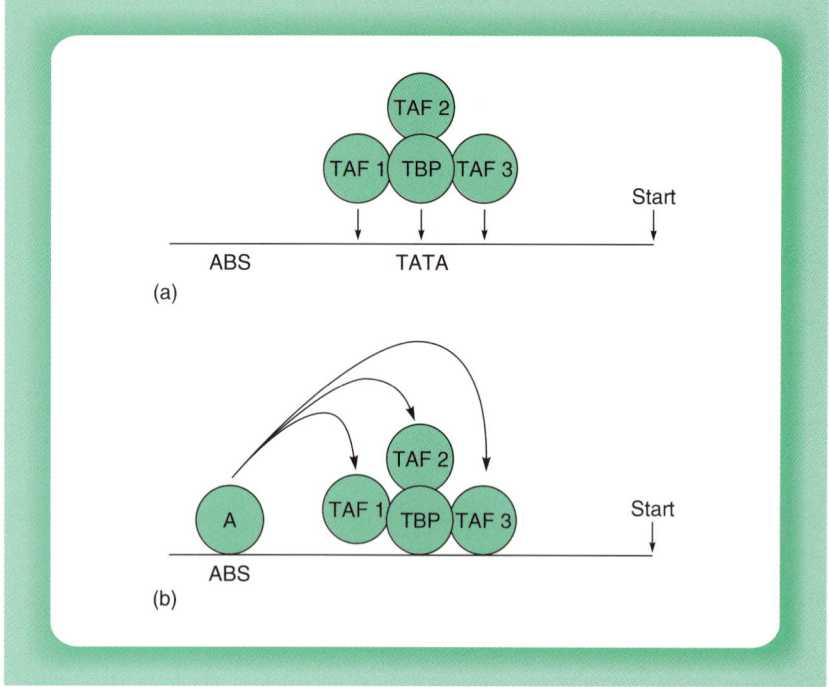

Figure 5.25

Mechanisms of TAF action. (a) The TAFs may act to enhance binding of TFIID to specific promoters by interacting with DNA sequences adjacent to the TATA box to which TBP binds. (b) The TAFs can mediate the response of TFIID to transcriptional activators.

domain with adjacent leucine zipper dimerization motif (Fig. 5.26) (see Chapter 4, section 4.5 for further discussion of this motif) and binds to the palindromic CRE as a dimer with each CREB monomer binding to one half of the palindrome (for review of CREB see de Cesare and Sassone-Corsi, 2000; Mayr and Montminy, 2001).

The CREB factor plays a key role in the activation of gene expression via the CRE following cyclic AMP treatment. The CREB factor is present in cells in an inactive form prior to exposure to the activating stimulus. Moreover, CREB is actually bound to the CRE prior to exposure to cyclic AMP but this DNA-bound CREB does not activate transcription. Elevated levels of cyclic AMP result in the activation of the protein kinase A enzyme which in turn phosphorylates CREB on the serine amino acid at position 133 in the molecule. This serine residue is located in a region of CREB known as the phosphorylation box (P-box), which is flanked on either side by regions rich in glutamine amino acids that act as transcriptional activation domains (see section 5.2) (Fig. 5.26). The

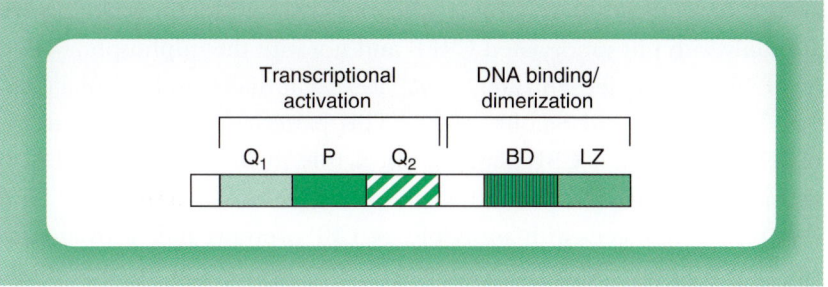

Figure 5.26

Structure of the CREB transcription factor indicating the glutamine-rich activation domains (Q_1 and Q_2), the phosphorylation box (P) containing the serine 133 residue, and the basic DNA binding domain (BD) with associated leucine zipper (LZ).

phosphorylation of CREB on serine 133 results in a change in the structure of the molecule which now allows it to activate transcription (Fig. 5.27).

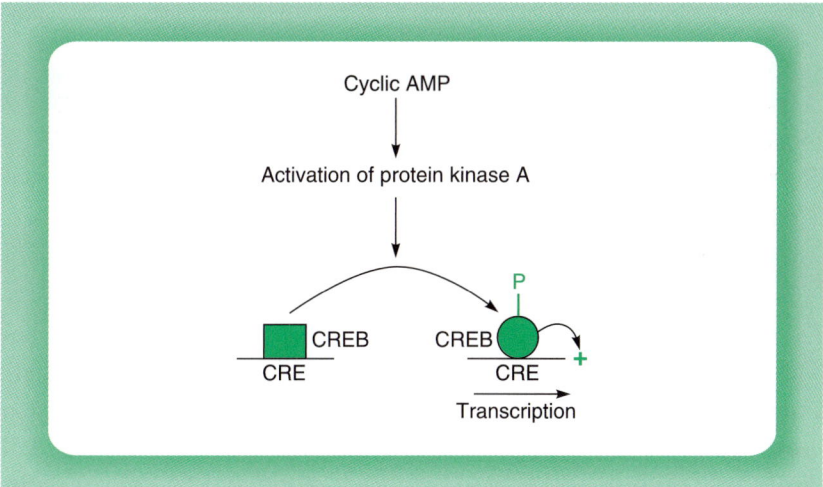

Figure 5.27

Activation of the CREB factor by cyclic AMP-induced phosphorylation. The ability of DNA-bound CREB to activate transcription is produced by the cyclic AMP-dependent activation of protein kinase A which phosphorylates the CREB protein resulting in its activation.

To identify the mechanism of this effect, Chrivia *et al.* (1993) screened a cDNA expression library with CREB protein phosphorylated on serine 133 to identify proteins which interact with phosphorylated CREB. This resulted in the isolation of cDNA clones encoding CBP

(CREB binding protein). CBP is a 265 kilo-dalton protein which associates only with phosphorylated CREB and not with the unphosphorylated form (for review see Shikama *et al.*, 1997; Giordano and Avantaggiati, 1999; Goodman and Smolik, 2000). This pattern of association immediately suggests that CBP plays a critical role in the ability of CREB to activate transcription only after phosphorylation. In agreement with this, injection of cells with antibodies to CBP prevents gene activation in response to cyclic AMP, indicating that CBP is essential for this effect. Hence, CBP is a co-activator molecule whose binding to phosphorylated CREB is essential for transcriptional activation to occur (Fig. 5.28).

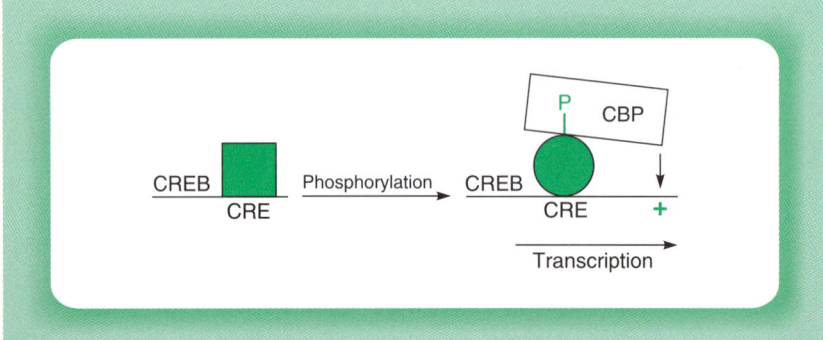

Figure 5.28
The phosphorylation of CREB on serine 133 allows it to bind the CBP co-activator which then stimulates transcription.

Although the CBP factor was originally defined as a co-activator essential for cyclic AMP stimulated transcription mediated via the CREB factor, it was subsequently shown that CBP and its close relative p300 are essential co-activators for a vast range of other factors such as the nuclear receptors (Chapter 4, section 4.4), MyoD (Chapter 7, section 7.2.1), AP1 (Chapter 9, section 9.3.1), p53 (Chapter 9, section 9.4.2) and a number of others (for review see Shikama *et al.*, 1997; Giordano and Avantaggiati, 1999; Goodman and Smolik, 2001) (Fig. 5.29).

This ability of CBP and p300 to interact with a vast array of transcription factors places them at the centre of a whole range of signalling pathways in the cell and they thus play a critical role in gene activation via these pathways. The relatively low abundance of CBP/p300 in the cell means that different signalling pathways compete for them and results in mutual antagonism between different competing pathways, such as the inflammation mediated by the AP1 pathway and the anti-inflammatory effects of glucocorticoids (see Chapter 6, section 6.5) or the growth promoting effects of the AP1 pathway compared to the

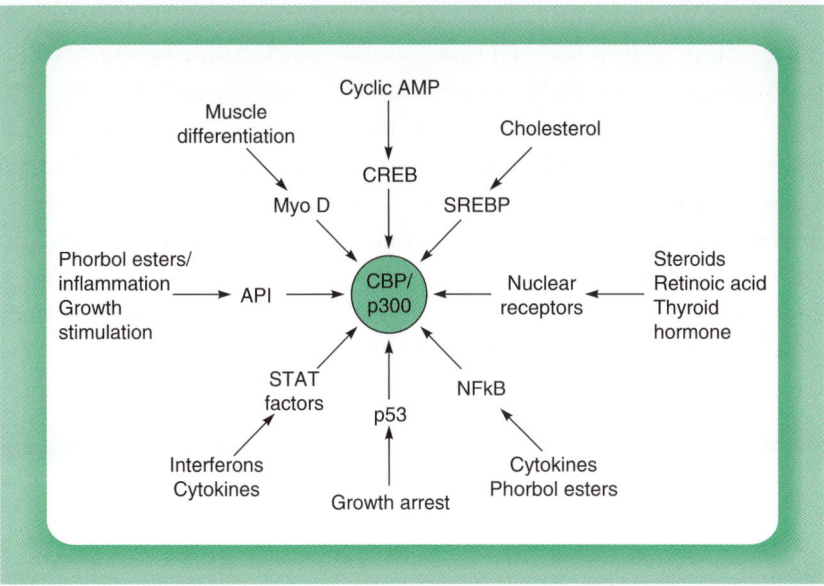

Figure 5.29

Some transcription factors which interact with the CBP/p300 co-activators and the signalling pathways which activate them.

growth arresting effects of the p53 pathway (see Chapter 9, section 9.4.2). Interestingly, the activation domain of CREB undergoes a structural transition from a coiled structure to form two α-helices when it interacts with CBP (Radhakrishnan *et al.*, 1997). This evidently parallels the change in the activation domain of VP16 when it interacts with TAF$_{II}$31 (see section 5.2.1), suggesting that the formation of a specific helical structure may be a general feature which occurs when many activation domains interact with their targets.

Although the p300/CBP proteins are the best defined co-activators, other co-activators have also been defined on the basis of their association with particular activators. Thus, for example, the nuclear receptors discussed in Chapter 4 (section 4.4) interact not only with CBP but also with a range of other co-activators such as TIF-1, TIF-2, SRC-1, SRC-3 and Sug1 (for review see Nagy and Schwabe, 2004; Lonard and O'Malley, 2006; O'Malley, 2006; Rosenfeld *et al.*, 2006). Moreover, several of these co-activators associate with the receptors only after they have been activated by binding their ligand, indicating that they are likely to play a key role in the ability of the receptors to activate transcription only following ligand binding (see Chapter 8, section 8.2.2 for a discussion of the mechanisms producing ligand-dependent activation of the nuclear receptors).

The key role of CBP/p300 and other co-activators obviously leads to the question of how they act. Two possible mechanisms by which CBP/p300 achieve their effects have been described. Thus, CBP/p300 have been shown to interact via a protein–protein interaction with several components of the basal transcriptional complex, such as TFIIB (see Chapter 3, section 3.2.4) and CBP/p300 have also been identified as part of the RNA polymerase II holoenzyme complex (which also contains RNA polymerase II, components of the basal transcriptional complex and other regulatory proteins). Hence, like the TAFs, CBP/p300 may serve as a bridge between CREB and the basal transcriptional complex, either interacting with components of the complex to enhance their activity or serving to recruit the RNA polymerase holoenzyme to the DNA by the CBP component binding to CREB (Fig. 5.30).

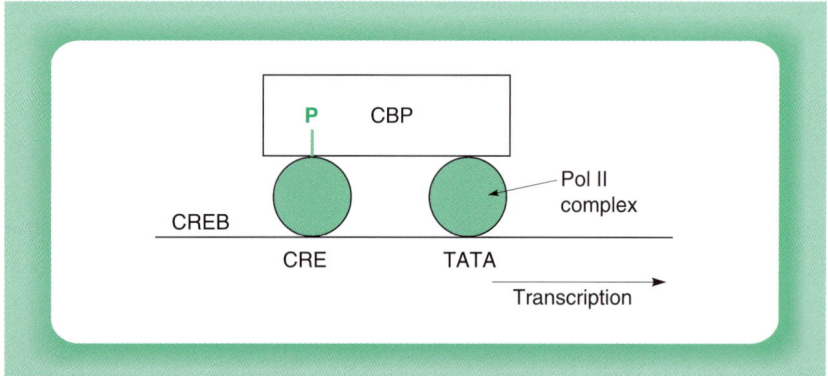

Figure 5.30
CBP can bind to both CREB and the basal transcriptional complex. It may therefore act as a bridge between CREB and the complex allowing transcriptional activation to occur.

As well as this mechanism, however, it is also possible that CBP/p300 act via a mechanism involving alterations in chromatin structure. Thus, several co-activators such as CBP/p300 and SRC-1 have been shown to have histone acetyltransferase activity (Ogryzko *et al.*, 1996). As discussed in Chapter 1 (section 1.2.3), acetylated histones are associated with the more open chromatin structure which is required for transcription. Hence, the binding of CBP to CREB, which recruits it to DNA, may then result in the acetylation of histones leading to a chromatin structure compatible with transcription (Fig. 5.31) (see section 5.5.2 for further discussion of this effect).

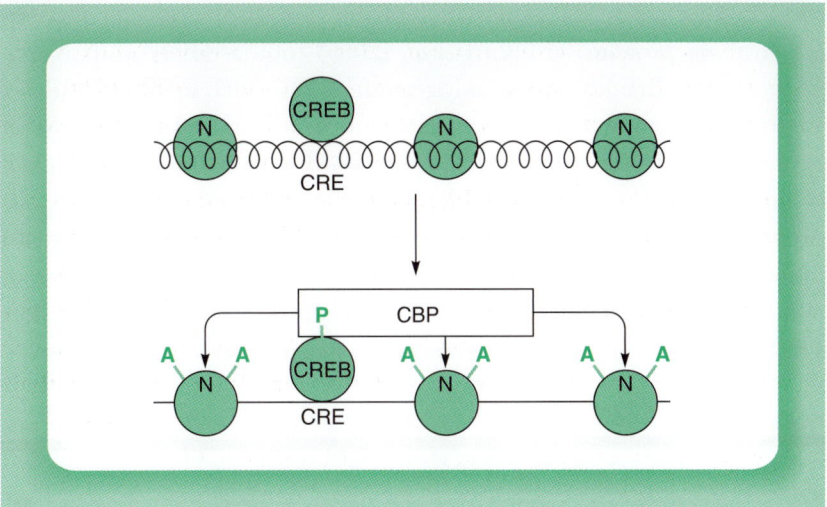

Figure 5.31
CBP has histone acetyltransferase activity. Therefore, following binding to phosphorylated CREB it can acetylate (A) histones within the nucleosome (N), resulting in a more open chromatin structure (wavy versus solid line) compatible with transcription.

5.4.4 A MULTITUDE OF TARGETS FOR TRANSCRIPTIONAL ACTIVATORS

There thus exists an array of target factors which are contacted by transcriptional activators and these include components of the basal complex such as TFIIB and TBP as well as the mediator, various TAFs and other co-activators, with some factors being contacted by activators of all classes and others by activators of only one class. Even when the finding that some of these targets such as individual TAFs can interact with only one class of activation domain is taken into account, there still remains a bewildering number of targets within the basal complex. Thus, for example, in the most extreme case described so far, the acidic activation domain of VP16 has been reported to interact with TFIIB, TFIIH, TBP, $TAF_{II}40$, $TAF_{II}31$ and the RNA polymerase holoenzyme (for review see Chang and Jaehning, 1997).

These interactions of VP16 were originally defined in the test tube, and not all interactions which can take place in the test tube will necessarily occur in the intact cell. However, several such interactions of VP16 have been confirmed in the intact cell (Hall and Struhl, 2002) using the ChIP assays described in Chapter 2 (section 2.4.3), suggesting that they are likely to be of functional importance.

To further probe the importance of the various different interactions of activating proteins, Fishburn *et al.* (2005) took a different approach. They used a chemical cross-linking method (in which adjacent proteins are linked together chemically in a stable manner) to find which proteins were in contact with the yeast acidic activator GCN4 in the transcriptional complex. Once these had been identified, they removed each of these target proteins individually from the transcriptional complex and investigated the effect on transcriptional activation (Fig. 5.32). They identified components of the mediator and SAGA complexes (see section 5.4.1) and of TFIID as being in contact with the activation domain of GCN4. However,

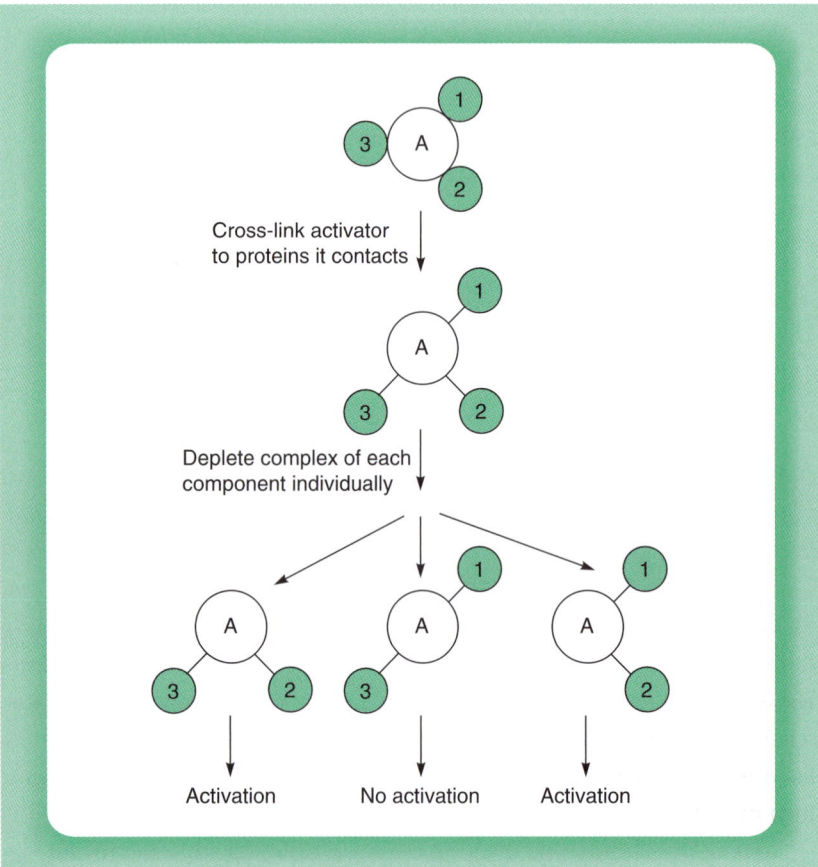

Figure 5.32

Chemical cross-linking can be used to identify the proteins which are in contact with an activator (A) in the transcriptional complex. The critical contacts for transcriptional activation can then be identified by removing each of these proteins individually and assessing the effect on transcriptional activation. In the case illustrated, only the contact of the activator with protein 2 is necessary for transcriptional activation.

only removal of the Tral component of SAGA prevented transcriptional activation, indicating it is of particular importance for activation by GCN4.

Hence, in some cases, only a small number of the protein contacts made by an activator may be of critical importance for transcriptional activation (for review see Green, 2005). It is likely, however, that in other situations the multiple contacts made by activators with different proteins are of importance for transcriptional activation. Indeed, the ability of different molecules of the same factor or different activating factors to interact with different components within the basal transcriptional complex is likely to be essential for the strong enhancement of transcription which is the fundamental aim of activating molecules (Fig. 5.33) (for review see Carey, 1998).

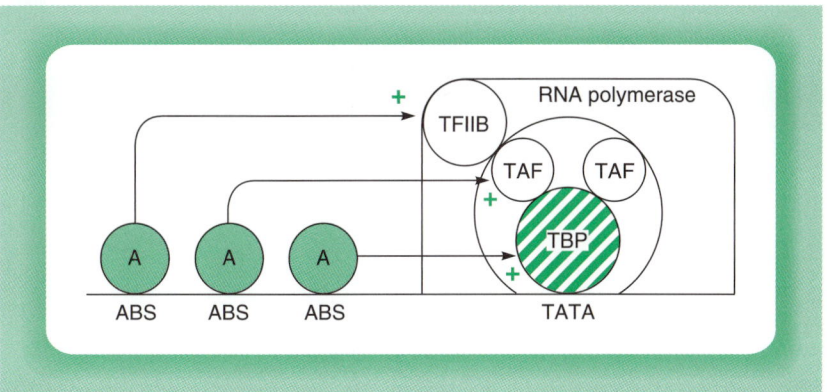

Figure 5.33
The ability of multiple activating molecules (A) to contact different factors allows strong activation of transcription.

The basal transcriptional complex which initiates transcription is therefore a critical target for transcriptional activators and co-activators. However, activators can also target at least two other stages of the transcriptional process, namely the alteration of chromatin structure required for transcription and transcriptional elongation. These will be discussed in turn in sections 5.5 and 5.6, respectively.

5.5 EFFECT OF TRANSCRIPTIONAL ACTIVATORS ON CHROMATIN STRUCTURE

5.5.1 EFFECT OF CHROMATIN REMODELLING FACTORS

As discussed in Chapter 1 (section 1.2), the DNA molecule is associated with histones and other proteins to form particles known as nucleosomes which are the basic unit of chromatin structure. Prior to the onset of transcription, the chromatin structure becomes altered, thus allowing the subsequent binding of the factors which actually stimulate transcription. This alteration in chromatin structure can itself be produced by the binding of a specific transcription factor. This results in a change in the nucleosome pattern of DNA/histone association, thereby allowing other activating factors access to their specific DNA binding sites (Fig. 5.34) (for reviews see Carey, 2005; Mellor, 2006).

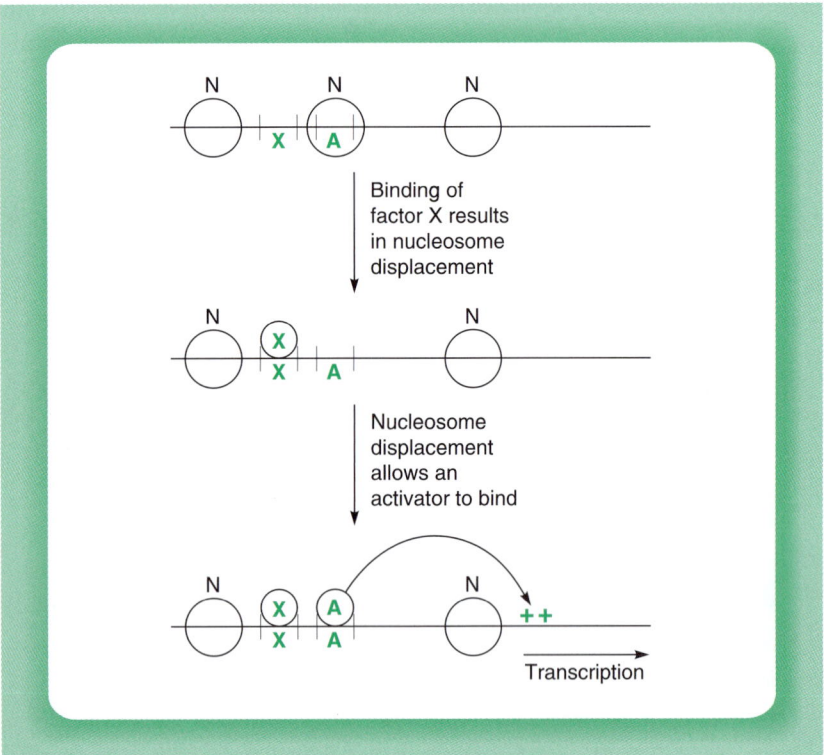

Figure 5.34
A transcription factor (X) can stimulate transcription by binding to DNA and displacing a nucleosome (N), so allowing a constitutively expressed activator (A) to bind and activate transcription.

Thus, as discussed in Chapter 1 (section 1.3.3), genes whose transcription is induced by elevated temperature share a common DNA sequence which, when transferred to another gene, can render the second gene heat inducible. This sequence is known as the heat shock element (HSE). The manner in which a *Drosophila* HSE, when introduced into mammalian cells, functioned at the mammalian rather than the *Drosophila* heat shock temperature suggested that this sequence acted by binding a protein rather than by acting directly as a thermosensor (see Chapter 1, Fig. 1.8).

Direct evidence that this was the case was provided by studying the proteins bound to the promoters of the hsp genes before and after heat shock. Thus, prior to heat shock, the TFIID complex (see Chapter 3, sections 3.5 and 3.6), is bound to the TATA box and another transcription factor, known as GAGA, is bound upstream (Fig. 5.35a; Wu, 1985; Tsukiyama *et al.*, 1994). Following heat shock, however, an additional factor is observed which is bound to the HSE (Fig. 5.35b) and it is this heat shock factor (HSF) which produces activation of the genes in response to the stimulus of elevated temperature.

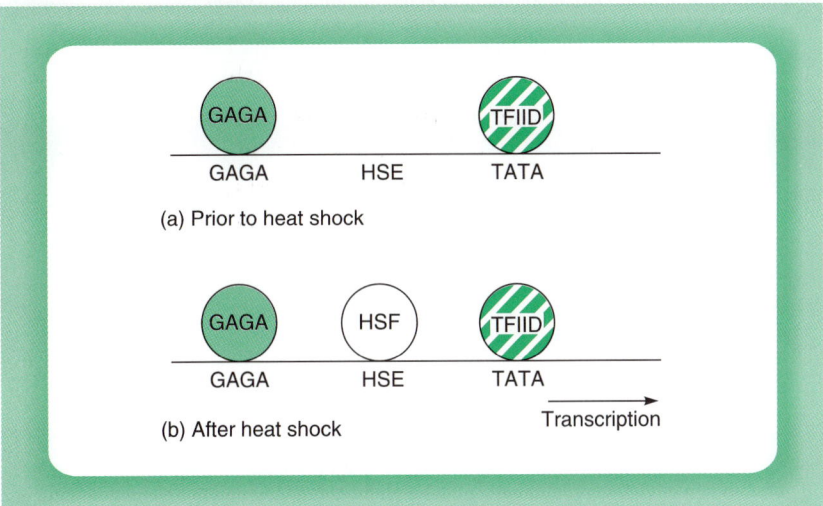

Figure 5.35
Proteins binding to the promoter of the hsp70 gene before (a) and after (b) heat shock.

However, prior to heat shock, the heat shock genes are poised for transcription. Thus, whilst the bulk of cellular DNA is associated with histone proteins to form a tightly packed chromatin structure, the binding of the GAGA factor to the heat shock gene promoters has resulted in the displacement of the histone-containing nucleosomes from the promoter region (for review see Wilkins and Lis, 1997; Simon and Tamkun, 2002; Lehmann, 2004). This opens up the chromatin and renders the promoter region exquisitely sensitive to digestion with the enzyme DNAseI.

Although such a DNAseI hypersensitive site marks a gene as poised for transcription (for review see Latchman, 2005), it is not in itself sufficient for transcription. The binding of the GAGA factor thus opens up the gene and renders it poised for transcription in response to a suitable stimulus. This role for the GAGA factor in chromatin remodelling is not confined to the heat shock genes. Thus, mutations in the gene encoding GAGA result in the *Drosophila* mutant trithorax in which a number of homeobox genes (which control the formation of the correct body plan – see Chapter 4, section 4.2) are not converted from an inactive to an active chromatin state and are hence not transcribed (for reviews see Simon and Tamkun, 2002; Lehmann, 2004). This mutation thus produces a fly with an abnormal body pattern and thus has a similar effect to the brahma mutation in the SWI 2 component of the SWI/SNF chromatin remodelling complex which was discussed in Chapter 1 (section 1.2.2). Indeed, the GAGA factor and other related members of the trithorax family have been shown to be associated with a multi-protein complex known as nucleosome remodelling factor (NURF) which, like SWI/SNF, can hydrolyse ATP and alter chromatin structure (Fig. 5.36).

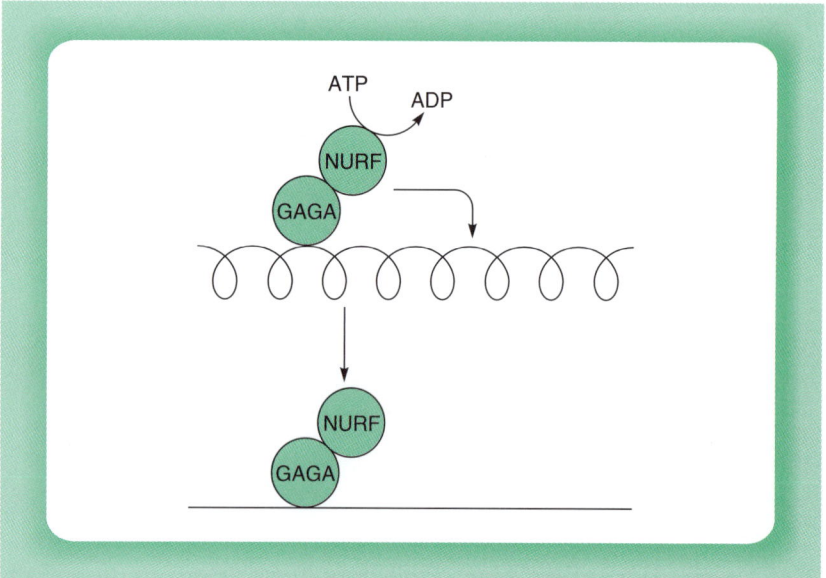

Figure 5.36

The GAGA factor can bind to DNA which is in a tightly packed chromatin structure (wavy line) and recruit the nucleosome remodelling factor (NURF). NURF then hydrolyses ATP and uses the energy to remodel the chromatin to a more open structure (solid line) to which other activating proteins can bind.

Hence, following binding of GAGA, the gene is in a state poised for the binding of an activating transcription factor which, in turn, will result in transcription of the gene. In the case of the heat shock genes, this is achieved following heat shock by the binding of the HSE to the HSF (Fig. 5.37). This factor then interacts with TFIID and other components of the basal transcription complex resulting in the activation of transcription. The manner in which HSF is activated in response to heat and can therefore mediate heat inducible transcription is discussed in Chapter 8 (section 8.3.1).

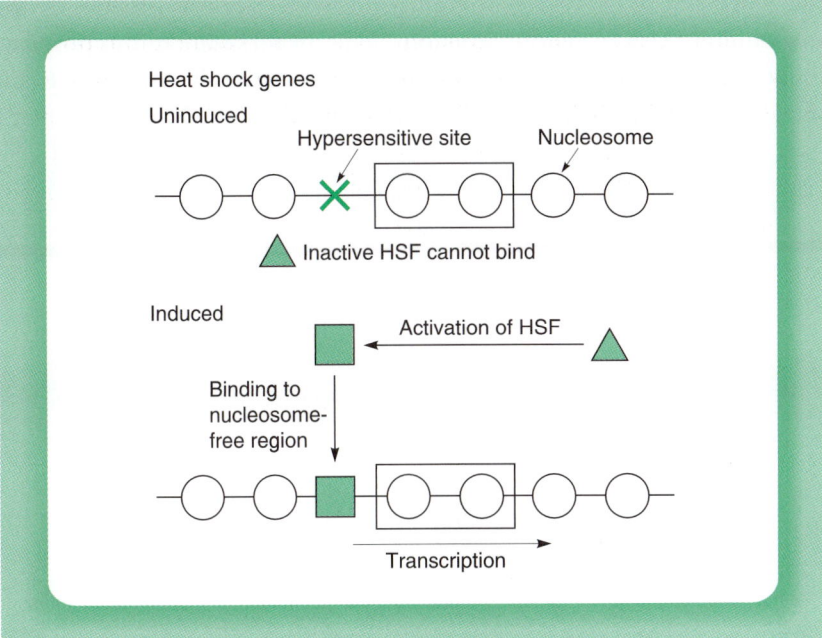

Figure 5.37
Activation of HSF by heat is followed by its binding to a pre-existing nucleosome free region in the heat shock gene promoters which is marked by a DNAseI hypersensitive site and was produced by the prior binding of the GAGA factor. Binding of HSF then results in the activation of heat shock gene transcription.

A similar modulation of chromatin structure to that produced by the GAGA factor is also seen in the case of members of the nuclear receptor family (see Chapter 4, section 4.4). In this case, the receptors are activated by treatment with the appropriate hormone and then bind to the DNA (see Chapter 8, section 8.2.2 for a discussion of the mechanism of this effect) allowing them to mediate hormone-inducible transcription.

In a number of cases, steroid hormone treatment has been shown to cause the induction of a DNAseI hypersensitive site located at the DNA sequence to which the receptor binds. Hence, the binding of the receptor may activate transcription by displacing or altering the structure of a nucleosome within the promoter of the gene creating the hypersensitive site. In turn, this would facilitate the binding of other transcription factors necessary for gene activation whose binding sites would be

exposed by the change in the position or structure of the nucleosome.
These factors would be present in the cell in an active form prior to steroid hormone treatment but could not bind to the gene because their
binding sites were masked by a nucleosome (Fig. 5.38) (for review see
Beato and Eisfeld, 1997). In agreement with this idea, the binding sites
for TFIID and CTF/NFI in the glucocorticoid-responsive mouse mammary tumour virus promoter are occupied only following hormone treatment, although these factors are present in an active DNA binding form
at a similar level in treated and untreated cells.

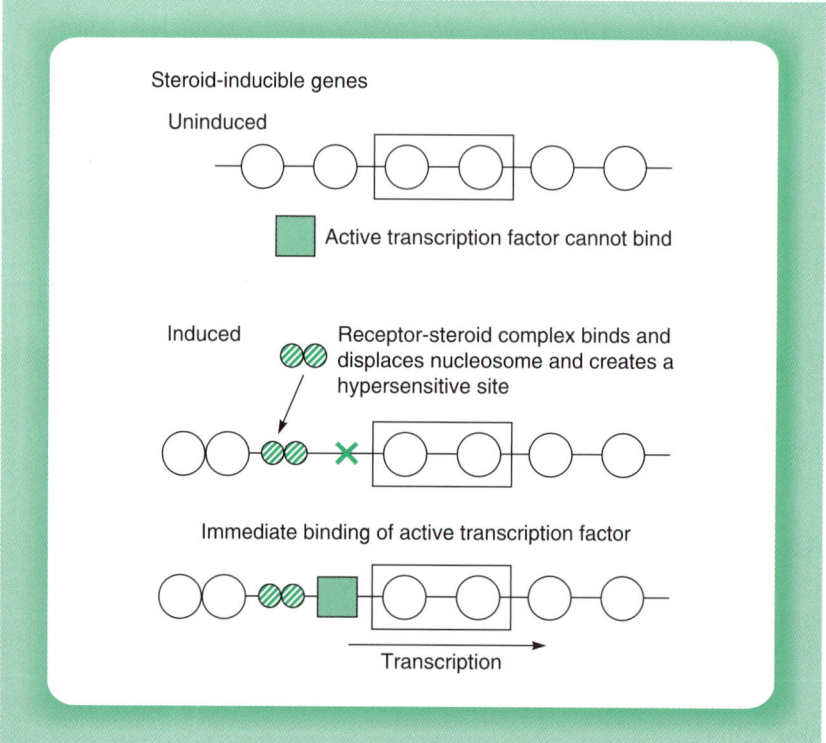

Figure 5.38

Binding of the steroid receptor–steroid hormone complex to the promoter of
a steroid-inducible gene results in a change in chromatin structure creating a
hypersensitive site and allowing pre-existing constitutively expressed transcription
factors to bind and activate transcription. Note that the binding of these
constitutive factors may occur because the receptor totally displaces a nucleosome
from the DNA as shown in the diagram or because it alters the structure of the
nucleosome so as to expose specific binding sites in the DNA.

This ability of the glucocorticoid receptor to alter chromatin structure
is likely to be linked to its ability to stimulate the activity of the SWI/SNF
complex (Inoue *et al.*, 2002), allowing it to fulfil its role of hydrolysing
ATP and unwinding chromatin (see Chapter 1, section 1.2.2) (Fig. 5.39).

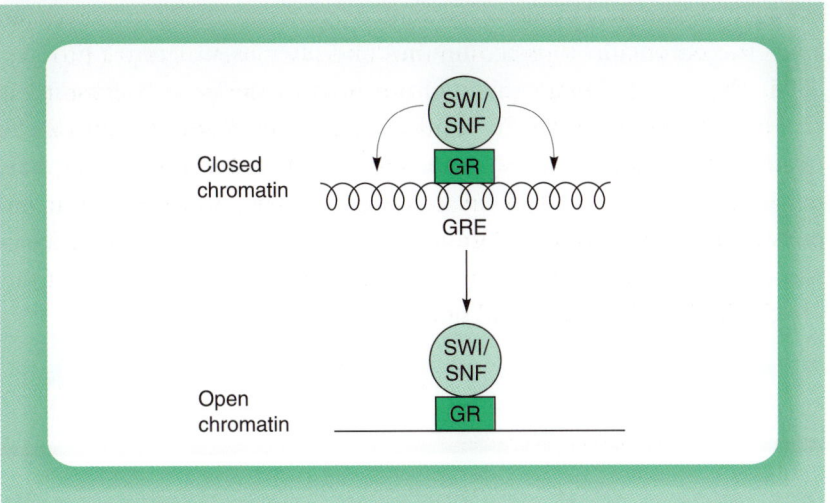

Figure 5.39

By binding chromatin remodelling factors such as SWI/SNF, the nuclear receptors can alter chromatin structure from a tightly packed (wavy line) to a more open (solid line) configuration.

As well as associating with the glucocorticoid receptor, the SWI/SNF complex has also been shown to be recruited to the DNA by other factors, including (as discussed in Chapter 1, section 1.2.2) the SATB1 protein, which is involved in looping of the chromatin. In addition, SWI/SNF has also been shown to be part of the RNA polymerase II holoenzyme which, as discussed in Chapter 3 (section 3.5.2), also contains the basal transcription factors TFIIB, TFIIF and TFIIH, as well as the mediator (see section 5.4.1). Moreover, when SWI/SNF is recruited as part of this complex, it is fully functional in opening up the chromatin and thus promoting transcription. In agreement with this, the TATA box which binds TFIID and hence recruits the RNA polymerase II holoenzyme complex has been shown to be of importance in the recruitment of chromatin remodelling complexes (Gui and Dean, 2003). The RNA polymerase II holoenzyme thus contains the polymerase and basal transcription factors, the mediator complex and a chromatin remodelling complex (see Chapter 3, section 3.5.2).

As well as the recruitment of SWI/SNF, it has recently been shown that the nuclear receptor family of transcription factors can induce chromatin structure changes in another way. Thus, during estrogen receptor-dependent gene activation of the pS2 gene promoter, the enzyme topoisomerase IIβ is recruited to the promoter and introduces a double-stranded break into the DNA. This effect of the topoisomerase has been shown to be essential for transcriptional activation by the estrogen receptor. The double-stranded break it produces is likely to allow the chromatin to relax, unwind and alter its structure (Ju *et al.*, 2006) (Fig. 5.40). This requirement for topoisomerase recruitment has also been shown to occur during the activation of the pS2 promoter by other nuclear receptors and by the unrelated transcription factor AP1, indicating that it is likely to be an important process involved in activation by a number of different transcription factors (for review see Haince *et al.*, 2006; Lis and Kraus, 2006).

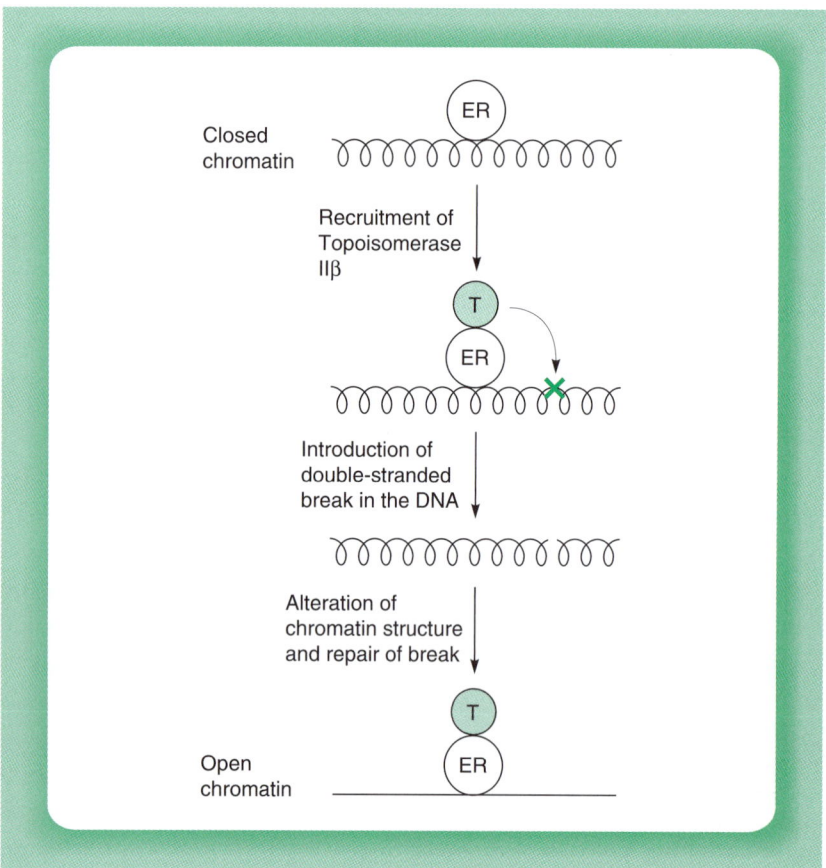

Figure 5.40

The estrogen receptor (ER) can recruit the enzyme topoisomerase IIβ (T) which can introduce a double-stranded break into the DNA. This allows the chromatin to relax and unwind, allowing its structure to be altered so that transcription can occur.

Hence, as described above, multiple mechanisms exist by which nuclear receptors and other transcription factors can produce remodelling of chromatin. In addition, however, these factors can also alter chromatin structure by modifying the histone proteins and these effects are discussed in the next section.

5.5.2 *EFFECT ON HISTONE MODIFICATION*

As discussed in Chapter 1 (section 1.2.3), changes in histone modifications have a key role in producing changes in chromatin structure. It is not surprising therefore that these modifications are targeted by transcriptional activators. Thus, as noted in section 5.4.3, CBP and other transcriptional co-activators have histone acetyltransferase (HAT) activity, allowing them to acetylate histones and open up the chromatin structure. Indeed, several of the co-activators which associate with the nuclear receptors, such as CBP, PCAF, SRC-1 and ACTR, have HAT activity, indicating that the nuclear receptors can interact both with chromatin remodelling complexes and histone modifying enzymes, thereby altering chromatin structure by both the mechanisms discussed in Chapter 1 (section 1.2) (Fig. 5.41).

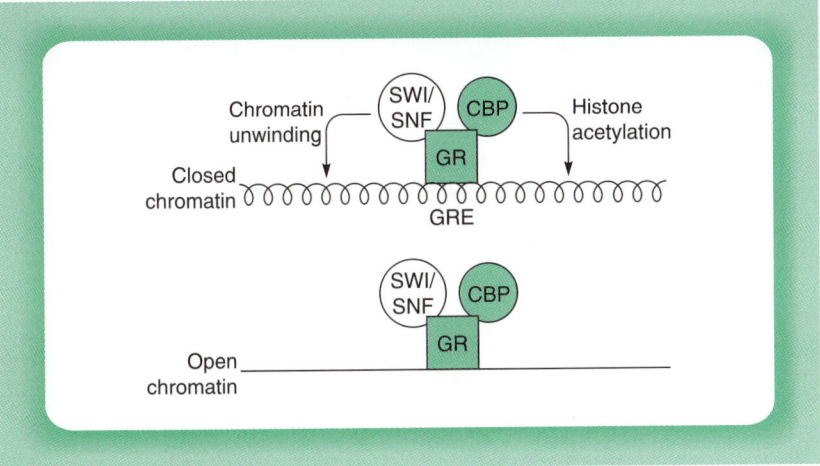

Figure 5.41

Nuclear receptors can alter chromatin structure both by recruiting chromatin remodelling complexes such as SWI/SNF and by recruiting co-activators such as CBP, with the ability to modify histones.

Interestingly, such HAT activity has also been demonstrated for a number of the factors which mediate transcriptional activation. Thus, the SAGA complex of proteins described in section 5.4.1 has HAT activity (for review see Hampsey, 1997; Grant *et al.*, 1998), as does TAF$_{II}$250 (for review see Struhl and Moqtaderi, 1998; Brown *et al.*, 2000) (see section 5.4.2). Such HAT activity has also been demonstrated for transcriptional activators which bind to DNA. Thus, the CLOCK protein which is involved in cellular timekeeping is a DNA binding transcription factor of the helix-loop-helix family (see Chapter 4, section 4.5.2) and has recently been shown to have HAT activity (Doi *et al.*, 2006; for reviews see Belden *et al.*, 2006; Hardin and Yu, 2006). Hence, the stimulation of histone acetylation plays a key role in transcriptional activation and a variety of different factors associated with this process have HAT activity (Fig. 5.42).

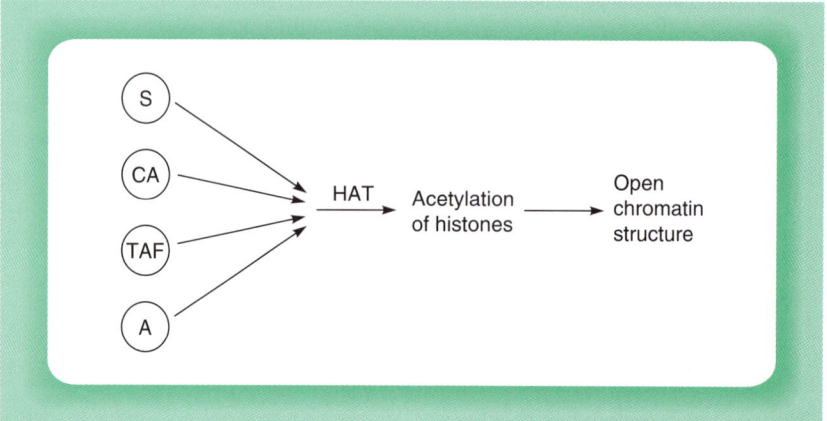

Figure 5.42
The SAGA complex (S), specific co-activators (CA), TAFs (TAF) and DNA binding transcriptional activators (A) can all have histone acetyltransferase (HAT) activity, allowing them to acetylate histones and promote a more open chromatin structure.

As noted in Chapter 1 (section 1.2.3), the histones are also modified in a variety of other ways, apart from acetylation. Some of these, such as the methylation of specific residues, have an inhibitory effect by inducing a more tightly packed chromatin structure (see Fig. 1.5) and this process can also be targeted in transcriptional activation. This has been demonstrated in the case of the androgen receptor, which induces gene transcription in response to androgen and is a member of the nuclear receptor family (see Chapter 4, section 4.4). Thus, following androgen treatment, the receptor binds a factor known as JHDM2A. This factor is a histone demethylase that is able to remove the inhibitory methyl groups on the lysine at position nine in histone H3 (Fig. 5.43) (Whetstine *et al.*, 2006; Yamane *et al.*, 2006) (for review of histone demethylation see Shi and Whetstine, 2007).

Evidently, this critical role for histone modifications as direct or indirect targets for transcriptional activators raises the question of how they promote a more open chromatin structure. In principle, there are at least two possible mechanisms by which this could occur. Thus, such modifications could affect the association of the histones with one another, thereby altering the structure of chromatin, which is dependent on the histones (Fig. 5.44a). Alternatively, they could affect the association of

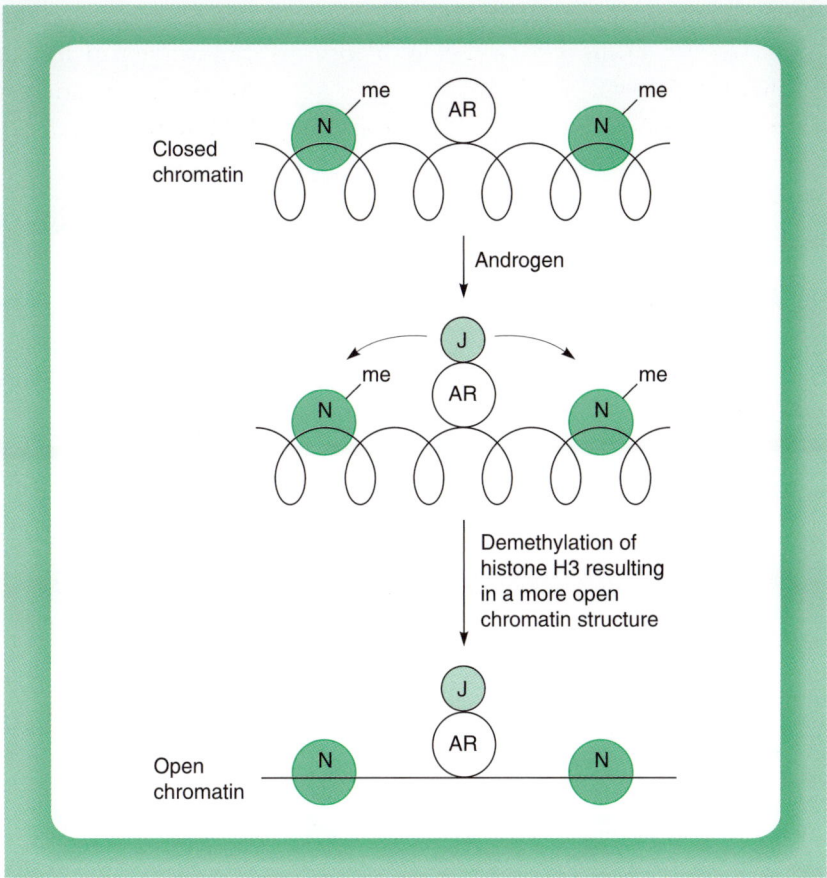

Figure 5.43

Following activation by androgen, the androgen receptor (AR) recruits the JHDM2A protein (J) which demethylates histone H3 within the nucleosome (N) producing a more open chromatin structure.

histones with other molecules, either promoting their association with proteins which remodel chromatin to a more open chromatin structure or inhibiting their association with proteins that promote a more closed chromatin structure (Fig. 5.44b).

In fact, there is evidence that both these mechanisms operate. Thus, for example, acetylation of histone H4 has been shown to promote the beads on a string structure of chromatin and prevent formation of the more compact solenoid structure (see Chapter 1, section 1.2.1), presumably by blocking interactions between adjacent nucleosomes (Shogren-Knaak *et al.*, 2006; for review see Marx, 2006). Interestingly, the same study found that acetylation of histone H4 also promoted the recruitment of a chromatin remodelling factor, which would then also act to promote a more open chromatin structure. Similarly, other studies have

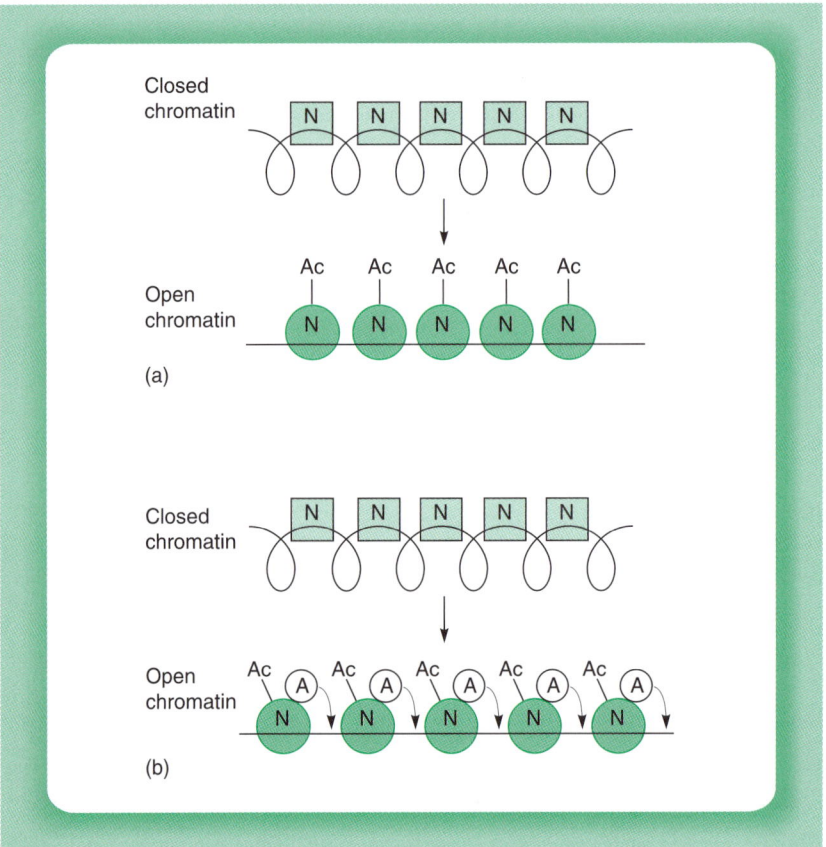

Figure 5.44

Acetylation (A) or other modifications of histones within the nucleosome (N) can produce chromatin opening either (a) by affecting the association of histones with each other (producing an altered nucleosome structure) or association between adjacent nucleosomes or (b) by affecting the association of the histones with a non-histone activator protein (A).

shown that acetylation of histone enhances binding by several activating factors which contain a region known as the bromodomain. Thus, this domain binds with greater affinity to histones when specific lysines are acetylated (for review see Loyola and Almouzni, 2004). Hence, a single modification such as acetylation can actually act by both the mechanisms illustrated in Fig. 5.44.

This effect of histone modifications in altering the recruitment of other regulatory proteins is not confined to acetylation. Thus, for example, it has recently been observed in the case of the methylation of lysine 4 on histone H3, which unlike modification on lysine 9 has a stimulatory effect on transcription (see Chapter 1, section 1.2.3 and Fig. 1.5) (for review see

Sims and Reinberg, 2006; Ruthenburg *et al.*, 2007). Such methylation on lysine 4 has been shown to promote the recruitment of the nucleosome remodelling factor, NURF (for review see Becker, 2006). As discussed in section 5.5.1, NURF is able to remodel chromatin. This example therefore links histone modification and chromatin remodelling factors, which are the two major means of altering chromatin structure.

5.5.3 TRANSCRIPTIONAL ACTIVATION BY CHROMATIN STRUCTURE CHANGES AND BY STIMULATION OF THE BASAL TRANSCRIPTION COMPLEX

Evidently, activation of transcription will require both the chromatin changes discussed in this section and the stimulation of transcriptional initiation by activators, which was discussed in earlier sections of this chapter. The two cases discussed in section 5.5.1 differ in how these mechanisms are combined.

Thus, as discussed in section 5.5.1, in the case of the glucocorticoid receptor, the receptor alters chromatin structure allowing constitutive factors access to their binding sites. This is clearly in contrast to the binding of HSF to a promoter which already lacks a nucleosome and contains bound GAGA factor and TFIID, as described in section 5.5.1. In this latter case, activation of transcription must occur not via alteration in chromatin structure but via interaction with the components of the constitutive transcriptional apparatus. It should be noted, however, that these two mechanisms are not exclusive. Thus, as discussed above (section 5.4.3), the CBP co-activator can also interact with components of the basal transcriptional complex to increase transcription. This finding indicates therefore that the nuclear receptors and their associated co-activators, such as CBP, promote transcription both by altering chromatin structure to allow constitutive factors to bind and also by interacting directly with other transcription factors such as components of the basal transcriptional complex (Fig. 5.45).

Activation by steroid hormones would therefore be a two-stage process involving, first, alteration of chromatin structure and, secondly, stimulation of the basal transcriptional complex (Jenster *et al.*, 1997). In agreement with this idea, chromatin disruption following binding of the thyroid hormone receptor to DNA is necessary but not sufficient for transcriptional activation to occur (Wong *et al.*, 1997). Similarly, recruitment of specific multi-protein complexes by nuclear receptors can result in chromatin opening, without transcription induction or both chromatin opening and transcriptional induction (King and Kingston, 2001).

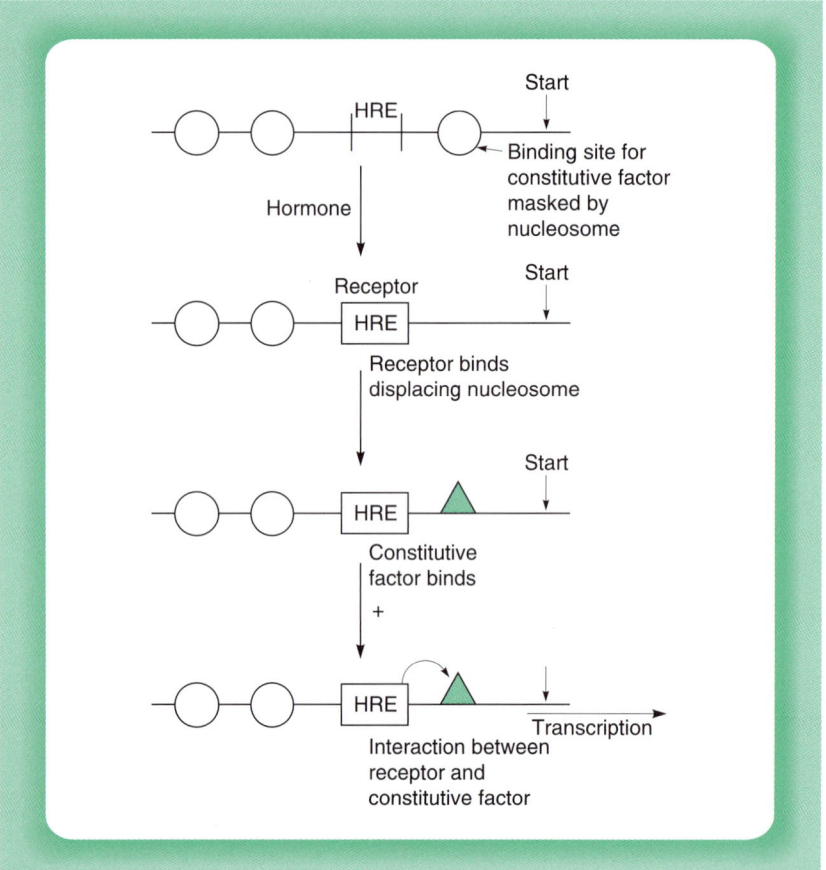

Figure 5.45

Activation of steroid-inducible genes by steroid receptors. As well as altering chromatin structure and allowing constitutive factors to bind, the hormone receptor and its associated co-factors is also able to interact directly with constitutive factors such as the basal transcriptional complex and directly activate transcription.

Hence, in the case of the nuclear receptors, a single factor and its associated co-factors can alter the chromatin structure and then activate transcription. In contrast, in heat shock gene activation, these functions are performed by separate factors, with the GAGA factor displacing a nucleosome, allowing HSE to bind and activate transcription following a subsequent heat shock.

Both cases illustrate, however, how factors which alter chromatin structure can prepare the way for the binding of the factors which actually activate transcription, with such binding occurring either immediately following the change in chromatin structure, as in the glucocorticoid receptor/NF-I case, or following a subsequent stimulus as in the GAGA/HSF case (Fig. 5.46).

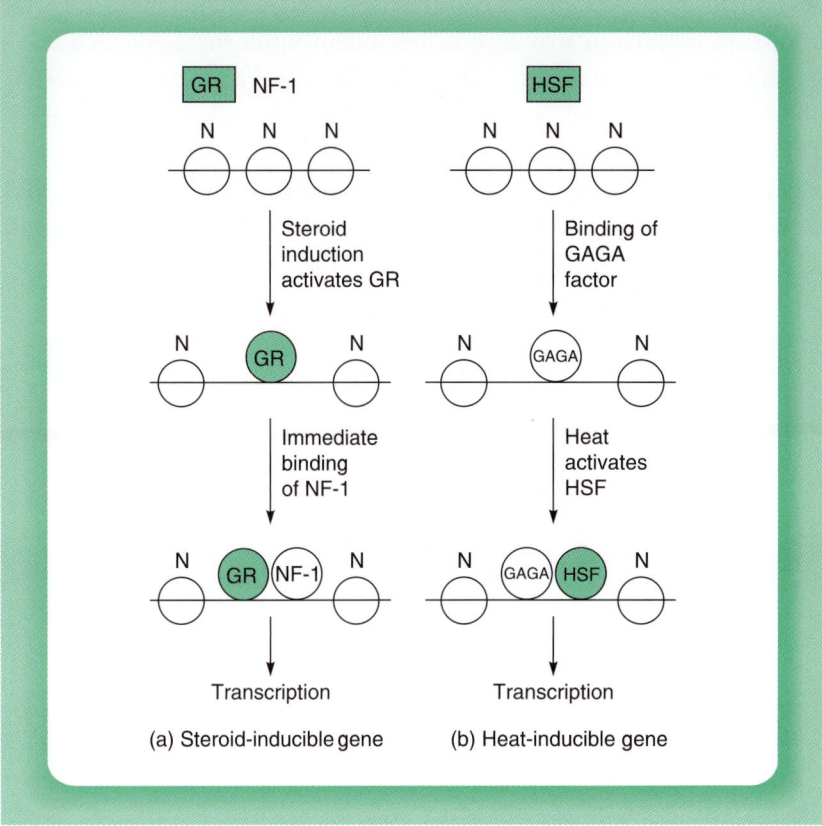

Figure 5.46

Two-stage induction of genes by steroid treatment or heat shock involving
alteration of chromatin structure and the subsequent binding of an activator
molecule. In the steroid case (a), the steroid activates the glucocorticoid receptor
(GR) resulting in a change in chromatin structure and immediate binding of the
constitutively expressed NF-1 factor. In contrast, in the heat-inducible case the
chromatin structure has already been altered prior to heat shock by binding of the
GAGA factor but transcription only occurs when heat shock activates the heat
shock factor (HSF) to a DNA binding form.

Interestingly, as well as interacting with the basal transcriptional apparatus, activation domains have also been shown to be involved in the
ability of specific factors to alter chromatin structure. Thus, the activation of the yeast PHO5 gene promoter following phosphate starvation
is mediated by the binding of the PHO4 factor to the PHO5 promoter
resulting in nucleosome displacement. Surprisingly, a truncated PHO4
molecule lacking the acidic activation domain but retaining the DNA
binding domain is incapable of nucleosome displacement, whilst this
ability can be restored by linking the truncated PHO4 molecule to the

acidic activation domain of VP16 (Fig. 5.47). Hence, the ability of PHO4 to disrupt chromatin structure is dependent upon the acidic activation domain which also interacts with the basal transcription complex to stimulate transcription (for reviews see Lohr, 1997; Svaren and Horz, 1997). This dual function is also seen in the case of the glucocorticoid receptor which, as well as altering chromatin structure thereby facilitating NF-1 binding and consequent transcriptional activation, can also itself directly stimulate transcription via interaction with other transcription factors.

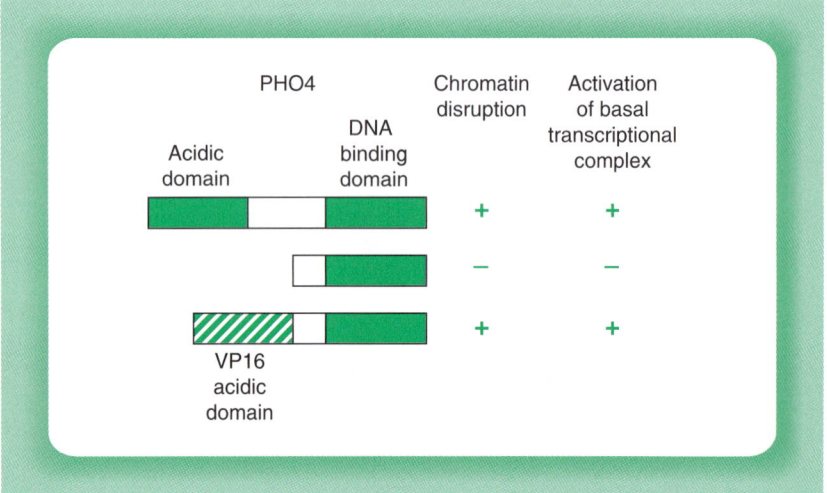

Figure 5.47
Both the disruption of chromatin and the activation of the basal transcription complex by the yeast PHO4 factor are dependent on its acidic activation domain. They are therefore lost when this domain is deleted but can be restored by addition of the acidic activation domain of VP16.

These findings indicate therefore that the remodelling of chromatin structure can be achieved both by specific factors such as the GAGA factor and by the activation domains of other factors which can modulate chromatin structure as well as activate transcription directly by interacting with the basal transcriptional apparatus. In both these types of cases, the ability to alter chromatin structure is likely to depend upon the ability of these factors to recruit other factors which then actually alter chromatin structure either via recruitment of ATP-dependent chromatin remodelling complexes such as the SWI/SNF complex (Fig. 5.48a) or via recruiting factors with histone acetyltransferase activity (Fig. 5.48b).

It is clear therefore that the alteration of chromatin structure by specific factors is of considerable importance in the control of transcription. Indeed, in some cases such chromatin remodelling may be the only

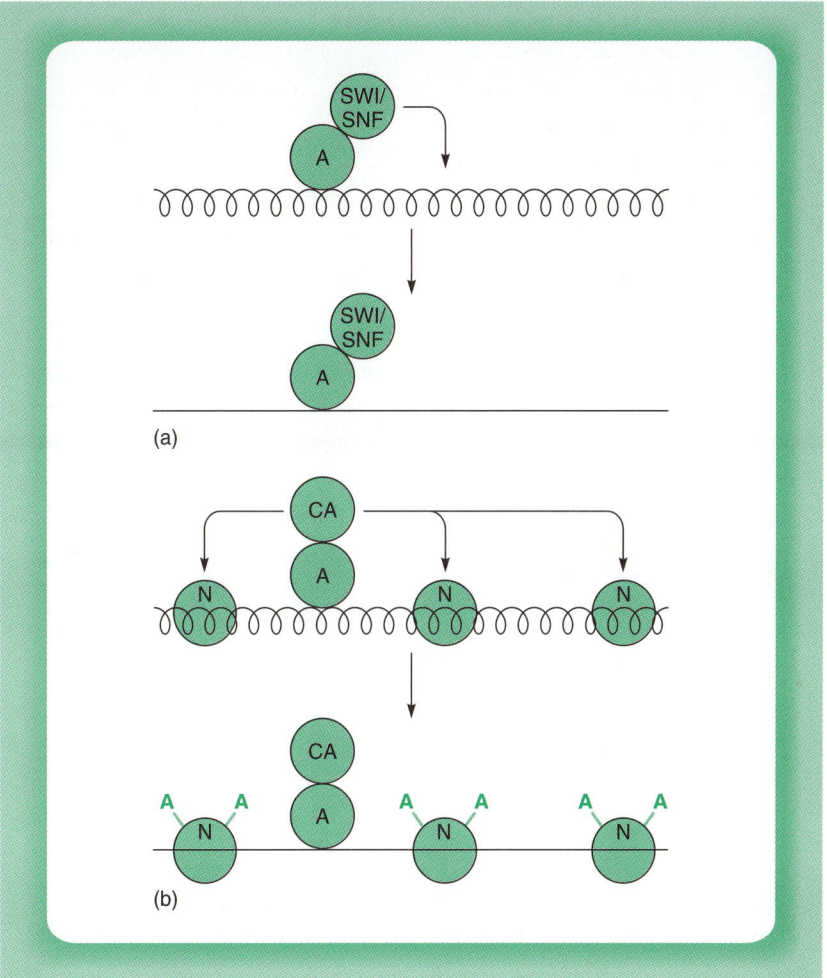

Figure 5.48

A transcriptional activator (A) can alter chromatin structure from a tightly packed (wavy line) to a more open (solid line) structure either (a) by recruiting the SWI/SNF chromatin remodelling complex or (b) by recruiting a co-activator (CA) with histone acetyltransferase activity. N = nucleosome.

requirement for transcriptional activation. Thus, recent studies in yeast have shown that transcriptional activators are not required for the activation of certain genes in a situation where nucleosomes are not present (for review see Davie and Dent, 2006). Hence, in such cases, the only role of transcriptional activators would be to disrupt chromatin structure. This example does illustrate therefore the importance of activators targeting chromatin structure, although in most cases this is likely to be combined with a subsequent step in which activators then stimulate the basal transcriptional complex, as discussed above.

5.6 STIMULATION OF TRANSCRIPTIONAL ELONGATION

As discussed in Chapter 3 (section 3.7), following transcriptional initiation, further steps occur to produce elongation of the nascent transcript. Clearly, therefore, transcriptional elongation represents a target for transcriptional activators, which could stimulate transcription at the level of transcriptional elongation by allowing full length transcripts to be produced (for review see Conaway *et al.*, 2000; Arndt and Kane, 2003; Sims *et al.*, 2004).

One case of this type involves the c-*myc* oncogene (see Chapter 9, section 9.3.3 for discussion of this oncogene). Thus, when the c-*myc* gene is transcribed in the pro-myeloid cell line HL-60, most transcripts terminate near the end of the first exon and do not produce a functional mRNA encoding the complete Myc protein. When the HL-60 cells are differentiated to form granulocytes, however, the majority of transcripts pass through this block and full length mRNA is produced (Fig. 5.49). Hence, in this case, an increased level of functional c-*myc* mRNA, able to produce the Myc protein, is obtained without an increase in transcriptional initiation (for review see Spencer and Groudine, 1990; Greenblatt *et al.*, 1993).

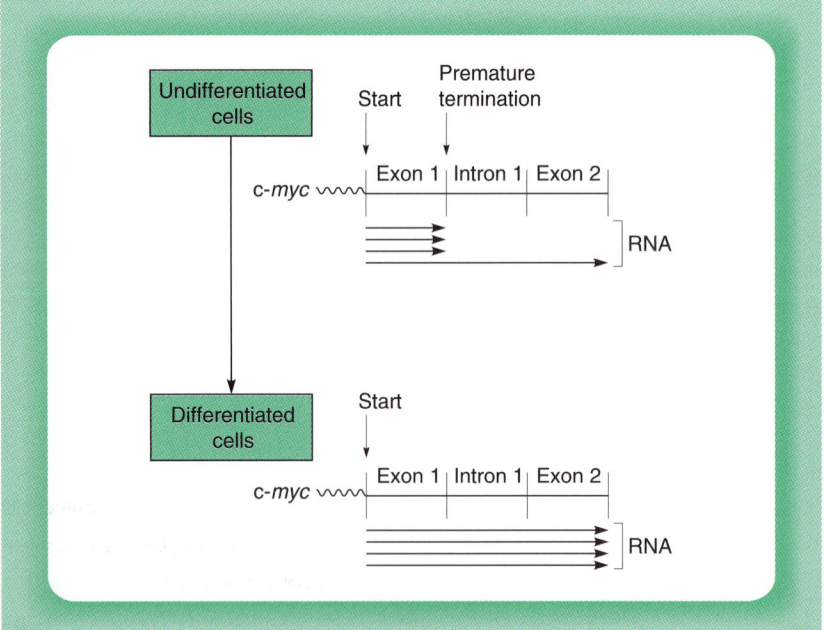

Figure 5.49

In undifferentiated HL-60 cells most transcripts from the c-*myc* gene terminate prematurely at the end of the first exon. In differentiated cells, however, this does not occur, resulting in an increase in full length functional transcripts without increased initiation.

Regulation at the level of transcriptional elongation has also been demonstrated in the human immunodeficiency virus (HIV). Thus, in this case, only short prematurely truncated transcripts are produced from the HIV promoter in the absence of the viral Tat protein. When Tat is present, however, it stimulates both the rate of initiation of transcription and also the proportion of full length transcripts which are produced, so overcoming the block to transcriptional elongation (Fig. 5.50) (for review see Greenblatt *et al.*, 1993; Jones, 1997). A similar role for HSF in stimulating transcriptional elongation in the hsp70 gene has also been proposed (for review see Lis and Wu, 1993). Hence activating factors can act to stimulate the proportion of full length RNA transcripts which are produced.

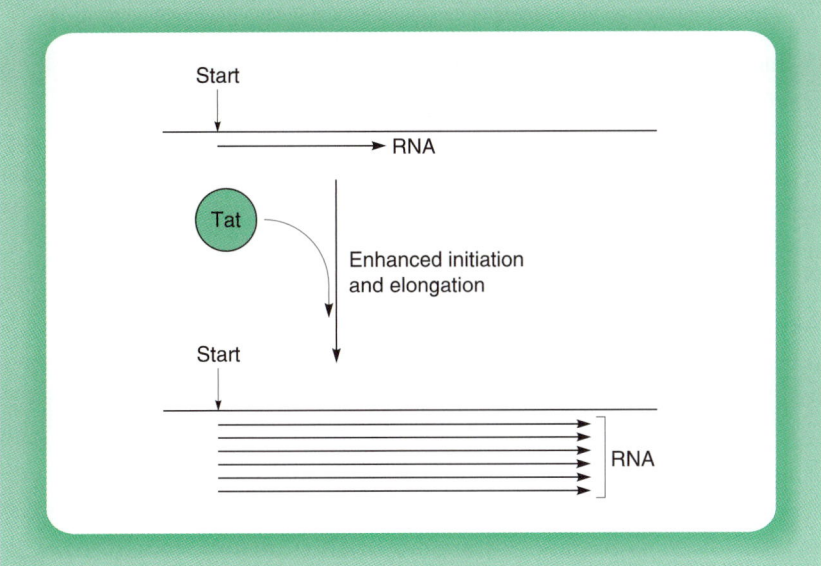

Figure 5.50
The human immunodeficiency virus Tat protein stimulates transcriptional initiation so that more RNA is made and also enhances the proportion of full length RNA species capable of encoding viral proteins.

It has been shown that Tat acts to stimulate elongation by recruiting the CDK9 kinase enzyme which phosphorylates the C-terminal domain (CTD) of RNA polymerase II (Kim *et al.*, 2002). As discussed in Chapter 3 (section 3.7), phosphorylation of this C-terminal domain on serine 2 is critical for transcriptional elongation by the polymerase. Hence, stimulation of phosphorylation in this way will enhance the production of full length transcripts (Fig. 5.51).

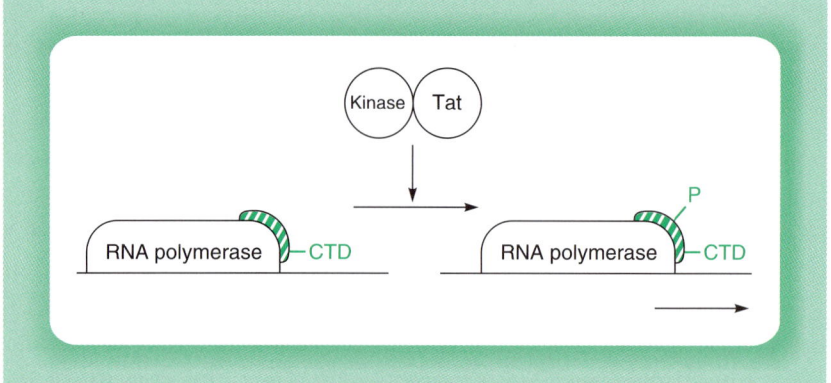

Figure 5.51
The HIV Tat protein can recruit a cellular kinase which phosphorylates the
C-terminal domain (CTD) of RNA polymerase, allowing transcription to proceed.

Such an effect on transcriptional elongation via the CTD of RNA polymerase II is not confined to the viral Tat protein but is observed also for cellular proteins. Thus, in yeast, the Fkh2p protein promotes transcriptional elongation by stimulating the phosphorylation of the CTD whilst its antagonist Fkh1p inhibits transcriptional elongation by blocking CTD phosphorylation (Morillon *et al.*, 2003).

Interestingly, as in the case of transcriptional initiation and the disruption of chromatin, there is evidence that the stimulation of transcriptional elongation involves activation domains. Thus, when binding sites for the yeast transcriptional activator GAL4 were placed upstream of the c-*myc* promoter, both the rate of initiation and the proportion of full length transcripts were greatly stimulated by the binding of hybrid transcription factors containing the DNA binding domain of GAL4 linked to an acidic or non-acidic activation domain (Yankulov *et al.*, 1994). This effect was not observed in the presence of the GAL4 DNA binding domain alone or when the GAL4 binding sites were deleted (Fig. 5.52). Hence, the ability of activating factors to act at the level of transcriptional elongation is dependent on the same activation domains which act to stimulate transcription at other stages. It is therefore possible for transcriptional activators to act by enhancing the proportion of full length transcripts which are produced, as well as by stimulating the number of transcripts which are initiated (for review see Bentley, 1995).

As well as involving activation domains, the stimulation of transcriptional elongation also involves the histone modifications which were described in Chapter 1 (section 1.2.3) and in section 5.5.2. Thus, as described in Chapter 3 (section 3.7), the initial phosphorylation of the CTD of RNA polymerase II on serine 5 is necessary for transcriptional

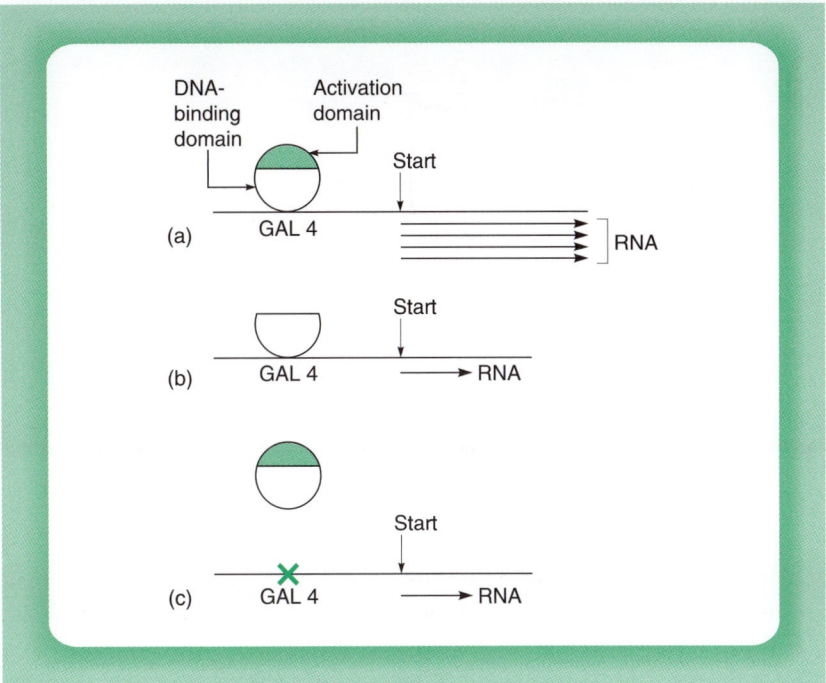

Figure 5.52

The binding of a chimaeric transcription factor containing the GAL4 DNA binding domain linked to an activation domain (shaded) stimulates both transcriptional initiation and the proportion of full length transcripts produced by a c-*myc* gene carrying a DNA binding site for GAL4 (a). This effect is not observed when only the DNA binding domain of GAL4 is present (b) or when the GAL4 binding sites are deleted so that the activator cannot bind (c).

initiation (Fig. 5.53a). However, such phosphorylation also leads to the recruitment of an enzyme complex, known as Set 1 which methylates histone H3 on lysine 4 (Fig. 5.53b). Subsequently, as described in Chapter 3 (section 3.7), serine 2 of the CTD is phosphorylated, allowing transcriptional elongation to occur (Fig. 5.53c). This results in the displacement of Set 1 by another methylation complex Set 2, which methylates H3 on lysine 36 (Fig. 5.53d). This Set 2 complex remains associated with the RNA polymerase, as it transcribes the gene and therefore methylates lysine 36 of H3 in all the nucleosomes it encounters (Fig. 5.53e).

Methylation of histone H3 at lysine 36, therefore, appears to have a key role in transcriptional elongation, as well as transcriptional initiation (for reviews see Gerber and Shilatifard, 2003; Sims *et al.*, 2003). Indeed, a similar association with transcriptional elongation has also been demonstrated for the modification of histone H2B by addition of the small protein ubiquitin, demonstrating that this link occurs for different histones and different modifications (Xiao *et al.*, 2005; Pavri *et al.*, 2006).

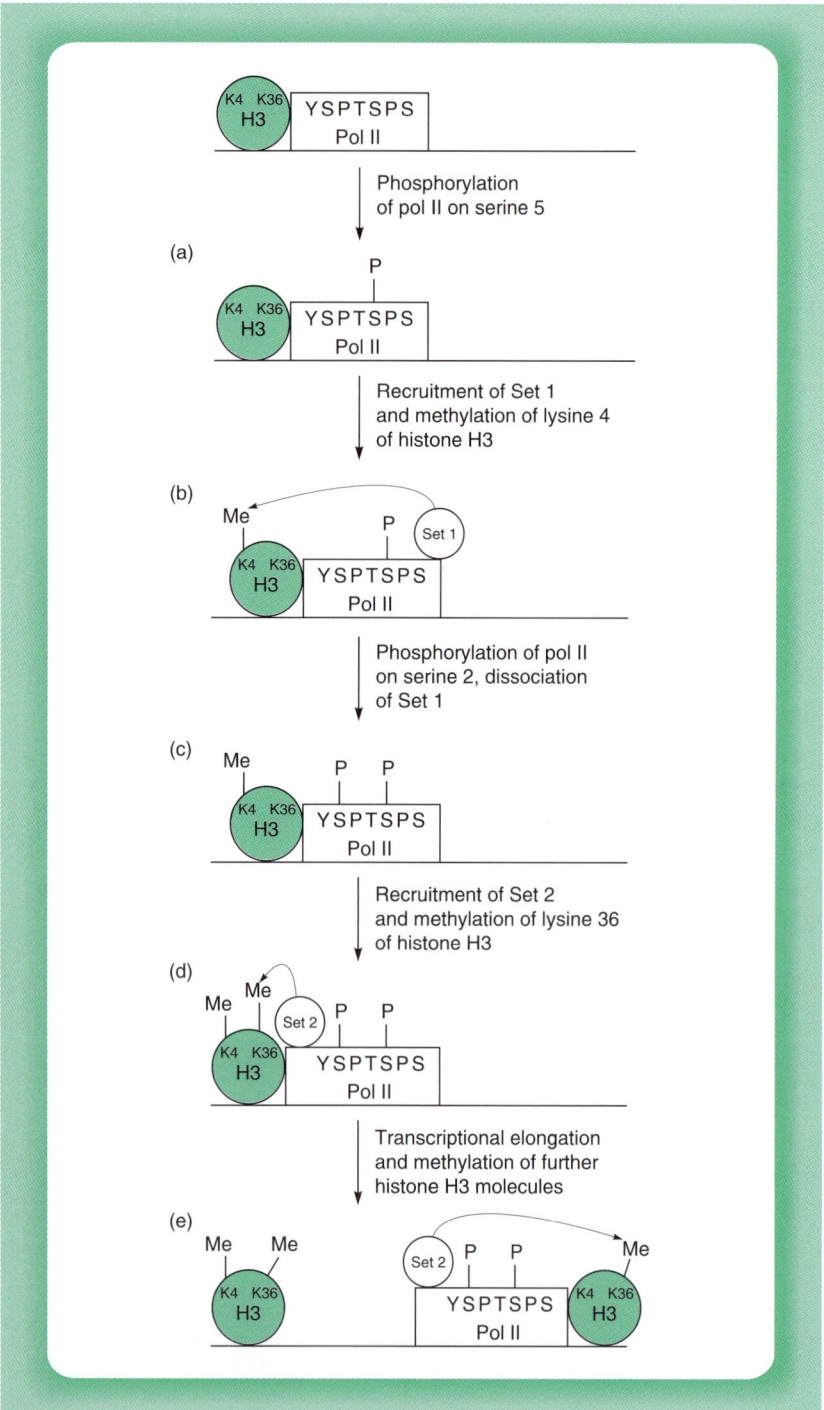

Figure 5.53

Histone methylation and transcriptional elongation. Phosphorylation of RNA polymerase II on serine 5 of its C-terminal domain (CTD) (panel a) stimulates recruitment of the Set 1 complex which methylates histone H3 on lysine 4 (K4)

Hence, transcriptional elongation is a process which can clearly be regulated by mechanisms that have common features with the regulation of transcriptional initiation and of chromatin structure, which were discussed earlier, such as the involvement of activation domains and of histone modifications. It is likely that such regulation of transcriptional elongation is of importance in a variety of organisms. Thus, inactivation of the Elongin A, transcriptional elongation factor in the fruit fly *Drosophila* blocks larval metamorphosis (Gerber *et al.*, 2004) whilst mutation of the Foggy protein, which as discussed in Chapter 6 (section 6.5) blocks transcriptional elongation in nematode worms, results in a failure to produce dopamine synthesizing neurones (Guo *et al.*, 2000).

5.7 CONCLUSIONS

It is clear therefore that activation of gene expression by transcription factors can occur at three distinct stages to stimulate transcription. Thus, activating factors can disrupt the chromatin structure to allow other activating factors to bind, stimulate the rate of transcriptional initiation so that more RNA transcripts are initiated and can stimulate transcriptional elongation (Fig. 5.54).

As described in section 5.5.3, these processes can be combined together in different ways, for example, in genes modulated by GAGA/HSF and the glucocorticoid receptor/NFI. An interesting example of this is provided by the activation of the interferon-β (IFN-β) gene by viral infection. Thus, as noted in Chapter 1 (section 1.3.6), the enhancer of the IFN-β gene binds a multi-protein enhanceosome complex. The binding of this complex (which includes the DNA binding protein HMGI(Y) and transcriptional activators such as NFκB) to the enhancer is the first stage in the activation of the IFN-β gene (for reviews see Cosma, 2002; Fry and Peterson, 2002) (Fig. 5.55a).

This enhanceosome complex then stimulates the recruitment of a histone acetyltransferase complex which acetylates nucleosomes (Fig. 5.55b), allowing the subsequent recruitment of the SWI/SNF complex (Fig. 5.55c). Further chromatin remodelling by this complex then displaces

Figure 5.53 (Continued)
(panel b). Subsequently, the CTD of polymerase II is phosphorylated on serine 2 (panel c). This results in the displacement of Set 1 and the recruitment of another complex, Set 2, which methylates histone H3 on lysine 39 (K9) (panel d). The Set 2 complex remains associated with the RNA polymerase, as it transcribes the DNA and methylates lysine 39 on downstream histone H3 molecules (panel e). Compare with Figure 3.13, which shows the same process but illustrates the enzymes which phosphorylate RNA polymerase II.

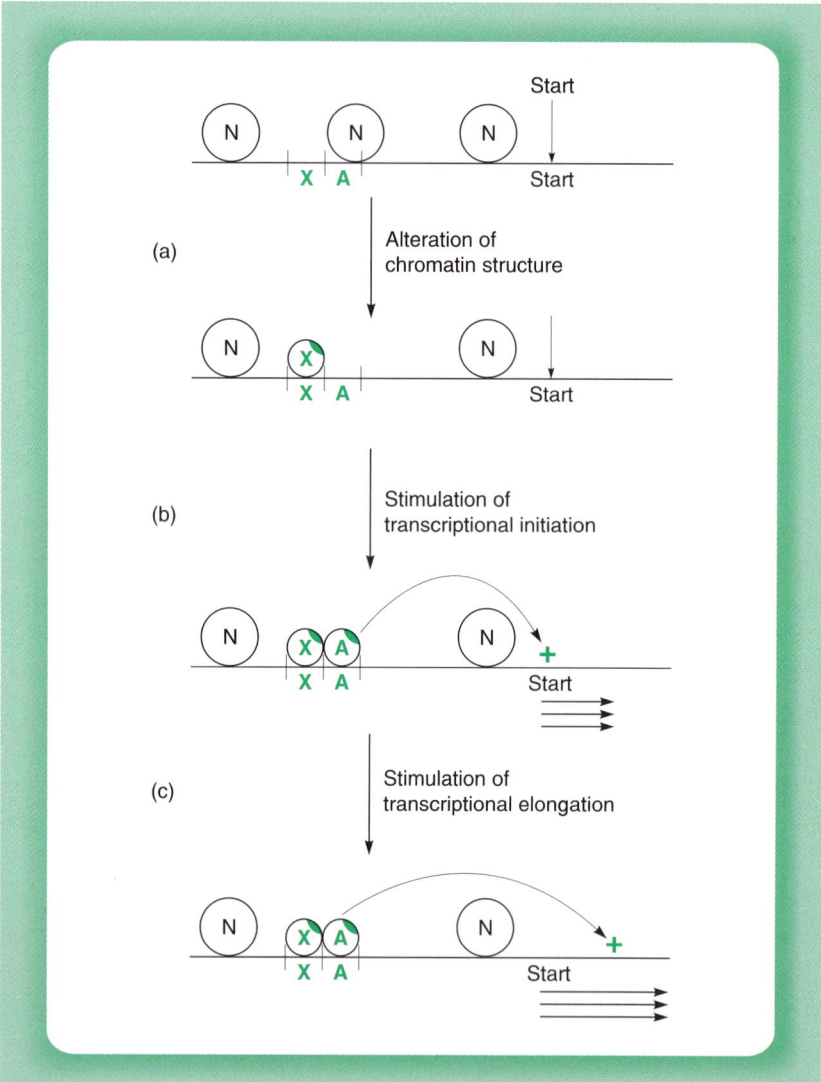

Figure 5.54

An activation domain (hatched) within an activating transcription factor can act (a) by disrupting the nucleosome (N) arrangement of the DNA so that an activating factor can bind; (b) by stimulating the rate of transcriptional initiation; and (c) by increasing the proportion of initiated transcripts which go on to produce a full length RNA.

a nucleosome exposing the core promoter, allowing the basal transcriptional complex to bind and transcription begins (Fig. 5.55d).

Interestingly, simple displacement of the core promoter nucleosome by artificial means is not sufficient to induce transcriptional activation, indicating that this multi-step process is required for correct transcriptional regulation. Similarly, distinct histone acetylation events mediate subsequent stages of activation on this promoter. Thus, acetylation of the lysine

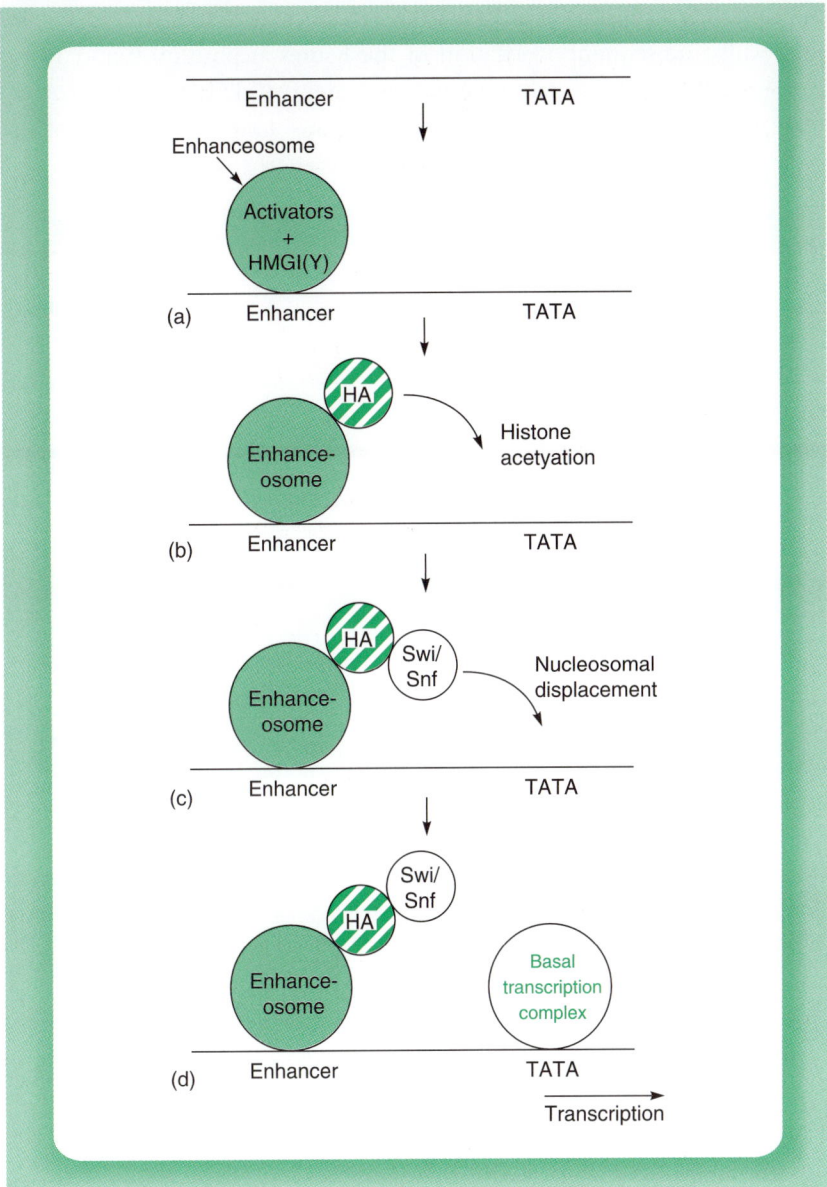

Figure 5.55

Multi-step activation of the interferon β promoter by viral infection. Binding of the enhanceosome multi-protein complex to the enhancer (panel a) is followed by successive recruitment of a histone acetylase (HA) (panel b), the SWI/SNF chromatin remodelling complex (panel c) and finally the recruitment of the basal transcriptional complex leading to transcriptional activation (panel d).

at position 8 on histone H4 allows the recruitment of the SWI/SNF complex whilst subsequent acetylation of the lysines at position 9 and 14 on histone H3 allows the subsequent recruitment of TFIID (Agalioti *et al.*, 2002) (Fig. 5.56). This case therefore illustrates how distinct parts of the

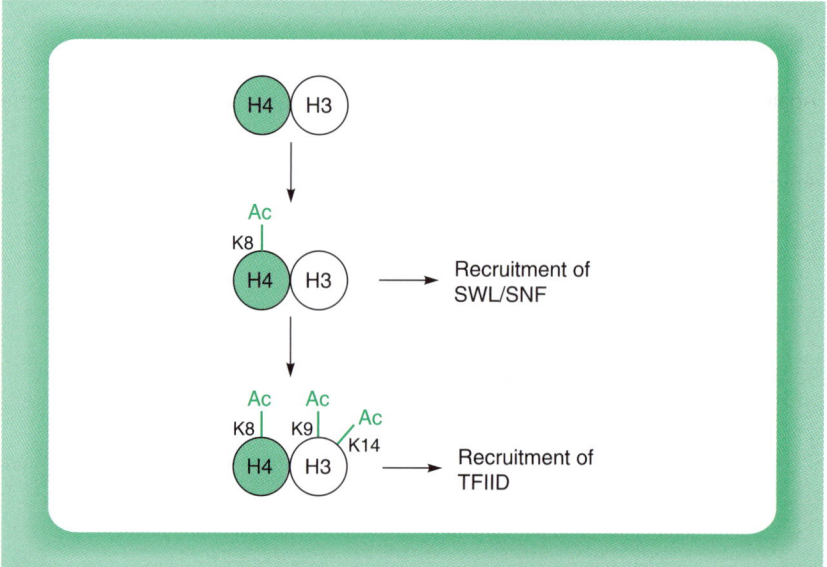

Figure 5.56

On the β-interferon gene promoter, acetylation (Ac) of histone H4 on the lysine (K) at position 8 facilitates recruitment of the chromatin remodelling complex SWI/SNF. Subsequent acetylation of histone H3 on the lysines at positions 9 and 14 facilitates recruitment of the basal transcription factor TFIID.

'histone code' (see Chapter 1, section 1.2.3) can mediate the recruitment of different activating factors, as well as how the two processes of histone modification and ATP-dependent chromatin remodelling, discussed in Chapter 1 (section 1.2) and in section 5.5 of this chapter, interact with one another to produce transcriptional activation.

Hence, the ordered recruitment of chromatin remodelling factors, histone modifiers, transcriptional activators and the basal transcriptional complex is likely to be crucial for the correct regulation of a variety of different genes, although the order in which these factors are recruited is likely to vary for different genes (for review see Cosma, 2002; Fry and Peterson, 2002).

A variety of means are therefore used to activate gene transcription involving modulation of transcriptional initiation, transcriptional elongation and chromatin structure with a number of factors including components of the basal transcriptional complex, co-activators, the mediator complex and chromatin remodelling factors being targeted by transcriptional activators. Taken together these effects allow transcriptional activators to fulfil their function and strongly stimulate transcription.

REFERENCES

Agalioti, T., Chen, G. and Thanos, D. (2002) Deciphering the transcriptional histone acetylation code for a human gene. Cell 111, 381–392.

Arndt, K.M. and Kane, C.M. (2003) Running with RNA polymerase: eukaryotic transcript elongation. Trends in Genetics 19, 543–550.

Barberis, A., Pearlberg, J., Simkovich, N., Farrell, S., Reinagel, P., Bamdad, C., Sigal, G. and Ptashne, M. (1995) Contact with a component of the polymerase II holoenzyme suffices for gene activation. Cell 81, 359–368.

Beato, M. and Eisfeld, K. (1997) Transcription factor access to chromatin. Nucleic Acids Research 25, 3559–3563.

Becker, P.B. (2006) A finger on the mark. Nature 442, 31–32.

Belden, W.J., Loros, J.J. and Dunlap, J.C. (2006) CLOCK leaves its mark on histones. Trends in Biochemical Sciences 31, 610–613.

Bentley, D. (1995) Regulation of transcriptional elongation by RNA polymerase II. Current Opinion in Genetics and Development 5, 210–216.

Brown, C.E., Lechner, T., Howe, L. and Workman, J.L. (2000) The many HATs of transcription co-activators. Trends in Biochemical Sciences 25, 15–19.

Buratowski, S. (1995) Mechanisms of gene activation. Science 270, 1773–1774.

Cantin, G.T., Stevens, J.L. and Berk, A.J. (2003) Activation domain-mediator interactions promote transcription preinitiation complex assembly on promoter DNA. Proceedings of the National Academy of Sciences, USA 100, 12003–12008.

Carey, M. (2005) Chromatin marks and machines, the missing nucleosome is a theme: gene regulation up and downstream. Molecular Cell 17, 323–330.

Carey, M. (1998) The enhanceosome and transcriptional synergy. Cell 92, 5–8.

Chadick, J.Z. and Asturias, F.J. (2005) Structure of eukaryotic Mediator complexes. Trends in Biochemical Sciences 30, 264–271.

Chang, M. and Jaehning, J.A. (1997) A multiplicity of mediators: alternative forms of transcription complexes communicate with transcriptional regulators. Nucleic Acids Research 25, 4861–4865.

Cheng, J.X., Gandolfi, M. and Ptashne, M. (2004) Activation of the Gal1 gene of yeast by pairs of 'non-classical' activators. Current Biology 14, 1675–1679.

Choy, B. and Green, M.R. (1993) Eukaryotic activators function during multiple steps of pre-initiation complex assembly. Nature 366, 531–536.

Chrivia, J.C., Kwok, R.P., Lamb, N., Hagiwara, M., Montminy, M.R. and Goodman, R.H. (1993) Phosphorylated CREB binds specifically to the nuclear protein CBP. Nature 365, 855–859.

Conaway, J.W., Shilatifard, A., Dvir, A. and Conaway, R.C. (2000) Control of elongation by RNA polymerase II. Trends in Biochemical Sciences 25, 375–380.

Conaway, R.C., Sato, S., Tomomori-Sato, C., Yao, T. and Conaway, J.W. (2005) The mammalian Mediator complex and its role in transcriptional regulation. Trends in Biochemical Sciences 30, 250–255.

Cosma, M.P. (2002) Ordered recruitment: gene-specific mechanism of transcription activation. Molecular Cell 10, 227–236.

Courey, A.J. and Tjian, R. (1988) Analysis of Sp1 *in vivo* reveals multiple transcriptional domains including a novel glutamine-rich activation motif. Cell 55, 887–898.

Davie, J.K. and Dent, S.Y. (2006) No Spt6, no nucleosomes, no activator required. Molecular Cell 21, 452–453.

De Cesare, D. and Sassone-Corsi, P. (2000) Transcriptional regulation by cyclic AMP-responsive factors. Progress in Nucleic Acids Research and Molecular Biology 64, 343–369.

Dikstein, R., Zhou, S. and Tjian, R. (1996) Human $TAF_{II}105$, is a cell type-specific TFIID subunit related to $TAF_{II}130$. Cell 87, 137–146.

Doi, M., Hirayama, J. and Sassone-Corsi, P. (2006) Circadian regulator CLOCK is a histone acetyltransferase. Cell 125, 497–508.

Ferreira, M.E., Hermann, S., Prochasson, P., Workman, J.L., Berndt, K.D. and Wright, A.P. (2005) Mechanism of transcription factor recruitment by acidic activators. Journal of Biological Chemistry 280, 21779–21784.

Fishburn, J., Mohibullah, N. and Hahn, S. (2005) Function of a eukaryotic transcription activator during the transcription cycle. Molecular Cell 18, 369–378.

Fry, C.J. and Peterson, C.L. (2002) Unlocking the gates to gene expression. Science 295, 1847–1848.

Gerber, H-P., Seipel, K., Georgiev, O., Hofferer, M., Hug, M., Rusioni, S. and Schaffner, W. (1994) Transcriptional activation modulated by homopolymeric glutamine and proline residues. Science 263, 808–811.

Gerber, M. and Shilatifard, A. (2003) Transcriptional elongation by RNA polymerase II and histone methylation. Journal of Biological Chemistry 278, 26303–26306.

Gerber, M., Eissenberg, J.C., Kong, S., Tenney, K., Conaway, J.W., Conaway, R.C. and Shilatifard, A. (2004) *In vivo* requirement of the RNA polymerase II elongation factor elongin A for proper gene expression and development. Molecular and Cellular Biology 24, 9911–9919.

Giordano, A. and Avantaggiati, M.L. (1999) p300 and CBP: partners for life and death. Journal of Cellular Physiology 181, 218–230.

Gonzalez-Couto, E.K., Klages, N. and Strubin, M. (1997) Synergistic and promoter-selective activation of transcription by recruitment of transcription factors TFIID and TFIIB. Proceedings of the National Academy of Sciences, USA 94, 8036–8041.

Goodman, R.H. and Smolik, S. (2000) CBP/p300 in cell growth, transformation and development. Genes and Development 14, 1553–1577.

Grant, P.A., Sterner, D.E., Duggan, L.J., Workman, J.L. and Berger, S.L. (1998) The SAGA unfolds: convergence of transcription regulators in chromatin-modifying complexes. Trends in Cell Biology 8, 193–197.

Green, M.R. (2000) TBP-associated factors (TAF$_{II}$s): multiple, selective transcriptional mediators in common complexes. Trends in Biochemical Sciences 25, 59–63.

Green, M.R. (2005) Eukaryotic transcription activation: right on target. Molecular Cell 18, 399–402.

Greenblatt, J., Nodwell, J.R. and Mason, S.W. (1993) Transcriptional anti-termination. Nature 364, 401–406.

Guarente, L. (1988) USAs and enhancers: common mechanism of transcriptional activation in yeast and mammals. Cell 52, 303–305.

Gui, C-Y. and Dean, A. (2003) A major role for the TATA box in recruitment of chromatin modifying complexes to a globin gene promoter. Proceedings of the National Academy of Sciences, USA 100, 7009–7014.

Guo, S., Yamaguchi, Y., Schillbach, S., Wada, T., Lee, J., Goddard, A., French, D., Handa, H. and Rosenthal, A. (2000) A regulator of transcriptional elongation controls vertebrate neuronal development. Nature 408, 366–369.

Hahn, S. (1998) The role of TAFs in RNA polymerase II transcription. Cell 95, 579–582.

Hahn, S. (1993a) Structure (?) and function of acidic transcription activators. Cell 72, 481–483.

Hahn, S. (1993b) Efficiency in activation. Nature 363, 672–673.

Haince, J.F., Rouleau, M. and Poirier, G.G. (2006) Gene expression needs a break to unwind before carrying on. Science 312, 1752–1753.

Hall, D.B. and Struhl, K. (2002) The VP16 activation domain interacts with multiple transcriptional components as determined by protein-protein cross-linking in vivo. Journal of Biological Chemistry 277, 46043–46050.

Hampsey, M. (1997) A SAGA of histone acetylation and gene expression. Trends in Genetics 13, 427–429.

Hardin, P.E. and Yu, W. (2006) Circadian transcription: passing the HAT to CLOCK. Cell 125, 424–426.

Hollenberg, S.M. and Evans, R.M. (1988) Multiple and cooperative trans-activation domains of the human glucocorticoid receptor. Cell 55, 899–906.

Hope, I.A. and Struhl, K. (1986) Functional dissection of a eukaryotic transcriptional activator, GCN4 of yeast. Cell 46, 885–894.

Horikoshi, M., Carey, M.F., Kakidani, H. and Roeder, R.G. (1988) Mechanism of action of a yeast activator: direct effect of GAL4 derivatives on mammalian TFIID promoter interactions. Cell 54, 665–669.

Inoue, H., Furukawa, T., Giannakopoulos, S., Zhou, S., King, D. S. and Tanese, N. (2002) Largest subunits of the human SWI/SNF chromatin-remodelling complex promote transcriptional activation by steroid hormone receptors. Journal of Biological Chemistry 277, 41674–41685.

Jenster, G., Spencer, T.H., Burcin, M., Tsai, S.Y., Tsai, M-J. and O'Malley, B.W. (1997) Steroid receptor induction of gene transcription: a two step model. Proceedings of the National Academy of Sciences, USA 94, 7879–7884.

Joliot, V., Demma, M. and Prywes, R. (1995) Interaction with RAP74 subunit of TFIIF is required for transcriptional activation by serum response factor. Nature 373, 632–635.

Jones, K. (1997) Taking a new TAK on Tat transactivation. Genes and Development 11, 2593–2599.

Ju, B.G., Lunyak, V.V., Perissi, V., Garcia-Bassets, I., Rose, D.W., Glass, C.K. and Rosenfeld, M.G. (2006) A topoisomerase IIbeta-mediated dsDNA break required for regulated transcription. Science 312, 1798–1802.

Kim, Y.K., Bourgeois, C.F., Isel, C., Churcher, M.J. and Karn, J. (2002) Phosphorylation of the RNA polymerase II carboxyl-terminal domain by CDK9 is directly responsible for human immunodeficiency virus type 1 Tat-activated transcriptional elongation. Molecular and Cellular Biology 22, 4622–4637.

King, I.F.G. and Kingston, R.E. (2001) Specifying transcription. Nature 414, 858–861.

Kinzler, M., Braus, G.H., Georgiev, O., Seipel, K. and Schaffner, W. (1994) Functional differences between mammalian transcriptional activation domains at the yeast GAL1 promoter. EMBO Journal 13, 641–645.

Kornberg, R.D. (2005) Mediator and the mechanism of transcriptional activation. Trends in Biochemical Sciences 30, 235–239.

Kuras, L., Kosa, P., Mencia, M. and Struhl, K. (2000) TAF-containing and TAF-independent forms of transcriptionally active TBP in vivo. Science 288, 1244–1248.

Latchman, D.S. (2005) Gene regulation–a eukaryotic perspective. Fifth edition. Taylor and Francis, Oxford.

Lehmann, M. (2004) Anything else but GAGA: a nonhistone protein complex reshapes chromatin structure. Trends in Genetics 20, 15–22.

Li, X-Y., Virbasius, A., Zhu, X. and Green, M.R. (1999) Enhancement of TBP binding by activators and general transcription factors. Nature 399, 605–609.

Lis, J. and Wu, C. (1993) Protein traffic on the heat shock promoter: parking, stalling and trucking along. Cell 74, 1–4.

Lis, J.T. and Kraus, W.L. (2006) Promoter cleavage: a TopoIIbeta and PARP-1 collaboration. Cell 125, 1225–1227.

Lohr, D. (1997) Nucleosome transactions on the promoters of the yeast GAL and PHO genes. Journal of Biological Chemistry 272, 26795–26798.

Lonard, D.M. and O'Malley, B.W. (2006) The expanding cosmos of nuclear receptor coactivators. Cell 125, 411–414.

Loyola, A. and Almouzni, G. (2004) Bromodomains in living cells participate in deciphering the histone code. Trends in Cell Biology 14, 279–281.

Malik, S. and Roeder, R.G. (2005) Dynamic regulation of pol II transcription by the mammalian Mediator complex. Trends in Biochemical Sciences 30, 256–263.

Marx, J. (2006) Protein tail modification opens way for gene activity. Science 311, 757.

Mayr, B. and Montminy, M. (2001) Transcriptional regulation by the phosphorylation-dependent factor CREB. Nature Reviews Molecular Cell Biology 2, 599–609.

Mellor, J. (2006) Dynamic nucleosomes and gene transcription. Trends in Genetics 22, 320–329.

Mermod, N., O'Neil, E.A., Kelley, T.J. and Tjian, R. (1989) The proline-rich transcriptional activator of CTF/NF-1 is distinct from the replication and DNA binding domain. Cell 58, 741–753.

Mitchell, P.J. and Tjian, R. (1989) Transcriptional regulation in mammalian cells by sequence specific DNA binding proteins. Science 245, 371–378.

Morillon, A., O'Sullivan, J., Azad, A., Proudfoot, N. and Mellor, J. (2003) Regulation of elongating RNA polymerase II by forkhead transcription factors in yeast. Science 300, 492–495.

Nagy, L. and Schwabe, J.W. (2004) Mechanism of the nuclear receptor molecular switch. Trends in Biochemical Sciences 29, 317–324.

Ogryzko, V.V., Schiltz, L., Russanova, V., Howard, B.H. and Nakatani, Y. (1996) The transcriptional coactivators p300 and CBP are histone acetyltransferases. Cell 87, 953–959.

O'Malley, B.W. (2006) Little molecules with big goals. Science 313, 1749–1750.

Ozer, J., Moore, P.A., Bolden, A.H., Lee, A., Rosen, C.A. and Lieberman, P.M. (1994) Molecular cloning of the small (α) sub-unit of human TFIIA reveals functions critical for activated transcription. Genes and Development 8, 2324.

Pavri, R., Zhu, B., Li, G., Trojer, P., Mandal, S., Shilatifard, A. and Reinberg, D. (2006) Histone H2B monoubiquitination functions cooperatively with FACT to regulate elongation by RNA polymerase II. Cell 125: 703–717.

Ptashne, M. (1988) How eukaryotic transcriptional activators work. Nature 335, 683–689.

Ptashne, M. and Gann, A. (1997) Transcriptional activation by recruitment. Nature 386, 569–577.

Radhakrishnan, I., Perez-Alvardo, G.C., Parker, D., Dyson, H.J., Montminy, M.R. and Wright, P. (1997) Solution structure of the KIX domain of CBP bound to the transactivation domain of CREB a model for activator: co-activator interactions. Cell 91, 741–752.

Roberts, S.G.E. and Green, M.R. (1994) Activator-induced conformational change in general transcription factor TFIIB. Nature 371, 717–720.

Rosenfeld, M.G., Lunyak, V.V. and Glass, C.K. (2006) Sensors and signals: a coactivator/corepressor/epigenetic code for integrating signal-dependent programs of transcriptional response. Genes and Development 20, 1405–1428.

Ruthenburg, A.J., Allis, C.D. and Wysocka, J. (2007) Methylation of lysine 4 on histone h3: intricacy of writing and reading a single epigenetic mark. Molecular Cell 25, 15–30.

Seipel, K., Georgiev, O. and Schaffner, W. (1992) Different activation domains stimulate transcription from the remote ('enhancer') and proximal ('promoter') positions. EMBO Journal 11, 4961–4968.

Shen, W-C. and Green, M.R. (1997) Yeast TAF$_{II}$145 functions as a core promoter selectivity factor, not a general co-activator. Cell 90, 615–624.

Shi, Y. and Whetstine, J.R. (2007) Dynamic regulation of histone lysine methylation by demethylases. Molecular Cell 25, 1–14.

Shikama, N., Lyon, J. and La Thangue, N.B. (1997) The p300/CBP family: integrating signals with transcription factors and chromatin. Trends in Cell Biology 7, 230–236.

Shogren-Knaak, M., Ishii, H., Sun, J.M., Pazin, M.J., Davie, J.R. and Peterson, C.L. (2006) Histone H4-K16 acetylation controls chromatin structure and protein interactions. Science 311, 844–847.

Simon, J.A. and Tamkun, J.W. (2002) Programming off and on states in chromatin: mechanisms of polycomb and trithorax group complexes. Current Opinion in Genetics and Development 12, 210–218.

Sims, R.J., III and Reinberg, D. (2006) Histone H3 Lys 4 methylation: caught in a bind? Genes and Development 20, 2779–2786.

Sims, R.J., III, Belotserkovskaya, R. and Reinberg, D. (2004) Elongation by RNA polymerase II: the short and long of it. Genes and Development 18, 2437–2468.

Sims, R.J., III, Nishioka, K. and Reinberg, D. (2003) Histone lysine methylation: a signature for chromatin function. Trends in Genetics 19, 629–639.

Spencer, C.A. and Groudine, M. (1990) Transcriptional elongation and eukaryotic gene regulation. Oncogene 5, 777–785.

Struhl, K. (2005) Transcriptional activation: mediator can act after preinitiation complex formation. Molecular Cell 17, 752–754.

Struhl, K. and Moqtaderi, Z. (1998) The TAFs in the HAT. Cell 94, 1–4.

Svaren, J. and Horz, W. (1997) Transcription factors vs nucleosomes: regulation of the PH05 promoter in yeast. Trends in Biochemical Sciences 22, 93–97.

Takagi, Y. and Kornberg, R.D. (2006) Mediator as a general transcription factor. Journal of Biological Chemistry 281, 80–89.

Triezenberg, S.J. (1995) Structure and function of transcriptional activation domains. Current Opinion in Genetics and Development 5, 190–196.

Tsukiyama, T., Becker, P.B. and Wu, C. (1994) ATP-dependant nucleosome disruption at a heat-shock promoter mediated by DNA binding of GAGA transcription factor. Nature 367, 525–532.

Uesugi, M., Nyanguile, O., La, H., Levine, A.J. and Verdine, G.L. (1997) Induced α-helix in the VP16 activation domain upon binding to a human TAF. Science 277, 1310–1313.

Verrijzer, C.P. (2001) Transcription factor IID – not so basal after all. Science 293, 2010–2011.

Wang, E.H. and Tjian, R. (1994) Promoter selective defect in cell cycle mutant ts13 rescued by hTAF$_{II}$250 in vitro. Science 263, 811–814.

Whetstine, J.R., Nottke, A., Lan, F., Huarte, M., Smolikov, S., Chen, Z., Spooner, E., Li, E., Zhang, G., Colaiacovo, M. and Shi, Y. (2006) Reversal of histone lysine trimethylation by the JMJD2 family of histone demethylases. Cell 125, 467–481.

Wilkins, R.C. and Lis, J.T. (1997) Dynamics of potentiation and activation: GAGA factor and its role in heat shock gene regulation. Nucleic Acids Research 25, 3963–3968.

Wong, J., Shi, Y-B. and Wolffe, A.P. (1997) Determinants of chromatin disruption and transcriptional regulation instigated by the thyroid hormone receptor: hormone regulated chromatin disruption is not sufficient for transcriptional activation. EMBO Journal 16, 3158–3171.

Wu, C. (1985) An exonuclease protection assay reveals heat shock element and TATA-box binding proteins in crude nuclear extracts. Nature 317, 84–87.

Wysocka, J. and Herr, W. (2003) The herpes simplex virus VP16-induced complex: the makings of a regulatory switch. Trends in Biochemical Sciences 28, 294–304.

Xiao, H., Pearson, A., Coulombe, B., Truant, R., Zhang, S., Regier, J.L., Triezenberg, S.J., Reinberg, D., Flores, O., Ingles, C.J. and Greenblatt, J. (1994) Binding of basal transcription factor TFIIH to the acidic activation domains of VP16 and p53. Molecular and Cellular Biology 14, 7013–7024.

Xiao, T., Kao, C.F., Krogan, N.J., Sun, Z.W., Greenblatt, J.F., Osley, M.A. and Strahl, B.D. (2005) Histone H2B ubiquitylation is associated with elongating RNA polymerase II. Molecular and Cellular Biology 25, 637–651.

Yamane, K., Toumazou, C., Tsukada, Y., Erdjument-Bromage, H., Tempst, P., Wong, J. and Zhang, Y. (2006) JHDM2A, a JmjC-containing H3K9 demethylase, facilitates transcription activation by androgen receptor. Cell 125, 483–495.

Yankulov, K., Blau, J., Purton, T., Roberts, S. and Bentley, D.L. (1994) Transcriptional elongation by RNA polymerase II is stimulated by transactivators. Cell 77, 748–759.

CHAPTER 6

REPRESSION OF GENE EXPRESSION BY TRANSCRIPTION FACTORS

6.1 REPRESSION OF TRANSCRIPTION

Although many transcription factors act in a positive manner, a number of cases have now been reported in which a transcription factor exerts an inhibitory effect on transcription initiation. Indeed, many key differentiation events are controlled in this manner. Thus, in many animal systems, the differentiation of mature nerve cells involves the loss of expression of the neuron-restrictive silencing factor (NRSF) which is expressed in non-neuronal cells and represses the expression of neural specific genes (for review see Lunyak and Rosenfeld, 2005) (Fig. 6.1a). Similarly, the differentiation of male germ cells in plants involves the loss of the germ-line-restrictive silencing factor (GRSF) which is expressed in germ cell precursors and stably represses the genes that are required for germ cell differentiation (Haerizadeh *et al.*, 2006) (Fig. 6.1b).

The inhibition of transcription by transcription factors can involve a variety of different mechanisms. This effect can occur by indirect repression in which the repressor interferes with the action of an activating factor, so preventing it stimulating transcription (Fig. 6.2a–d). Alternatively, it can occur via direct repression in which the factor reduces the activity of the basal transcriptional complex (Fig. 6.2e). These two mechanisms will be discussed in sections 6.2 and 6.3, respectively. As with transcriptional activation, transcriptional repression can also occur via the alteration of chromatin structure or at the level of transcriptional elongation and these effects are discussed in sections 6.4 and 6.5, respectively (for reviews of transcriptional repression see Hanna-Rose and Hansen, 1996; Latchman, 1996; Maldonado *et al.* 1999; Courey and Jia, 2001).

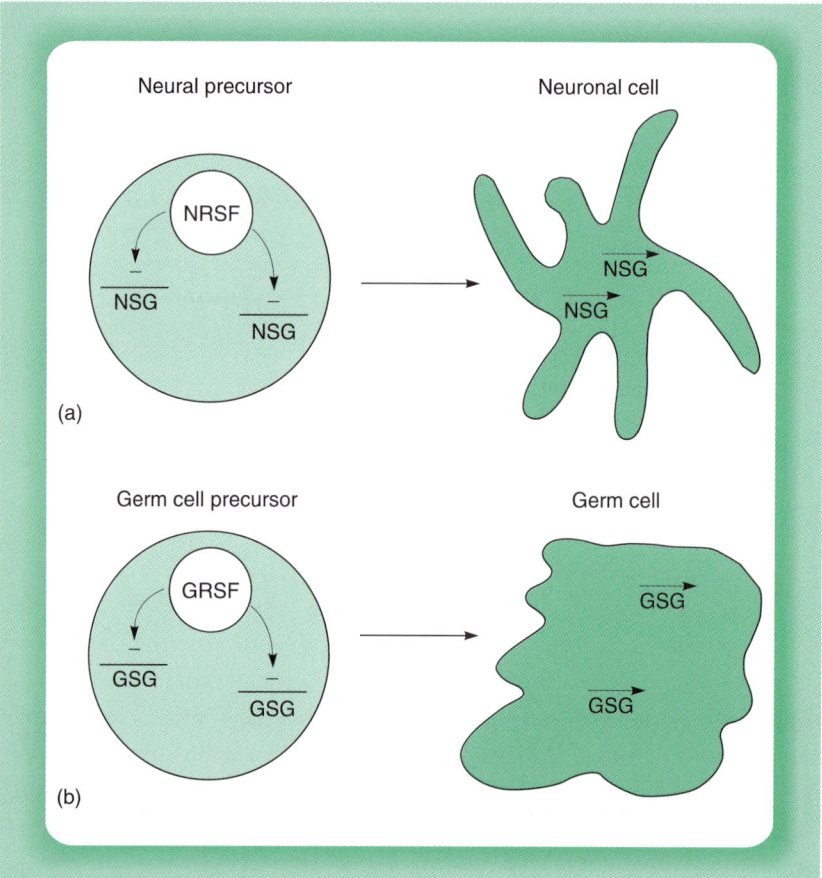

Figure 6.1

Key role for negatively acting transcription factors in specific differentiation events. Panel (a): In animals, neuronal differentiation involves the loss of the NRSF factor which inhibits the expression of neuronal specific genes in neuronal precursors and non-neuronal cells. Panel (b): In plants, the GRSF factor similarly represses the expression of male germ cell specific genes (GSG) in germ cell precursors.

6.2 INDIRECT REPRESSION

Several mechanisms exist by which an inhibitor can interfere with the action of an activator and these will be discussed in turn.

6.2.1 INHIBITION OF ACTIVATOR BINDING BY MASKING OF ITS DNA BINDING SITE

One means by which repression can occur is by the masking of the DNA binding site for the factor so preventing it binding to the DNA and

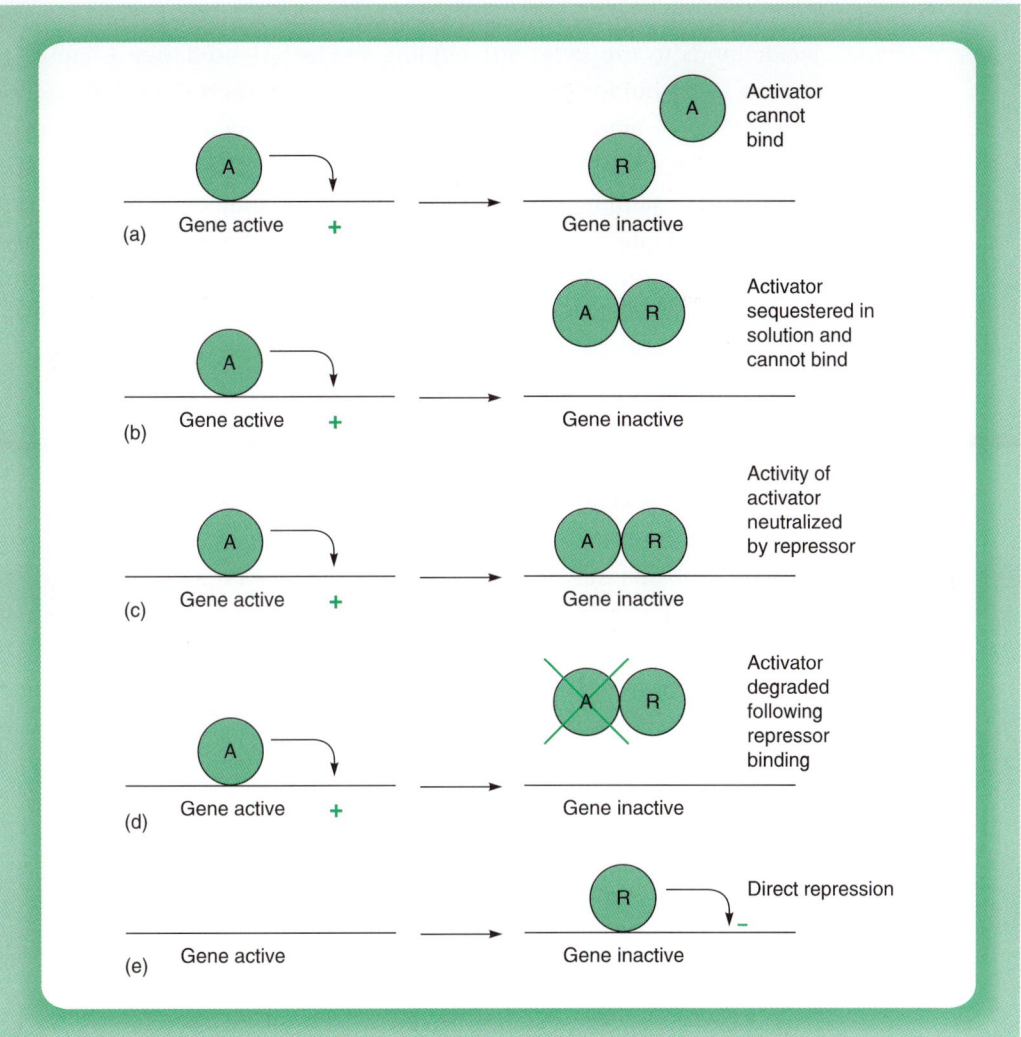

Figure 6.2

Potential mechanisms by which a transcription factor can repress gene expression. This can occur by the repressor (R) binding to DNA and preventing an activator (A) from binding and activating gene expression (a), by the repressor interacting with the activator in solution and preventing its DNA binding (b), by the repressor binding to DNA with the activator and neutralizing its ability to activate gene expression (c), by the repressor promoting degradation of the activator (d) or by direct repression by an inhibitory transcription factor (e).

activating transcription. By preventing the binding of the positively acting factor, the negatively acting factor effectively inhibits gene activation.
This masking of the binding site can be achieved simply by the negatively
acting factor binding to the same site as the positively acting factor but
failing to activate transcription (Fig. 6.2a). This is seen, for example, in

the case of the Sp3 factor, a factor related to the Sp1 factor. Thus the Sp3 factor binds to the same Sp1 binding site as Sp1 itself (see Chapter 1, section 1.3.2) but unlike Sp1 it cannot activate transcription. It therefore blocks the Sp1 binding site, preventing Sp1 binding and activating transcription (for review see Lania *et al.*, 1997).

A similar example is seen in the case of the homeobox proteins, discussed in Chapter 4 (section 4.2). The homeobox proteins engrailed (eng), Fushi-tarazu (Ftz), paired (prd) and zerknult (zen) can all bind to the sequence TCAATTAAAT. When plasmids expressing each of these genes are co-transfected with a target promoter carrying multiple copies of this binding site, the Ftz, prd and zen proteins can activate transcription of the target promoter (Han *et al.*, 1989). In contrast, the eng protein has no effect on the transcription of such a promoter. It does, however, interfere with the ability of the activating proteins to induce transcription, presumably by blocking the binding of the activating factor. Thus, for example, whilst Ftz can stimulate the target promoter when co-transfected with it, it cannot do so in the presence of eng since eng prevents binding of Ftz to its binding site (Jaynes and O'Farrell, 1988). Hence, the expression of Ftz alone in a cell would activate particular genes whereas its expression in a cell also expressing engrailed would not have any effect (Fig. 6.3).

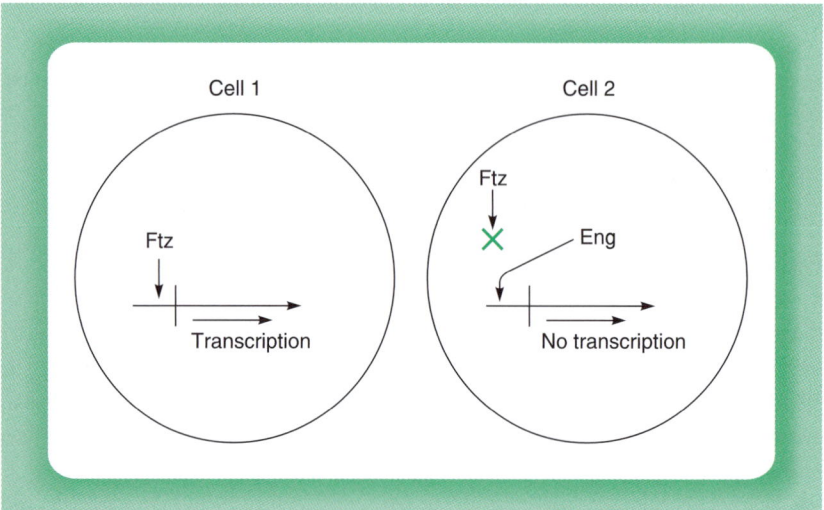

Figure 6.3
Blockage of gene induction by Ftz in cells expressing the engrailed (Eng) protein which binds to the same sequence as Ftz but does not activate transcription.

Hence, both Sp3 and eng act purely as transcriptional repressors by blocking binding of activators to their binding sites. Interestingly, the glucocorticoid receptor can also repress transcription of specific genes in this way, even though, as discussed in Chapter 4 (section 4.4) it acts primarily as an activator of target gene expression. Thus, treatment with glucocorticoid and activation of the glucocorticoid receptor inhibits expression of the genes encoding bovine prolactin and human pro-opiomelanocortin. The inhibitory effect observed in these cases is mediated by binding to DNA of the identical receptor/hormone complex which activates glucocorticoid-inducible genes (see Chapter 4, section 4.4 and Chapter 8, section 8.2.2) and these genes are therefore repressed by glucocorticoid. However, the DNA sequence element to which the complex binds when mediating its negative effect (nGRE) is distinct from the glucocorticoid response element (GRE) to which it binds when inducing gene expression, although the two are related (Fig. 6.4).

| Binding site for positive regulation | **R G R A C A N N N T G T Y C Y** |
| Binding site for negative regulation | **A T Y A C A N N N T G A T C W** |

Figure 6.4

Relationship of the sites in DNA which mediate gene activation or repression by binding the glucocorticoid receptor. Note that the sites are related but distinct.

This has led to the suggestion that the sequence difference causes the receptor/hormone complex to bind to the nGRE in a configuration in which its activation domain cannot interact with other transcription factors to activate transcription, as occurs following binding to the positive element (Fig. 6.5). In agreement with this idea, the glucocorticoid receptor has been shown to bind to the nGRE in the POMC gene as a trimer rather than the dimer form which binds to the GRE and stimulates transcription (for review see Latchman, 2001). The receptor bound in this configuration to the negative element apparently acts by preventing binding of a positive acting factor to this or an adjacent site thereby preventing gene induction. In agreement with this idea, the nGRE in the human glycoprotein hormone alpha subunit gene which overlaps a cyclic AMP response element (CRE) is only able to inhibit gene expression when the CRE is left intact. Hence, it is likely that receptor bound at the negative element prevents binding of a transcriptional activator to the CRE and thereby inhibits gene expression (Fig. 6.6). Interestingly, the glucocorticoid receptor can also inhibit gene expression at the level of transcriptional elongation and this is discussed in section 6.5.

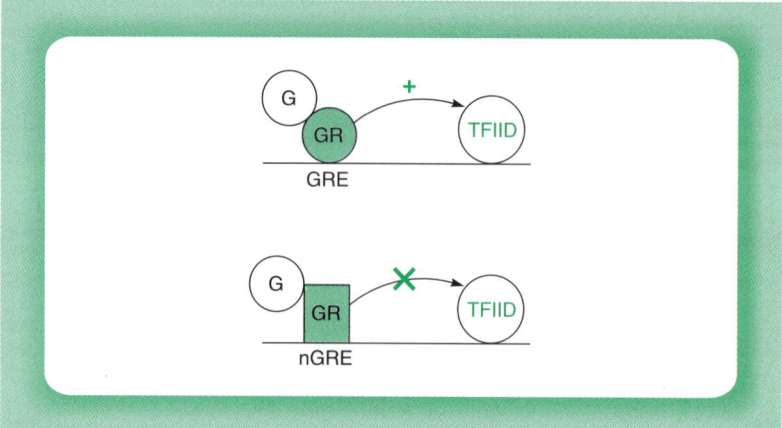

Figure 6.5

Consequences of glucocorticoid receptor binding to the DNA binding sites which mediate gene activation (GRE) or repression (nGRE). Note that the receptor is likely to bind in a different configuration to the two different sequences resulting in its ability to activate transcription only following binding to the GRE.

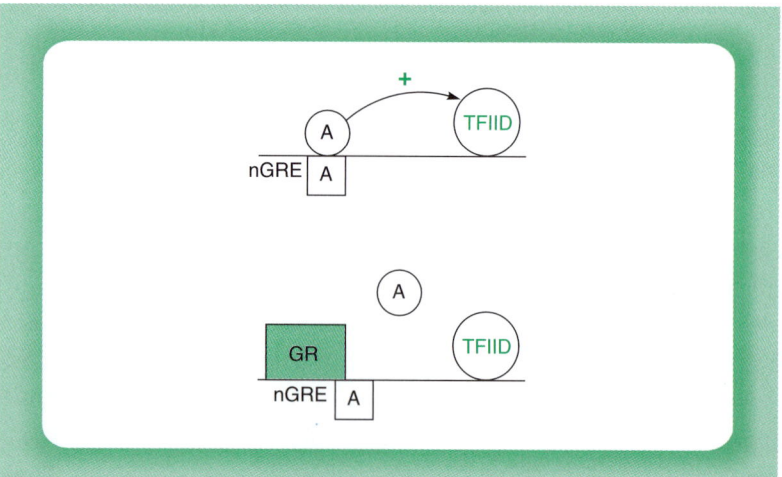

Figure 6.6

Inhibition of gene expression by glucocorticoid receptor binding to an nGRE is likely to be mediated by preventing the binding of a positively acting activator protein (A) to a site adjacent to or overlapping the nGRE.

Hence, the inhibition of DNA binding by a specific activator via masking of its binding site is a major method of transcriptional inhibition. This can involve either factors such as Sp3 or eng which function only as transcriptional inhibitors, as well as factors such as the glucocorticoid receptor which either repress via this mechanism or activate transcription depending on the nature of their DNA binding site in a specific target gene.

6.2.2 INHIBITION OF ACTIVATOR BINDING BY FORMATION OF A NON-DNA BINDING COMPLEX

As well as preventing activator binding to DNA via masking its binding site, an inhibitor can also inhibit transcription via the formation of a non-DNA binding complex with an activating factor (Fig. 6.2b). Thus, as discussed in Chapter 7 (section 7.2.1), the MyoD transcription factor is specifically expressed in skeletal muscle cells and plays a key role in activating skeletal muscle genes. Its expression can be induced in the 10T½ fibroblast cell line by treatment with 5-azacytidine and this converts the fibroblast cells into muscle cell precursors, known as myoblasts. However, activation of muscle-specific genes and the production of differentiated myotubes requires these cells to be incubated in the absence of serum (Fig. 6.7).

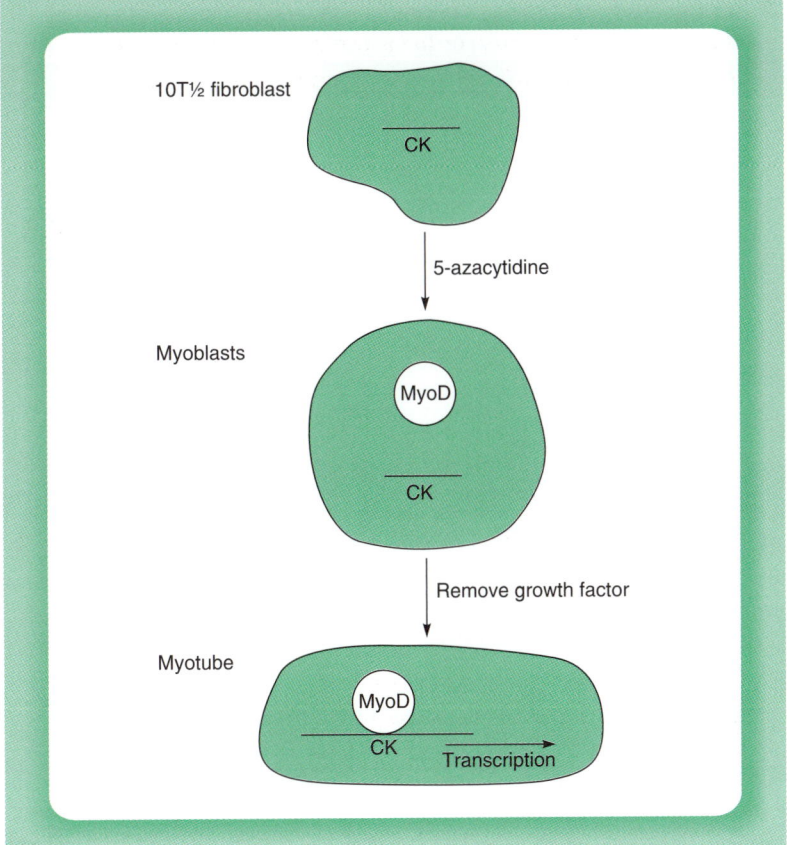

Figure 6.7

Differentiation of 10T½ cells into myoblasts by 5-azacytidine and then into myotubes by removal of growth factors. Note that the MyoD-dependent induction of genes encoding terminal differentiation markers such as creatine kinase (CK), which occurs in myotubes, occurs without an increase in MyoD concentration.

Paradoxically, however, MyoD levels do not change in this transition from myoblast to myotubes and yet MyoD-dependent muscle-specific genes are activated.

The explanation of this paradox was provided by the identification of the Id protein (Benezra *et al.*, 1990), which like MyoD contains a helix-loop-helix motif but lacks the basic domain mediating DNA binding (see Chapter 4, section 4.5 for a discussion of these motifs). Because the helix-loop-helix motif mediates dimerization of proteins containing it, Id can dimerize with other helix-loop-helix proteins, such as MyoD and inhibit their DNA binding since the resulting heterodimer lacks the necessary pair of DNA binding motifs (Fig. 6.8). When 10T½-derived myoblasts are induced to form myotubes, Id levels decline, indicating that this second stage of myogenesis is mediated by a decline in the inhibitory protein rather than an increase in the activator, MyoD.

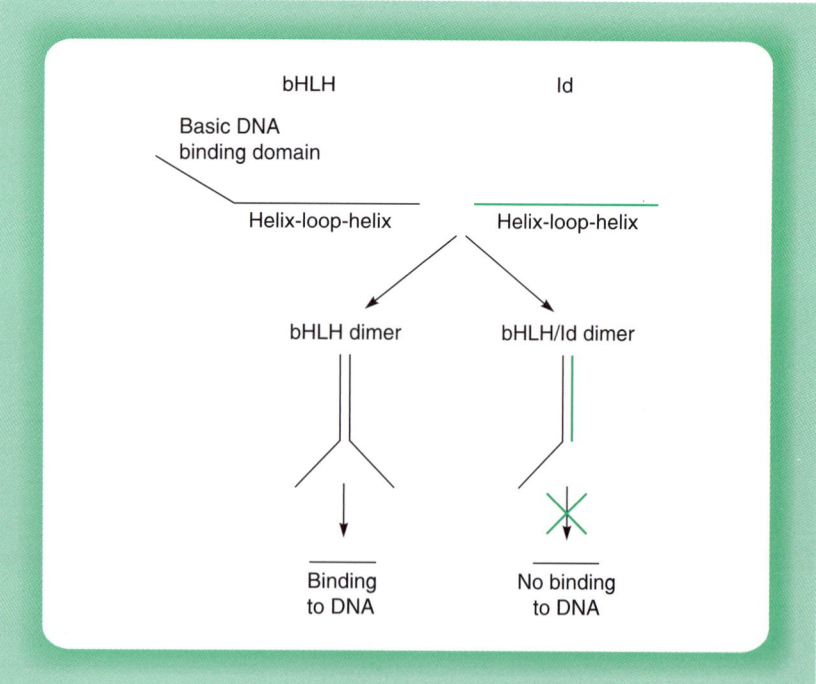

Figure 6.8

Dimerization of functional basic helix-loop-helix proteins (bHLH) with Id. Note that whilst Id can dimerize with other proteins via the helix-loop-helix domain, it lacks the basic DNA domain and hence the Id-containing heterodimer cannot bind to DNA.

The role of inhibitory helix-loop-helix proteins is not confined to myogenesis. Thus, the Id2 protein, which is another member of the Id family, binds to the E12 and E47 members of the basic helix-loop-helix family (see Chapter 4, section 4.5). This binding of Id2 blocks DNA binding by E12/E47 and thereby blocks the activation of genes which are essential for the differentiation of neuronal cells and which require E12/E47 for their expression. Moreover, inactivation of Id proteins in knock out mice results in abnormal development of the brain and the blood vessels, providing direct evidence for the key role of proteins of this type in development (Lyden *et al.*, 1999; for review see Carmeliet, 1999). Similarly, the product of the *emc* gene, which regulates neurogenesis in *Drosophila*, also contains a helix-loop-helix motif and lacks a basic DNA binding domain. Hence, this form of repression is not confined to mammalian cells.

6.2.3 QUENCHING OF AN ACTIVATOR

The cases of repression described so far all involve the inhibition of DNA binding either by blocking the binding site for a factor (Fig. 6.2a) or by forming a non-DNA binding protein–protein complex (Fig. 6.2b). Since DNA binding is a prerequisite for gene activation, this constitutes an effective form of repression. In addition, however, inhibition of transcription can also be achieved by interfering with transcriptional activation by a DNA-bound factor in a phenomenon known as quenching (Fig. 6.2c).

This effect can occur via an inhibitory factor, binding to a DNA-bound factor and masking its activation domain. A case of this type, involving repression of E2F by Rb-1 which involves, at least in part, masking of the activation domain is discussed in Chapter 9 (section 9.4.3). Similarly, the MDM4 protein binds to the p53 anti-oncogenic, transcription factor and inhibits its activity via this quenching mechanism (for review see Toledo and Wahl, 2006) (see Chapter 9, section 9.4.2 for further discussion of p53).

A related example of quenching in which the inhibitory factor binds to a DNA sequence adjacent to the quenched factor, rather than only to the factor itself, is seen in the case of the c-*myc* promoter. Thus, an inhibitory transcription factor myc-PRF binds to a site adjacent to that occupied by an activating factor myc-CF1 and interferes with its ability to activate c-*myc* gene transcription (Kakkis *et al.*, 1989). Hence quenching can occur either by an inhibitory factor binding to the positively acting factor (Fig. 6.9a) or by the inhibitory factor binding to DNA adjacent to the positive factor (Fig. 6.9b). In both cases, however, this effect involves the inhibitor interfering with the ability of the activator's activation domain to stimulate transcription.

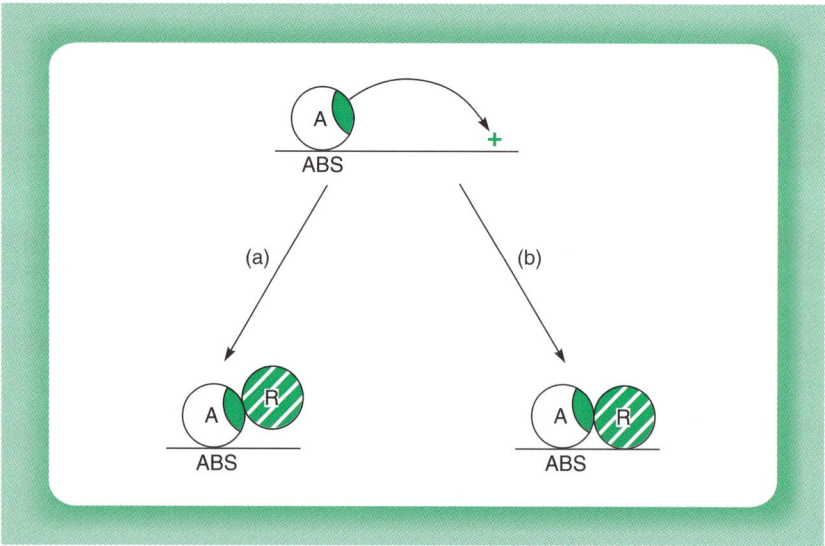

Figure 6.9

The ability of a bound activator (A) to stimulate transcription via its activation domain (hatched) can be inhibited by quenching of the activation domain by inhibitory factors (R) which either bind to the activator without binding to DNA (a) or which bind to a site adjacent to the activator (b).

6.2.4 DEGRADATION OF AN ACTIVATOR

As well as interfering functionally with the action of the activator by preventing its DNA binding or quenching its activation domain, an indirect repressor can act by targeting an activator for degradation (Fig. 6.2d). This is seen in the case of MDM2, which targets the p53 anti-oncoprotein for degradation (see Chapter 9, section 9.4.2).

In the case of p53, MDM2 is likely to induce degradation of p53 by stimulating the recognition of p53 by protease enzymes which are present in the cell (Fig. 6.10a). However, the AEBP1 transcription factor, which regulates adipocyte differentiation, actually itself has the ability to degrade other proteins and it is therefore likely to act directly by degrading activators to which it binds (Fig. 6.10b). Hence this factor combines the ability to bind to DNA with the ability to degrade other factors with which it comes into contact (He *et al.*, 1995). A similar mechanism is also used to maintain the undifferentiated state of embryonic stem cells by rapidly degrading any initiation complexes of RNA polymerase II and associated factors which bind to genes that should only be transcribed in differentiated cells (for review see Zwaka, 2006).

Thus, the degradation of activators, or of the initiation complex itself, appears to be an important mechanism of transcriptional repression.

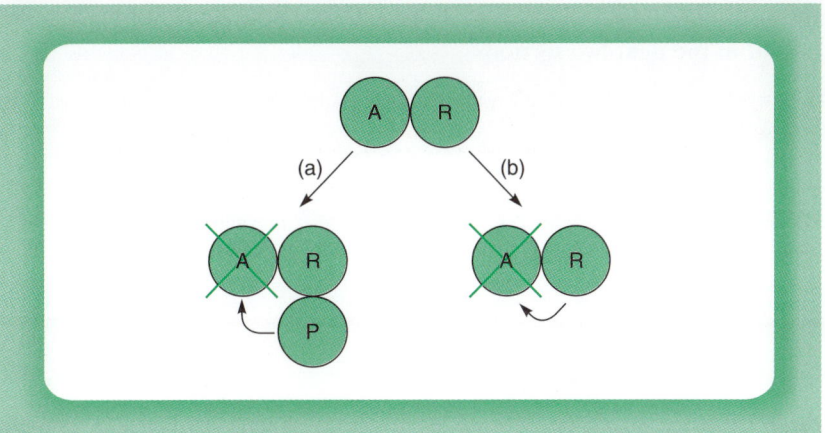

Figure 6.10

A repressor (R) can promote the degradation of an activator (A) either (a) indirectly by making it a target for a protease (P) or (b) directly by itself degrading the activator.

Interestingly, however, it is also possible for the reverse to occur, with a factor activating transcription because it directs the degradation of a repressor. Thus, in *Drosophila* the PHYL and SINA factors cause the degradation of the TTK88 transcription repressor and thereby produce activation of transcription (Li *et al.*, 1997; Tang *et al.*, 1997).

6.3 DIRECT REPRESSION

6.3.1 MECHANISMS OF TRANSCRIPTIONAL REPRESSION

In the cases described so far, a negative factor exerts its effect by neutralizing the action of a positively acting factor by preventing either its DNA binding (Figs. 6.2a and b) or inhibiting its activation of transcription following such binding (Fig. 6.2c) or promoting its degradation (Fig. 6.2d). In other cases, however, the inhibitory effect of a particular factor can be observed in the absence of any activating factors. This indicates that these inhibitory factors inhibit transcription directly by interacting with the basal transcriptional complex to reduce its activity (Fig. 6.2e). Several factors of this type have now been shown to bind to specific DNA binding sites within their target genes and reduce the activity of the basal transcriptional complex (Fig. 6.11a). This effect is evidently similar in nature but opposite in effect to the stimulation of the basal complex by the binding of activating factors to specific DNA sequences in the promoter. Alternatively, a direct repressor may actually join the basal transcriptional complex via a protein–protein interaction, without binding to DNA and then inhibit transcription (Fig. 6.11b). These two mechanisms are discussed in the next two sections.

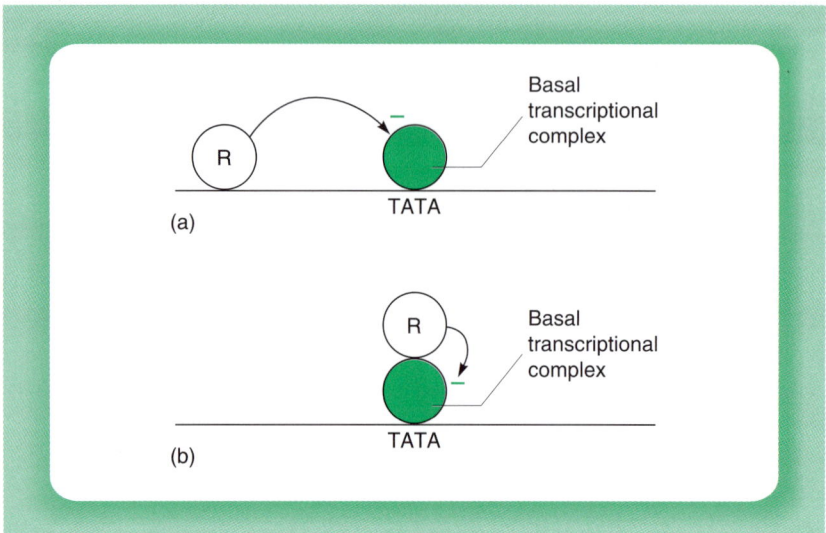

Figure 6.11

An inhibitory factor (R) can reduce the activity of the basal transcriptional complex either by binding to DNA and then interacting with the complex (a) or by binding directly to the complex by protein–protein interaction (b).

6.3.2 DIRECT REPRESSION BY DNA BINDING TRANSCRIPTION FACTORS

One factor capable of inhibiting the basal initiation complex following binding to the DNA is the *Drosophila* eve protein which is a member of the homeobox family discussed in Chapter 4 (section 4.2) and can act, for example, to repress the gene encoding the Ubx protein which is also a member of the homeobox family.

Thus, if the Ubx promoter linked to a marker gene is added to a suitable cell free extract, transcription of the marker gene driven by the promoter can be observed. Addition of the purified protein even-skipped (eve) to this extract inhibits Ubx promoter activity, however, and this inhibition is dependent upon binding sites for the eve protein within the Ubx promoter. Such findings parallel the ability of a vector expressing eve to repress the Ubx promoter following co-transfection into cultured cells and the genetic evidence which originally led to the definition of eve as a repressor of Ubx (Fig. 6.12). This case thus represents an interesting example of the transcription of the gene encoding one homeobox transcription factor (Ubx) being repressed by another (eve).

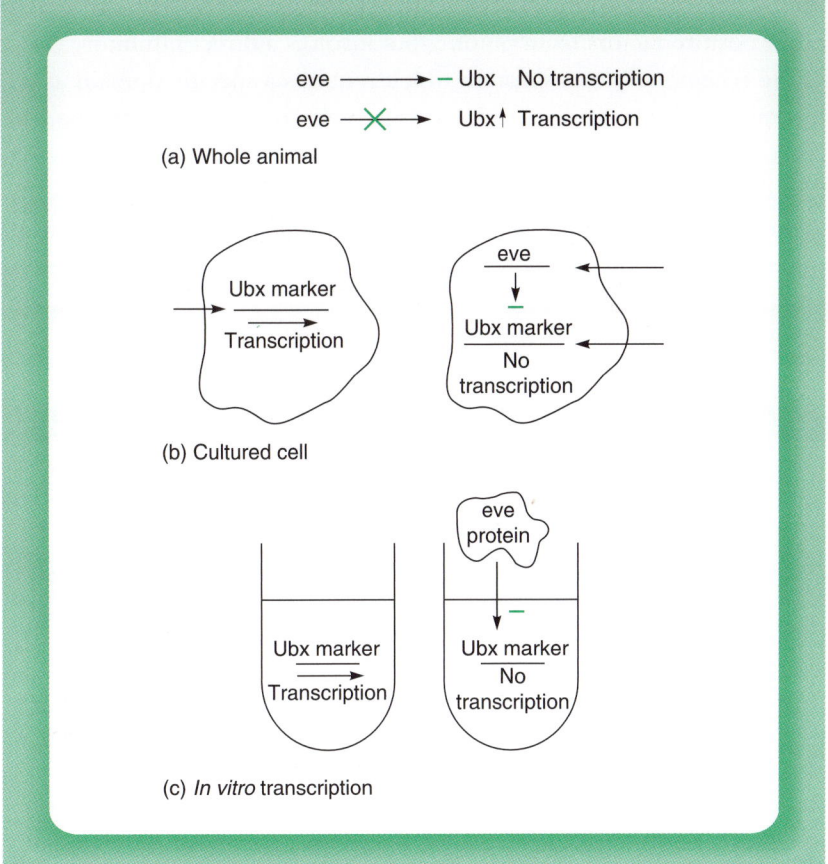

Figure 6.12

Inhibitory effect of the eve protein on expression of the Ubx gene. This inhibitory effect can be observed in the whole animal where mutation of the eve gene enhances Ubx expression (top panel); in cultured cells where introduction of a plasmid expressing the eve gene represses a co-transfected Ubx promoter driving a marker gene (middle panel) and in a test tube *in vitro* transcription system where addition of purified eve protein represses transcription of a marker gene driven by the Ubx promoter (bottom panel).

Interestingly, the number of such directly inhibitory factors is growing steadily and now includes some which were previously thought to function only in an indirect manner. Thus, the Rb-1 protein, which was originally thought to function solely by masking the activation domain of E2F, is now known to also act as a direct repressor (see Chapter 9, section 9.4.3).

An interesting example of a directly acting repressor is provided by the thyroid hormone receptor which is a member of the nuclear receptor family discussed in Chapter 4 (section 4.4). Thus, this receptor can bind to its response element (TRE) in the absence of thyroid hormone and inhibit gene expression. This effect is not due to the receptor preventing other positive factors from binding but involves a direct inhibitory effect of the receptor on transcription which requires a specific domain at the C-terminus of the molecule. In the presence of thyroid hormone, the receptor undergoes a conformational change which exposes its activation domain and converts it from a repressor to an activator (Fig. 6.13). Hence, in this case, gene activation or repression can be mediated from the same DNA binding site with the effect depending on the presence or absence of the hormone.

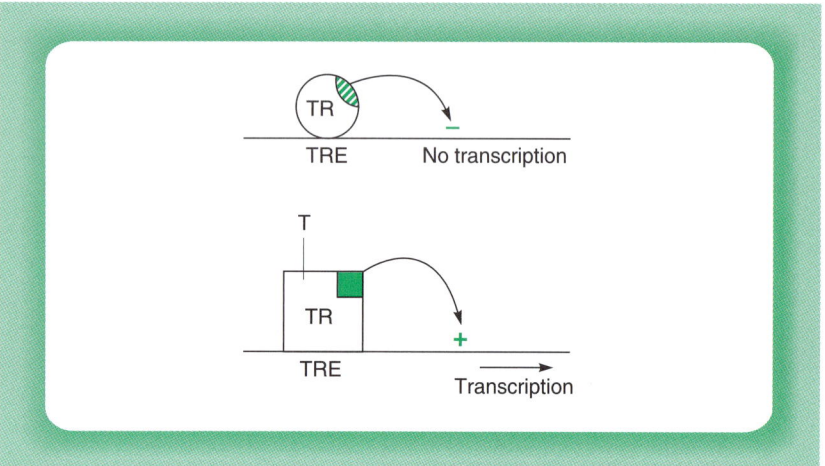

Figure 6.13

In the absence of thyroid hormone (T) the thyroid hormone receptor inhibits gene expression via a discreet inhibitory domain (hatched box). Binding of thyroid hormone (T) exposes the activation domain of the receptor (solid box) and allows it to activate transcription.

Although the thyroid hormone receptor can therefore act as either a transcriptional activator or repressor, the mechanism differs from that observed with the glucocorticoid receptor (which is also a member of the nuclear receptor family) and, as discussed in section 6.2.1, can also act as either an activator or a repressor. Thus, in the case of the glucocorticoid receptor, both activation or repression are dependent upon activation of the receptor by glucocorticoid and it is the nature of the binding site which determines whether activation or repression is observed. Moreover, repression is indirect, being achieved by preventing an activator from binding. In contrast, in the case of the thyroid receptor, the activation/repression decision is controlled by thyroid hormone and inhibition of gene expression involves direct repression (for review see Latchman, 2001).

Interestingly, in addition to the thyroid binding form of the thyroid hormone receptor, alternative splicing generates another form (alpha 2) lacking a part of the hormone binding domain and therefore unable to bind hormone (Koenig *et al.*, 1989; Fig. 6.14a). Both the alpha 2 form and the hormone binding alpha 1 form can bind to DNA. However, binding of alpha 2 to the thyroid response element (TRE) sequence prevents binding of alpha 1 and thereby prevents gene induction in response to thyroid hormone (Fig. 6.14b). As discussed in Chapter 9 (section 9.3.2), a similar non-hormone binding form of the thyroid hormone receptor is encoded by the v-*erbA* oncogene, which produces cancer by inhibiting the expression of thyroid hormone responsive genes involved in erythroid differentiation.

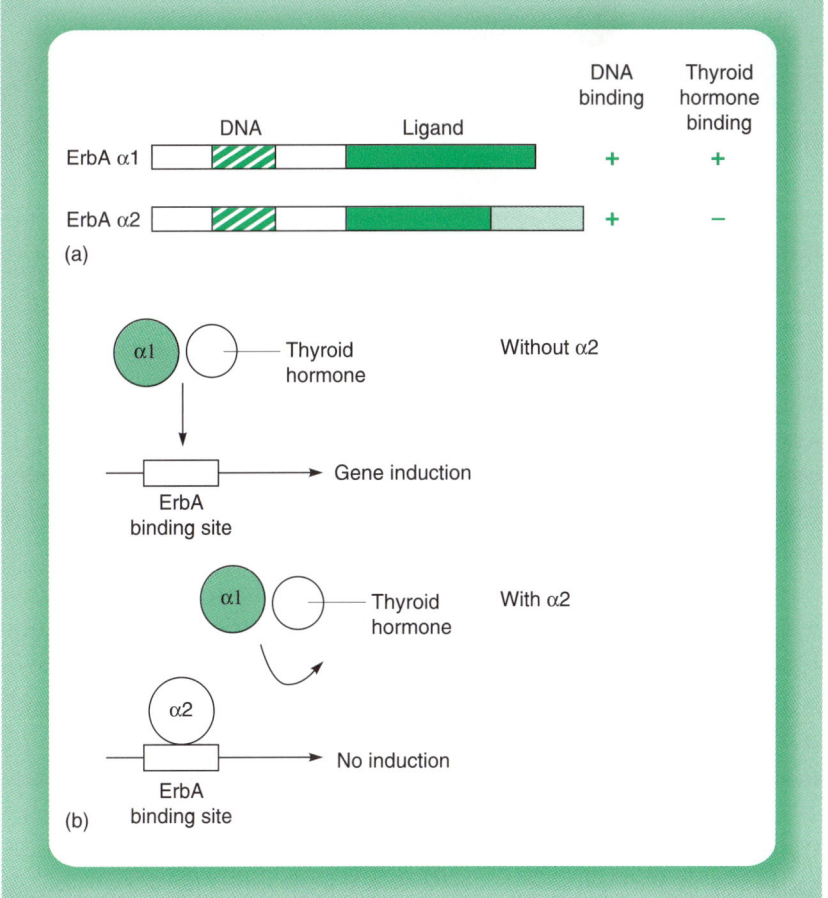

Figure 6.14

(a) Relationship of the ErbA alpha 1 and alpha 2 proteins. Note that only the alpha 1 protein has a functional thyroid hormone binding domain. (b) Inhibition of ErbA alpha 1 binding and of gene activation in the presence of the alpha 2 protein.

Inhibitory factors of the directly acting type such as the thyroid hormone receptor or the eve factor generally contain a small domain which can confer the ability to repress gene expression upon the DNA binding domain of another factor when the two are artificially linked (see for example, Han and Manley, 1993; Lillycrop *et al.*, 1994) (Fig. 6.15). Hence, these directly repressing factors contain specific inhibitory domains paralleling the existence of specific activation domains in activating transcription factors.

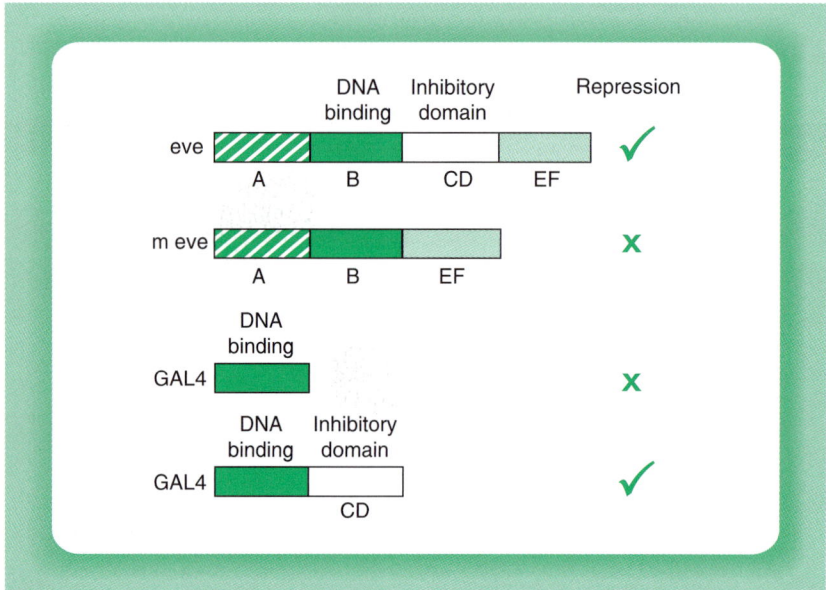

Figure 6.15

A specific region (CD) of the eve factor which is distinct from the DNA binding domain (B) acts as a transferable inhibitory domain. Thus its deletion from the eve protein results in a loss of the ability to repress transcription whilst its linkage to the DNA binding domain of GAL4 generates a functional repressor.

Interestingly, the inhibitory domain in the human Wilms tumour anti-oncogene product and those from several *Drosophila* inhibitory factors, including eve, appear to share the common features of proline richness and an absence of charged residues (Han and Manley, 1993) suggesting that these factors have a common inhibitory domain. However, other inhibitory domains such as those in the mammalian factors Oct-2 (Lillycrop *et al.*, 1994) and E4BP4 (Cowell and Hurst, 1994) are distinct both from this common domain and from each other, indicating that, as with activation domains, several types of inhibitory domain may exist.

By analogy with activation domains, inhibitory domains are likely to inhibit either the assembly of the basal transcriptional complex or reduce its activity and/or stability after it has assembled. In agreement with this

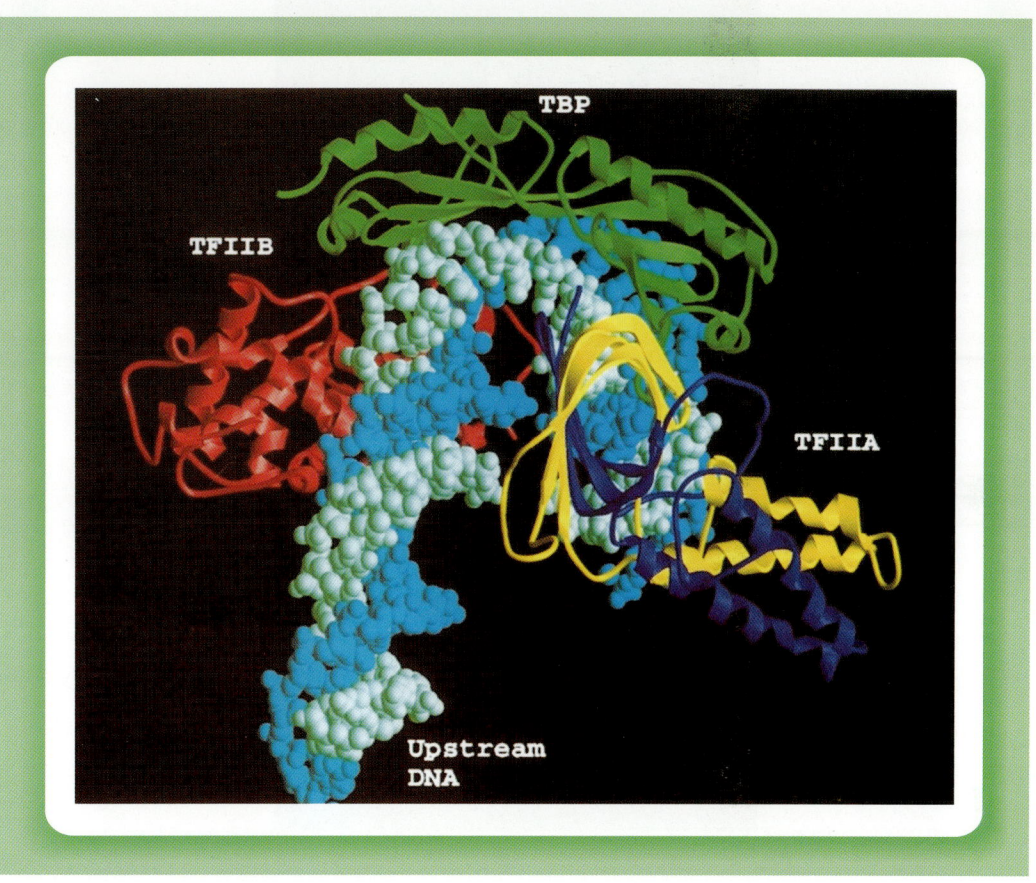

Plate 1 Schematic diagram illustrating the structure of the TFIIB/TBP/TFIIA complex bound to DNA. Note the bending of the DNA induced by TBP binding and the positions of TFIIB and TFIIA relative to TBP.

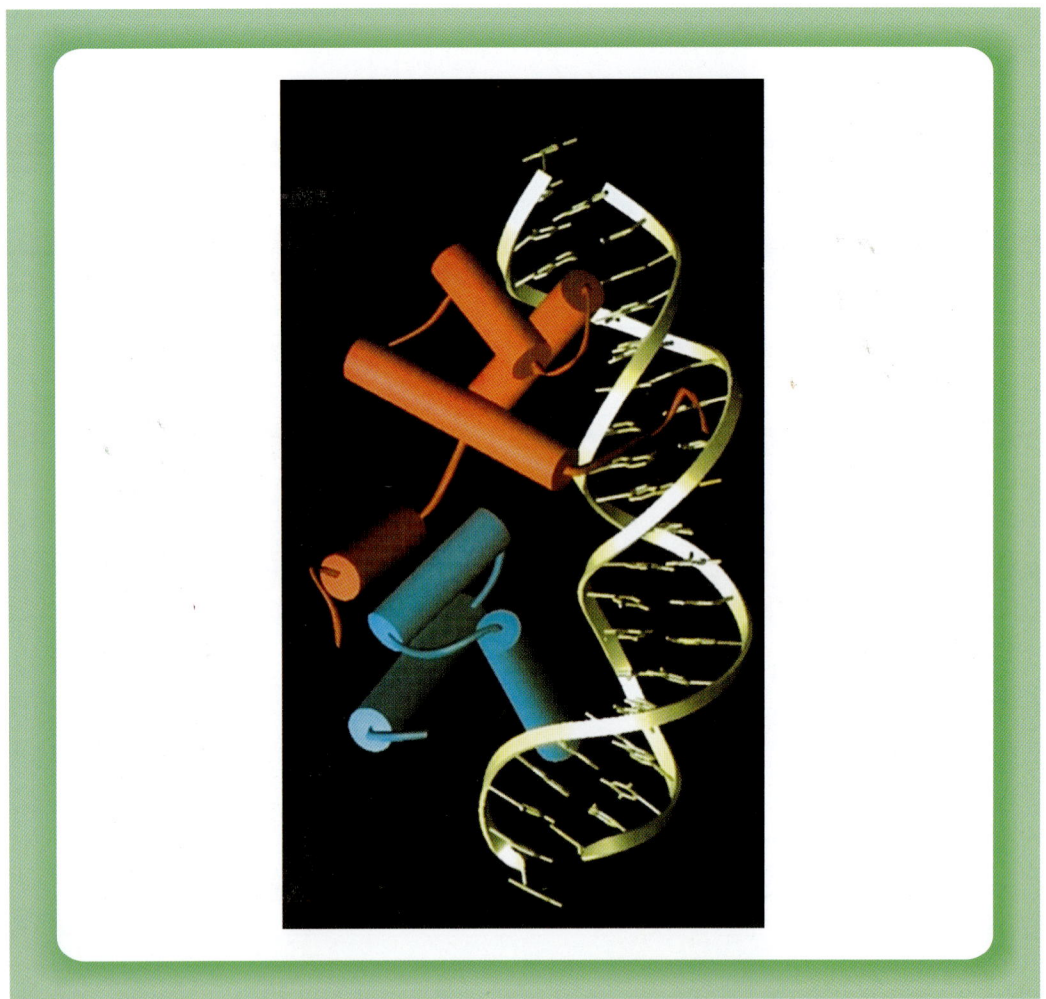

Plate 2 Binding of the αl (blue)/ α2 (red) homeodomain heterodimer to DNA. α-helices are shown as cylinders. Note the three helical structure of the homeodomains of a1 and α2 and the C-terminal region of α2, which forms an additional α-helix in the presence of a1 and packs against the α2 homeodomain forming the dimerization interface.

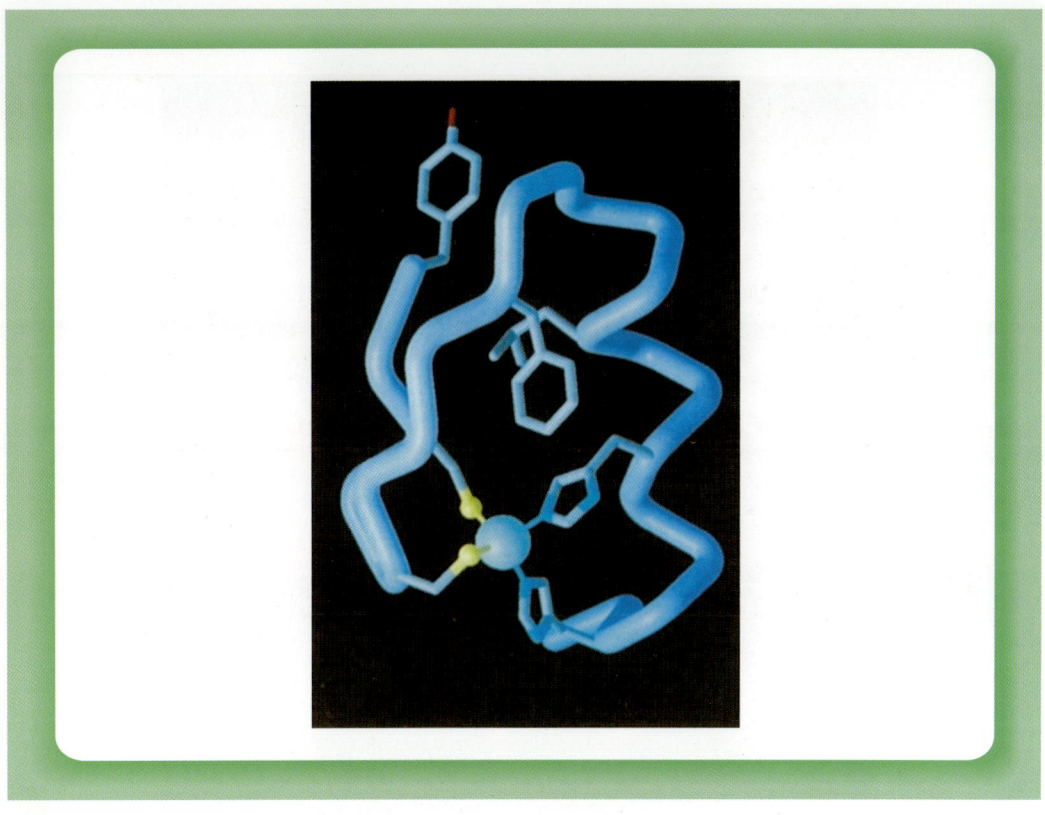

Plate 3 Structure of the Cys_2 His_2 Zinc finger from Xfin. The Cys residues are shown in yellow and the His residues in dark blue.

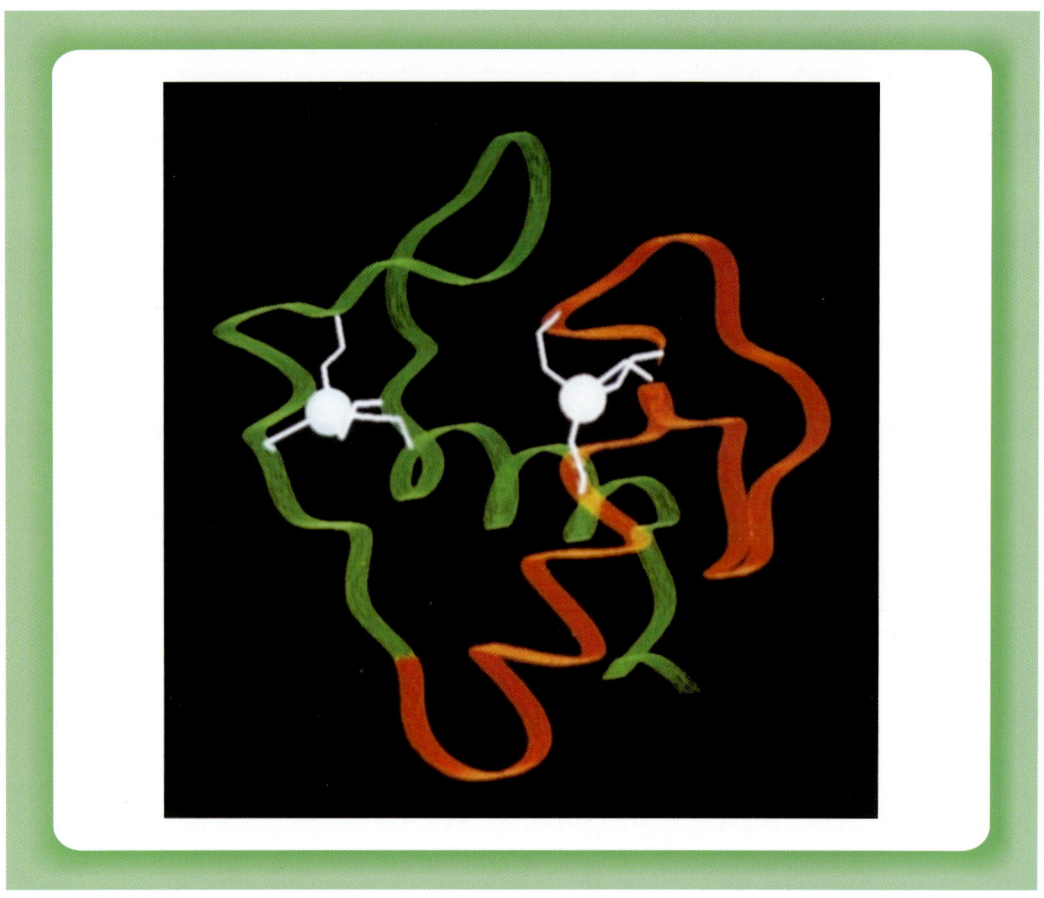

Plate 4 Structure of the two Cys_4 zinc fingers in a single molecule of the glucocorticoid receptor. The first finger is shown in red and the second finger in green with the zinc atoms shown white.

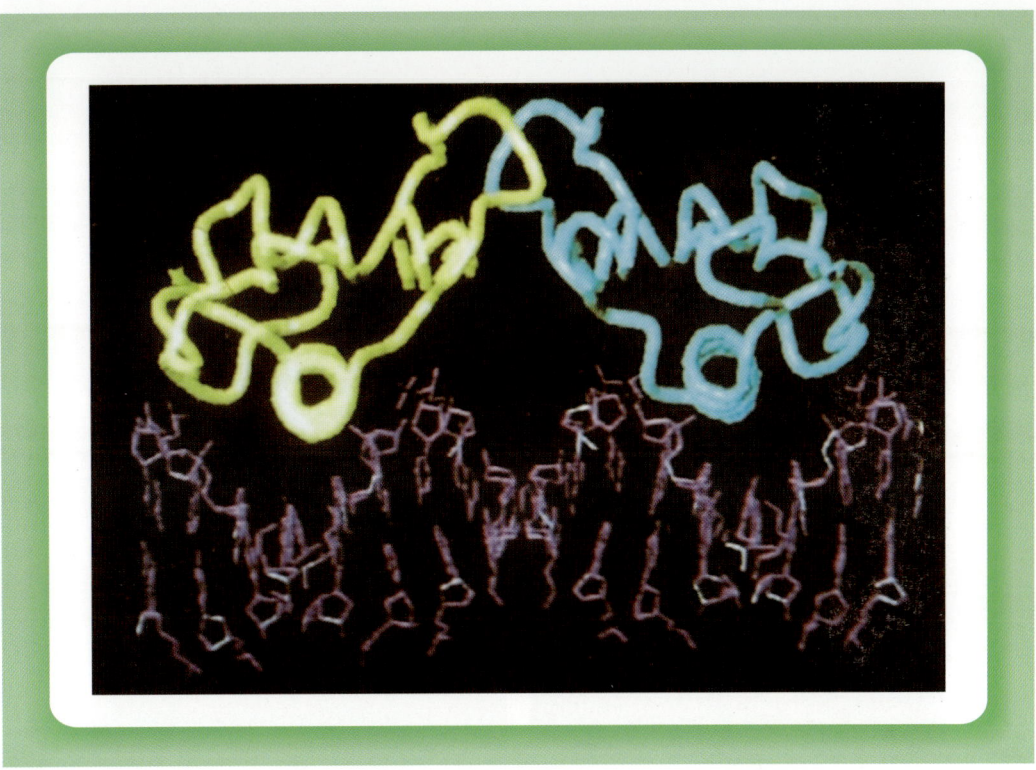

Plate 5 Structure of the estrogen receptor dimer consisting of two receptor molecules bound to DNA. The two molecules of the receptor are shown yellow and blue respectively and the DNA is shown in purple.

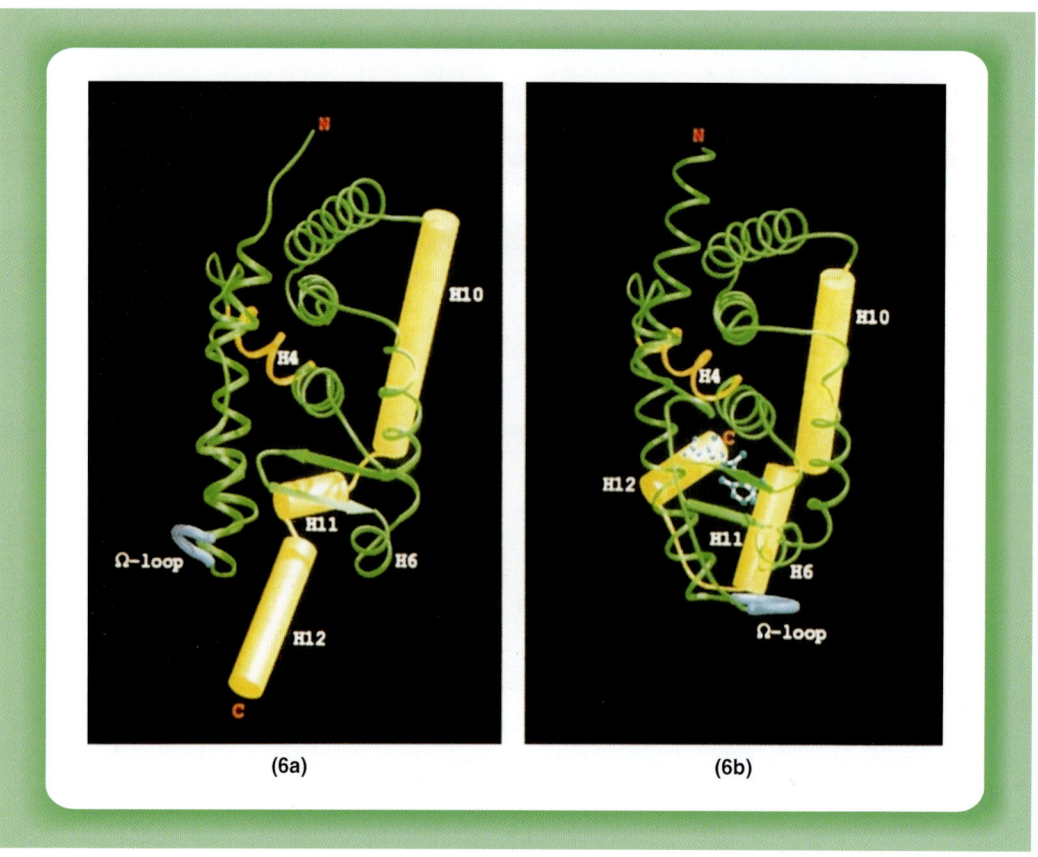

(6a)   (6b)

Plate 6 Structure of (a) the RXRα receptor in the absence of ligand and (b) the closely related RARα receptor following binding of ligand (light blue atoms joined by white bonds). Note the structural change induced by the binding of ligand involving the movement of the H12 helix towards the ligand binding core so creating a sealed pocket in which the ligand is trapped.

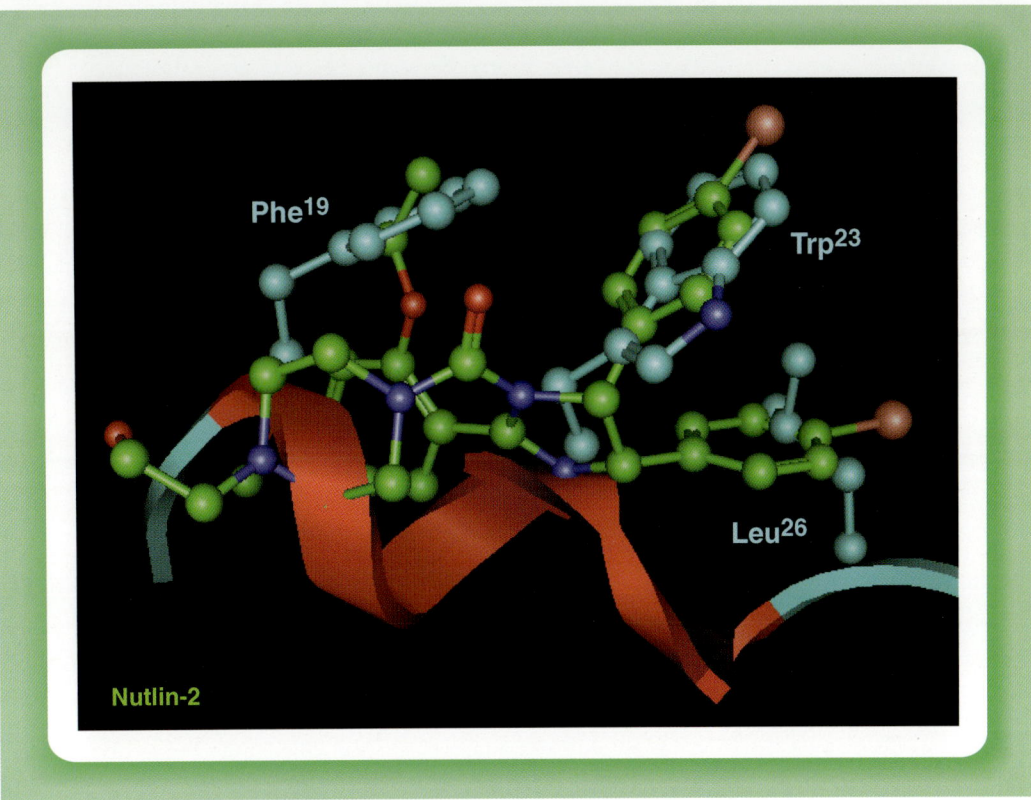

Phe19

Trp23

Leu26

Nutlin-2

Plate 7 Structure of the nutlin-2 inhibitor compared to that of amino acids in p53 which are critical for its interaction with MDM2. Nutlin-2 carbon atoms are shown in green, nitrogen in dark blue, oxygen in red and bromine in brown. The peptide side-chains of the key p53 amino acids Phe[19], Trp[23] and Leu[26] are shown in light blue. Note the structural similarity of nutlin-2 to the amino acids in p53, which bind to MDM2. This allows nutlin-2 to bind to MDM2 and block its interaction with p53.

idea, the inhibitory domain of the Kruppel repressor in *Drosophila* has been shown to interact with a component of the basal transcriptional complex, TFIIEβ (Sauer *et al.*, 1995). Interestingly, Kruppel can also act as an activator by interacting with TFIIB to stimulate its activity. This interaction with TFIIB is seen in the monomeric Kruppel factor which hence acts as an activator, whereas the Kruppel dimer which forms at high concentrations inhibits transcription by interacting with TFIIEβ. Hence, Kruppel can act as activator or repressor depending on its concentration in the cell, which results in it being present as an activating monomer or an inhibitory dimer (Fig. 6.16).

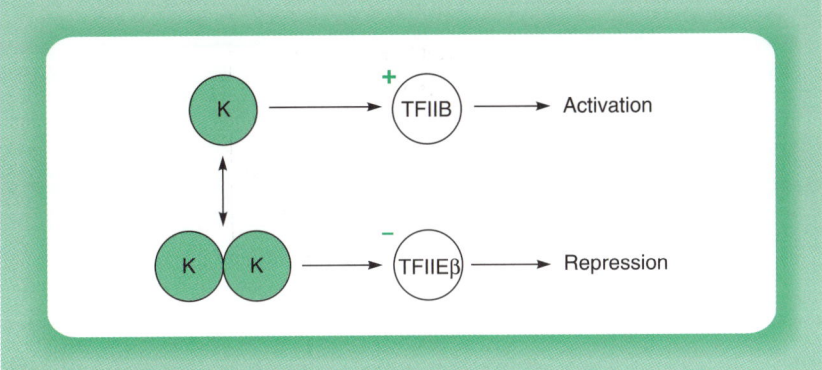

Figure 6.16
The Kruppel factor (K) when present as a monomer can interact with TFIIB to stimulate transcription. At high concentrations when it forms a dimer it interacts with TFIIEβ to repress transcription.

Although repressors may therefore interact directly with the basal transcriptional complex, in many cases they do so via a non-DNA binding co-repressor molecule which then actually represses transcription (Fig. 6.17). Such non-DNA binding co-repressors have been observed in a range of organisms from yeast to man (for reviews see Knoepfler and Eisenman, 1999; Smith and Johnson, 2000; Chinnadurai, 2002). Thus, the inhibitory effect of the thyroid hormone receptor discussed above involves co-repressor molecules such as N-CoR (nuclear receptor co-repressor) which bind to the receptor in the absence of hormone and produce its inhibitory effect on transcription (for reviews see Rosenfeld and Glass, 2001; Nagy and Schwabe, 2004; Rosenfeld *et al.*, 2006).

Interestingly, studies on mice lacking N-CoR have shown multiple defects in the development of numerous organs and cell types (Jepsen *et al.*, 2000; Hermanson *et al.*, 2002). Hence, co-repressors play critical roles in the regulation of gene expression with N-CoR, for example, being involved both in responses to thyroid hormone and in embryonic development.

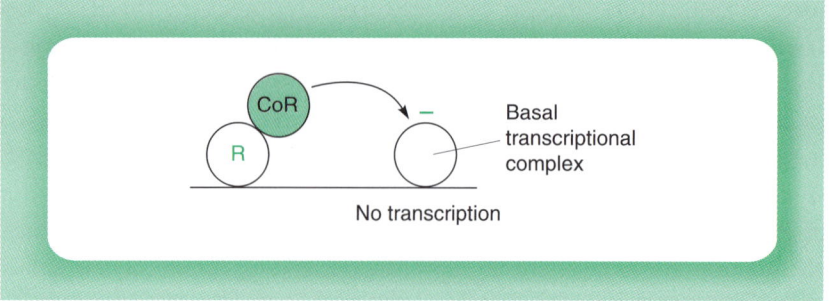

Figure 6.17
A DNA binding repressor (R) can recruit a non-DNA binding co-repressor (CoR) which then represses transcription.

In the case of the thyroid hormone receptor, following treatment with thyroid hormone, the conformation of the receptor changes and the co-repressor can no longer bind. This conformational change allows co-activator molecules such as CBP to bind and induce transcriptional activation. Hence, in this case, treatment with thyroid hormone causes a conformational change in the receptor resulting in the removal of co-repressors and the binding of co-activators (Fig. 6.18).

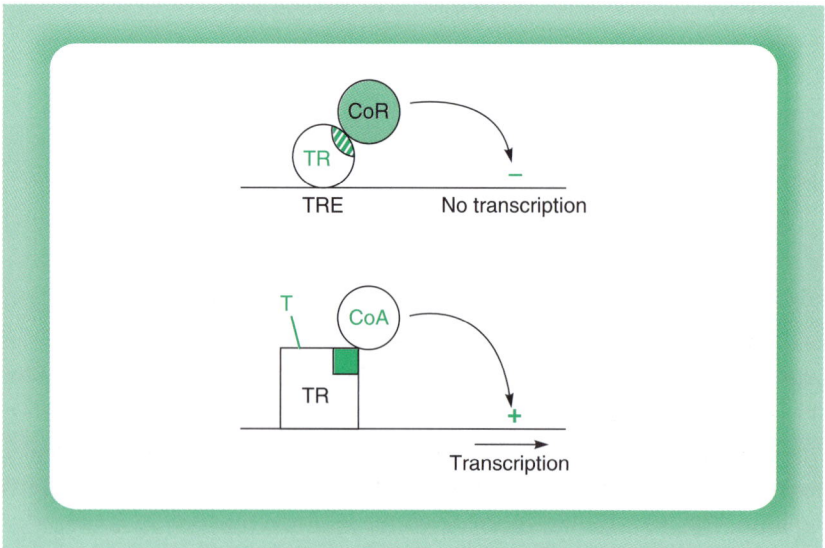

Figure 6.18
In the absence of thyroid hormone, the inhibitory domain of the thyroid hormone receptor (TR) bound to its response element (TRE) can recruit a co-repressor (CoR) which then inhibits transcription. In the presence of thyroid hormone (T), the conformation of the receptor changes, exposing its activation domain (solid), allowing recruitment of co-activator molecules (CoA), thereby producing activation of transcription in response to thyroid hormone.

An interesting example of such a co-activator/co-repressor interchange has been described in the case of the LIM homeodomain transcription factors (Ostendorff *et al.*, 2002). Thus, these factors bind both the RLIM co-repressor molecule and the CLIM co-activator molecule. In a novel mechanism, the CLIM co-activator promotes degradation of RLIM, removing the repressor and allowing activation to occur (Fig. 6.19).

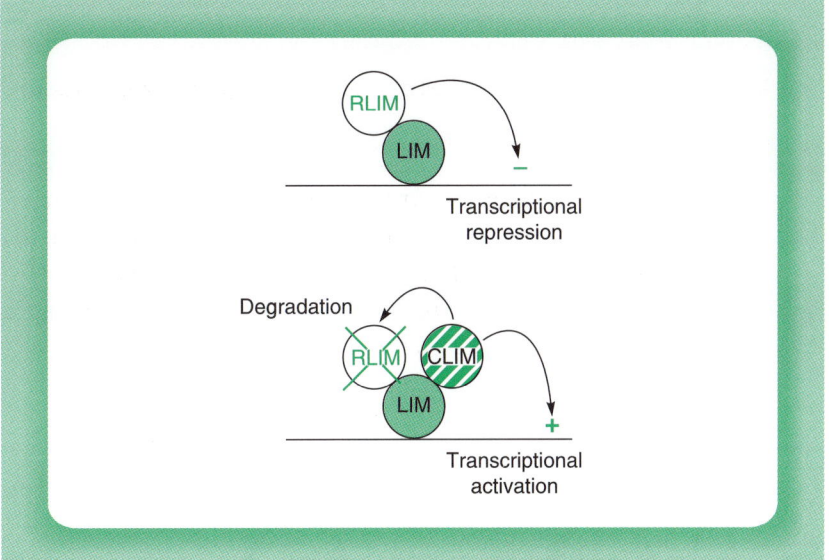

Figure 6.19
The LIM homeodomain protein binds the RLIM co-repressor producing transcriptional repression. However, following binding of the CLIM activator protein, CLIM digests the RLIM protein, allowing transcriptional activation to occur.

It is clear therefore that a number of factors can inhibit transcription by binding to upstream DNA sequences and inhibiting the activity of the basal transcriptional complex, either directly or via co-repressor molecules. The binding of such factors is likely to be of vital importance in producing the inhibitory effect of many of the silencer elements which were described in Chapter 1 (section 1.3.5), the silencer element in the chicken lysozyme gene, for example, having been shown to act by binding the inhibitory thyroid hormone receptor.

6.3.3 DIRECT REPRESSION BY FACTORS BINDING TO THE BASAL TRANSCRIPTIONAL COMPLEX

As well as interfering with the basal complex by binding to distinct DNA binding sites (Fig. 6.11a), it is also possible for inhibitory factors to bind

to the complex itself by protein–protein interaction and thereby interfere with its activity or assembly (Fig. 6.11b) (for review see Maldonado
et al., 1999). An example of this is provided by the Dr1 protein which
inhibits the assembly of the basal transcriptional complex by binding
to TBP and preventing TFIIB from binding (Fig. 6.20) (Inostroza *et al.*,
1992). As the recruitment of TFIIB to the promoter by interaction with
the TBP component of TFIID is an essential step in the assembly of the
basal transcriptional complex, this effectively inhibits transcription (see
Chapter 3, section 3.5.1).

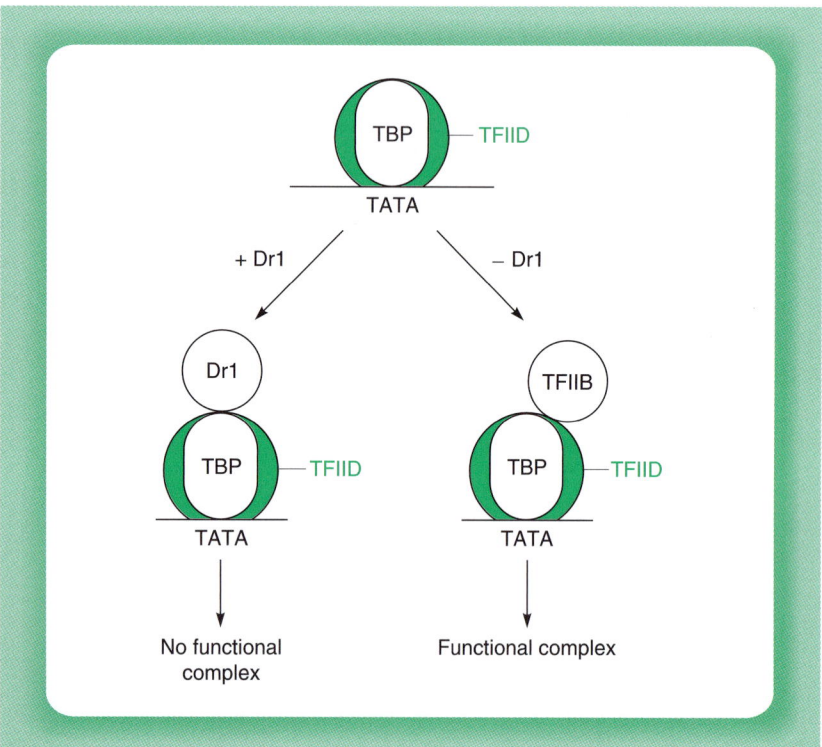

Figure 6.20
The Dr1 inhibitory factor can interact with the TBP component of TFIID, thereby
preventing it binding TFIIB and thus inhibiting the assembly of the basal
transcriptional complex.

As noted in Chapter 3 (section 3.6), TBP is a component of the initiation complexes of all three polymerases and in each case acts by recruiting other factors to the promoter. It has been shown (White *et al.*, 1994)
that Dr1 can inhibit this ability of TBP to recruit other factors within
the RNA polymerase II and III initiation complexes but not within the
RNA polymerase I complex. It therefore inhibits transcription by RNA

polymerase II and III but not by RNA polymerase I. Thus, Dr1 may play a critical role in regulating the balance of transcriptional activity between the ribosomal genes which are the only genes transcribed by RNA polymerase I and all the other genes in the cell.

As well as this potential role for Dr1, there is evidence that it can also alter the balance between transcription of different types of promoter by RNA polymerase II. Thus, although Dr1 inhibits transcription from TATA box-containing promoters, it actually stimulates transcription from polymerase II promoters lacking a TATA box and containing an initiator element and an associated downstream promoter element (Willy *et al.*, 2000) (see Chapter 1, section 1.3.2 for discussion of promoters with or without a TATA box). Hence, Dr1 may switch transcription between different genes transcribed by RNA polymerase II but containing different core promoter elements.

In addition to Dr1, other factors which bind to TBP and inhibit the assembly of the RNA polymerase II basal complex have been described and are likely to be important in controlling the rate of transcription (Chitikila *et al.*, 2002; for review see Maldonado *et al.*, 1999). Thus, for example, the Mot 1 factor also targets TBP but rather than preventing TFIIB binding, it displaces TBP from the DNA, thereby inhibiting transcription (Fig. 6.21).

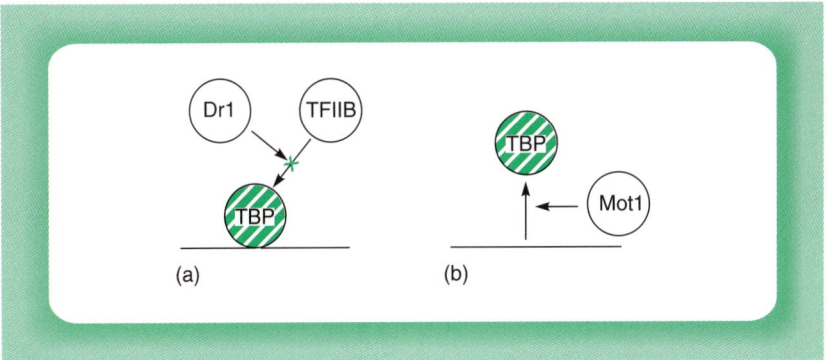

Figure 6.21

The Dr1 repressor interacts with TBP to prevent binding of TFIIB (panel a) whilst the Mot1 repressor displaces TBP from the DNA (panel b).

As well as interacting with activating factors (see Chapter 5, section 5.3.3), the TFIIA factor also appears to be able to bind to TBP, preventing these inhibitors from binding, thus preventing Mot 1 from inhibiting TBP binding to DNA or allowing the recruitment of TFIIB in the presence of Dr1 (Fig. 6.22). This indicates that the activity of inhibitory molecules which act by interacting with the basal transcriptional complex can be regulated by activating factors. Moreover, it illustrates that the balance between transcriptional activators and repressors is of central importance in the control of transcription, with directly acting repressors playing a key role either by binding to upstream DNA sequences or by joining the basal transcriptional complex.

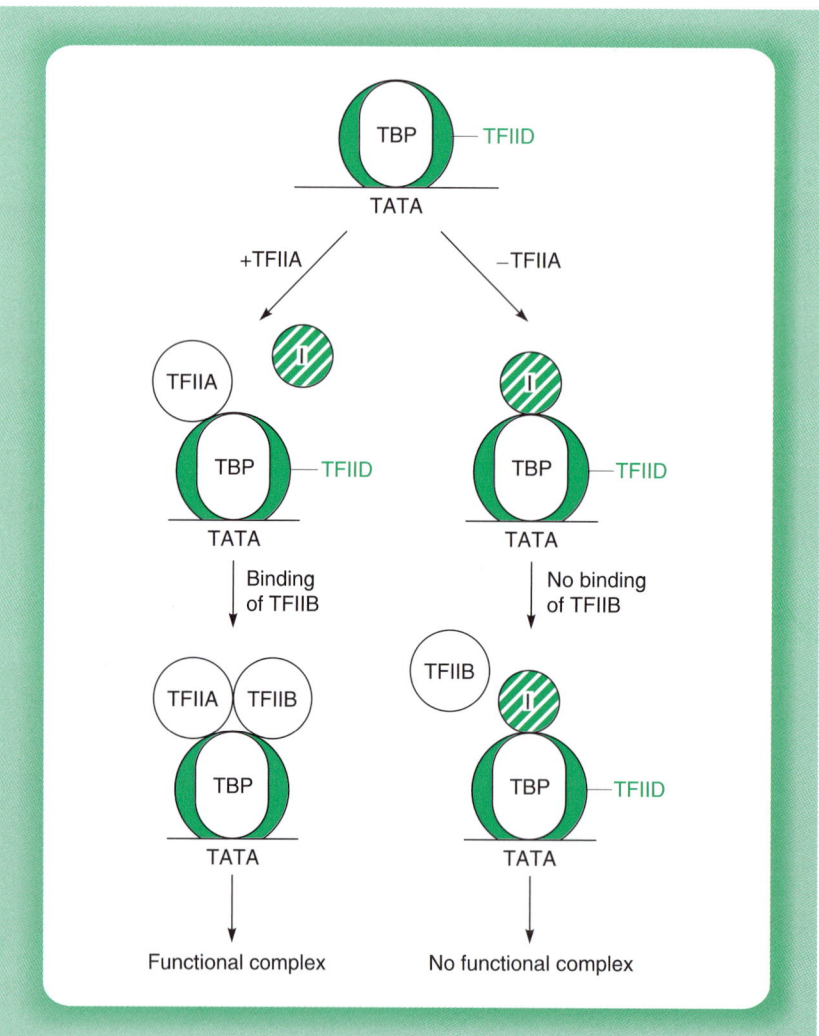

Figure 6.22
Binding of TFIIA to TBP prevents inhibitory molecules (I) from binding but still allows the binding of TFIIB and thereby promotes the assembly of the basal transcriptional complex.

Interestingly, these inhibitory effects on the basal transcriptional complex are not confined to RNA polymerase II. Thus, as noted above, Dr1 can target TBP within the RNA polymerase III basal transcriptional complex, whilst another inhibitory factor, pTEN, has been shown to target the SLI factor, which is part of the RNA polymerase I basal transcriptional complex, discussed in Chapter 3 (section 3.3) (Zhang *et al.*, 2005).

6.4 INHIBITION BY ALTERATION OF CHROMATIN STRUCTURE

6.4.1 EFFECT OF REPRESSORS ON CHROMATIN

Evidently, in the same way as an activating factor can activate transcription, by opening up the chromatin (see Chapter 5, section 5.5.1), an inhibitory factor can produce repression by directing a more tightly packed chromatin structure (for reviews see Tyler and Kadonaga, 1999; Courey and Jia, 2001) (Fig. 6.23). This could occur either via ATP-dependent remodelling of nucleosomes or via altering the modification of histones (see Chapter 1, section 1.2).

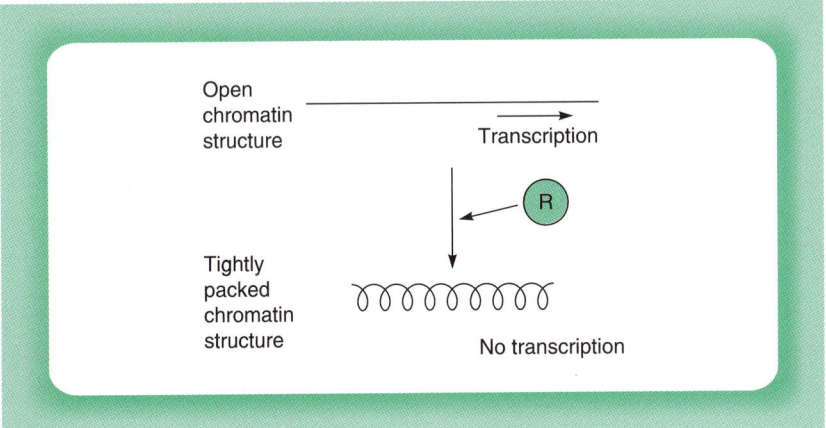

Figure 6.23

A repressor of transcription (R) can act by inducing a tightly packed chromatin structure incompatible with transcription.

An example of a factor which acts in this way is the polycomb repressor of *Drosophila* which normally represses inappropriate expression of several homeotic genes by modulating their chromatin structure so that activating molecules cannot bind. When this factor is inactive, inappropriate expression of these genes in the wrong cell type is observed, leading to dramatic transformations in the nature of specific parts of the body (for review see Orlando, 2003; Levine *et al.*, 2004; Mohd-Sarip and Verrijzer, 2004; Marx, 2005; Schwartz and Pirrotta, 2007). By directing the tight packing of specific genes and thereby preventing transcription, the polycomb factor evidently has the opposite effect to that of the trithorax factors which, as discussed in Chapter 5 (section 5.5.1), direct an open chromatin structure allowing activator binding (Fig. 6.24).

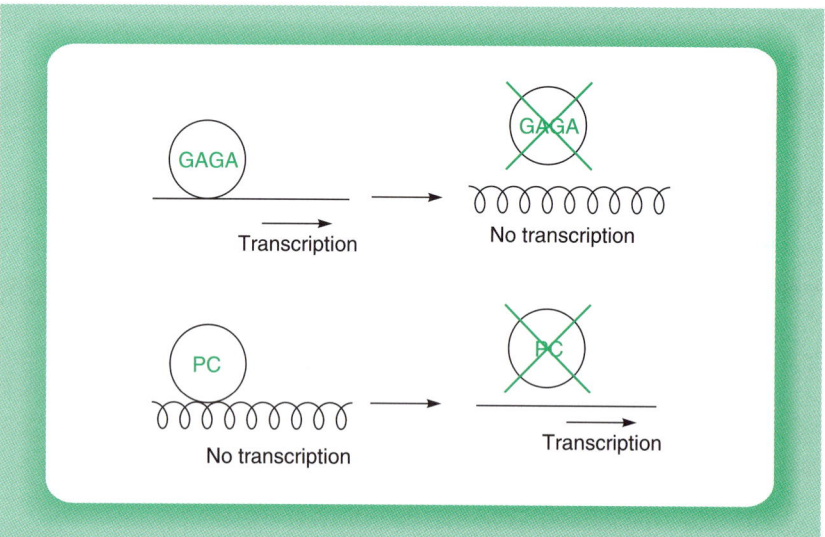

Figure 6.24

The GAGA/trithorax factor induces an open chromatin structure. Its inactivation by mutation produces an inappropriate tightly packed chromatin structure of its target genes producing a failure of gene transcription. In contrast, the polycomb factor (PC) induces a tightly packed chromatin structure. Its inactivation by mutation produces inappropriate chromatin opening and transcription.

Interestingly, both polycomb and trithorax factors have been shown to affect the function of insulator elements (see Chapter 1, section 1.3.5) with mutation of polycomb preventing the blocking effects of insulators on enhancer function, whilst mutation of trithorax enhances the blocking effects of insulators. Hence, insulator elements represent one target for the antagonistic effects of polycomb and trithorax factors (Gerasimova and Corces, 1998).

Although polycomb was originally shown to be an important transcriptional repressor in *Drosophila*, it is clear that members of the polycomb family of proteins play a critical role in a wide range of organisms. Thus, for example, in mammals, polycomb has been shown to be essential to maintain the undifferentiated nature of embryonic stem (ES) cells and prevent them differentiating prematurely (for reviews see Buszczak and Spradling, 2006; Holden, 2006; Sparmann and van Lohuizen, 2006). Thus, in these cells, the chromatin immunoprecipitation assay (see Chapter 2, section 2.4.3) was used to show that polycomb proteins were associated with hundreds of genes. A number of these are likely to play critical roles in inducing the formation of specific differentiated cell types, and hence their repression prevents differentiation (Fig. 6.25). Polycomb therefore plays a central role in keeping stem cells in an undifferentiated state by repressing the genes whose activity is required to induce particular states of differentiation. Obviously, this repression needs to be relieved to allow differentiation to occur and this is likely to involve factors such as the trithorax factors (see above) which have the opposite effect to polycomb and open the chromatin.

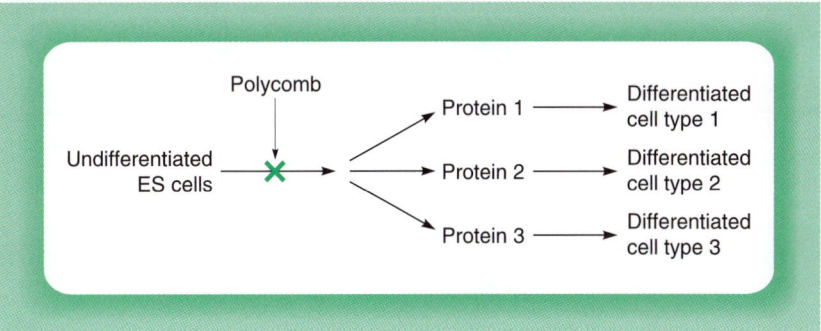

Figure 6.25

The polycomb factor maintains embryonic stem (ES) cells in an undifferentiated state by preventing the transcription of the genes encoding proteins, which are required for differentiation into specific cell types.

During the differentiation of male germ cells in the testis, a novel mechanism is used to antagonize the inhibitory effects of polycomb and achieve opening of the chromatin. Thus, in this case, tissue specific TAFs (TBP-associated factors, see Chapter 5, section 5.4.2), which are found only in the testis, bind to the promoters of the genes to be activated and reduce polycomb binding. In turn, this is likely to allow binding of trithorax factors and opening of the chromatin (Chen *et al.*, 2005).

The key role of polycomb in these effects leads to the question of how it achieves its inhibitory effect on chromatin structure. There is considerable evidence that this involves effects on the histones, producing modifications which promote tight packing of the chromatin (see Chapter 1, section 1.2.3).

Thus, the polycomb factor is part of a multi-protein complex which has histone methyltransferase activity (Czermin *et al.*, 2002; Müller *et al.*, 2002). Moreover, genes which are repressed by polycomb in ES cells show enhanced methylation of histone H3 on lysine 27 (for reviews see Buszczak and Spradling, 2006; Schwartz and Pirrotta, 2007), which is known to produce a more tightly packed chromatin structure (see Chapter 1, Fig. 1.5). Such methylation of histone H3 at lysine 27 (and also at another inhibitory residue: lysine 9) has also been observed at sites where polycomb has bound in other cell types (see for example Ringrose *et al.*, 2004). Hence, binding of the polycomb complex to the DNA results in methylation of histones and thereby produces a closed chromatin structure (Fig. 6.26). This effect is opposed by trithorax proteins, which direct demethylation of histone H3 at lysines 9 and 27, thereby producing a more open chromatin structure (Fig. 6.27) (Papp and Muller, 2006).

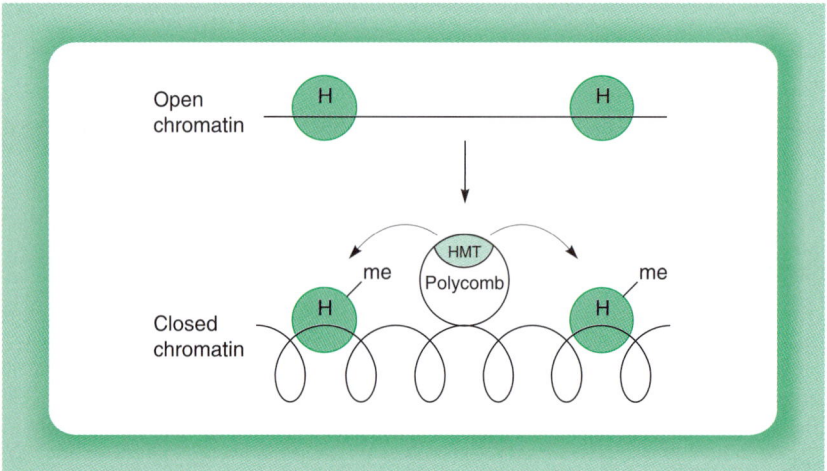

Figure 6.26
The polycomb protein complex contains a histone methyltransferase enzyme which methylates histones and thereby produces a more tightly packed chromatin structure.

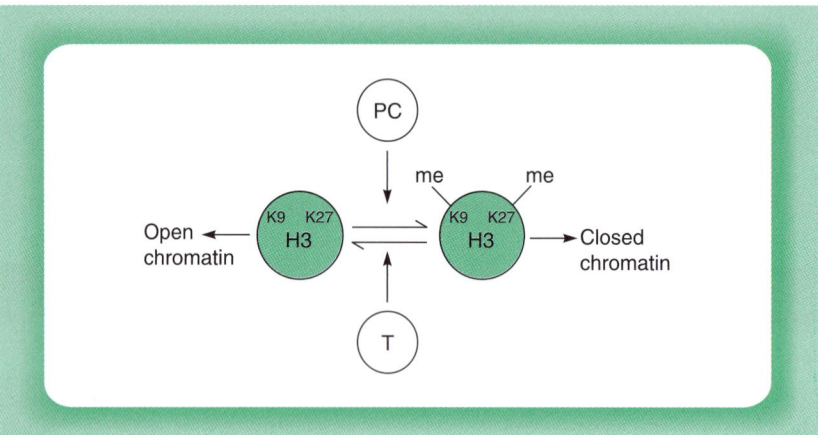

Figure 6.27
Polycomb proteins (PC) promote a closed chromatin structure by stimulating the methylation of histone H3 on lysines (K) 9 and 27. This effect is opposed by the trithorax proteins (T) which promote demethylation at these positions and hence a more open chromatin structure.

Such involvement of histone methylation in transcriptional repression is also seen in the case of the MyoD transcription factor which plays a key role in muscle differentiation. Thus, as discussed in section 6.2.2, the Id factor inhibits the binding of the MyoD factor to DNA and hence MyoD can only bind to DNA, once Id levels decline during differentiation of muscle cells. However, recent findings have demonstrated that MyoD initially binds to DNA in association with a methyltransferase enzyme (Mal, 2006). This enzyme methylates histone H3 on lysine 9, producing an inactive chromatin structure. It is only when expression of this methyltransferase decreases that MyoD can induce muscle cell differentiation. Hence, MyoD is inhibited successively by a factor which blocks its binding to DNA and then by a factor which produces an inactive chromatin structure (Fig. 6.28).

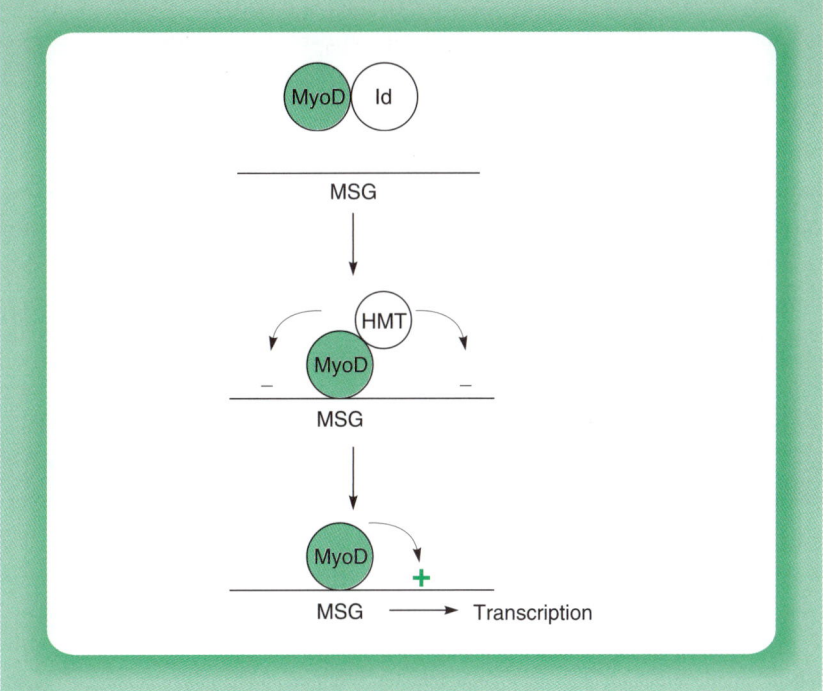

Figure 6.28

The ability of the MyoD transcription factor to induce muscle-specific gene expression is repressed both by the Id factor which prevents it binding to DNA and a histone methyltransferase (HMT) which associates with it and produces a more tightly packed chromatin structure.

As noted in Chapter 1 (section 1.2.3 and Fig. 1.5), methylation of histones at some positions, other than lysine 9 or lysine 27, can produce an open rather than a closed chromatin structure. Such methylation events are also a target for inhibitory molecules but, in this case, they produce a demethylation of the amino acid concerned. Thus, the CoRest co-repressor has been shown to demethylate lysine 4 of histone H3 (Lee *et al.*, 2005), whilst the JHDM3A repressor is a demethylase which targets lysine 36 (Klose *et al.*, 2006).

As well as targeting histones at the level of methylation, transcriptional co-repressors such as the NuRD and SIN3 complexes can also produce a more tightly packed chromatin structure by acting as histone deacetylases, removing the acetyl groups which promote chromatin opening (for reviews see Ahringer, 2000; Ng and Bird, 2000; Yang and Gregoire, 2005) (see Chapter 1, section 1.2.3). This effect evidently parallels the ability of transcriptional co-activators to produce a more open chromatin structure by acetylating histones (see Chapter 5, section 5.5.2).

Hence, the state of histone acetylation and the structure of chromatin can be controlled by the balance of deacetylases and acetylases which are bound to the DNA (Fig. 6.29). In most cases the acetylating and deacetylating factors will be respectively co-activators and co-repressors and will be brought to the DNA via interactions with distinct activating and inhibiting transcription factors, respectively. In the case of the thyroid hormone receptor, however, both types of factors bind to the same molecule. Thus, the N-CoR co-repressor which binds to the thyroid hormone receptor prior to exposure to hormone also binds other co-repressor complexes such as mRPD3 and SIN3, which have histone deacetylase

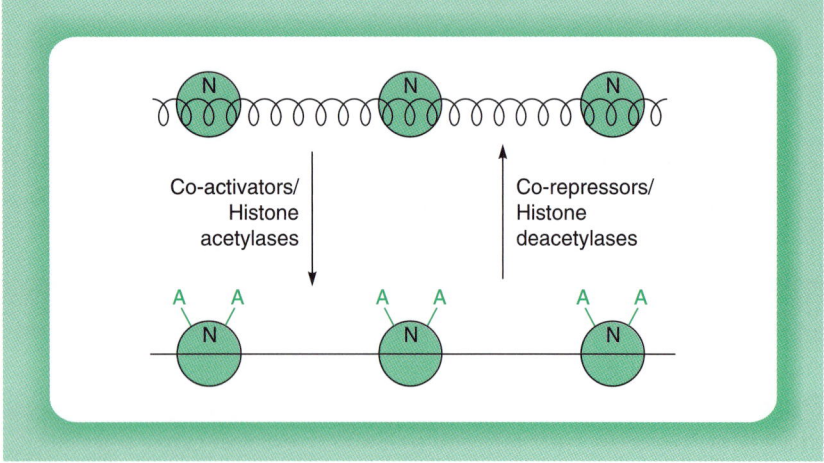

Figure 6.29

The balance between tightly packed chromatin (wavy line) and open chromatin (solid line) can be controlled by the balance between co-activating molecules which acetylate histones and co-repressors which deacetylate histones in the nucleosome (N).

activity. Conversely, following hormone binding, these factors dissociate and are replaced by co-activators such as CBP which acetylate histones and allow the receptor to activate transcription (Fig. 6.30).

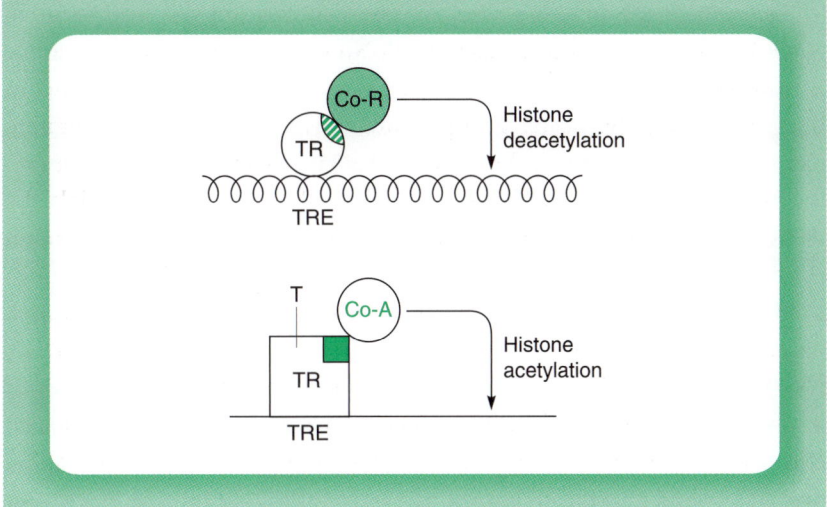

Figure 6.30

The inhibitory domain (hatched semi-circle) of the thyroid hormone receptor binds a co-repressor (Co-R) complex which deacetylates histones inducing a closed chromatin structure (wavy line). Binding of ligand results in the release of the co-repressor and binding of co-activator (Co-A) molecules to the exposed activation domain (hatched square). The co-activators have histone acetyltransferase activity and produce a more open chromatin structure (solid line) compatible with transcription.

Hence, transcriptional repressors can modulate chromatin structure, acting at least in part by altering the histone code at the level of both methylation and acetylation.

6.4.2 SMALL RNAs AND TRANSCRIPTIONAL INHIBITION

The last few years have seen an explosion of information on the role of small RNA molecules between nineteen and twenty-five bases in length. These inhibitory RNAs (RNAi) are produced in a range of organisms from yeast to man and act to inhibit gene expression. In most cases, this is achieved by post-transcriptional mechanisms in which the small RNA binds to a messenger RNA by complementary base-pairing and either induces degradation of the mRNA or blocks its translation into protein.

Such post-transcriptional processes are beyond the scope of this book (for further details, see Latchman, 2005).

However, in some cases, small RNAs can produce an inhibitory effect at the level of transcription. This effect appears to be of particular importance in plants, although it has also been shown to occur, for example, in *Drosophila* and in yeast (for reviews see Buratowski and Moazed, 2005; Ronemus and Martienssen, 2005; Sontheimer and Carthew, 2005; Wassenegger, 2005; Rana, 2007). When a gene is targeted in this way, an RNAi binds to it by complementary base-pairing and this then results in the gene being organized into a tightly packed chromatin structure, producing inhibition of transcription.

As expected, from the previous section (section 6.4.1), this shift to a closed chromatin structure induced by the RNAi can involve alterations in histone methylation. Thus, the methylation of histone H3 on lysine 9 has been shown to be induced by RNAi in *Drosophila* (Pal-Bhadra *et al.*, 2004). Such methylation on lysine 9 in turn induces the recruitment of inhibitory proteins such as HP1 which organize a more tightly packed chromatin structure (Bannister *et al.*, 2001; Lachner *et al.*, 2001). These proteins contain a region, known as a chromodomain, which binds with greater efficiency to histone H3 molecules in which lysine 9 has been methylated. This evidently parallels (but with the opposite effect) the ability of activating proteins, containing a different protein domain, known as the bromodomain, to bind to acetylated histones with greater affinity (see Chapter 5, section 5.5.2).

As well as inducing histone modification, in a number of studies RNAi has been shown to induce a more tightly packed chromatin structure by inducing methylation of cytosine residues in the DNA itself, which is normally associated with inactive tightly packed chromatin (for discussion of this modification of the DNA see Latchman, 2005). Hence, RNAi appears to repress gene expression by binding to the gene and promoting the binding of proteins which modify histones and methylate the DNA (Fig. 6.31).

Such repression via a combination of histone modification and DNA modification is not unique to RNAi. Thus, the polycomb repressor complex which, as discussed in section 6.4.1, induces histone methylation, also induces methylation of C residues in DNA (for review see Taghavi and van Lohuin, 2006). Indeed, there is evidence that RNAi and polycomb can co-operate to produce transcriptional repression in some circumstances (for reviews see Kavi *et al.*, 2006; Lei and Corces, 2006).

Evidently, the small inhibitory RNAs which inhibit gene transcription in this way must themselves be transcribed. In plants, this is achieved by a fourth RNA polymerase enzyme, which is not found in animals and

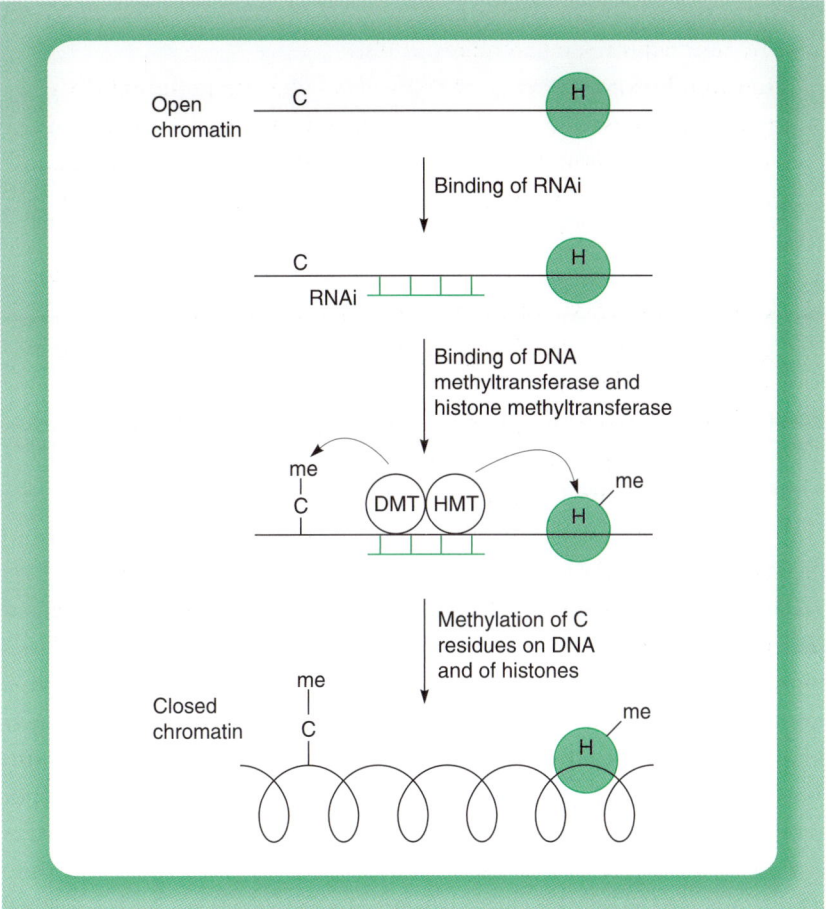

Figure 6.31
Binding of a small inhibitory RNA (RNAi) to a gene results in the binding of proteins
with DNA methyltransferase (DMT) and histone methyltransferase (HMT) activity.
These enzymes respectively methylate C residues in the DNA and histone proteins
(H) to produce a tightly packed chromatin structure. Methylation will target many
C residues and histone molecules but only one of each is shown for clarity.

which is distinct from the three well-characterized RNA polymerases
discussed in Chapter 3 (for review see Vaucheret, 2005; Vaughn and
Martienssen, 2005) (Fig. 6.32).

6.5 INHIBITION OF TRANSCRIPTIONAL ELONGATION

Just as activators can stimulate transcriptional elongation, as well as
transcriptional initiation (see Chapter 5, section 5.5.2), repressors can
inhibit transcription by blocking transcriptional elongation. Thus, the

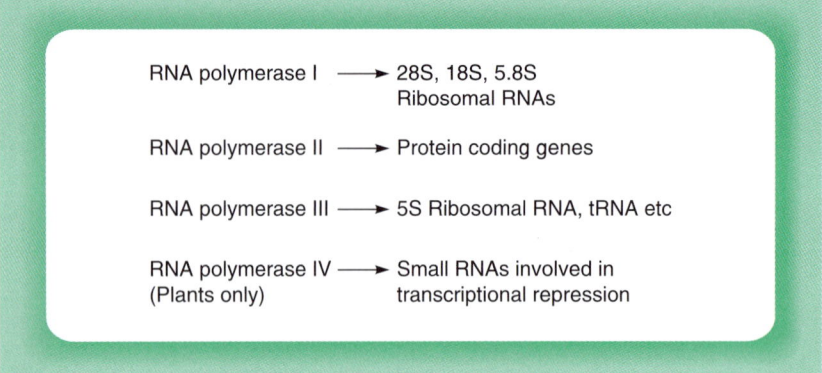

Figure 6.32

In addition to the three RNA polymerases found in all eukaryotes, plants contain a fourth RNA polymerase which is involved in transcribing the small inhibitory RNAs involved in transcriptional repression.

zebrafish Foggy protein acts by interacting with the non-phosphorylated form of RNA polymerase and prevents it from catalysing transcriptional elongation. When the polymerase is phosphorylated on its C-terminal domain (see Chapter 3, section 3.7), it is no longer inhibited by Foggy and transcriptional elongation proceeds (Guo *et al.*, 2000) (Fig. 6.33). Importantly, when Foggy is mutated so that it cannot block transcriptional elongation, the development of the zebrafish nervous system is severely disrupted. This indicates that the correct regulation of transcriptional elongation by proteins such as Foggy is necessary for normal development.

Inhibition of transcriptional elongation is also produced by the von Hippel–Lindau protein (VHL). However, this factor targets the phosphorylated form of RNA polymerase. Thus, VHL forms part of a complex which adds the small protein ubiquitin to the large subunit of RNA polymerase II (see Chapter 8, section 8.4.5 for further details of this protein modification). This ubiquitination, occurs only for the phosphorylated form of RNA polymerase II and targets it for degradation, thereby blocking transcriptional elongation (Kuznetsova *et al.*, 2003) (Fig. 6.34). As with Foggy, the action of VHL is of critical importance for normal cell function. Thus, as discussed in Chapter 9 (section 9.4.4), VHL is an anti-oncogene and cancers result when the function of VHL is disrupted by mutations.

Hence, transcriptional elongation can be repressed both by Foggy, which inhibits the non-phosphorylated form of RNA polymerase II, and by VHL, which targets the phosphorylated form of the polymerase for degradation.

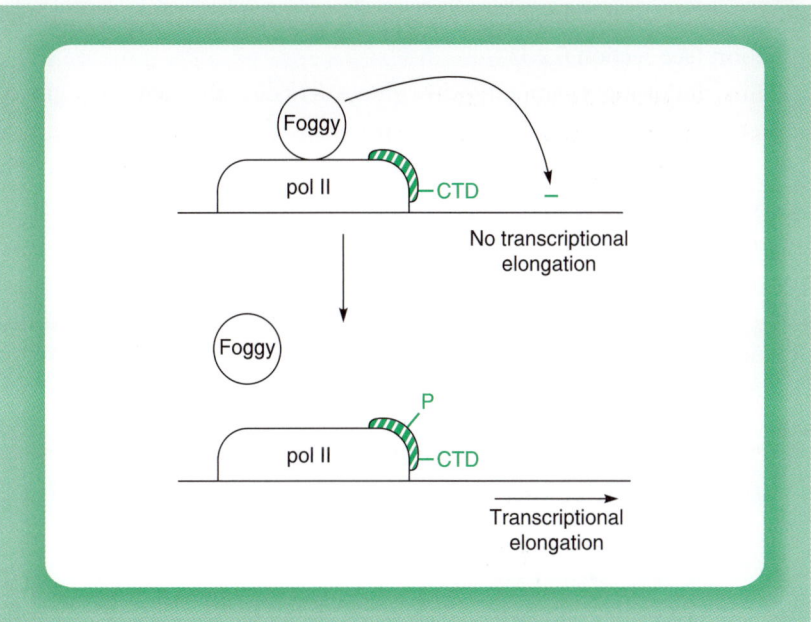

Figure 6.33
Phosphorylation of the C-terminal domain (CTD) of RNA polymerase prevents binding of the Foggy protein which would otherwise inhibit transcriptional elongation.

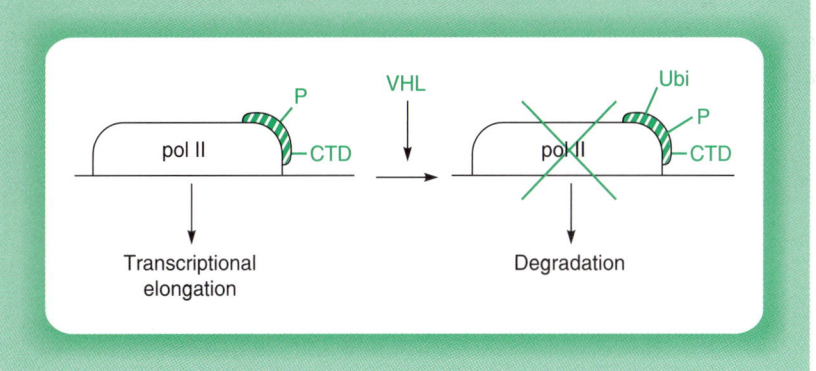

Figure 6.34
The von Hippel–Lindau gene product (VHL) adds ubiquitin (Ubi) to the phosphorylated C-terminal domain (CTD) of RNA polymerase. This promotes degradation of the polymerase inhibiting transcriptional elongation.

As well as targeting the polymerase itself, inhibitors of transcriptional elongation can also act by blocking the modification of the polymerase which is necessary for such elongation to occur. This is seen in the case of the glucocorticoid receptor which can act as an inhibitor of transcriptional elongation, as well as blocking the binding of activators of transcriptional initiation (see section 6.2.1).

Thus, following treatment with glucocorticoid, the activated glucocorticoid receptor can inhibit the expression of the interleukin-8 gene. This is achieved by the receptor blocking the binding of the P-TEFb kinase to the promoter (Luecke and Yamamoto, 2005). This kinase is required for the phosphorylation of serine 2 in the RNA polymerase II C-terminal domain (see Chapter 3, section 3.7). Since such phosphorylation is necessary for transcriptional elongation to occur, this effect of the glucocorticoid receptor results in the inhibition of transcriptional elongation (Fig. 6.35).

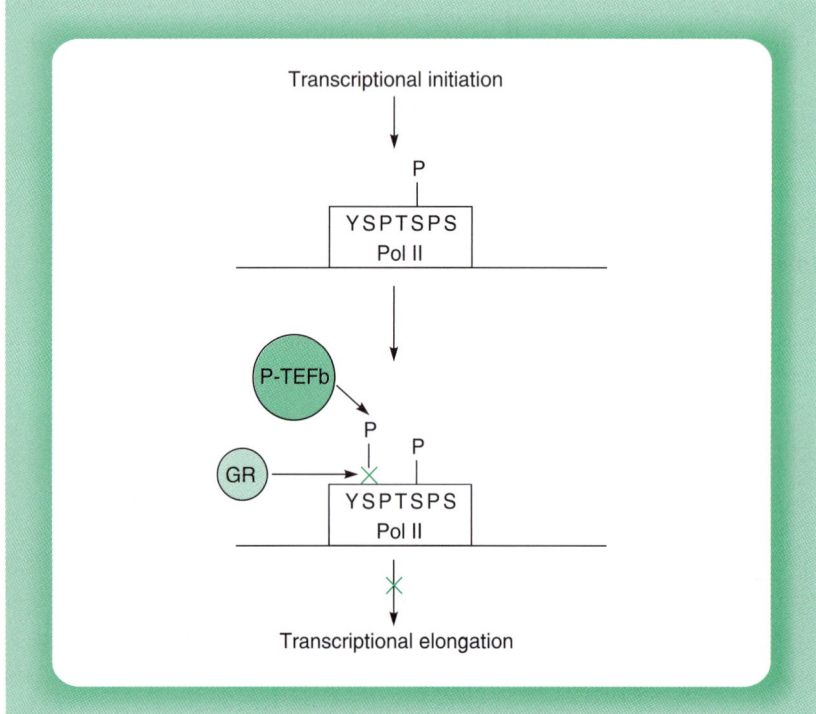

Figure 6.35
The glucocorticoid receptor (GR) blocks the phosphorylation of serine 2 in the RNA polymerase II C-terminal domain and hence blocks transcriptional elongation. Compare with Figure 3.13.

Hence, transcriptional elongation is a target for inhibitory factors, including factors such as Foggy which act specifically at transcriptional elongation and others such as the glucocorticoid receptor which also target transcriptional initiation. As with transcriptional initiation, the correct regulation of these inhibitory processes and their activity relative to processes which stimulate transcriptional elongation (see Chapter 5, section 5.6) is likely to be critical for the correct regulation of cellular function.

6.6 CONCLUSIONS

In this chapter we have discussed how the repression of transcription can be produced by the neutralization of a positively acting factor, or by direct repression of the basal transcriptional complex, as well as by alteration of chromatin structure or the inhibition of transcriptional elongation. These properties offer ample scope for gene regulation in different cell types or in different tissues. Thus, in addition to the simple activation of gene expression by a positively acting factor present in only one cell type, the effect of a positively acting factor present in several different cell types can be affected by the presence or absence of a negatively acting factor, which is active in only one cell type and which inhibits its activity. Similarly, a single factor may act either positively or negatively depending on the gene involved (as in the case of the glucocorticoid receptor) or depending on whether a specific hormone is present (as in the case of the thyroid hormone receptor).

Interestingly, two positive factors can also repress one another if they compete for the same co-factor. Thus, glucocorticoid hormones have been known for some time to be a potent inhibitor of the induction of the collagenase gene by phorbol esters resulting in their having an anti-inflammatory effect. This inhibition is mediated by the glucocorticoid receptor which inhibits the activity of the Jun and Fos proteins that normally activate the collagenase gene via the AP-1 sites in its promoter (for discussion of Fos, Jun and AP-1 see Chapter 9, section 9.3.1). This effectively inhibits collagenase gene activation. Unlike the examples of repression by the glucocorticoid receptor discussed in section 6.2.1, however, the collagenase promoter does not contain any binding sites for the receptor adjacent to the AP-1 sites nor does the receptor apparently bind to the collagenase promoter.

Interestingly, however, like the glucocorticoid receptor, the Fos/Jun complex requires the CBP protein as a co-activator to activate transcription. Hence, the glucocorticoid receptor may compete with Fos/Jun for limited quantities of the CBP co-activator which are present in the cell, resulting in a failure of Fos/Jun to activate transcription in the presence of activated glucocorticoid receptor (Kamei *et al.*, 1996) (Fig. 6.36). Clearly, such competition between Fos/Jun and the glucocorticoid receptor for limited quantities of CBP will also result in inhibition of glucocorticoid-dependent genes in response to hormone in the presence of high concentrations of Fos and Jun and this is indeed observed (Fig. 6.36).

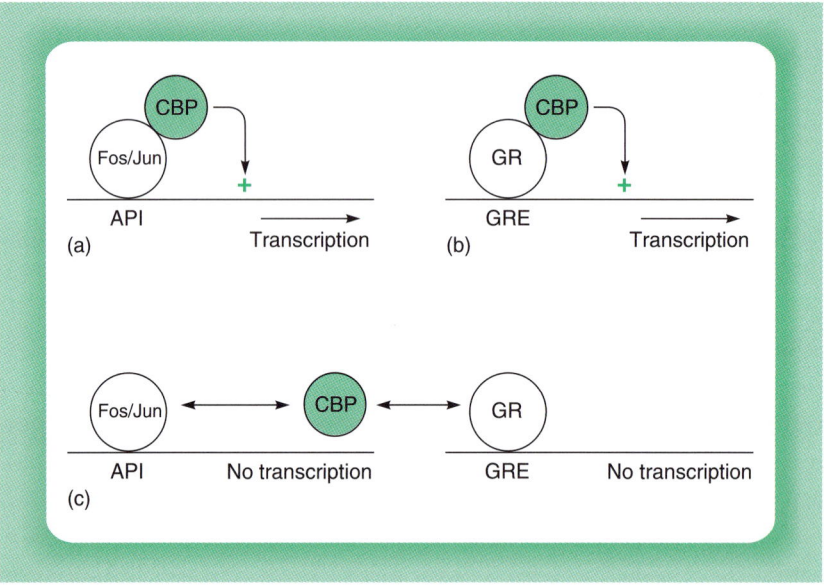

Figure 6.36

Mutual transrepression by Fos/Jun and the glucocorticoid receptor. Competition between Fos/Jun and the glucocorticoid receptor for the CBP co-activator inhibits the expression of genes containing binding sites for either Fos/Jun (AP-1 sites) or for the glucocorticoid receptor (GRE).

Hence, mutual transrepression of two different activating proteins can be achieved by competition for a co-activator (Fig. 6.36). Moreover, this mutual repression illustrates how different cellular signalling pathways which are activated respectively by phorbol esters and glucocorticoid hormones can interact with one another, resulting in cross talk between the pathways (for review see Janknecht and Hunter, 1996).

Hence, as well as being able to activate gene expression, the glucocorticoid receptor also illustrate three mechanisms by which repression of gene expression can be achieved. These are the neutralization of a positive factor either by preventing its binding to DNA by masking of its site (see section 6.2.1) or by competing for a co-activator (see above) or by the inhibition of transcriptional elongation (see section 6.5) (Fig. 6.37). As noted above (section 6.3.6), the thyroid hormone receptor, which like the glucocorticoid receptor is a member of the nuclear receptor family, inhibits transcription by a different mechanism, namely, the direct inhibition of transcription. It should be noted, however, that this case differs in that the glucocorticoid receptor needs to be activated by steroid before it can inhibit gene expression by any of its three distinct modes of action, whereas the thyroid hormone receptor directly inhibits by binding to its response element in the absence of hormone.

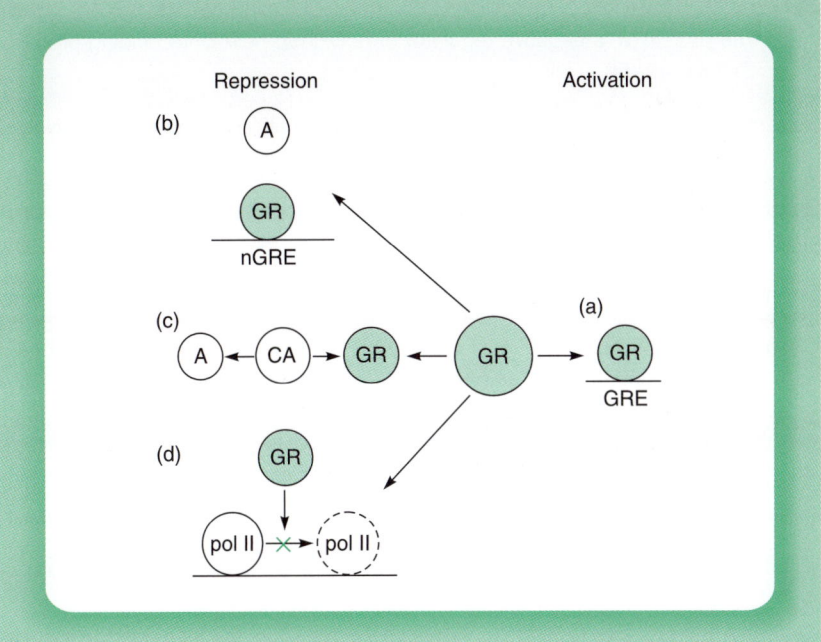

Figure 6.37

The glucocorticoid receptor (GR) can activate transcription by binding to a GRE sequence (panel a). Alternatively, it can repress gene expression by (i) binding to a distinct nGRE sequence and preventing an activator (A) from binding (panel b), (ii) by competing with an activator for a co-activator (CA) (panel c) or (iii) by blocking transcriptional elongation (panel d).

Hence, numerous methods of transcriptional repression exist paralleling the different mechanisms of transcriptional activation discussed in Chapter 5. Ultimately, however, as with transcriptional activation, all such potential mechanisms involving the inhibition of gene expression in response to specific stimuli or in specific cell types are dependent upon mechanisms which control the synthesis or activity of specific transcription factors in different cell types or in response to specific stimuli. These mechanisms are discussed in the next two chapters.

REFERENCES

Ahringer, J. (2000) NuRD and SIN3 histone deacetylase complexes in development. Trends in Genetics 16, 351–356.

Bannister, A.J., Zegerman, P., Partridge, J.F., Miska, E A., Thomas, J.O., Allshire, R.C. and Kouzarides, T. (2001) Selective recognition of methylated lysine 9 on histone H3 by the HP1 chromo domain. Nature 410, 120–124.

Benezra, R., Davis, R.L., Lockshon, D., Turner, D.L. and Weintraub, H. (1990) The protein Id: a negative regulator of helix-loop-helix DNA binding proteins. Cell 61, 49–59.

Buratowski, S. and Moazed, D. (2005) Expression and silencing coupled. Nature 435, 1174–1175.

Buszczak, M. and Spradling, A.C. (2006) Searching chromatin for stem cell identity. Cell 125, 233–236.

Carmeliet, P. (1999) Controlling the cellular brakes. Nature 401, 657–658.

Chen, X., Hiller, M., Sancak, Y. and Fuller, M.T. (2005) Tissue-specific TAFs counteract Polycomb to turn on terminal differentiation. Science 310, 869–872.

Chinnadurai, G. (2002) CtBP, an unconventional transcriptional corepressor in development and oncogenesis. Molecular Cell 9, 213–224.

Chitikila, C., Huisinga, K.L., Irvin, J.D., Basehoar, A.D. and Pugh, B.F. (2002) Interplay of TBP inhibitors in global transcriptional control. Molecular Cell 10, 871–882.

Courey, A.J. and Jia, S. (2001) Transcriptional repression: the long and the short of it. Genes and Development 15, 2786–2796.

Cowell, I.G. and Hurst, H.C. (1994). Transcriptional repression by the human bZIP factor E4BP4: definition of a minimal repressor domain. Nucleic Acids Research 22, 59–65.

Czermin, B., Melfi, R., McCabe, D., Seitz, V., Imhof, A. and Pirrotta, V. (2002) *Drosophila* enhancer of Zeste/ESC complexes have a histone H3 methyltransferase activity that marks chromosomal polycomb sites. Cell 111, 185–196.

Gerasimova, T.I. and Corces, V.G. (1998) Polycomb and trithorax group proteins mediate the function of a chromatin insulator. Cell 92, 511–521.

Guo, S., Yamaguchi, Y., Schillbach, S., Wada, T., Lee, J., Goddard, A., French, D., Handa, H. and Rosenthal, A. (2000) A regulator of transcriptional elongation controls vertebrate neuronal development. Nature 408, 366–369.

Haerizadeh, F., Singh, M.B. and Bhalla, P.L. (2006) Transcriptional repression distinguishes somatic from germ cell lineages in a plant. Science 313, 496–499.

Han, K. and Manley, J.L. (1993). Transcriptional repression by the *Drosophila* even skipped protein: definition of a minimal repressor domain. Genes and Development 7, 491–503.

Han, K., Levine, M.S. and Manley, J.L. (1989) Synergistic activation and repression of transcription by *Drosophila* homeobox proteins. Cell 56, 573–583.

Hanna-Rose, W. and Hansen, U. (1996) Active repression mechanisms of eukaryotic transcriptional repressors. Trends in Genetics 12, 229–234.

He, G-P., Muise, A., Li, A.W. and Ro, H-S. (1995) A eukaryotic transcriptional repressor with carboxypeptidase activity. Nature 378, 92–96.

Hermanson, O., Jepsen, K. and Rosenfeld, M.G. (2002) N-CoR controls differentiation of neural stem cells into astrocytes. Nature 419, 934–939.

Holden, C. (2006) Gene-suppressing proteins reveal secrets of stem cells. Science 312, 349.

Inostroza, J.A., Mermelstein, F.H., Ha, I., Lane, W.S. and Reinberg, D. (1992) Dr1, a TATA-binding protein-associated phosphoprotein and inhibitor of class II gene transcription. Cell 70, 477–489.

Janknecht, R. and Hunter, T. (1996) A growing co-activator network. Nature 383, 22–23.

Jepsen, K., Hermanson, O., Onami, T.M., Gleiberman, A.S., Lunyak, V., McEvilly, R.J., Kurokawa, R., Kumar, V., Liu, F., Seto, E., Hedrick, S.M., Mandel, G., Glass, C.K., Rose, D.W. and Rosenfeld, M.G. (2000) Combinatorial roles of the nuclear receptor corepressor in transcription and development. Cell 102, 753–763.

Kakkis, E., Riggs, K.J., Gillespie, W. and Calame, K. (1989) A transcriptional repressor of C-myc. Nature 339, 718–721.

Kamei, Y., Heinzel, T., Torchia, J., Kurokawa, R., Gloss, B., Lin, S.C., Heyman, R.A., Rose, D.W., Glass, C.K. and Rosenfeld, M.G. (1996) A CBP integrator complex mediates transcriptional activation and AP-1 inhibition by nuclear receptors. Cell 85, 403–414.

Kavi, H.H., Fernandez, H.R., Xie, W. and Birchler, J.A. (2006) Polycomb, pairing and P. Trends in Biochemical Sciences 31, 485–487.

Klose, R.J., Yamane, K., Bae, Y., Zhang, D., Erdjument-Bromage, H., Tempst, P., Wong, J. and Zhang, Y. (2006) The transcriptional repressor JHDM3A demethylates trimethyl histone H3 lysine 9 and lysine 36. Nature 442, 312–316.

Knoepfler, P.S. and Eisenman, R.N. (1999) Sin meets NuRD and other tails of repression. Cell 99, 447–450.

Koenig, R.G., Lazar, M.A., Hoden, R.A., Brent, G.A., Larsen, P.R., Chin, W.W. and Moore, D.D. (1989) Inhibition of thyroid hormone action by a non hormone binding c-erb A protein generated by alternative RNA splicing. Nature 337, 659–661.

Kuznetsova, A.V., Meller, J., Schnell, P.O., Nash, J.A., Ignacak, M.L., Sanchez, Y., Conaway, J.W., Conaway, R.C. and Czyzyk-Krzeska, M.F. (2003) von Hippel–Lindau protein binds hyperphosphorylated large subunit of RNA polymerase II through a proline hydroxylation motif and targets it for ubiquitination. Proceedings of the National Academy of Sciences, USA 100, 2706–2711.

Lachner, M., O'Carroll, D., Rea, S., Mechtler, K. and Jenuwein, T. (2001) Methylation of histone H3 lysine 9 creates a binding site for HP1 proteins. Nature 410, 116–120.

Lania, L., Majello, B. and DeLuca, P. (1997) Transcriptional regulation by the Sp family proteins. International Journal of Biochemistry and Cell Biology 29, 1313–1323.

Latchman, D.S. (1996). Inhibitory factors. International Journal of Biochemistry and Cell Biology 28, 965–974.

Latchman, D.S. (2001) Transcription factors: bound to activate or repress. Trends in Biochemical Sciences 26, 211–213.

Latchman, D.S. (2005) Gene regulation: a eukaryotic perspective. Fifth edition. Taylor and Francis, Oxford and New York.

Lee, M.G., Wynder, C., Cooch, N. and Shiekhattar, R. (2005) An essential role for CoREST in nucleosomal histone 3 lysine 4 demethylation. Nature 437, 432–435.

Lei, E.P. and Corces, V.G. (2006) A long-distance relationship between RNAi and Polycomb. Cell 124, 886–888.

Levine, S.S., King, I.F.G. and Kingston, R.E. (2004) Division of labour in Polycomb group repression. Trends in Biochemical Sciences 29, 478–485.

Li, S., Li, Y., Carthew, R.W. and Lai, Z-C. (1997) Photoreceptor cell differentiation requires regulated proteolysis of the transcriptional repressor tramtrack. Cell 90, 469–478.

Lillycrop, K.A., Dawson, S.J., Estridge, J.K., Gester, T., Matthias, P. and Latchman, D.S. (1994) Repression of a herpes simplex virus immediate–early promoter by the Oct-2 transcription factor is dependent upon an inhibitory region at the N-terminus of the protein. Molecular and Cellular Biology 14, 7633–7642.

Luecke, H.F. and Yamamoto, K.R. (2005) The glucocorticoid receptor blocks P-TEFb recruitment by NFκB to effect promoter-specific transcriptional repression. Genes and Development 19, 1116–1127.

Lunyak, V.V. and Rosenfeld, M.G. (2005) No rest for REST: REST/NRSF regulation of neurogenesis. Cell 121, 499–501.

Lyden, D., Young, A.Z., Zagzag, D., Yan, W., Gerald, W., O'Reilly, R., Bader, B.L., Hynes, R. O., Zhuang, Y., Manova, K. and Benezra, R. (1999) Id1 and Id3 are required for neurogenesis, angiogenesis and vascularization of tumour xenografts. Nature 401, 670–677.

Mal, A.K. (2006) Histone methyltransferase Suv39h1 represses MyoD-stimulated myogenic differentiation. EMBO Journal 25, 3323–3334.

Maldonado, E., Hampsey, M. and Reinberg, D. (1999) Repression: targeting the heart of the matter. Cell 99, 455–458.

Marx, J. (2005) Combing over the Polycomb group proteins. Science 308, 624–626.

Mohd-Sarip, A. and Verrijzer, C.P. (2004) A higher order of silence. Science 306, 1484–1485.

Müller, J., Hart, C.M., Francis, N.J., Vargas, M.L., Sengupta, A., Wild, B., Miller, E.L., O'Connor, M.B., Kingston, R.E. and Simon, J.A. (2002) Histone methyltransferase activity of a *Drosophila* polycomb group repressor complex. Cell 111, 197–208.

Nagy, L. and Schwabe, J.W. (2004) Mechanism of the nuclear receptor molecular switch. Trends in Biochemical Sciences 29, 317–324.

Ng, H-H. and Bird, A. (2000) Histone deacetylases: silencers for hire. Trends in Biochemical Sciences 25, 121–126.

Orlando, V. (2003) Polycomb, epigenomes and control of cell identity. Cell 112, 599–606.

Ostendorff, H.P., Peirano, R.I., Peters, M.A., Schlüter, A., Bossenz, M., Scheffner, M. and Bach, I. (2002) Ubiquitination-dependent cofactor exchange on LIM homeodomain transcription factors. Nature 416, 99–103.

Pal-Bhadra, M., Leibovitch, B.A., Gandhi, S.G., Rao, M., Bhadra, U., Birchler, J.A. and Elgin, S.C. (2004) Heterochromatic silencing and HP1 localization in *Drosophila* are dependent on the RNAi machinery. Science 303: 669–672.

Papp, B. and Muller, J. (2006) Histone trimethylation and the maintenance of transcriptional ON and OFF states by trxG and PcG proteins. Genes and Development 20, 2041–2054.

Rana, T.M. (2007) Illuminating the silence: understanding the structure and function of small RNAs. Nature Reviews Molecular Cell Biology 8, 23–36.

Ringrose, L., Ehret, H. and Paro, R. (2004) Distinct contributions of histone H3 lysine 9 and 27 methylation to locus-specific stability of polycomb complexes. Molecular Cell 16, 641–653.

Ronemus, M. and Martienssen, R. (2005) RNA interference: methylation mystery. Nature 433, 472–473.

Rosenfeld, M.G. and Glass, C.K. (2001) Coregulator codes of transcriptional regulation by nuclear receptors. Journal of Biological Chemistry 276, 36865–36868.

Rosenfeld, M.G., Lunyak, V.V. and Glass, C.K. (2006) Sensors and signals: a coactivator/corepressor/epigenetic code for integrating signal-dependent programs of transcriptional response. Genes and Development 20, 1405–1428.

Sauer, F., Fondell, J.D., Ohkuma, Y., Roeder, R. and Jackle, H. (1995) Control of transcription by Kruppel through interactions with TFIIB and TFIIEβ. Nature 375, 162–164.

Schwartz, Y.B. and Pirrotta, V. (2007) Polycomb silencing mechanisms and the management of genomic programmes. Nature Reviews Genetics 8, 9–22.

Smith, R.L. and Johnson, A.D. (2000) Turning genes off by Ssn6-Tup1: a conserved system of transcriptional repression in eukaryotes. Trends in Biochemical Sciences 25, 325–330.

Sontheimer, E.J. and Carthew, R.W. (2005) Silence from within: endogenous siRNAs and miRNAs. Cell 122, 9–12.

Sparmann, A. and van Lohuizen, M. (2006) Polycomb silencers control cell fate, development and cancer. Nature Reviews Cancer 6, 846–856.

Taghavi, P. and van Lohuizen, M. (2006) Two paths to silence merge. Nature 439, 794–795.

Tang, A.H., Neufeld, T.P., Kwan, E. and Rubin, G.M. (1997) PHYL acts to down regulate TTK88, a transcriptional repressor of neuronal cell fates by a SINA-dependent mechanism. Cell 90, 459–467.

Toledo, F. and Wahl, G.M. (2006) Regulating the p53 pathway: in vitro hypotheses, in vivo veritas. Nature Reviews Cancer 6, 909–923.

Tyler, J.K. and Kadonaga, J.T. (1999) The 'dark side' of chromatin remodelling: repressive effects on transcription. Cell 99, 443–446.

Vaucheret, H. (2005) RNA polymerase IV and transcriptional silencing. Nature Genetics 37, 659–660.

Vaughn, M.W. and Martienssen, R.A. (2005) Finding the right template: RNA Pol IV, a plant-specific RNA polymerase. Molecular Cell 17, 754–756.

Wassenegger, M. (2005) The role of the RNAi machinery in heterochromatin formation. Cell 122, 13–16.

White, R.J., Khoo, B.C-E., Inostroza, J.A., Reinberg, D. and Jackson, S.P. (1994) Differential regulation of RNA polymerases I, II and III by the TBP-binding repressor Dr1. Science 266, 448–450.

Willy, P.J., Kobayashi, R. and Kadonaga, J.T. (2000) A basal transcription factor that activates or represses transcription. Science 290, 982–984.

Yang, X.J. and Gregoire, S. (2005) Class II histone deacetylases: from sequence to function, regulation, and clinical implication. Molecular and Cellular Biology 25, 2873–2884.

Zhang, C., Comai, L. and Johnson, D.L. (2005) PTEN represses RNA Polymerase I transcription by disrupting the SL1 complex. Molecular and Cellular Biology 25, 6899–6911.

Zwaka, T.P. (2006) Keeping the noise down in ES cells. Cell 127, 1301–1302.

CHAPTER 7

REGULATION OF TRANSCRIPTION FACTOR SYNTHESIS

7.1 TRANSCRIPTION FACTOR REGULATION

Transcription factors play a central role in a number of biological processes, producing, for example, the induction of specific genes in response to particular stimuli as well as controlling the cell type specific or developmentally regulated expression of other genes. The ability to bind to DNA (Chapter 4) and influence the rate of transcription either positively (Chapter 5) or negatively (Chapter 6) are clearly features of many transcription factors which regulate gene expression in response to specific stimuli or in specific cell types. Most importantly, however, such factors must also have their activity regulated such that they only become active in the appropriate cell type or in response to the appropriate stimulus, thereby producing the desired pattern of gene expression.

Two basic mechanisms by which the action of transcription factors can be regulated have been described. These involve either controlling the synthesis of the transcription factor so that it is made only when necessary (Fig. 7.1a) or, alternatively, regulating the activity of the factor so that pre-existing protein becomes activated when required (Fig. 7.1b). This chapter considers the regulation of transcription factor synthesis whilst Chapter 8 considers the regulation of transcription factor activity.

7.2 REGULATED SYNTHESIS OF TRANSCRIPTION FACTORS

Regulating the synthesis of transcription factors such that they are only made when the genes which they regulate are to be activated is an obvious mechanism of ensuring that specific genes become activated only at the appropriate time and place. This mechanism is widely used therefore, particularly for transcription factors which regulate the expression of cell type specific or developmentally regulated genes.

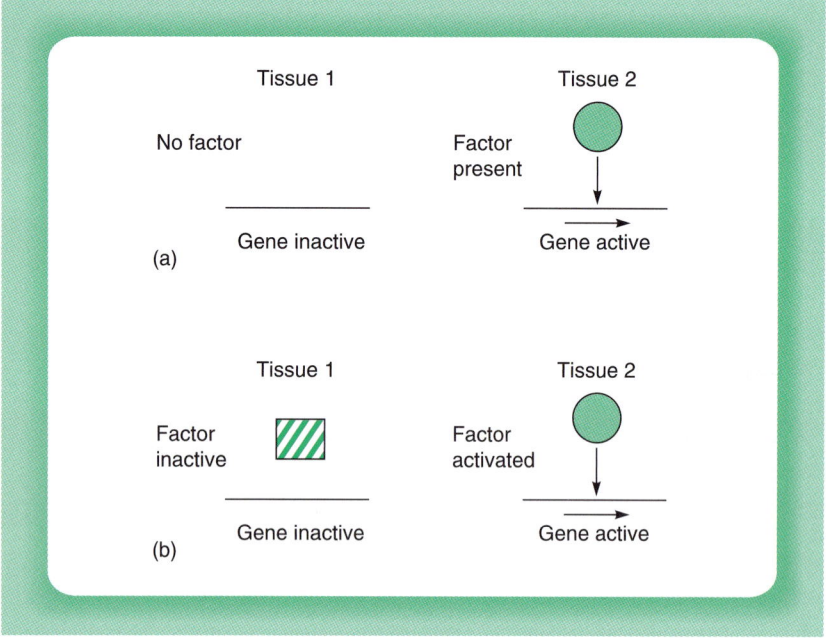

Figure 7.1

Gene activation mediated by the synthesis of a transcription factor only in a specific tissue (a) or its activation in a specific tissue (b).

Thus, for example, as discussed in Chapter 6 (section 6.4.1), the poly-comb transcriptional repressor maintains embryonic stem cells in an undifferentiated state by repressing the transcription of other genes whose expression is required to produce different types of differentiated cells. A number of these polycomb-repressed genes have been shown to encode transcription factors which can induce the stem cells to differentiate into particular types of cells. These polycomb-repressed transcription factors include members of several different families, discussed in Chapter 4, such as homeobox-containing factors, POU factors, Pax factors and helix-loop-helix proteins. Hence, polycomb maintains the undifferentiated state by preventing the synthesis of specific transcription factors (Fig. 7.2).

Two specific examples of the regulated synthesis of particular tran-scription factors which illustrate the role of this mechanism in regulating cell type specific or developmental gene expression are discussed in the following sections.

7.2.1 THE MYoD TRANSCRIPTION FACTOR

Probably the most novel approach to the cloning of the gene encoding a transcription factor was taken by Davis *et al.* (1987), who isolated cDNA

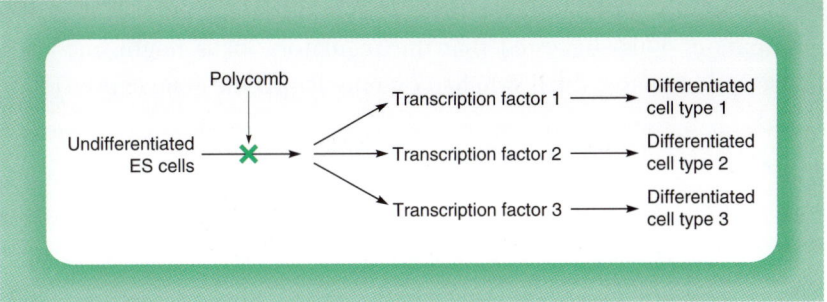

Figure 7.2

The polycomb transcription repressor maintains embryonic stem cells in an undifferentiated state by preventing the transcription of genes encoding transcription factors which are required for differentiation into specific cell types. Compare with Figure 6.25 and note that a number of the polycomb-regulated genes shown in that diagram have been shown to encode transcription factors.

clones encoding MyoD, a factor which plays a critical role in skeletal muscle-specific gene regulation. They used an embryonic muscle fibroblast cell line known as C3H 10T½. Although these cells do not exhibit any differentiated characteristics, they can be induced to differentiate into myoblast cells, expressing a number of muscle-lineage genes, upon treatment with 5-azacytidine (Constantinides *et al.*, 1977). This agent is a cytidine analogue having a nitrogen instead of a carbon atom at position 5 on the pyrimidine ring and is incorporated into DNA instead of cytidine. Unlike cytidine, however, it cannot be methylated at this position and hence its incorporation results in demethylation of DNA. As methylation of DNA at C residues is thought to play a critical role in transcriptional silencing of gene expression (for review see Latchman, 2005), this artificial demethylation can result in the expression of particular genes which were previously silent.

In the case of 10T½ cells, this demethylation was thought to result in the expression of, previously silent, regulatory loci which are necessary for differentiation into muscle myoblasts. Several experiments also suggested that the activation of only one key regulatory locus might be involved. Thus, 5-azacytidine induces myoblasts at very high frequency consistent with only the demethylation of one gene being required, whilst DNA prepared from differentiated cells can also induce differentiation in untreated cells at a frequency consistent with the transfer of only one activated locus.

Hence differentiation is thought to occur via the activation of one regulatory locus (gene X in Fig. 7.3),whose expression in turn switches on the expression of genes encoding muscle lineage markers, which is observed in the differentiated 10T½ cells and thereby induces their differentiation. This suggested that the regulatory locus might encode a transcription factor which switched on muscle-specific gene expression.

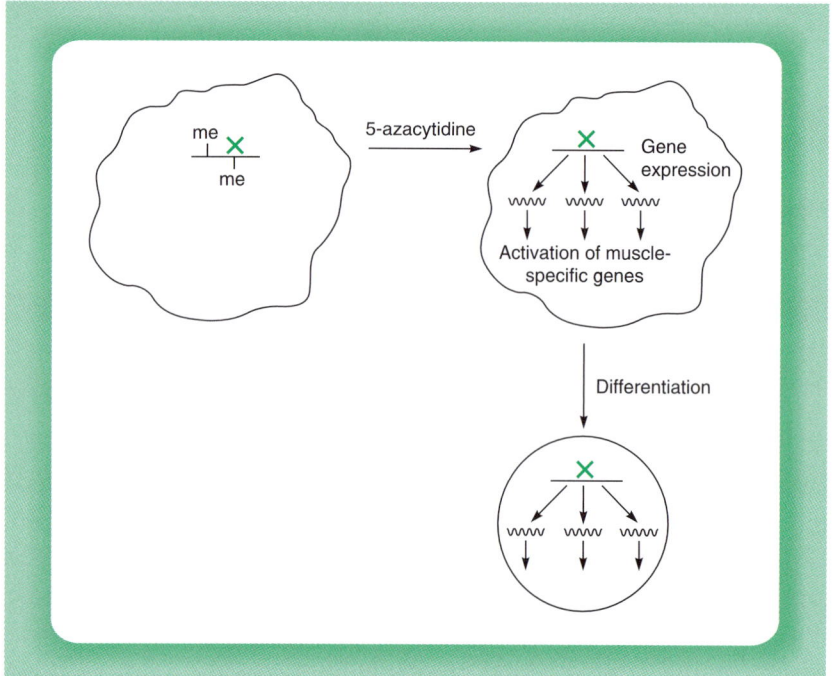

Figure 7.3

Model for differentiation of 10T½ cells in response to 5-azacytidine. Activation of a master locus (x) by demethylation allows its product to activate the expression of muscle-specific genes thereby producing differentiation.

To isolate the gene encoding this factor, Davis *et al.* (1987) reasoned that it would continue to be expressed in the myoblast cells but would evidently not be expressed in the undifferentiated cells. They therefore prepared RNA from the differentiated cells and removed from it by subtractive hybridization all the RNAs which were also expressed in the undifferentiated cells. After various further manipulations to exclude RNAs characteristic of terminal muscle differentiation such as myosin and others induced non-specifically in all cells by 5-azacytidine, the enriched probe was used to screen a cDNA library prepared from differentiated 10T½ cells.

This procedure (Fig. 7.4) resulted in the isolation of three clones, MyoA, MyoD and MyoH whose expression was specifically activated when 10T½ cells were induced to form myoblasts with 5-azacytidine. When each of these genes was artificially expressed in 10T½ cells, MyoA and MyoH had no effect. However, artificial expression of MyoD was able to

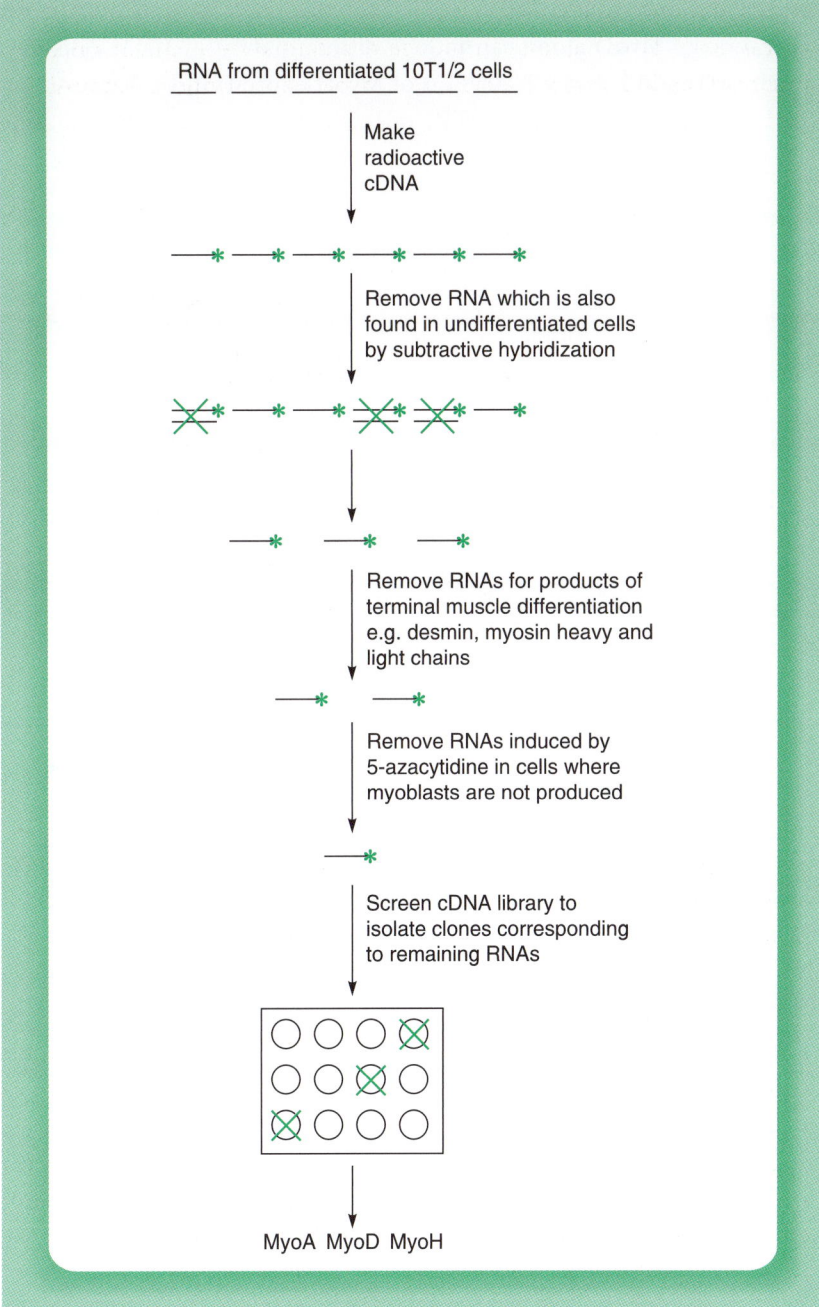

Figure 7.4

Strategy for isolating the master regulatory locus expressed in 10T½ cells after but not before treatment with 5-azacytidine. Subtractive hybridization was used to isolate all RNA molecules which are present in 10T½ cells only following treatment with 5-azacytidine. After removal of RNAs for terminal differentiation products of muscle and RNAs induced in non-muscle-producing cells by 5-azacytidine, the remaining RNAs were used to screen a cDNA library. Three candidates for the master regulatory locus MyoA, MyoD and MyoH were isolated in this way.

convert undifferentiated 10T½ cells into myoblasts (Fig. 7.5). Hence, expression of MyoD alone can induce differentiation of 10T½ cells into muscle cells and it is the induction of MyoD expression by 5-azacytidine which is responsible for the ability of this compound to induce muscle differentiation.

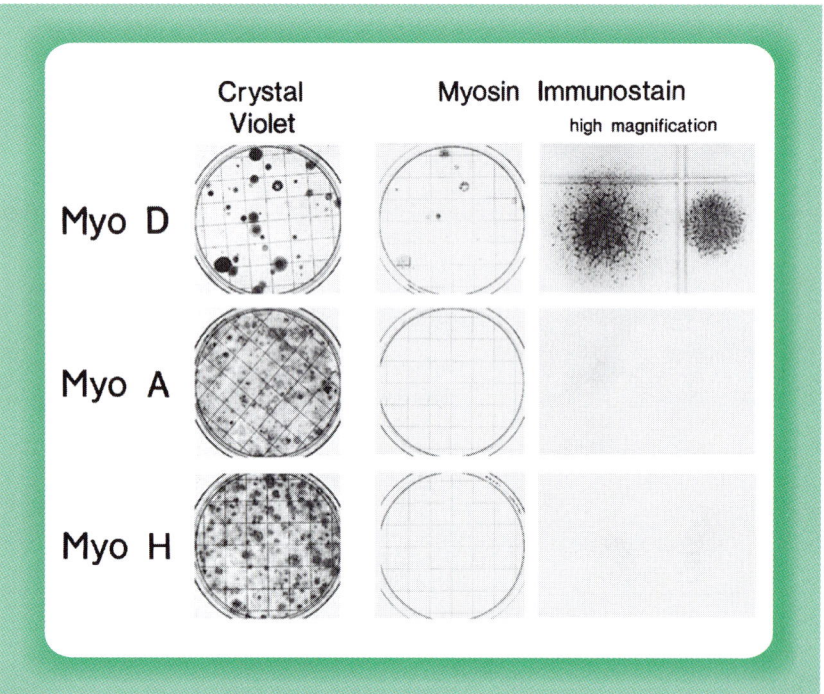

Figure 7.5

Test of each of the putative master regulatory loci MyoA, MyoD and MyoH. Each of the genes was introduced into 10T½ cells and tested for the ability to induce the cells to differentiate into muscle cells. Note that whilst MyoA and MyoH have no effect, introduction of MyoD results in the production of muscle cells which contain the muscle protein myosin. The differentiated muscle cells induced by MyoD cease to divide on differentiation resulting in less cells being detectable by staining with crystal violet compared to the MyoA and MyoH treated cells which continue to proliferate. Hence only MyoD has the capacity to cause 10T½ cells to differentiate into non-proliferating muscle cells producing myosin, identifying it as a master regulatory locus for muscle differentiation.

The differentiated 10T½ cells produced by artificial expression of MyoD, like those induced by 5-azacytidine, express a variety of muscle lineage markers and indeed also switch on both MyoA and MyoH, as well as the endogenous MyoD gene itself. This suggests that MyoD is a transcription factor which switches on genes expressed in muscle cells. In agreement with this, MyoD was shown to bind to a region of the creatine kinase gene upstream enhancer, which was known to be necessary for its muscle-specific gene activity.

Moreover, it has been shown that MyoD can actually bind to its binding sites within target genes when they are in the tightly packed chromatin structure characteristic of genes which are inactive in a particular lineage (Gerber *et al.*, 1997). This binding results in the remodelling of the chromatin to a more open form and is then followed by enhanced transcription stimulated by MyoD (Fig. 7.6). This alteration in chromatin structure is likely to be dependent on the ability of MyoD to interact with the p300 co-activator protein (Puri *et al.*, 1997) (see Chapter 5, section 5.4.3). Like CBP, p300 has histone acetyltransferase activity and is therefore able to alter chromatin to the more open structure associated with acetylated histones (see Chapter 1, section 1.2.3).

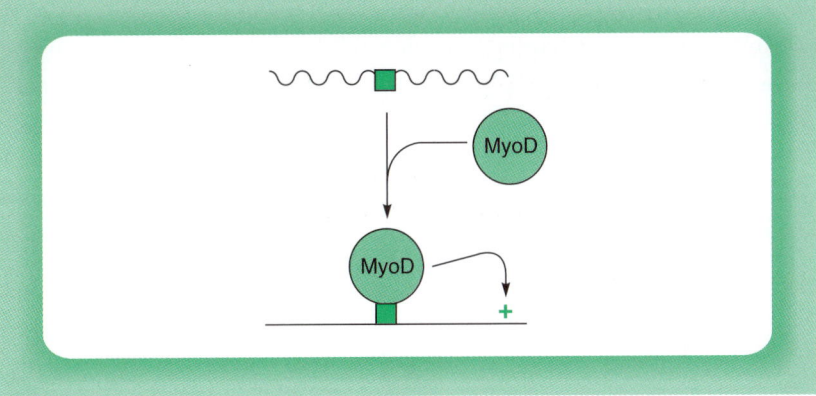

Figure 7.6
MyoD binding to its binding site (solid box) both converts the chromatin structure from a closed (wavy line) to a more open (solid line) structure compatible with transcription and also directly enhances the rate of transcription (arrow).

Hence, MyoD is capable of activating transcription by two distinct means, namely the remodelling of chromatin and the direct stimulation of enhanced transcription (see Chapter 5, for a discussion of the mechanisms of transcriptional activation). This is particularly important since it allows enhanced synthesis of MyoD to induce the development of myogenic cells from non-differentiated precursors, in which the genes that must be switched on are in an inactive closed chromatin structure that is inaccessible to many transcriptional activators.

Interestingly, as well as stimulating muscle-specific genes, MyoD also promotes differentiation by modulating gene expression so as to inhibit cellular proliferation, thereby producing the non-dividing phenotype characteristic of muscle cells. Thus, MyoD has been shown to activate the gene encoding the p21 inhibitor of cyclin-dependent kinases (Halevy *et al.*, 1995). This results in the inhibition of these kinases whose activity is necessary for cell division (see Chapter 9, section 9.4.2). In addition, MyoD can also repress the promoter of the c-*fos* gene whose protein product is important for cellular proliferation (see Chapter 9, section 9.3.1) indicating that MyoD can also act by repressing genes whose products are not required in non-dividing muscle cells (Trouche *et al.*, 1993).

Like gene activation by MyoD, repression of the c-*fos* promoter is dependent on DNA binding, which in this case prevents the binding of a positively acting factor to a site known as the serum response element that overlaps the MyoD binding site in the c-*fos* promoter (Fig. 7.7). Obviously, in contrast to its binding to the creatine kinase enhancer, MyoD must bind to its binding site in the c-*fos* promoter in a form which cannot activate transcription. Hence, like the glucocorticoid receptor, MyoD can have different effects on gene expression depending on the nature of its binding site (see Chapter 6, section 6.2.1 for discussion of the mechanism of transcriptional repression by the glucocorticoid receptor).

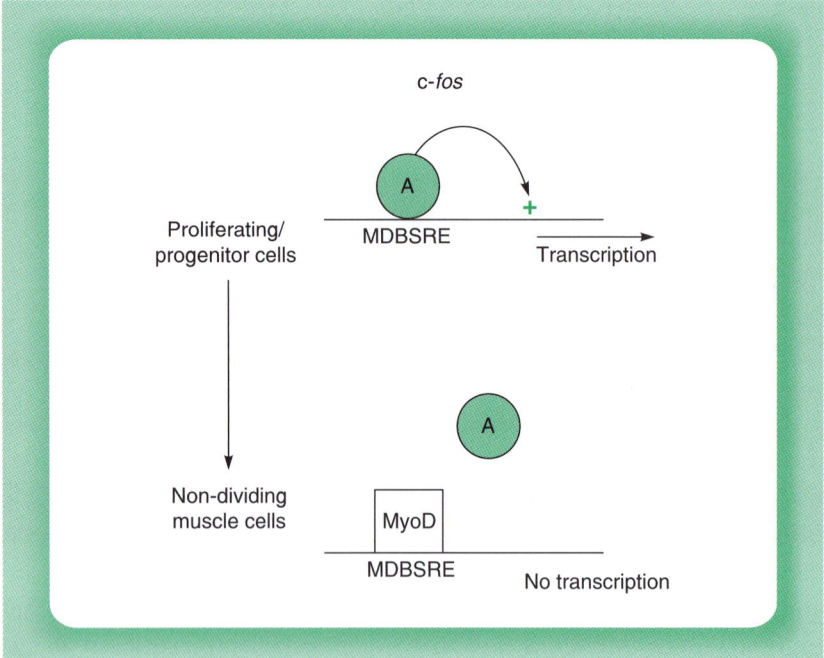

Figure 7.7
MyoD binds to its binding site (MDBS) in the c-*fos* promoter in a configuration which does not activate transcription and prevents binding of an activating factor (A) to the overlapping serum response element (SRE). This therefore results in the repression of c-*fos* transcription in MyoD-expressing muscle cells.

In both cases, however, DNA binding by MyoD is dependent upon a basic region of the protein which binds directly to the DNA and an adjacent region which can form a helix-loop-helix structure and is essential for dimerization of MyoD (see Chapter 4, section 4.5 for further discussion of these motifs).

Hence, MyoD can induce muscle differentiation by both activating and repressing the expression of protein-coding genes. Interestingly, it has recently been shown that MyoD can also induce the expression of specific microRNAs which are expressed only in muscle differentiation. As described in Chapter 6 (section 6.4.2), such small RNAs repress the expression of other genes. Hence, in this case, it is likely that MyoD indirectly represses the expression of non-muscle genes by activating the transcription of specific microRNAs (Rao *et al.*, 2006). MyoD therefore utilizes multiple mechanisms to induce the muscle phenotype (Fig. 7.8).

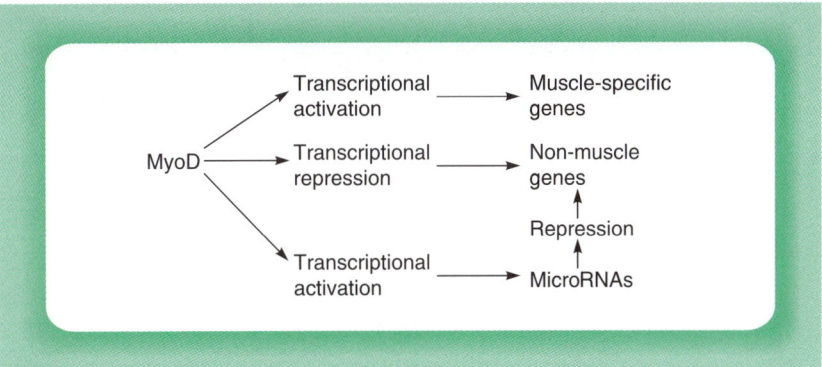

Figure 7.8

The MyoD transcription factor induces muscle cell differentiation by activating the expression of muscle-specific genes and by inhibiting the expression of non-muscle genes, both directly by transcriptional repression and indirectly by inducing the expression of inhibitory microRNAs.

Hence, synthesis of MyoD results in the production of the skeletal muscle phenotype by activating and repressing the expression of specific target genes. As expected, in view of the critical role which MyoD plays in the development of muscle cells, the MyoD mRNA is present in skeletal muscle tissue taken from a variety of different sites in the body but is absent in all other tissues, including cardiac muscle (Davis *et al.*, 1987) (Fig. 7.9). The MyoD mRNA and protein therefore accumulate only in a specific cell type where it is required and the activation of the MyoD gene during myogenesis is likely to be of central importance in switching on the expression of muscle-specific genes. In turn, this suggests that other developmentally regulated transcription factors will be involved in switching on MyoD expression during myogenesis. In agreement with this, the paired-type homeobox factor Pax3 (Chapter 4, section 4.2.7) has been shown to activate MyoD expression and myogenic differentiation in a variety of non-muscle cell types, whilst the classical homeobox factor MSX1 can repress the transcription of the MyoD gene.

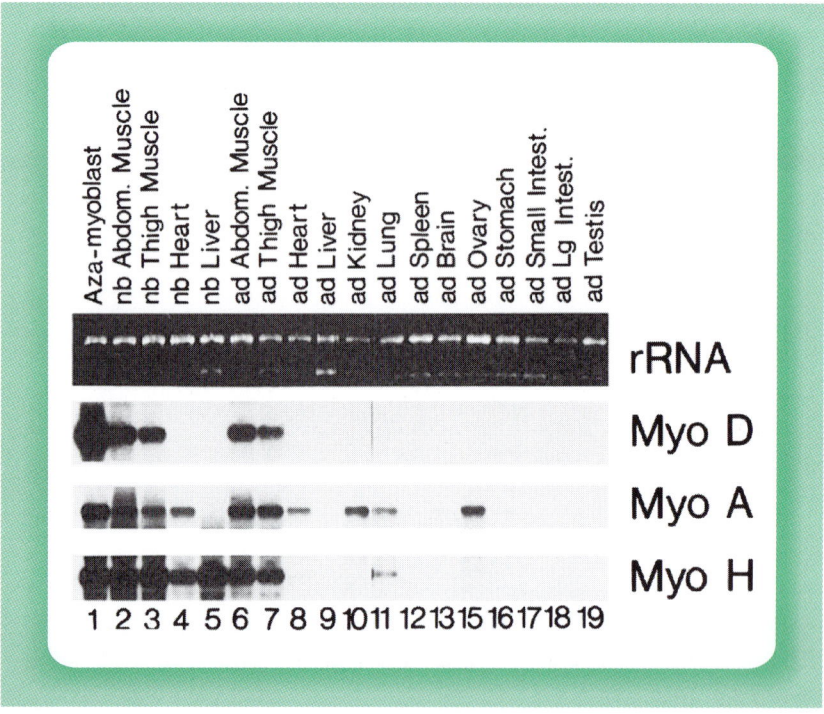

Figure 7.9

Northern blotting experiment to detect the mRNAs encoding MyoA, MyoD and MyoH in different muscle and non-muscle tissues. Note that the MyoD mRNA is present only in skeletal muscle as expected in view of its ability to produce muscle differentiation whereas the MyoA and MyoH mRNAs are more widely distributed. nb, indicates new born; ad, indicates adult; rRNA, indicates the ribosomal RNA control used to show that all samples contain intact RNA.

Thus, MyoD is a transcription factor whose regulated synthesis results in the activation of muscle-specific gene expression and the production of skeletal muscle cells. Interestingly, the observation that the introduction of MyoD into cells switches on the endogenous MyoD gene (see above) suggests that a positive feedback loop normally regulates MyoD expression, so that once the gene is initially expressed, expression is maintained, producing commitment to the myogenic lineage (Fig. 7.10). This is of importance since MyoD appears to be essential for the repair of damaged muscle in adult animals, indicating that its expression must be maintained throughout life.

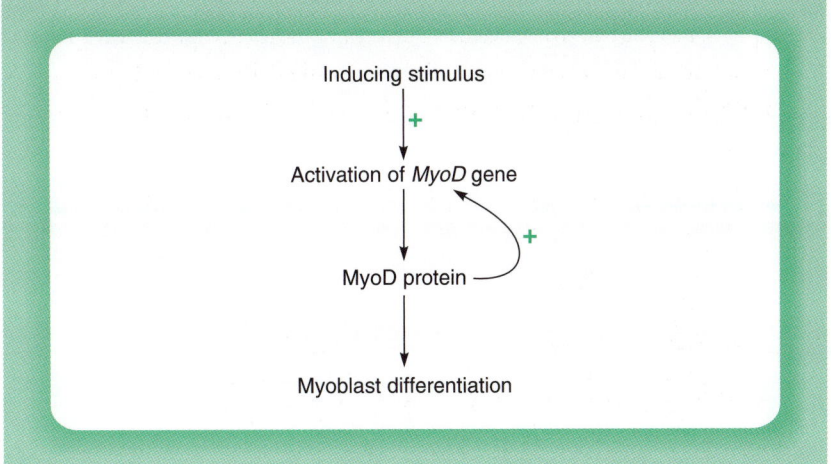

Figure 7.10
Ability of MyoD protein to activate expression of its own gene creating a positive feedback loop which ensures that following an initial stimulus, the MyoD protein is continuously produced and hence maintains myoblast differentiation.

MyoD therefore offers a classical example of the role of transcription factor synthesis in regulating cell type specific gene expression. It should be noted, however, that as discussed in Chapter 6 (section 6.2.2), the activity of MyoD is also regulated by its interaction with the Id inhibitor protein. Hence, MyoD is regulated both by regulating its synthesis and by regulating its activity.

7.2.2 HOMEOBOX TRANSCRIPTION FACTORS

In addition to its role in controlling cell type specific gene expression, regulation of transcription factor synthesis is also widely used in the control of developmentally regulated gene expression. Thus, numerous studies of the *Drosophila* homeobox transcription factors discussed in Chapter 4 (section 4.2) using both immunofluorescence with specific antibodies and *in situ* hybridization, have revealed highly specific expression patterns for individual factors and the mRNAs which encode them, indicating that their role in regulating gene expression in development is dependent, at least in part, on the regulation of their synthesis (Fig. 7.11).

Moreover, such regulated synthesis of specific transcription factors can specifically determine the nature of the cell types which are produced during development. Thus, the LIM homeobox factors Lhx3 and Lhx4

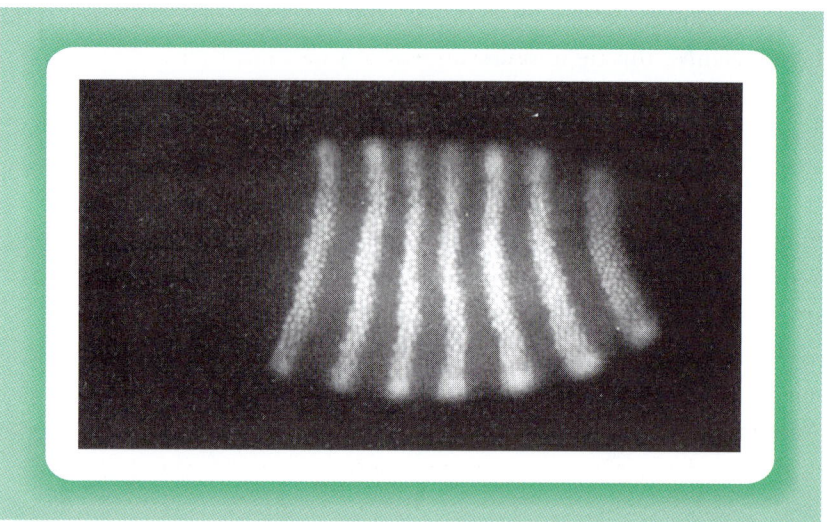

Figure 7.11

Localization of the Ftz protein in the *Drosophila* blastoderm embryo using a fluorescent antibody which reacts specifically with the protein. The anterior end of the embryo is to the left and the dorsal surface to the top of the photograph. Note the precise pattern of seven stripes of Ftz-expressing cells around the embryo.

are expressed transiently in motor neurones whose axons extend ventrally (v-MN) but not in those which extend dorsally (d-MN). In knock out mice lacking both Lhx3 and Lhx4, cells that should become v-MN cells instead become d-MN cells (Fig. 7.12). In contrast, misexpression

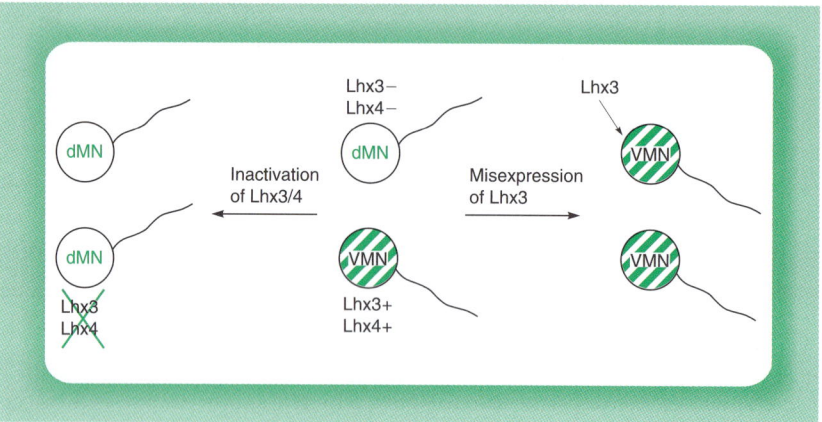

Figure 7.12

The homeobox transcription factors Lhx3 and Lhx4 are expressed in motor neurones whose axons project ventrally (vMN) but not in those which project dorsally (dMN). Inactivation of Lhx3 and Lhx4 in knock out mice converts vMN cells into dMN cells whereas artificial expression of Lhx3 in dMN cells converts them into vMN cells.

of Lhx3 is sufficient to convert d-MN cells into v-MN cells (Sharma *et al.*, 1998). Hence, the regulated synthesis of these two homeodomain proteins results in cells which express them becoming one type of motor neurone (v-MN), whereas the cells which do not express them, form a different type of motor neurone (d-MN).

In an even more dramatic example of this effect, the Pitx2 homeobox factor is expressed only on the left side of the developing embryo in the mouse or chicken. Expressing Pitx2 on the right side of the embryo affects the normal pattern of asymmetry between the left and right sides of the embryo (Logan *et al.*, 1998; Piedra *et al.*, 1998), indicating that the appropriate regulation of its synthesis is required for the embryo to develop distinct left and right sides.

Hence, the regulated expression of homeobox factors is essential for their role in regulating gene transcription and cell fate in development. As discussed in Chapter 4 (section 4.2), the mouse homeobox genes are found in clusters containing a number of different genes (Fig. 7.13). Most interestingly, in both *Drosophila* and mammals, the position of a gene within a cluster is related to its expression pattern during embryogenesis. Thus, in the mouse Hoxb cluster, all the genes are expressed in the developing central nervous system of the embryo. However, in moving from the 5′ to the 3′ end of the cluster (i.e., from Hoxb-9 (2.5) to Hoxb-1 (2.9) in Fig. 7.13), each successive gene is expressed earlier in

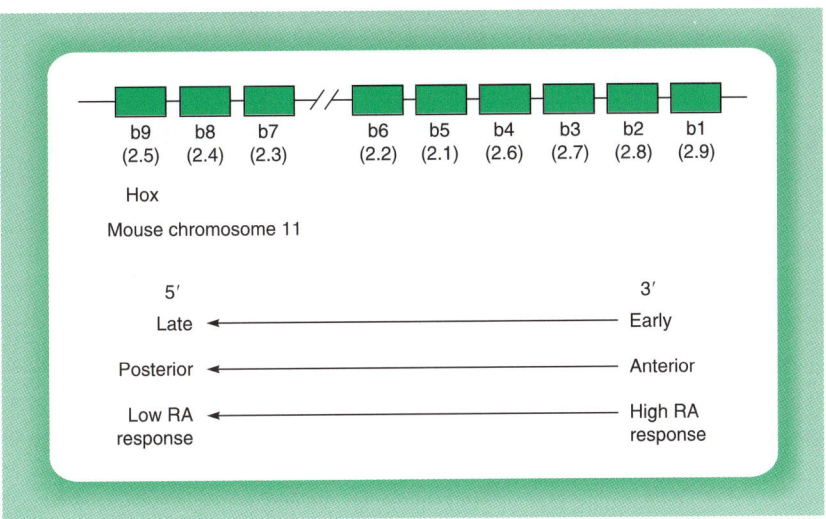

Figure 7.13

Hoxb gene cluster on mouse chromosome 11. Note that in moving from the left to the right of the mouse complex, the genes are expressed progressively earlier in development, have a more anterior boundary of expression and a greater responsiveness to retinoic acid.

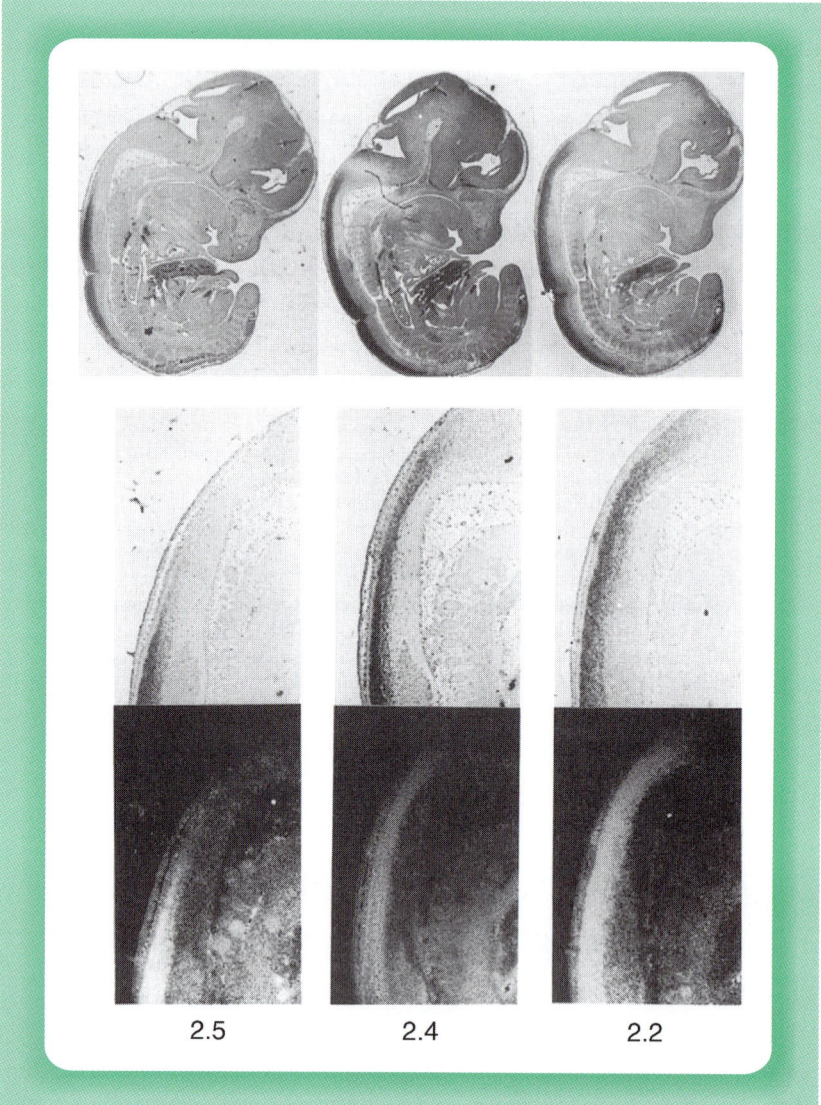

2.5   2.4   2.2

Figure 7.14

Comparison of the expression pattern of the Hox b-9 (2.5), b-8 (2.4) and b-6 (2.2) genes in the 12.5 day mouse embryo. The top panel shows *in situ* hybridization with the appropriate gene probe to a section of the entire embryo, whilst the middle row shows a high power view of the region in which the anterior limit of gene expression occurs. In these panels, which show the sections in bright field, hybridization of the probe and therefore gene expression is indicated by the dark areas. In the lower panel, which shows the same area in dark field, hybridization is indicated by the bright areas. Note the progressively more anterior boundary of expression of Hox b-6 (2.2) compared to Hox b-8 (2.4) and to Hox b-9 (2.5) and compare with their positions in the Hox b (Hox 2) complex in Figure 7.11.

development and also displays a more anterior boundary of expression within the central nervous system (Figs. 7.14, 7.15). Similar expression patterns have also been observed in *Drosophila*, where each successive gene in the Bithorax and Antennapedia clusters is expressed more anteriorly and affects progressively more anterior segments when it is mutated. Indeed, studies in which regulatory elements from the invertebrate *Amphioxus* were tested in mouse and chick embryos, have indicated that the elements regulating homeobox gene expression have been highly conserved in evolution with the *Amphioxus* elements functioning in these very different species (Manzanares *et al.*, 2000).

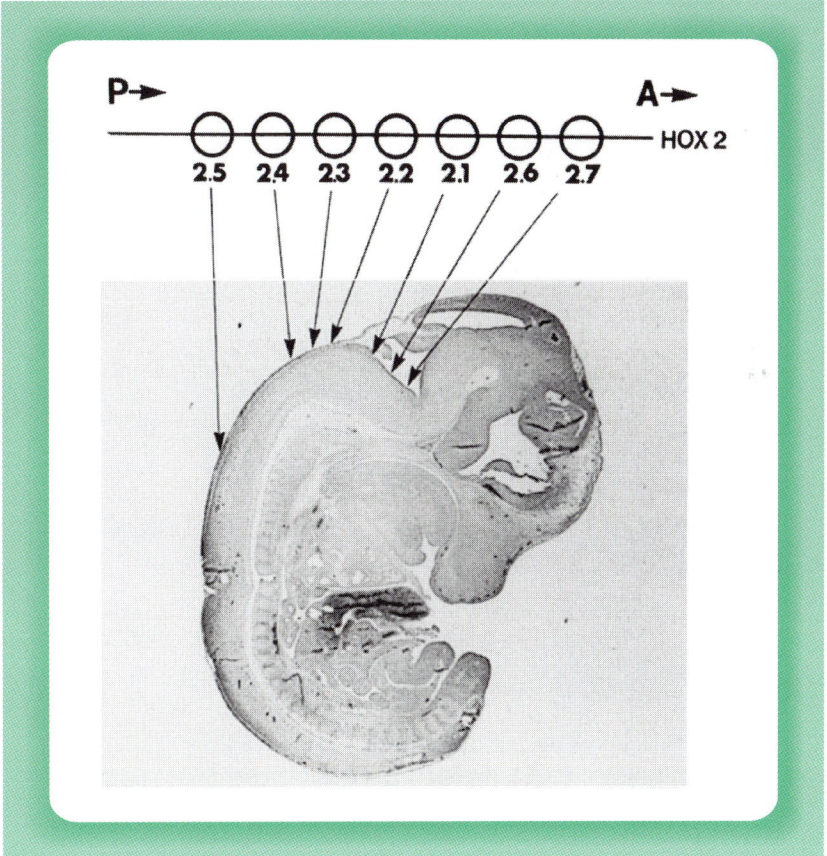

Figure 7.15

Summary of the anterior boundary of expression of the genes in the Hox b (2) complex indicated on a section of a 12.5 day mouse embryo and compared to the position of the gene in the Hox b (2) cluster. Note the progressively more anterior boundary of expression from the 5′ to the 3′ end of the Hox b (2) cluster.

In the case of the mouse genes, a possible molecular mechanism for the differential expression pattern across a cluster is provided by the finding that genes in the 3' half of the Hoxb cluster are activated in cultured cells by treatment with low levels of retinoic acid, whereas genes in the 5' half of the cluster require much higher levels of retinoic acid for their activation. Considerable evidence exists that retinoic acid can act as a morphogen in vertebrate development and it has been suggested that a gradient of retinoic acid concentration may exist across the developing embryo. Hence, the observed difference in expression of the Hoxb genes could be controlled by a retinoic acid gradient (Fig. 7.16). In turn, the Hoxb genes, like their *Drosophila* counterparts, would switch on other genes required in cells at particular positions in the embryo accounting for the morphogenetic effects of retinoic acid.

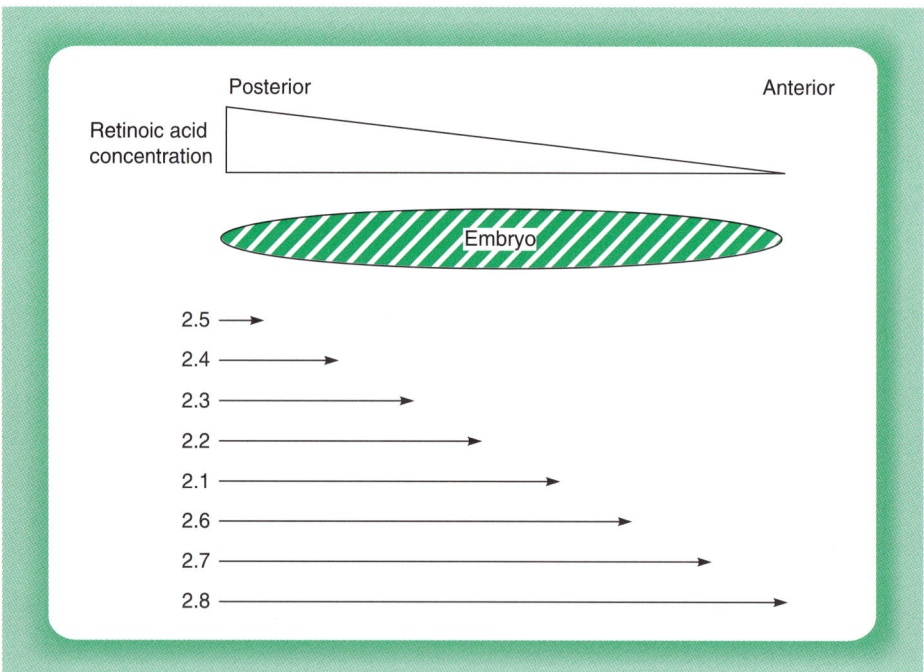

Figure 7.16

Model for the progressively more anterior expression of the genes in the Hox b (2) cluster in which expression is controlled by a posterior to anterior gradient in retinoic acid concentration and the increasing sensitivity to induction by retinoic acid which occurs from the 5' to the 3' end of the cluster. Thus because genes at the 3' end of the cluster are inducible by very low levels of retinoic acid they will be expressed in anterior points of the embryo where the retinoic acid level will be too low to induce the genes at the 5' end of the cluster which require a much higher level of retinoic acid to be activated.

Retinoic acid functions by binding to and activating specific receptors which are members of the nuclear receptor super family and which in turn bind to specific sequences within retinoic acid responsive genes, activating their expression (see Chapter 4, section 4.4, and Chapter 8, section 8.2.2). Hence, the activation of regulatory genes and the initiation of a regulatory cascade can be achieved by the activation of specific receptors/transcription factors by an inducing stimulus.

This illustrates therefore how the synthesis of one set of transcription factors (the homeobox proteins) can be regulated by the activation of another set of transcription factors (the retinoic acid receptors). In agreement with this idea, the treatment of mouse embryos with retinoic acid results, for example, in changes in the expression pattern of the Hoxb-1 gene, which contains a retinoic acid response element in its 3' regulatory region. Moreover, the inactivation of this element so that it no longer binds the retinoic acid receptors, abolishes expression of Hoxb-1 in the neuroectoderm of the early embryo, providing direct evidence that the retinoic acid response element is necessary to produce the expression pattern of this gene observed in the developing embryo (for review see Stern and Foley, 1998).

Interestingly, the regulation of Hox gene expression by such DNA response elements located adjacent to the individual genes appears to interact with other regulatory processes which operate over the whole gene cluster. Thus, in experiments where individual Hox genes (with their adjacent control elements) were moved to a different position within the gene cluster, their pattern of expression was altered so that they behaved similarly to genes normally located at that position in the cluster, for example, in terms of the time at which they were switched on during development (van der Hoeven et al., 1996) (Fig. 7.17).

In the case of the HoxD cluster, this effect appears to involve the order of the genes relative to a distant enhancer element located at least 100,000 bases away. Thus, in this cluster the first gene, HoxD13, is expressed most anteriorly and at the highest level with each successive gene being expressed at lower levels and more posteriorly. If HoxD13 is deleted, the next gene in the cluster HoxD12 is now expressed in the manner typical of HoxD13 even though it remains in its normal position (for review see Zeller and Deschamps, 2002). In this case, therefore, the genes appear to compete to interact with the distant enhancer element, so that the closest gene is expressed in a particular pattern and so on (Fig. 7.18). This effect is evidently reminiscent of the locus control region (LCR) in the β-globin gene cluster (see Chapter 1, section 1.3.4) where the globin genes were expressed in a specific order in development, which is determined by their position relative to the LCR.

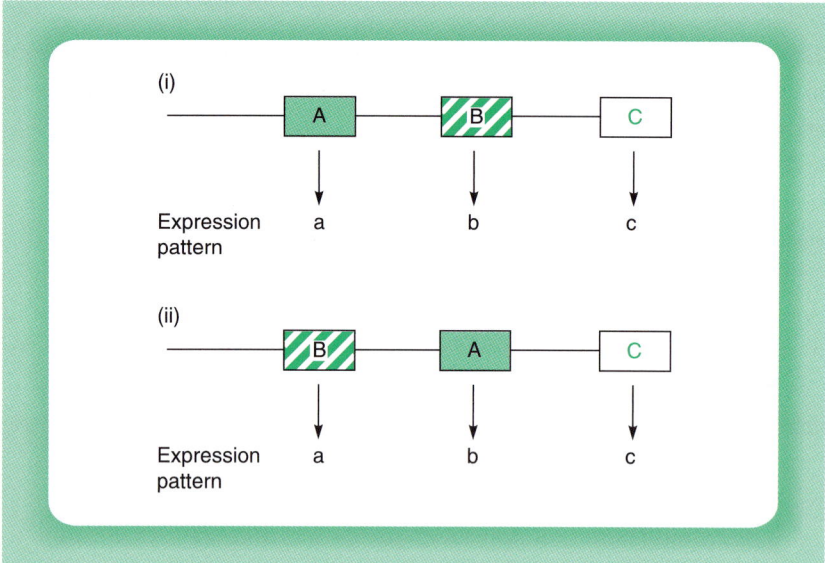

Figure 7.17

Each gene in a Hox cluster has its own specific expression pattern (panel i).
Moving a particular gene to a new position in the cluster results in it having the
expression pattern of the gene which is normally located at that position (panel ii).

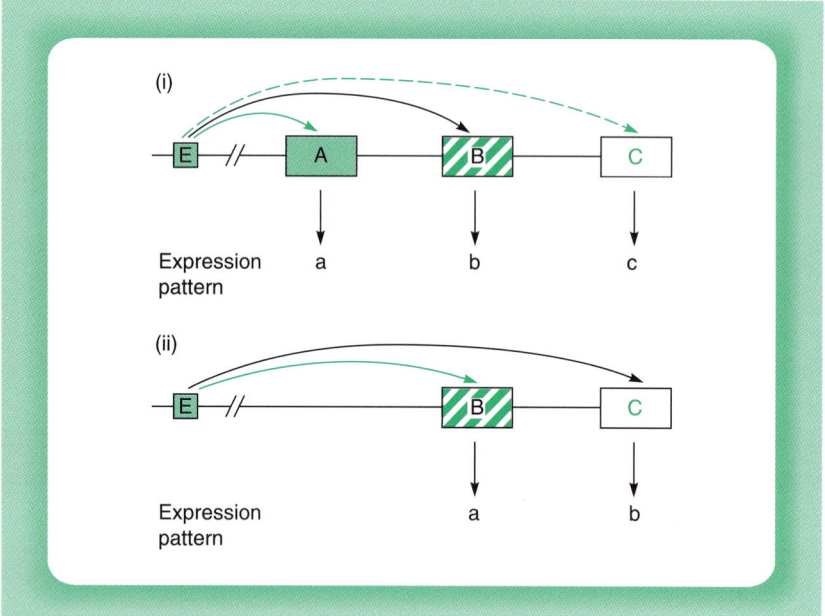

Figure 7.18

In the HoxD cluster, the expression of the genes is affected by their order relative
to a distant enhancer element (E) (panel i). Deletion of gene A results in gene B
being the closest to the enhancer. It is therefore expressed in the normal pattern
for gene A even though its physical location is unchanged (panel ii).

The specific pattern of expression of individual homeobox genes, which is determined by their position in the cluster, is absolutely critical to their function. Indeed, it appears that it is the different patterns of regulation rather than the different proteins which they encode, that determine the different roles of specific genes in a cluster. Thus, if an individual gene in a cluster is deleted, the other genes in the cluster cannot substitute for it and an abnormal animal results (Fig. 7.19). However, if the deleted gene is replaced by a further copy of another gene in the

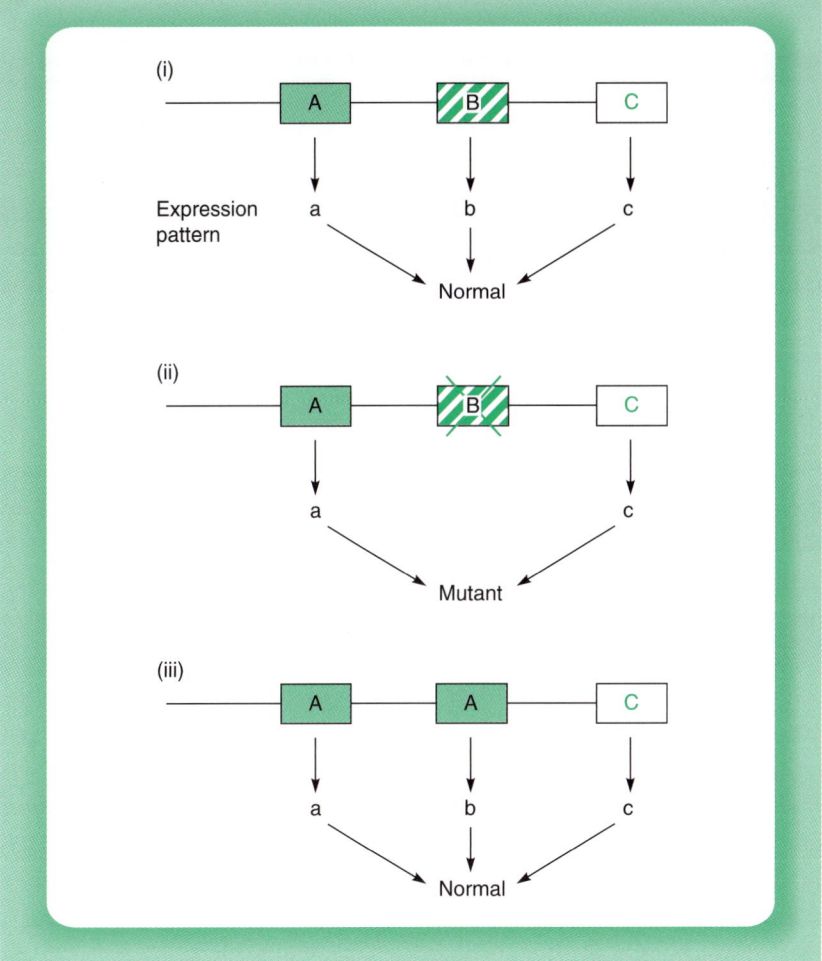

Figure 7.19

The normal expression pattern of Hox genes results in the production of a normal animal (panel i). Inactivation of a specific gene results in a mutant animal being produced (panel ii). However, if the mutant gene is replaced with a further copy of another gene in the cluster, this gene is expressed in the same way as the deleted gene and a normal animal results (panel iii).

cluster, then a normal animal results (for review see Duboule, 2000) (Fig. 7.19). This occurs because the expression of the inserted gene is now determined by its position in the cluster and it is therefore expressed in the manner characteristic of the deleted gene. Hence, the products of different genes in a cluster can functionally substitute for one another but only if they are expressed in the appropriate pattern, as determined by their position in the cluster. This illustrates the critical role of the regulated synthesis of transcription factors in allowing them to produce their functional effects.

The manner in which the regulated synthesis of multiple homeobox factors can regulate the production of several different cell types has been analysed in detail in the ventral neural tube. In this case, the system is regulated by a gradient in the concentration of a protein signalling molecule known as sonic hedgehog (Shh) rather than via a retinoic acid gradient. The expression of several homeobox factors (Dbx1, Dbx2, Irx3 and Pax6) is repressed by Shh, but their sensitivity to such repression differs so that Pax6, for example, is expressed at higher Shh concentrations than Irx3 and so on (Fig. 7.20). In contrast, two other homeobox genes are activated by Shh, but their sensitivity differs so that Nkx6.1 is expressed at lower levels of Shh than Nkx2.2 (Fig. 7.20).

The different expression patterns of these genes, therefore, convert the gradient of Shh expression into a homeobox code, in which each region has a unique pattern of expression of the different homeobox genes. In turn, this results in five different neuronal types forming at different positions in the ventral neural tube (Fig. 7.20) (Briscoe *et al.*, 2000). Hence, in this case, the precise combination of specific home-obox genes expressed in each position controls the precise cell type which is formed (for review see Marquardt and Pfaff, 2001).

In our discussion so far, it has been assumed that a homeobox factor is either present in a particular cell or is entirely absent. In fact, however, a further level of complexity exists, since many homeobox factors are not expressed in a simple on/off manner but rather show a concentration gradient ranging from high levels in one part of the embryo via inter-mediate levels to low levels in another part. For example, in *Drosophila*, the bicoid protein (bcd), whose absence leads to the development of a fly without head and thoracic structures, is found at high levels in the anterior part of the embryo and declines progressively posteriorly, being absent in the posterior one third of the embryo (Fig. 7.21).

Genes which are activated in response to bicoid, contain binding sites in their promoters which have either high affinity or low affinity for the bicoid protein. If these sites are linked to a marker gene, it can be dem-onstrated that genes with low affinity binding sites are only activated at

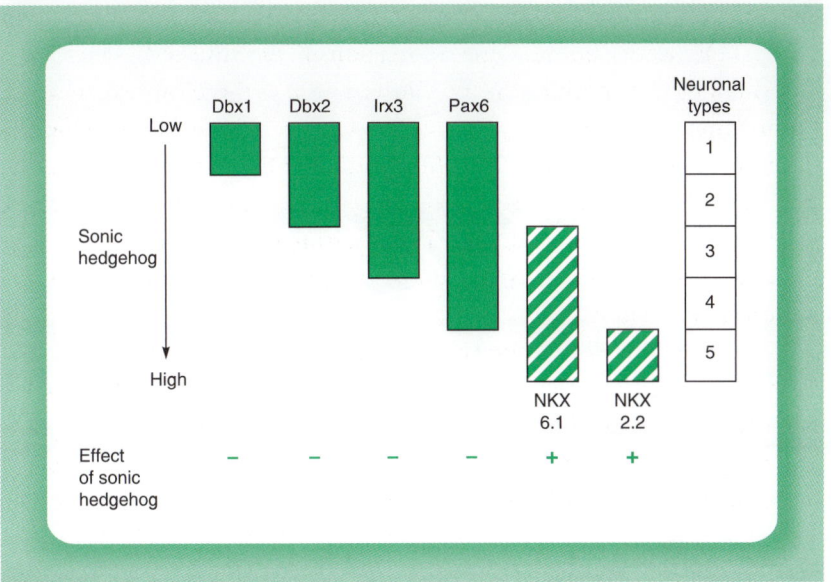

Figure 7.20

In the ventral neural tube, a gradient of sonic hedgehog regulates the expression of several homeobox genes. Dbx1, Dbx2, Irx3 and Pax6 are repressed by sonic hedgehog but differ in their sensitivity to repression. Thus, Pax6, which is the least sensitive to repression, is expressed at higher sonic hedgehog concentrations than Irx3 and so on. Conversely, Nkx6.1 and Nkx2.2 are activated by sonic hedgehog with Nkx6.1 being activated at lower concentrations than Nkx2.2. Together these effects create a homeodomain code in which each region has a different pattern of expression of the six genes and hence different neuronal types (1–5) form at each point.

high concentrations of bicoid and are therefore expressed only at the extreme anterior end of the embryo. In contrast, genes which have higher affinity binding sites are active at much lower protein concentrations and will be active both at the anterior end and more posteriorly. Moreover, the greater the number of higher affinity sites the greater the level of gene expression which will occur at any particular point in the gradient (Driever *et al.*, 1989; Fig. 7.21).

The gradient in bicoid expression can be translated therefore into the differential expression of various bicoid-dependent genes along the anterior part of the embryo. Each cell in the anterior region will be able to 'sense' its position within the embryo and respond by activating specific genes. One of the genes activated by bicoid is the homeobox-containing segmentation gene hunchback. In turn, this protein regulates the expression of gap genes, encoding the transcription factors Kruppel and giant (Struhl *et al.*, 1992). All four of these proteins then act on the eve gene,

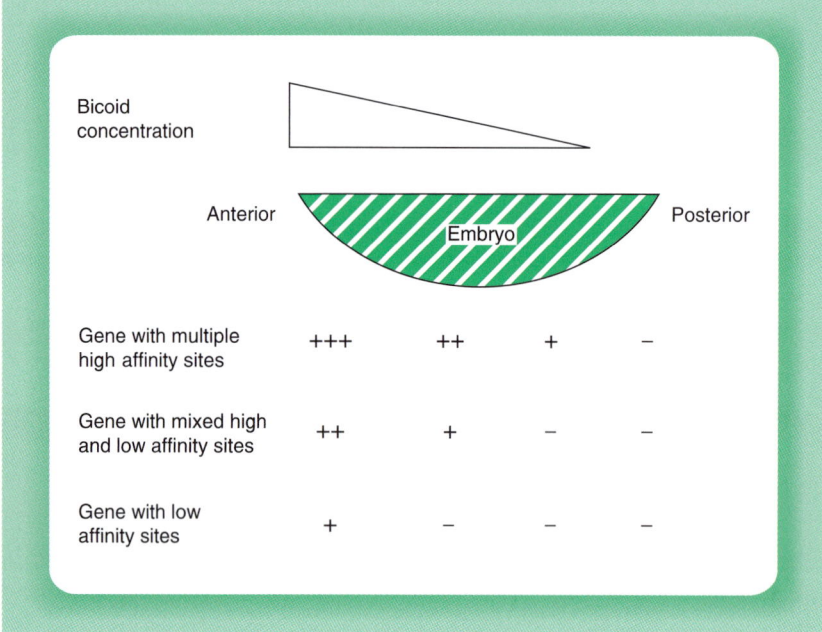

Figure 7.21

The gradient in Bicoid concentration from the anterior to the posterior point of the embryo results in bicoid-dependent genes with only low affinity binding sites for the protein being active only at the extreme anterior part of the embryo, whereas genes with high affinity binding sites are active more posteriorly. Note that in addition to the different posterior boundaries in the expression of genes with high and low affinity binding sites, genes with high affinity binding sites will be expressed at a higher level than genes with low affinity binding sites at any point in the embryo.

with bicoid and hunchback activating its expression whilst Kruppel and giant repress it. The concentration gradients of these four transcription factors thus result in the spatial localization of eve gene expression in a defined region of the embryo where it exerts its inhibitory effects on gene expression (Small *et al.*, 1991; Fig. 7.22). Hence, the gradient in bicoid gene expression results in changes in the expression of other genes encoding regulatory proteins, leading to the activation of regulatory networks involving the controlled synthesis of multiple transcription factors.

The bicoid factor therefore has all the properties of a morphogen whose concentration gradient determines position in the anterior part of the embryo. This idea is strongly supported by the results of genetic experiments in which the bicoid gradient was artificially manipulated. Thus, cells containing artificially increased levels of bicoid assume a phenotype characteristic of more anterior cells which normally contain

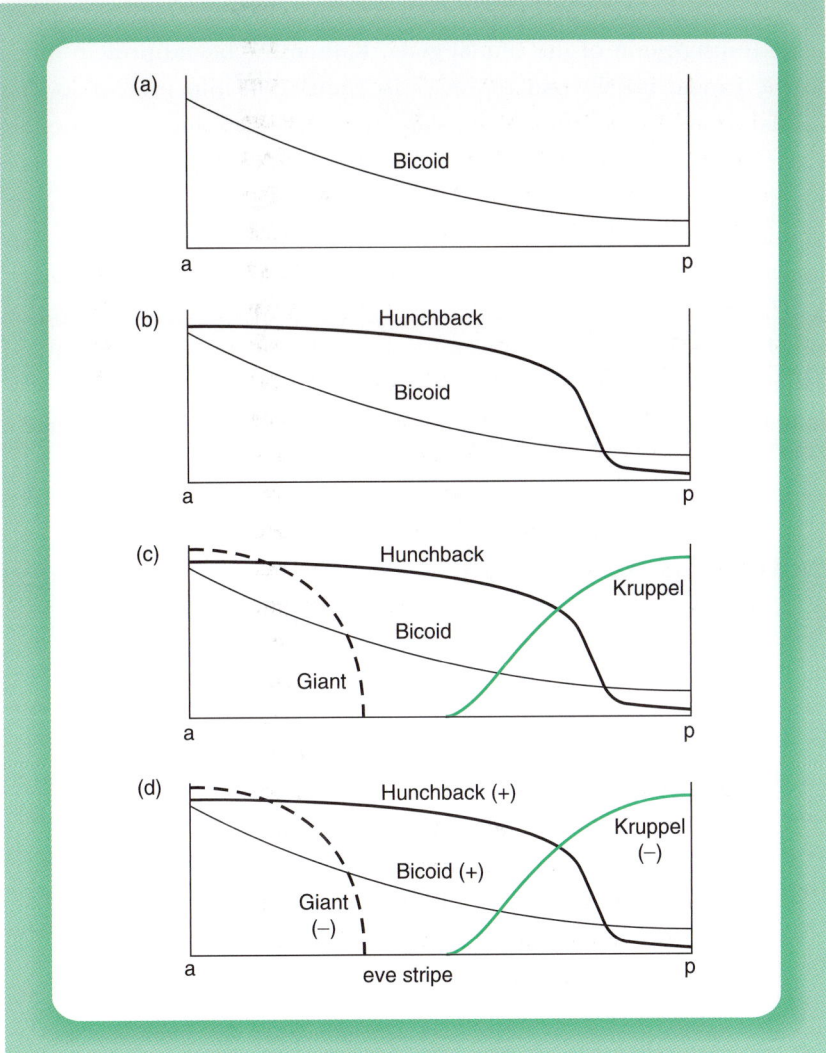

Figure 7.22

Model illustrating how the concentration gradients of the activators bicoid and hunchback and the repressors Kruppel and giant produce a stripe of eve gene expression. Note that the bicoid gradient (a) affects hunchback expression (b) which in turn affects giant and Kruppel expression (c). Eve gene activation (+) by hunchback and bicoid and its repression (−) by giant and Kruppel then produces a specific stripe or region of the embryo in which eve is expressed (d).

the new level of bicoid and vice versa (Driever and Nusslein-Volhard, 1988).

The anterior to posterior gradient in bicoid levels is required to produce the opposite posterior to anterior gradient in the level of another protein, caudal. However, the caudal mRNA is equally distributed throughout the embryo, indicating that the bicoid gradient does not regulate transcription of the caudal gene. Rather, the bicoid protein binds to the caudal mRNA and represses its translation into protein so that caudal protein is not produced when bicoid levels are high (reviewed by Carr, 1996; Chan and Struhl, 1997). As well as providing further evidence for the key role of the bicoid factor, this finding also shows that homeodomain proteins can bind to RNA as well as to DNA and that they may therefore act at the post-transcriptional level as well as at transcription.

The bicoid case clearly illustrates therefore how the regulated synthesis of an individual factor, resulting in a gradient in its concentration, can alter the expression of a regulatory network of other genes and ultimately control the differentiation of specific cells during development.

7.3 MECHANISMS REGULATING THE SYNTHESIS OF TRANSCRIPTION FACTORS

The cases discussed in the previous section illustrate therefore that where a factor must be active in a particular cell type or at a specific point in development, this is frequently achieved by the factor being present only in the particular cells where it is required. Clearly, such regulated synthesis of a specific transcription factor could be achieved by any of the methods which are normally used to regulate the production of individual proteins such as the regulation of gene transcription, RNA splicing or translation of the mRNA (Fig. 7.23, for review of the levels at which gene regulation can occur see Latchman, 2005). Several of these mechanisms of gene regulation are utilized in the case of individual transcription factors and these will be discussed in turn.

7.3.1 REGULATION OF TRANSCRIPTION

As discussed above, a number of cases where the cell type specific expression of a transcription factor is paralleled by the presence of its corresponding mRNA in the same cell type have now been described. In turn this cell type specific expression of the transcription factor mRNA is likely to result from the regulated transcription of the gene encoding the transcription factor. Unfortunately, the low abundance of many transcription factors has precluded the direct demonstration of the regulated transcription of the genes which encode them. This has been achieved, however, in the case of the CCAAT box binding factor C/EBP which regulates the transcription of several different liver-specific genes such as transthyretin and alpha-1 anti-trypsin. Thus, by using nuclear run

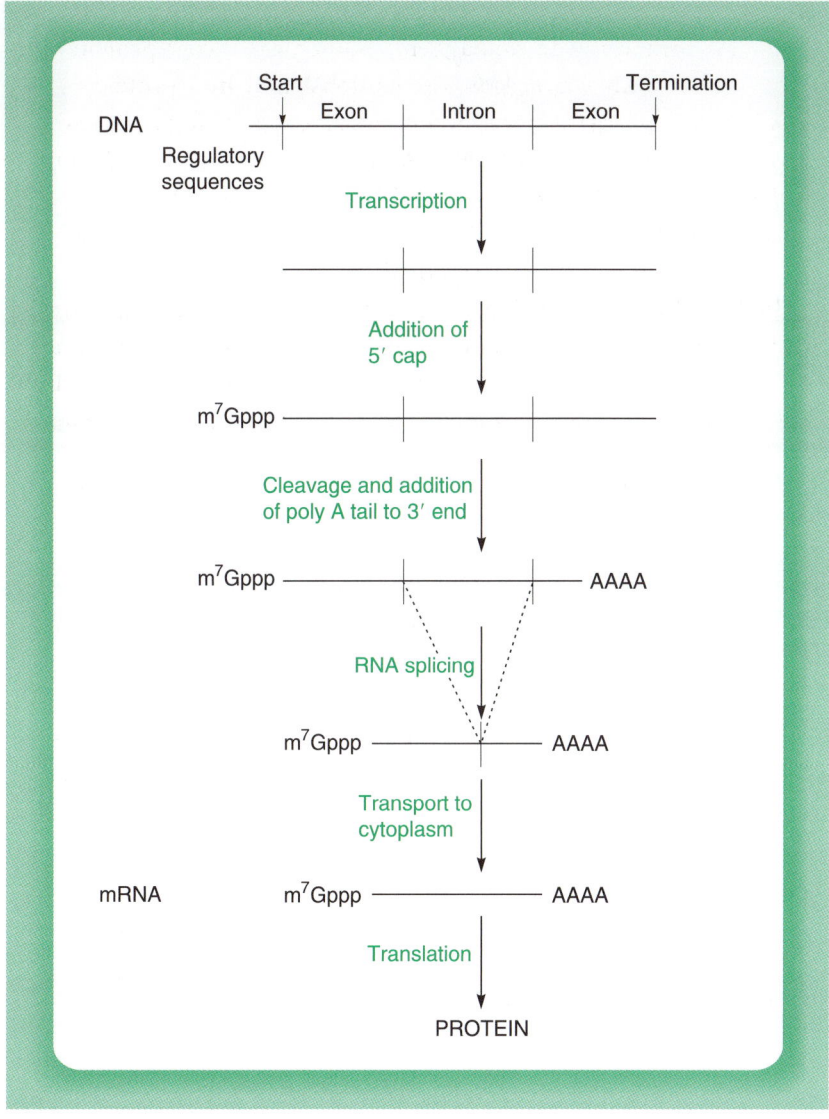

Figure 7.23
Potential regulatory stages in the expression of a gene encoding a transcription factor.

on assays to directly measure transcription of the gene encoding C/EBP, Xanthopoulos *et al.* (1989) were able to show that this gene is transcribed at high levels only in the liver, paralleling the presence of C/EBP itself and the mRNA encoding it at high levels only in this tissue (Fig. 7.24). Hence the regulated transcription of the C/EBP gene, in turn, controls the production of the corresponding protein, which, in turn, directly controls the liver specific transcription of other genes such as alpha-1 anti-trypsin and transthyretin.

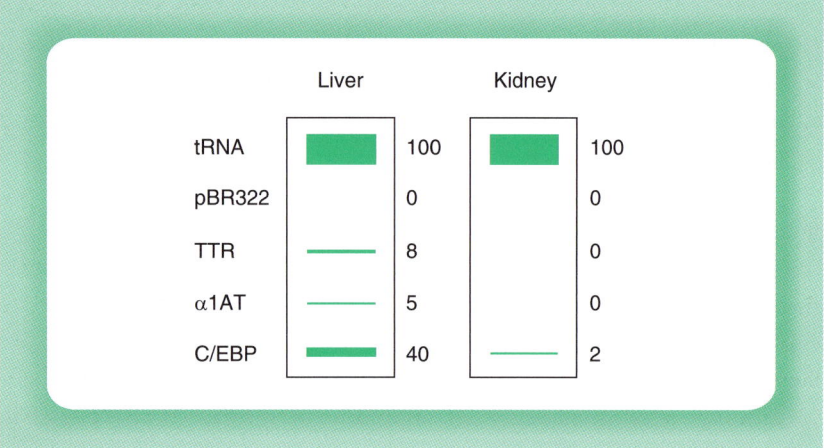

Figure 7.24

Nuclear run on assay of transcription in the nuclei of kidney and liver. Values indicate the degree of transcription of each gene in the two tissues. Note the enhanced transcription in the liver of the gene encoding the transcription factor C/EBP as well as of the genes encoding the liver-specific proteins transthyretin (TTR) and alpha-1 anti-trypsin (alpha 1AT). The positive control transfer RNA gene is, as expected, transcribed at equal levels in both tissues whilst the negative control, pBR322 bacterial plasmid, does not detect any transcription.

Interestingly, as well as being used to regulate the relative amounts of a particular factor produced by different tissues, transcriptional control can also be used to regulate factor levels within a specific cell type. Thus, the levels of the liver specific transcription factor DBP are highest in rat hepatocytes in the afternoon and evening, with the protein being undetectable in the morning. This fluctuation is produced by regulated transcription of the gene encoding DBP which is highest in the early evening and undetectable in the morning, whereas the C/EBP gene is transcribed at equal levels at all times. In turn, the alterations in DBP level produced in this way produce similar diurnal fluctuations in the transcription of the albumin gene which is dependent on DBP for its transcription (Wuarin and Schibler, 1990).

Although regulated transcription of the genes encoding the transcription factors themselves is likely therefore to constitute an important means of regulating their synthesis, it is clear that this process simply sets the problem of gene regulation one stage further back. Thus, it will be necessary to have some means of regulating the specific transcription of the gene encoding the transcription factor itself, which in turn will require other transcription factors.

Indeed, a number of such cases where specific transcription factors regulate the transcription of the gene encoding another transcription factor have already been discussed. Thus, as discussed in section 7.2, in embryonic stem cells, the polycomb factor represses the transcription of the genes encoding several other transcription factors, whilst several transcription factors regulating the muscle-specific transcription of the MyoD transcription factor have been identified.

Clearly, to achieve their effect such transcription factors will need to themselves be synthesized or be in an active form only in the appropriate cell type. It is not surprising therefore that the synthesis of transcription factors is often modulated by post-transcriptional control mechanisms not requiring additional transcription factors. These mechanisms will now be discussed.

7.3.2 REGULATION OF RNA SPLICING

Numerous examples have now been described in eukaryotes, where a single RNA species transcribed from a particular gene can be spliced in two or more different ways to yield different mRNAs encoding proteins with different properties (for review see Latchman, 2005). This process is also used in several cases of genes encoding specific transcription factors, for example, in the case of the *era*-1 gene which encodes a transcription factor that mediates the induction of gene expression in early embryonic cells in response to retinoic acid. In this case, two alternatively spliced mRNAs are produced, one of which encodes the active form of the molecule, whilst the other produces a protein lacking the homeobox region. As the homeobox mediates DNA binding by the intact protein (see Chapter 4, section 4.2.3), this truncated form of the protein is incapable of binding to DNA and activating gene expression (Larosa and Gudas, 1988). A similar use of alternative splicing to create mRNAs encoding proteins with and without the homeobox has also been reported for the Hoxb6 (2.2) gene (Shen *et al.*, 1991).

Hence, in these cases where one of the two proteins encoded by the alternatively spliced mRNAs is inactive, alternative splicing can be used in the same way as the regulation of transcription in order to control the amount of functional protein which is produced.

Interestingly, however, unlike transcriptional regulation, alternative splicing can also be used to regulate the relative production of two distinct functional forms of a transcription factor which have different properties. This is seen in the case of the Pax8 factor, which is a member of the Pax family (see Chapter 4, section 4.2.7). In this case, alternative splicing results in the insertion of a single serine residue in the recognition helix of the paired domain which is critical for DNA binding (Fig. 7.25). This alters the DNA binding properties of the factor so that it recognizes different DNA sequences to the form of Pax8 which lacks this residue (Kozmik *et al.*, 1997). Hence, alternative splicing can introduce a subtle, single amino acid, change in a transcription factor which results in the existence of two forms of the factor with different DNA binding specificities.

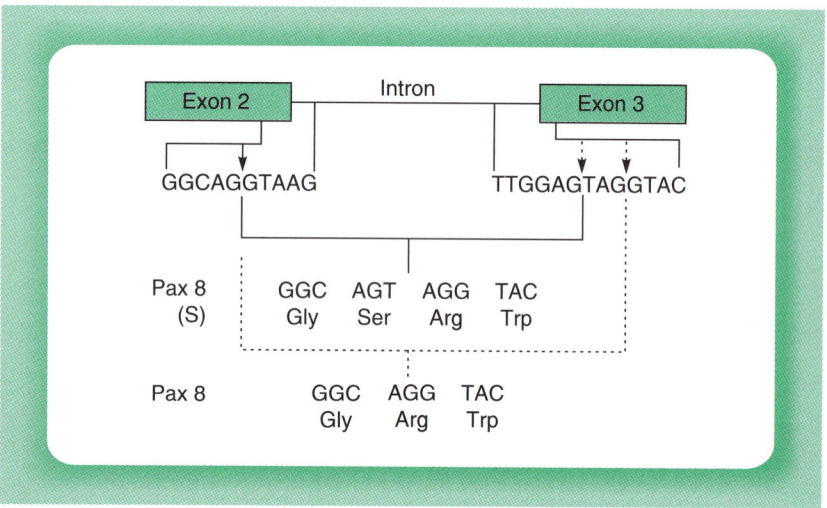

Figure 7.25

Alternative splicing in the Pax8 gene involving the use of different splice sites in exon 3 (dotted arrows), together with the same splice site in exon 2 (solid arrow), generates different forms of the protein with and without an additional serine residue and thus having different DNA binding specificities.

As well as affecting DNA binding specificity, alternative splicing can also produce forms of a transcription factor with distinct effects on transcription. This is seen in the case of the CREM factor, which is related to the CREB factor discussed in Chapter 5 (section 5.4.3). Thus, CREM resembles CREB in being phosphorylated following cyclic AMP treatment at a site located between two glutamine-rich activation domains. Like CREB, it can therefore bind to the CRE and activate transcription in response to cyclic AMP by binding the co-activator CBP (for reviews see de Cesare and Sassoni-Corsi, 2000; Mayr and Montminy, 2001).

Interestingly, however, alternative splicing produces distinct forms of the CREM factor which lack the activation domains, although they retain the leucine zipper and basic DNA binding domain (Fig. 7.26a) (see Chapter 4, section 4.5 for a discussion of these motifs). These forms can therefore bind to DNA but cannot activate transcription since they lack an activation domain. They therefore inhibit transcription by competing

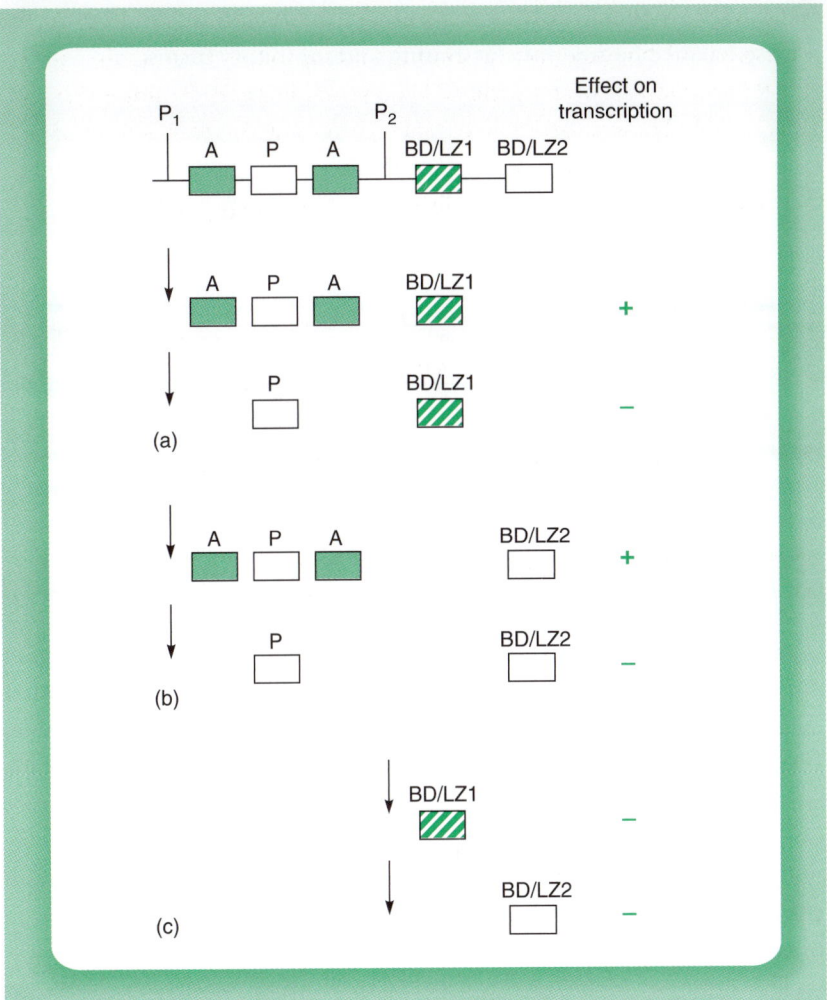

Figure 7.26

The CREM protein contains two transcriptional activation domains (A), a region containing a site for cyclic AMP-induced phosphorylation (P) and two DNA binding domains containing a basic domain and leucine zipper (BD/LZ). After transcription from the P_1 promoter, alternative splicing can result in forms with or without the activation domains (a) or having either of the DNA binding domains (b). In addition, cyclic AMP-inducible transcription from the P_2 promoter can produce forms (inducible cyclic AMP early repressors: ICERs) containing only one or other of the DNA binding domains but lacking the activation domains and the phosphorylated region (c). Arrows indicate the transcriptional start sites used in each case.

for binding to the CRE with the activating forms (Fig. 7.27) (see Chapter 6, section 6.2.1 for a discussion of indirect repression of this type). Since the proportion of the activating and inhibitory forms of CREM varies in different cell types, the level of transcription directed by a CRE following cyclic AMP treatment will be different in these cells depending on the precise balance between the activating and inhibitory forms.

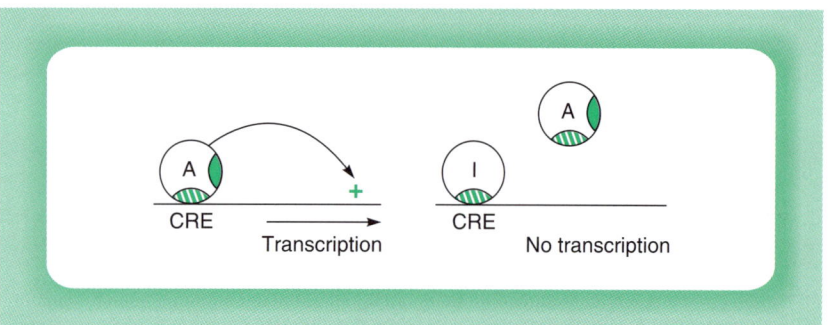

Figure 7.27
Gene activation by the activating forms of the CREM protein (A) can be inhibited by forms (I) which contain the DNA binding domain (light shading) but lack the activation domain (heavy shading). They therefore bind to the CRE and prevent binding by the activating forms.

As well as producing distinct forms with and without the activation domain, the CREM factor also undergoes alternative splicing in another manner. Thus, two distinct exons in the CREM gene contain two distinct DNA binding domains. Alternative splicing results in the proteins which either do or do not contain the activation domains, also having one or other of the DNA binding domains (Fig. 7.26b). As the relative usage of the two DNA binding domains is different in different cell types, this effect is likely to have biological significance but its precise role is at present unclear.

The different forms of the CREM factor which have been discussed so far are all produced by alternative splicing of a single RNA transcript whose rate of production is unaffected by cyclic AMP. The ability of the CREB factor and the activating forms of CREM to switch on gene expression is then stimulated post-translationally by their phosphorylation following cyclic AMP treatment, hence allowing them to switch on gene expression in response to cyclic AMP (see Chapter 5, section 5.4.3).

In contrast to such post-translational regulation, the CREM gene also contains a promoter which is activated in response to cyclic AMP. This promoter produces transcripts encoding short proteins, which contain one or other of the DNA binding domains and the phosphorylated region but lack the activation domain (Fig. 7.26c). These proteins can therefore bind to the cyclic AMP response element and repress transcriptional activation by the activating forms, exactly as described above for the alternatively spliced forms lacking the activation domain. These forms are therefore known as ICERs (inducible cyclic AMP early repressors).

As they are inducible by cyclic AMP, these forms are likely to play a key role in making the cyclic AMP response self-limiting. Thus, following cyclic AMP treatment, CREB and CREM will become phosphorylated and will then activate the expression of promoters containing a CRE, including that which produces the ICERs. The ICERs produced in this manner will then bind to the CRE and switch off the inducible genes by preventing the binding of CREB and CREM (Fig. 7.28) thereby making the cyclic AMP response a transient one.

The regulation of cyclic AMP-inducible transcription by the CREB and CREM factor is therefore extraordinarily complex with both alternative splicing and the use of two different promoters in the CREM gene. It illustrates therefore how the combination of transcriptional and post-transcriptional control of synthesis can be used to produce multiple forms of transcription factors with different functional roles.

Alternative splicing can also occur in factors which contain a specific inhibitory domain and which can therefore function as direct repressors interfering with the activity of the basal transcriptional complex (see Chapter 6, section 6.3.2). Thus, although it is transcribed in B cells and not in most other cell types, the gene encoding the Oct-2 transcription factor (which is a member of the POU family, discussed in Chapter 4, section 4.2.6), is also transcribed in neuronal cells. In neuronal cells, the Oct-2 RNA is spliced so that the protein it encodes does not contain the C-terminal activation domain which allows it to activate transcription. It does, however, retain the N-terminal inhibitory domain discussed in Chapter 6 (section 6.3.2) as well as the DNA binding domain and can therefore act as a direct inhibitor of gene expression (Lillycrop *et al.*, 1994). In contrast, in B cells, alternative splicing produces an mRNA which encodes a protein containing both the inhibitory domain and the stronger activation domain and which therefore activates transcription (Fig. 7.29). Hence, in this case alternative splicing produces different forms of a factor in different cell types which have opposite effects on the activity of their target promoters.

Such alternative splicing is also seen in the case of another transcription factor containing an inhibitory domain, namely the thyroid hormone receptor. Thus, as discussed in Chapter 6 (section 6.3.2), alternative splicing produces two forms of the receptor, one of which lacks the ligand binding domain and therefore cannot bind thyroid hormone (Fig. 6.14). Although it cannot therefore respond to thyroid hormone, this alpha 2 form of the protein still contains the DNA binding domain and can therefore bind to the specific binding site for the receptor in hormone responsive genes. By doing so, it acts as a dominant repressor of gene activation mediated by the normal receptor in response to hormone binding.

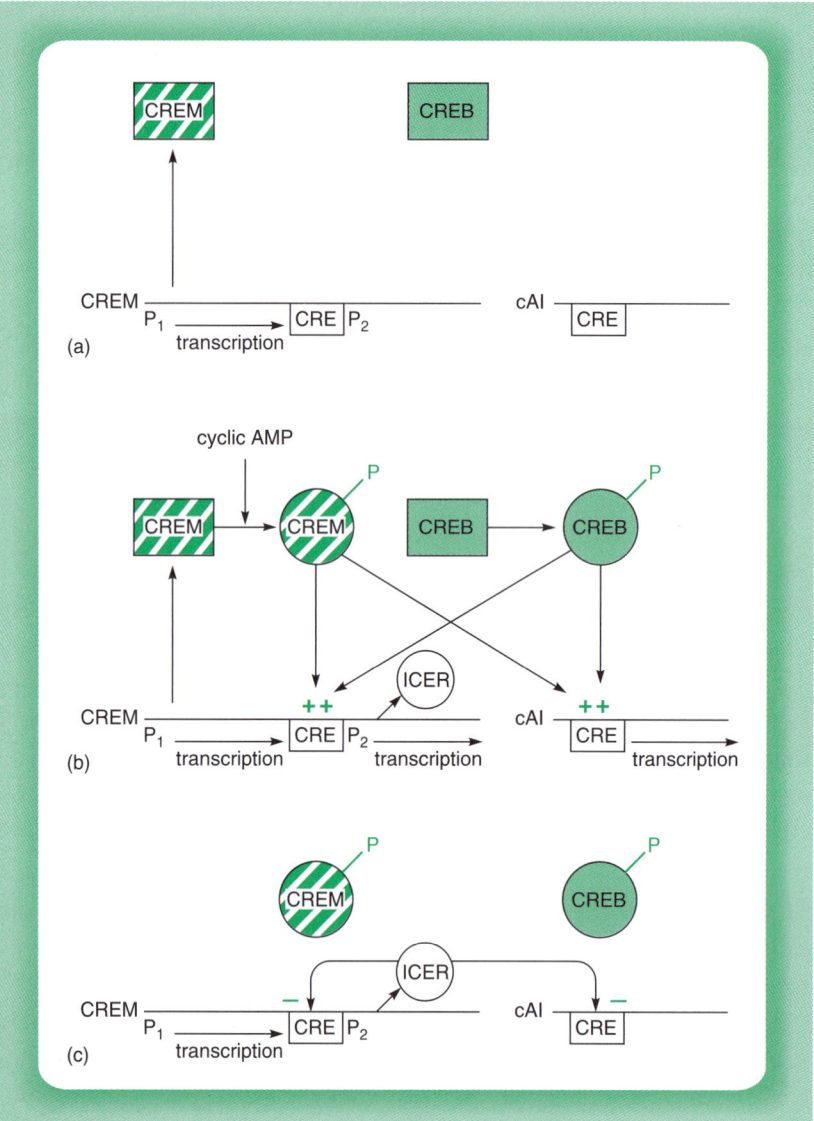

Figure 7.28

(a) In the absence of cyclic AMP, the CREM gene is transcribed from the P_1 promoter. However, neither the CREM produced in this way or the CREB protein can activate transcription until they are activated post-translationally. (b) Following cyclic AMP treatment, the CREB and CREM proteins become activated post-translationally by phosphorylation. They therefore activate the cyclic AMP-inducible genes (cAI) which contain a cyclic AMP response element (CRE) in their promoters. In addition they also activate the P_2 promoter of CREM which also contains a CRE. (c) The ICERs (inducible cyclic AMP early repressors) produced by the CREM P_2 promoter bind to the CREs and prevent activation by CREB and CREM thereby repressing transcription.

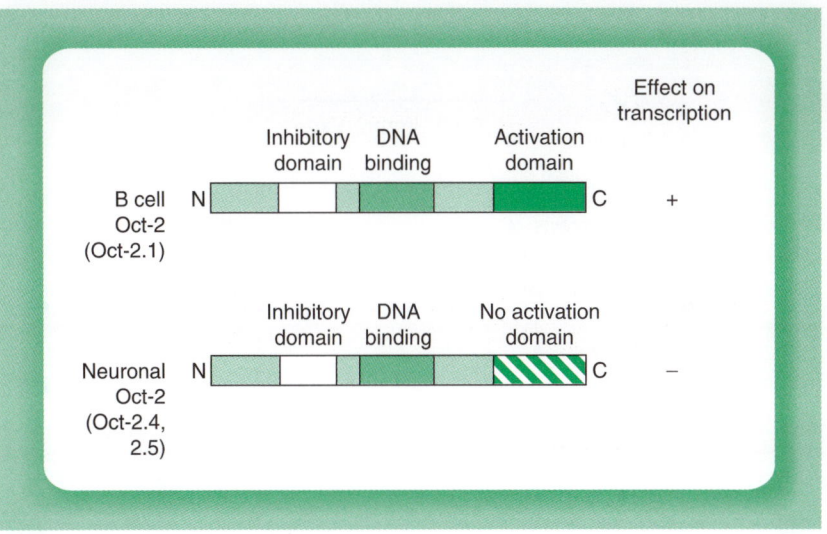

Figure 7.29

In B lymphocytes the predominant form of Oct-2 (Oct-2.1) contains the C-terminal activation domain as well as the DNA binding domain and an inhibitory domain. As the activation domain overcomes the effect of the inhibitory domain, this form is able to activate transcription. In contrast the predominant neuronal forms of Oct-2 (Oct-2.4 and 2.5) contain different C-terminal regions and lack the activation domain. As they retain both the inhibitory domain and the DNA binding domain, however, they can bind to specific DNA binding sites and inhibit gene expression.

Hence, these two alternatively spliced forms of the transcription factor, which are made in different amounts in different tissues, mediate opposing effects on thyroid hormone-dependent gene expression.

As well as affecting the actual properties of a transcription factor, regulation of splicing can also be used to determine how much of the protein accumulates. This is seen in the case of the Haclp protein, which is a member of the basic-leucine zipper transcription factor family discussed in Chapter 4 (section 4.5). This factor accumulates at an increased level in the presence of unfolded proteins in the cell and then activates the expression of genes which assist other proteins to fold properly. This increased accumulation of Haclp is controlled by a splicing event which removes an intron from the Haclp transcript. When this intron is present, the RNA forms a folded structure which cannot be translated to produce Haclp protein. When the intron is removed by splicing, this folded structure no longer forms and the Haclp mRNA is translated (Rüegsegger and Lebb, 2001) (Fig. 7.30). Hence, in this case the regulation of splicing alters the amount of the transcription factor produced rather than its activity.

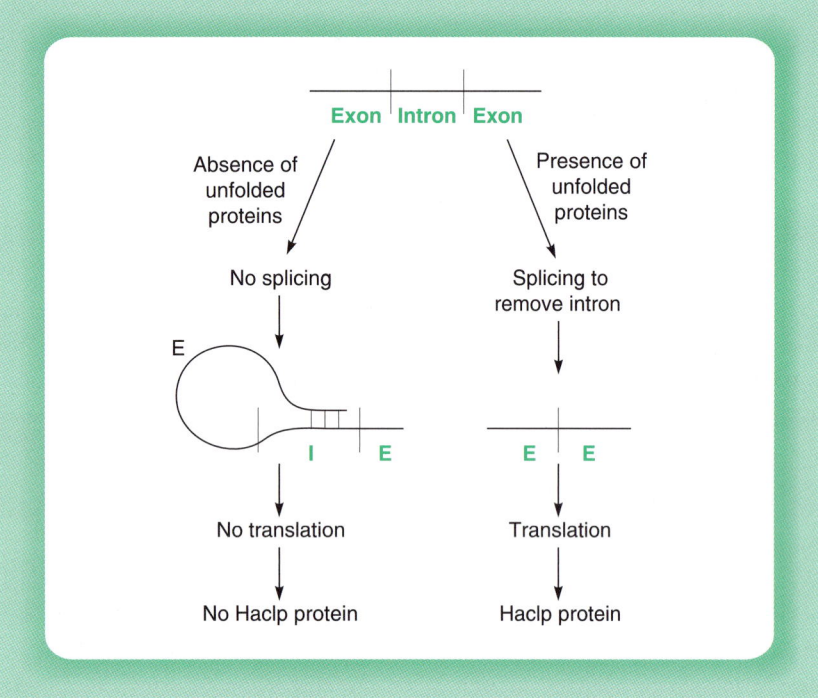

Figure 7.30

Regulated splicing of the RNA encoding Haclp results in its enhanced synthesis in response to the presence of unfolded proteins in the cell. In the absence of unfolded proteins, the intron is not removed from the RNA and base-pairing between the first exon and the intron prevents the RNA from being translated into protein. In the presence of unfolded protein, the intron is removed and the unfolded mRNA is translated to produce functional Haclp protein.

The examples of regulated splicing discussed above illustrate the potential of this process. Thus, it can control the level of functional transcription factor which is produced or it can generate different forms of a particular transcription factor which, because of differences in the regions which mediate DNA binding or transcriptional activation, have different properties that result in differences in their effects on gene expression.

7.3.3 REGULATION OF TRANSLATION

The final stage in the expression of a gene is the translation of its corresponding mRNA into protein. In theory, therefore, the regulation of synthesis of a particular transcription factor could be achieved by producing its mRNA in all cell types but translating it into active protein only in the particular cell type where it was required. However, the observed parallels between the cell type specific expression of a particular transcription factor and the cell type specific expression of its corresponding mRNA discussed above (section 7.2) indicate that this cannot be the case for the majority of transcription factors. Nonetheless, this mechanism is used to control the synthesis of at least one transcription factor in yeast.

Thus, the yeast GCN4 transcription factor controls the activation of several genes in response to amino acid starvation and the factor itself is synthesized in increased amounts following such starvation, allowing it to mediate this effect. This increased synthesis of GCN4 following amino acid starvation is mediated via increased translation of pre-existing GCN4 mRNA (for reviews see Hinnebusch, 1997; Morris and Geballe, 2000). This translational regulation is dependent upon short sequences within the 5' untranslated region of the GCN4 mRNA, upstream of the start point for translation of the GCN4 protein.

Most interestingly, such sequences are capable of being translated to produce short peptides of two or three amino acids (Fig. 7.31). Under conditions when amino acids are plentiful, these short peptides are synthesized and the ribosome fails to reinitiate at the start point for GCN4 production, resulting in this protein not being synthesized. Following amino-acid starvation, however, the production of the small peptides is suppressed and the production of GCN4 is correspondingly enhanced. Hence, this mechanism ensures that GCN4 is synthesized only in response to amino-acid starvation and then activates the genes encoding the

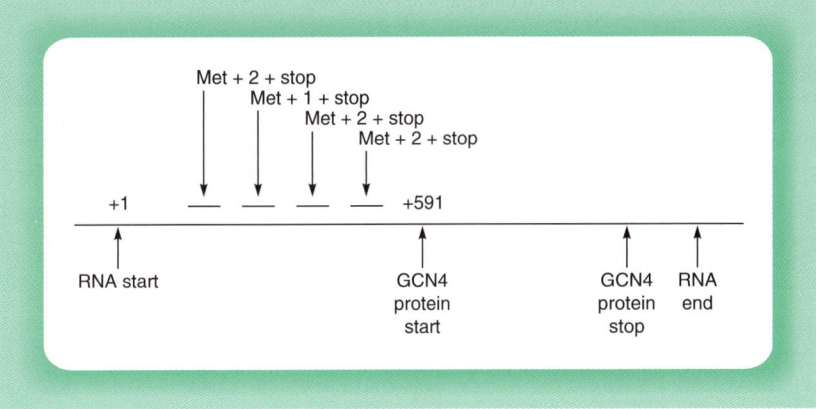

Figure 7.31

Presence of short open reading frames capable of producing small peptides in the 5' untranslated region of the yeast GCN4 RNA. Translation of the RNA to produce these small proteins suppresses translation of the GCN4 protein. The position of the methionine residue beginning each of the small peptides is indicated together with the number of additional amino acids incorporated before a stop codon is reached.

enzymes required for the biosynthetic pathways necessary to make good this deficiency.

Interestingly, the use of distinct translational start sites is also seen in the case of the C/EBP transcription factors expressed in the mammalian liver. In this case, however, the two start sites of translation result in two different forms of the C/EBP proteins. The longer form contains an activation domain as well as a basic DNA binding domain and leucine zipper. The other is produced by translational initiation from a downstream start site and therefore lacks the activation domain, although it retains both the basic domain and the leucine zipper (Fig. 7.32). This shorter protein can bind to the same sites as the longer form and since it cannot activate transcription, acts as an inhibitor of gene activation by the longer form (Descombes and Schibler, 1991).

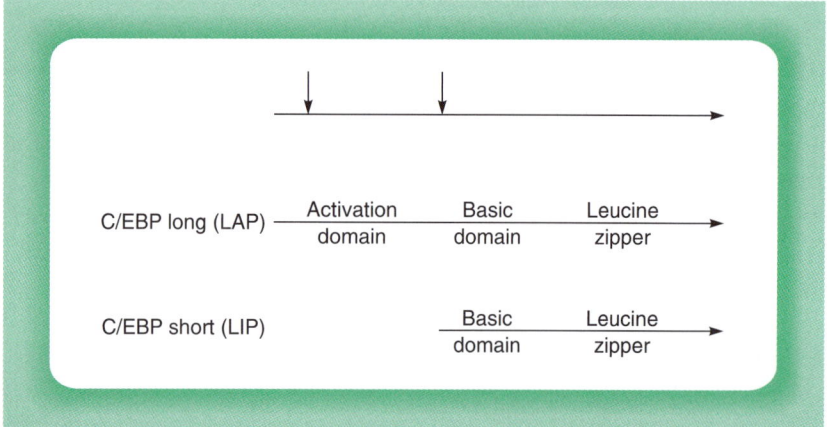

Figure 7.32
The use of different translational initiation codons (vertical arrows) in the mRNA encoding the C/EBP transcription factors produces the longer LAP (liver activator protein) form of the protein which possesses an activation domain and the shorter LIP (liver inhibitor protein) form of the protein which lacks this domain and therefore inhibits gene activation by LAP.

Interestingly, the balance between the long and short forms of C/EBP is controlled by the level in the cell of factors required for the translation of all mRNAs. Thus, when a low level of these translation factors is present in the cell, the upstream start site of translation is used preferentially and the full length protein predominates. In contrast, when higher levels of the translation factors are present, the shorter form of C/EBP is produced in increasing amounts (Calkhover *et al.*, 2000). Moreover, it has been shown that the shorter form of C/EBP promotes cellular proliferation whereas the longer form promotes growth arrest and terminal differentiation. Hence, in this case the regulated translation of a transcription factor produces two distinct forms with opposite effects on cellular proliferation and differentiation (Fig. 7.33).

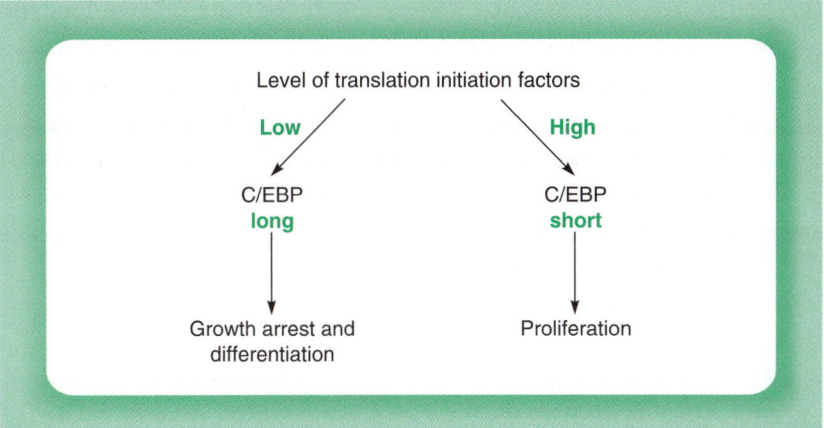

Figure 7.33
The level of translation initiation factors controls the balance between the activating long form of C/EBP which induces growth arrest and differentiation and the inhibitory short form which induces cellular proliferation.

As with the regulation of splicing, the regulation of translation can therefore be used to control the amount of an active factor which is produced, as well as to regulate the balance between two functionally antagonistic factors encoded by the same gene.

7.4 CONCLUSIONS

Regulating the synthesis of a transcription factor constitutes a metabolically inexpensive way of controlling its activity. Thus, in situations where the activity of a particular factor is not required, no energy is expended on making it in an inactive form. Such regulation probably takes place predominantly at the level of transcription so that no energy is expended on the production of an RNA, its splicing, transport, etc. However, even in cases where regulation occurs at later stages such as splicing or translation, the system is relatively efficient in terms of energy usage, since the step in gene expression which requires the most energy is the final one of translation.

In view of its metabolic efficiency, it is not surprising therefore that the regulation of their synthesis is widely used to control the activity of the factors which mediate cell type specific gene regulation, where differences in the activity of a given factor in different cell types are maintained for long periods of time. Similarly, alternative splicing or use of different translational initiation codons is used to produce different forms of the same factor which often have antagonistic effects on gene expression.

The regulation of factor activity by regulating its synthesis does suffer, however, from the defect that a change in the level of activity of a factor which is controlled purely by a change in its actual amount can take some time to occur. Thus, in response to a signal which induces new transcription of the gene encoding a particular factor, it is necessary to go through all the stages illustrated in Fig. 7.23, before the production of active factor, which is capable of activating the expression of other genes in response to the inducing signal.

It is not surprising therefore that, although some factors such as GCN4 which mediate inducible gene expression are regulated by the regulation of their synthesis, the majority of such factors are regulated by post-translational mechanisms, which activate pre-existing transcription factor protein in response to the inducing signal. Thus, although mechanisms of this type are metabolically expensive in that they require the synthesis of the factor in situations where it is not required, they have the necessary rapid response time required for the regulation of inducible gene expression. Moreover, unlike transcriptional regulation, they constitute an independent method of gene regulation rather than requiring the activation of other transcription factors in order to activate the transcription of the gene encoding the factor itself. The regulation of transcription factor activity and the manner in which it is achieved is discussed in the next chapter.

REFERENCES

Briscoe, J., Pierani, A., Jessell, T.M. and Ericson, J. (2000) A homeodomain protein code specifies progenitor cell identity and neuronal fate in the ventral neural tube. Cell 101, 435–445.

Calkhoven, C.F., Müller, C. and Leutz, A. (2000) Translational control of C/EBPa and C/EBPb isoform expression. Genes and Development 14, 1920–1932.

Carr, K. (1996) RNA bound to silence. Nature 379, 676.

Chan, S.K. and Struhl, G. (1997) Sequence specific RNA binding by bicoid. Nature 388, 634.

Constantinides, P.G., Jones, P.A. and Gevers, W. (1977) Functional striated muscle cells from non-myoblast precursors following 5-azacytidine treatment. Nature 267, 364–366.

Davis, H.L., Weintraub, H. and Lassar, A.B. (1987) Expression of a single transfected cDNA converts fibroblasts to myoblasts. Cell 51, 987–1000.

De Cesare, D. and Sassoni-Corsi, P. (2000) Transcriptional regulation by cyclic AMP-responsive factors. Progress in Nucleic Acids Research and Molecular Biology 64, 343–369.

Descombes, P. and Schibler, U. (1991) A liver transcriptional activator protein LAP and a transcriptional inhibitory protein LIP are translated from the same mRNA. Cell 67, 569–580.

Driever, W. and Nusslein-Volhard, C. (1988) The bicoid protein determines position in the *Drosophila* embryo in a concentration-dependent manner. Cell 54, 95–104.

Driever, W., Thoma, G. and Nusslein-Volhard, C. (1989) Determination of spatial domains of zygotic gene expression in the *Drosophila* embryo by the affinity of binding sites for the bicoid morphogen. Nature 340, 363–367.

Duboule, D. (2000) A Hox by any other name. Nature 403, 607–610.

Gerber, A.N., Klesert, T.R., Bergstrom, D.A. and Tapscott, S.J. (1997) Two domains of MyoD mediate transcriptional activation of genes in repressive chromatin: a mechanism for lineage determination in myogenesis. Genes and Development 11, 436–450.

Halevy, O., Novitch, B.G., Spicer, D.B., Skapek, S.X., Rhee, J., Hannon, G.J., Beach, D. and Lassar, A.B. (1995) Correlation of terminal cell cycle arrest of skeletal muscle with induction of p21 by MyoD. Science 267, 1018–1021.

Hinnebusch, A.G. (1997) Translational regulation of GCN4. Journal of Biological Chemistry 272, 21661–21664.

Kozmik, Z., Czerny, T. and Busslinger, M. (1997) Alternatively spliced insertions in the paired domain restrict the DNA sequence specificity of Pax6 and Pax8. EMBO Journal 16, 6793–6803.

Larosa, G.J. and Gudas, L.J. (1988) Early retinoic acid-induced F9 teratocarcinoma stem cell gene ERA-1: alternative splicing creates transcripts for a homeobox-containing protein and one lacking the homeobox. Molecular and Cellular Biology 8, 3906–3917.

Latchman, D.S. (2005) *Gene regulation: a eukaryotic perspective.* Fifth edition. Taylor and Francis, Oxford and New York.

Lillycrop, K.A., Dawson, S.J., Estridge, J.K., Gerster, T., Matthias, P. and Latchman, D.S. (1994). Repression of a herpes simplex virus immediate–early promoter by the Oct-2 transcription factor is dependent upon an inhibitory region at the N-terminus of the protein. Molecular and Cellular Biology 14, 7633–7642.

Logan, M., Pagán-Westphal, S.M., Smith, D.M., Paganessi, L. and Tabin, C.J. (1998) The transcription factor Pitx2 mediates situs-specific morphogenesis in response to left-right asymmetric signals. Cell 94, 307–317.

Manzanares, M., Wada, H., Itasaki, N., Trainor, P.A., Krumlauf, R. and Holland, P.W.H. (2000) Conservation and elaboration of *Hox* gene regulation during evolution of the vertebrate head. Nature 408, 854–857.

Marquardt, T. and Pfaff, S.L. (2001) Cracking the transcriptional code for cell specification in the neural tube. Cell 106, 651–654.

Mayr, B. and Montminy, M. (2001) Transcriptional regulation by the phosphorylation-dependent factor CREB. Nature Reviews Molecular Cell Biology 2, 599–6091.

Morris, D.R. and Geballe, A.P. (2000) Upstream open reading frames as regulators in mRNA translation. Molecular and Cellular Biology 20, 8635–8642.

Piedra, M.E., Icardo, J.M., Albajar, M., Rodriguez-Rey, J.C. and Ros, M.A. (1998) *Pitx2* participates in the late phase of the pathway controlling left-right asymmetry. Cell 94, 319–324.

Puri, P.L., Avantaggiati, M.L., Balsone, C., Sang, N., Graessmann, A., Giordano, A. and Leureo, M. (1997) p300 is required for MyoD-dependent cell cycle arrest and muscle-specific gene transcription. EMBO Journal 6, 369–383.

Rao, P.K., Kumar, R.M., Farkhondeh, M., Baskerville, S. and Lodish, H.F. (2006) Myogenic factors that regulate expression of muscle-specific microRNAs. Proceedings of the National Academy of Sciences, USA 103, 8721–8726.

Rüegsegger, U. and Leber, J.H. (2001) Block of *HAC1* mRNA translation by long-range base pairing is released by cytoplasmic splicing upon induction of the unfolded protein response. Cell 107, 103–114.

Sharma, K., Sheng, H.Z., Lettieri, K., Li, H., Karavanov, A., Potter, S., Westphal, H. and Pfaff, S.L. (1998) LIM homeodomain factors Lhx3 and Lhx4 assign subtype identities for motor neurones. Cell 95, 817–828.

Shen, W-F., Detmer, K., Simonitch-Easton, T., Lawrence, H.J. and Largman, C. (1991) Alternative splicing of the Hox 2.2 homeobox gene in human hematopoetic cells and murine embryonic and adult tissues. Nucleic Acids Research 19, 539–545.

Small, S., Krant, R., Hoey, T., Warrior, R. and Levine, M. (1991) Transcriptional regulation of a pair-rule stripe in *Drosophila*. Genes and Development 5, 827–839.

Stern, C.D. and Foley, A.C. (1998) Molecular dissection of *Hox* gene induction and maintenance in the hindbrain. Cell 94, 143–145.

Struhl, G., Johnston, P. and Lawrence, P.A. (1992) Control of *Drosophila* body pattern by the hunchback morphogen gradient. Cell 69, 237–249.

Trouche, D., Grigoriev, M., Lenormard, J.C., Robin, P., Leibovitch, S.A., Sassone-Corsi, P. and Harel-Bellan, A. (1993) Repression of c-*fos* promoter by MyoD on muscle cell differentiation. Nature 363, 79–82.

Van der Hoeven, F., Zakany, J. and Duboule, D. (1996) Gene Transpositions in the HoxD complex reveal a hierarchy of regulatory controls. Cell 85, 1025–1035.

Wuarin, J. and Schibler, U. (1990) Expression of the liver-enriched transcriptional activator protein DBP follows a stringent circadian rhythm. Cell 63, 1257–1269.

Xanthopoulos, K.G., Mirkovitch, J., Decker, T., Kuo, C.G. and Darnell, J.E. Jr. (1989) Cell-specific transcriptional control of the mouse DNA binding protein mC/EBP. Proceedings of the National Academy of Sciences, USA 86, 4117–4121.

Zeller, R. and Deschamps, J. (2002) First come, first served. Nature 420, 138–139.

CHAPTER 8

REGULATION OF TRANSCRIPTION FACTOR ACTIVITY

8.1 EVIDENCE FOR THE REGULATED ACTIVITY OF TRANSCRIPTION FACTORS

In a number of cases, it has been shown that a particular transcription factor pre-exists in an inactive form prior to its activation and the consequent switching on of the genes which depend on it for their activity. Thus, as discussed in Chapter 5 (section 5.5.1) and in section 8.3.1 of this chapter, the activation of heat inducible genes by elevated temperature is dependent on the activity of the heat shock transcription factor (HSF). However, this induction can be achieved in the presence of the protein synthesis inhibitor cycloheximide (Zimarino and Wu, 1987; for review see Morimoto, 1998). Hence, this process cannot be dependent on the synthesis of HSF in response to heat but rather must depend on the heat-induced activation of pre-existing inactive HSF (for further details see section 8.3.1).

Although for the reasons discussed in Chapter 7 (section 7.4) the activation of pre-existing transcription factors is predominantly used to modulate transcription factors involved in controlling inducible rather than cell type specific gene expression, it has also been reported for factors involved in regulating cell type specific gene expression. Thus, the transcription factor NFκB (which is a heterodimer of two subunits p50 and p65) plays an important role in the B cell specific expression of the immunoglobulin κ gene (for reviews see Hayden *et al.*, 2006; Hoffman and Baltimore, 2006; Hoffman *et al.*, 2006; Tergaonkar, 2006). However, both subunits of NFκB are expressed in a wide variety of cell types and the factor is present in an inactive form both in pre B cells and in a wide variety of other cell types, such as T cells and HeLa cells, which do not express the immunoglobulin genes. This pre-existing form of NFκB can be activated by treatment of pre-B cells with substances such as lipopoly-saccharides. As in the case of HSF, this activation can take place in the presence of inhibitors of protein synthesis, indicating that it does not require *de novo* synthesis of NFκB protein. These treatments therefore activate pre-existing NFκB and thus result in the activation of the immunoglobulin κ gene in pre-B cells, which do not normally express it.

Interestingly, the inactive form of NFκB is widely distributed in different cell types and can be activated in both T cells and HeLa cells by treatment with phorbol ester. Although in these cases NFκB activation does not result in immunoglobulin light chain gene expression since the gene has not rearranged and is tightly packed within inactive chromatin, it does play a role in gene regulation. Thus, the activation of NFκB by agents which activate T cells results in the active transcription factor inducing increased expression of cellular genes, such as that encoding the interleukin-2 α receptor and is also responsible for the increased activity of the human immunodeficiency virus promoter in activated T cells. NFκB therefore plays a role not only in B cell specific gene activity but also in gene activity specific to activated T cells. Indeed, further work has suggested an additional role for NFκB in bone development, indicating that it plays a key role in a number of different cell types (for review see Abu-Amer and Tondravi, 1997; Dixit and Mak, 2002).

The process in which pre-existing NFκB becomes activated both during B cell differentiation and by agents such as phorbol esters which activate T cells therefore allows NFκB to play a dual role both in B cell specific gene expression and in the expression of particular genes in response to T cell activation by various agents (Fig. 8.1). This effect would otherwise require a complex pattern of regulation in which NFκB was synthesized both in response to B cell maturation and to agents which activate T cells.

Hence, modulating the activity of a transcription factor represents a rapid and flexible means of activating a particular factor. Moreover, unlike transcriptional control, such mechanisms allow a direct linkage between the inducing stimulus and the activation of the factor rather than requiring the regulated activity of other transcription factors which in turn activate transcription of the gene encoding the regulated factor. Hence, they represent a highly efficient means of allowing specific cellular signalling pathways to produce changes in cellular transcription factor activity and hence affect gene expression (for reviews see Barolo and Posakony, 2002; Brivanlou and Darnell, 2002).

In the most extreme example of the linkage between signalling pathways and transcription factors, the signalling molecule and the transcription factor are identical. Thus, in response to microbial infection,

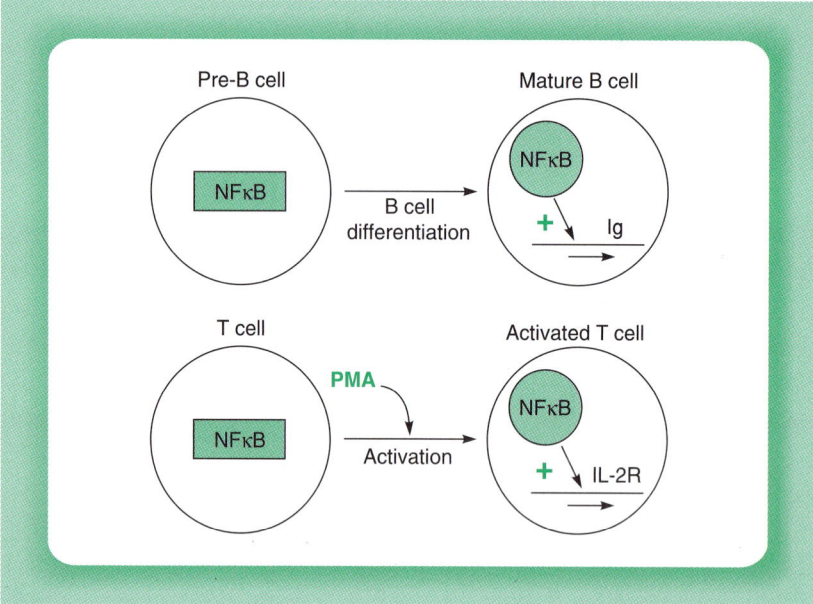

Figure 8.1

Activation of NFκB during B cell differentiation or by agents such as PMA which activate T cells allows it to activate expression of the immunoglobulin κ chain gene in B cells and the interleukin 2 receptor gene in activated T cells.

mammalian neutrophils secrete the protein lactoferrin into the medium. It has been shown that the lactoferrin protein can be taken up by other cells of the immune system. It then enters the nucleus of the cells and binds to specific DNA sequences activating genes whose protein products are required for the cells to neutralize the microbial infection (He and Furmanski, 1995). Hence, in this case, the signalling factor and the transcription factor are the same protein (for discussion see Baeuerle, 1995). In most cases, however, the signalling molecule acts indirectly to produce a change in the activity of a distinct transcription factor which pre-existed within the cell in an inactive form prior to exposure to the signal. Four basic means by which such mechanisms can regulate factor activity have been described (Fig. 8.2) and these will be discussed in turn.

8.2 REGULATION BY PROTEIN–LIGAND BINDING

8.2.1 EXAMPLES OF REGULATION BY LIGAND BINDING

As discussed above, one of the principal advantages of regulating the activity of a factor in response to an inducing stimulus is that it allows a direct interaction between the inducing stimulus and the activation of

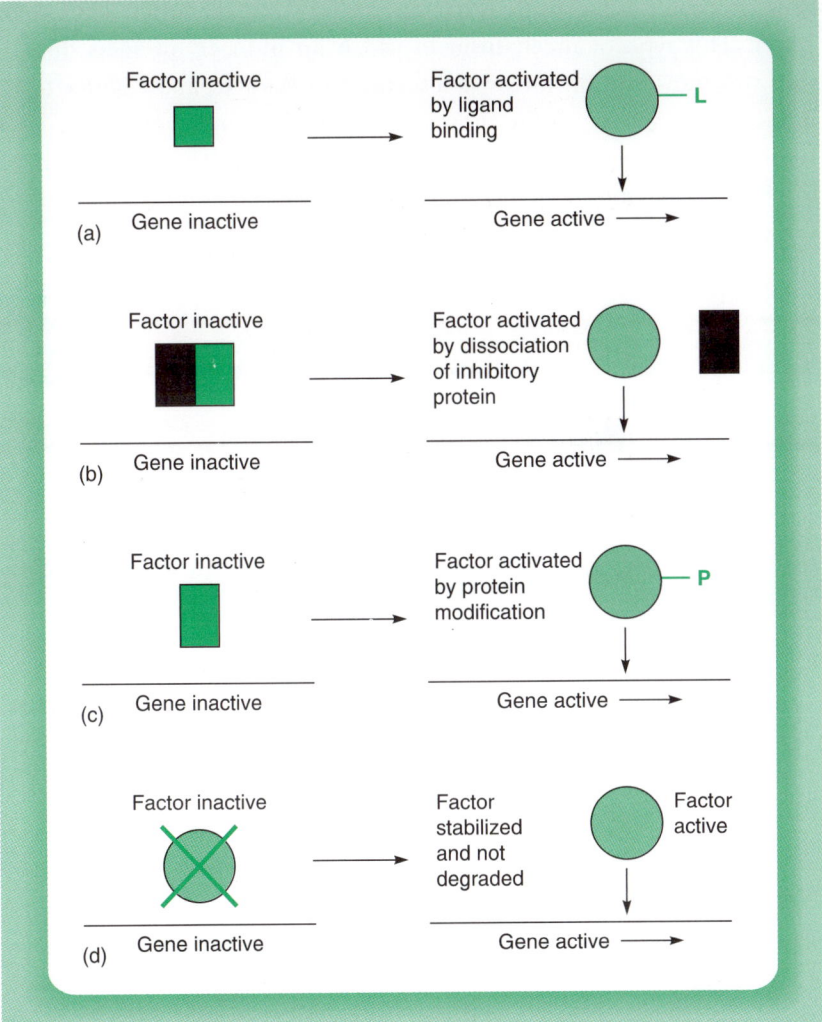

Figure 8.2
Methods of activating a transcription factor in response to an inducing stimulus.
This can occur by a ligand-mediated conformational change (a), by removal of an
inhibitory protein (b), by a modification to the protein such as phosphorylation
(c) or by stabilizing the factor so that it is not degraded (d).

the factor, ensuring a rapid response. The simplest method for this is for
an inducing ligand to bind to the transcription factor and alter its structure so that it becomes activated (Fig. 8.2a).

An example of this effect is seen in the case of the ACE1 factor, which
mediates the induction of the yeast metallothionein gene in response to
copper. In this case, the transcription factor undergoes a major conformational change in the presence of copper. This converts it to an active form
which is able to bind to its appropriate binding sites in the metallothionein

gene promoter and activate transcription (Fig. 8.3; for review see Thiele, 1992). This type of mechanism in which an inducer interacts directly with the transcription factor is also used in yeast to allow genes to be induced in response to the presence of particular nutrients in the environment (for review see Sellick and Reece, 2005). Thus, for example, the Put3 transcription factor undergoes a conformational change in the presence of proline, allowing it to induce the genes required to use this amino acid as a nitrogen source.

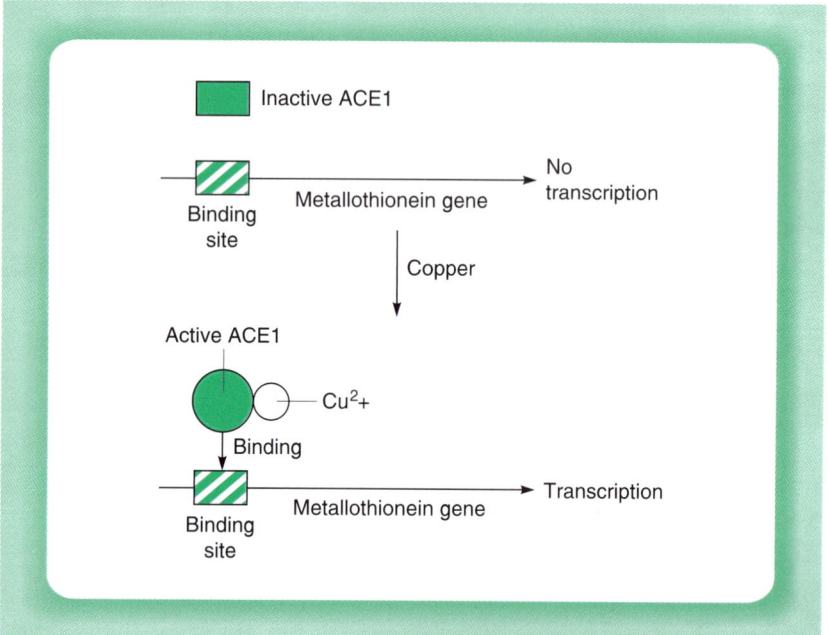

Figure 8.3
Activation of the ACE1 factor in response to copper results in transcription of the metallothionein gene.

An interesting variant on this theme is seen in the case of the Yap 1 factor, which stimulates the expression of genes encoding anti-oxidant proteins in response to high oxygen levels. Under high oxygen conditions, the Yap 1 factor contains internal disulphide bonds between cysteine amino acids. The resulting folded structure masks a region of the protein containing a nuclear export signal and so the protein remains in the nucleus, and can regulate its target genes. When oxygen levels are low, the disulphide bonds are reduced and therefore break, leading to exposure of the nuclear export signal. Hence, Yap 1 is exported to the cytoplasm and so can no longer activate its target genes (Wood *et al.*, 2004) (Fig. 8.4).

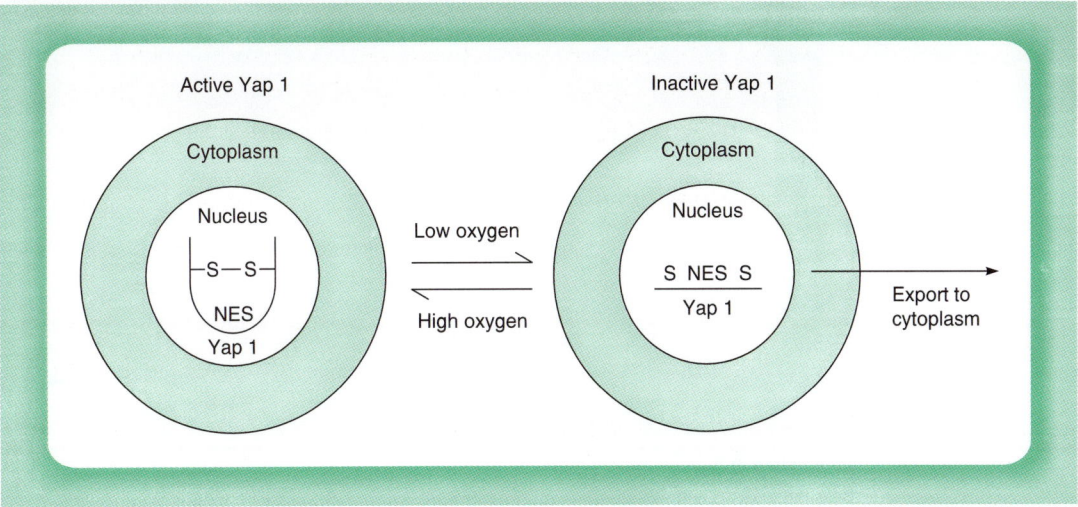

Figure 8.4

Under conditions of high oxygen, disulphide bonds (S–S) in the Yap 1 transcription factor mask its nuclear export signal (NES) allowing it to remain in the nucleus and activate its target genes. In low oxygen, the disulphide bonds are reduced, the NES is exposed and Yap 1 is exported to the cytoplasm.

Although widely used in yeast cells, which are in close contact with their environment, direct regulation by an inducing stimulus also occurs in higher eukaryotes. Thus, for example, in mammalian cells, the DREAM transcription factor represses the transcription of the dynorphin gene and thereby enhances the response to painful stimuli (for review see Costigan and Woolf, 2002). The activity of this factor is directly modulated by the level of calcium, which binds directly to the DREAM protein and reduces its ability to bind to its binding site in the dynorphin gene. Hence, the repression of the gene by DREAM is relieved and the dynorphin gene is transcribed (for review see Mandel and Goodman, 1999) (Fig. 8.5).

8.2.2 THE NUCLEAR RECEPTORS

A more complex example of regulation by ligand binding to a transcription factor is provided by the steroid hormone receptors. These receptors are members of the nuclear receptor or steroid-thyroid receptor family discussed in Chapter 4 (section 4.4) and mediate gene activation in response to steroids such as glucocorticoid or estrogen (for review see Weatherman *et al.*, 1999).

Following identification of the steroid hormone receptors, it was very rapidly shown that the receptors were only found associated with DNA after hormone treatment. These early studies were subsequently

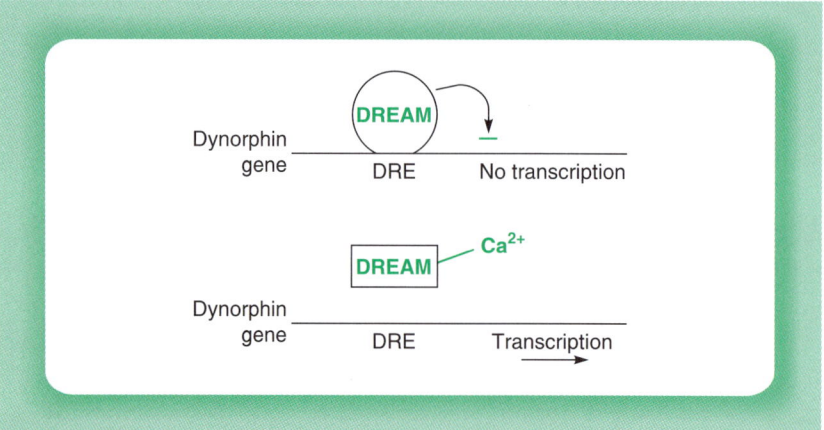

Figure 8.5

In the absence of calcium, the DREAM repressor binds to its response element (DRE) in the dynorphin gene and represses its transcription. When calcium is present, it binds to the DREAM factor and changes its conformation so that it does not bind to the DRE. This relieves the repression and allows transcription of the dynorphin gene.

confirmed by using DNAseI footprinting on whole chromatin to show that the receptor was only bound to the hormone response sequence following hormone treatment (Becker *et al.*, 1986). These studies were therefore consistent with a model in which the hormone induces a conformational change in the receptor activating its ability to bind to DNA and thereby activate transcription.

Subsequent studies have suggested that the situation is more complex, however. Thus, although in the intact cell the receptor binds to DNA only in the presence of the hormone, purified receptor can bind to DNA *in vitro* in a band shift or footprinting assay regardless of whether hormone is present or not (Wilmann and Beato, 1986; Figs. 8.6 and 8.7).

This discrepancy led to the suggestion that the receptor is inherently capable of binding to DNA, but is prevented from doing so in the absence of steroid because it is anchored to another protein. The hormone acts to release it from this association and allow it to fulfil its inherent ability to bind to DNA. In agreement with this possibility, in the absence of hormone, the glucocorticoid receptor protein is found in the cytoplasm complexed to a 90,000 molecular weight heat-inducible protein (hsp90) in an 8S complex. This complex is dissociated upon steroid treatment releasing the 4S receptor protein (for review see Pratt, 1997). The released receptor is free to dimerize and move into the nucleus. Since these processes have been shown to be essential for DNA binding

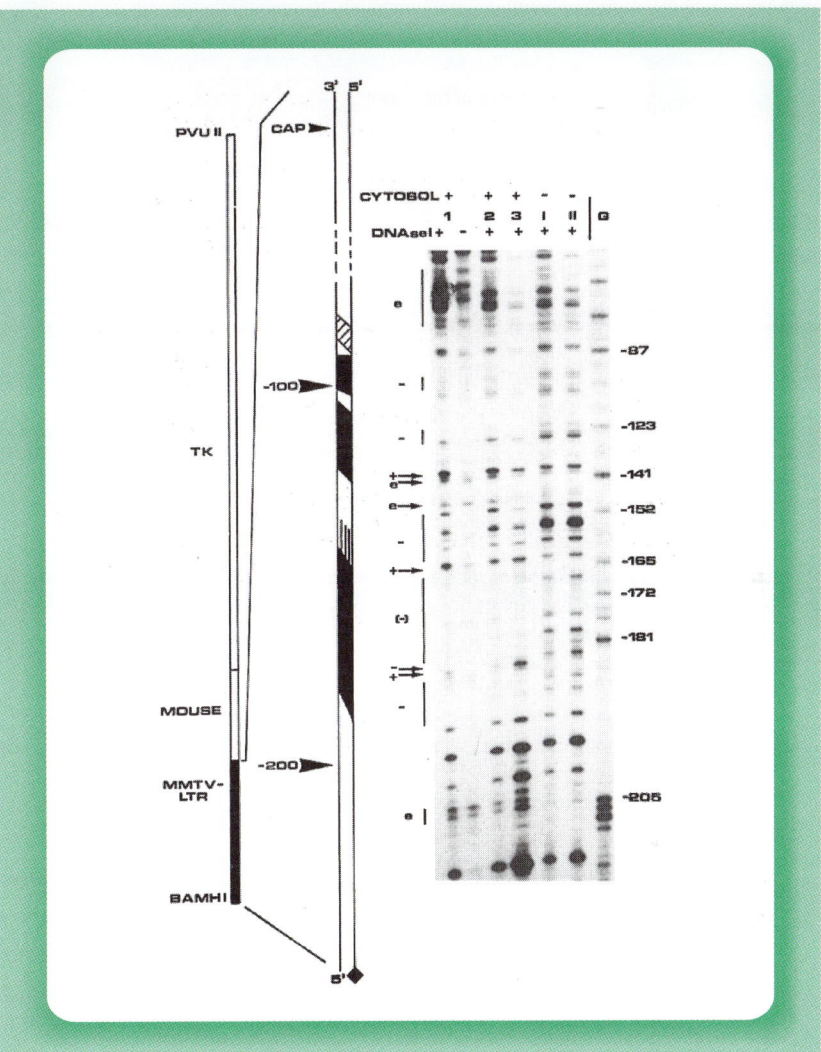

Figure 8.6

DNAseI footprint analysis of the binding of the glucocorticoid receptor to the glucocorticoid-inducible mouse mammary tumour virus long terminal repeat promoter (MMTV–LTR). In tracks I and II the DNAseI digestion has been carried out without any added receptor. In tracks 1–3, glucocorticoid receptor has been added prior to DNAseI digestion either alone (track 1+), with the glucocorticoid hormone corticosterone (track 2) or with the anti-hormone RU486 which inhibits steroid-induced activation of the receptor (track 3). Track 1[−] shows the result of adding receptor to the DNA in the absence of DNAseI addition in which some cleavage by endogenous nucleases (e) occurs, whilst track G is a marker track produced by cleaving the same DNA at each guanine residue. Minus signs indicate footprinted regions protected by receptor, plus signs are hypersensitive sites at which cleavage is increased by the presence of the receptor. The DNA fragment used and position of the radioactive label (diamond) are shown together with the distances upstream from the initiation site for transcription. Note that the identical footprint is produced by the receptor either alone or in the presence of hormone or anti-hormone. Hence *in vitro* the receptor can bind to DNA in the absence of hormone.

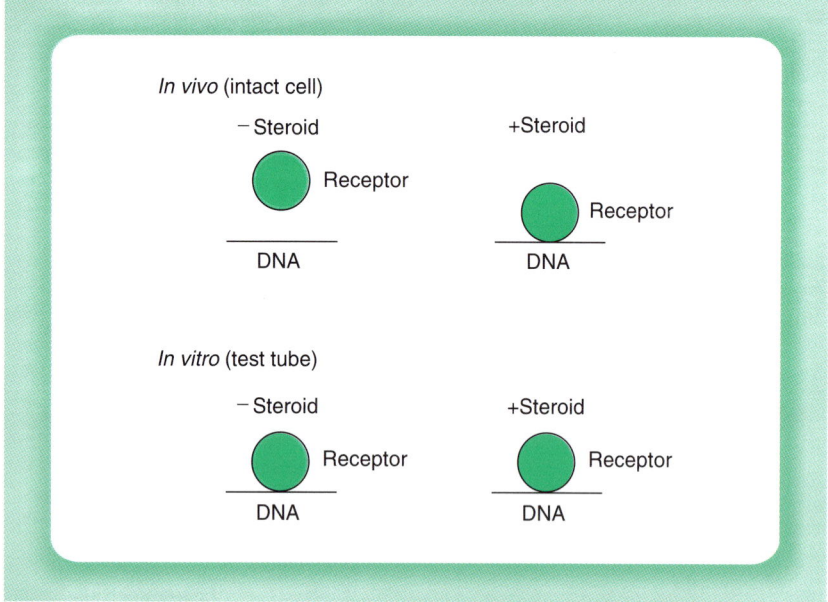

Figure 8.7

Comparison of steroid receptor binding to DNA in the presence or absence of hormone *in vivo* and *in vitro*. Note that whilst *in vivo* DNA binding can occur only in the presence of hormone, *in vitro*, it can occur in the presence or absence of hormone.

and transcriptional activation by steroid hormone receptors, dissociation of the receptor from hsp90 is essential if gene activation is to occur. In agreement with this, antiglucocorticoids which inhibit the positive action of glucocorticoids have been shown to stabilize the 8S complex of hsp90 and the receptor.

Similar complexes with hsp90 have also been reported for the other steroid hormone receptors. Thus the activation of the different steroid receptors such as the glucocorticoid and estrogen receptors by their specific hormones is likely to involve disruption of the protein–protein interaction with hsp90 (Fig. 8.8).

Most interestingly, the association of hsp90 with the glucocorticoid receptor occurs via the C-terminal region of the receptor, which also contains the steroid binding domain. It has been suggested therefore that by associating with the C-terminal region of the receptor, hsp90 masks adjacent domains whose activity is necessary for gene activation by the receptor, for example, those involved in receptor dimerization or subsequent DNA binding, thereby preventing DNA binding from occurring. Following steroid treatment, however, the steroid binds to the C-terminus of the receptor, displacing hsp90 and thereby unmasking these

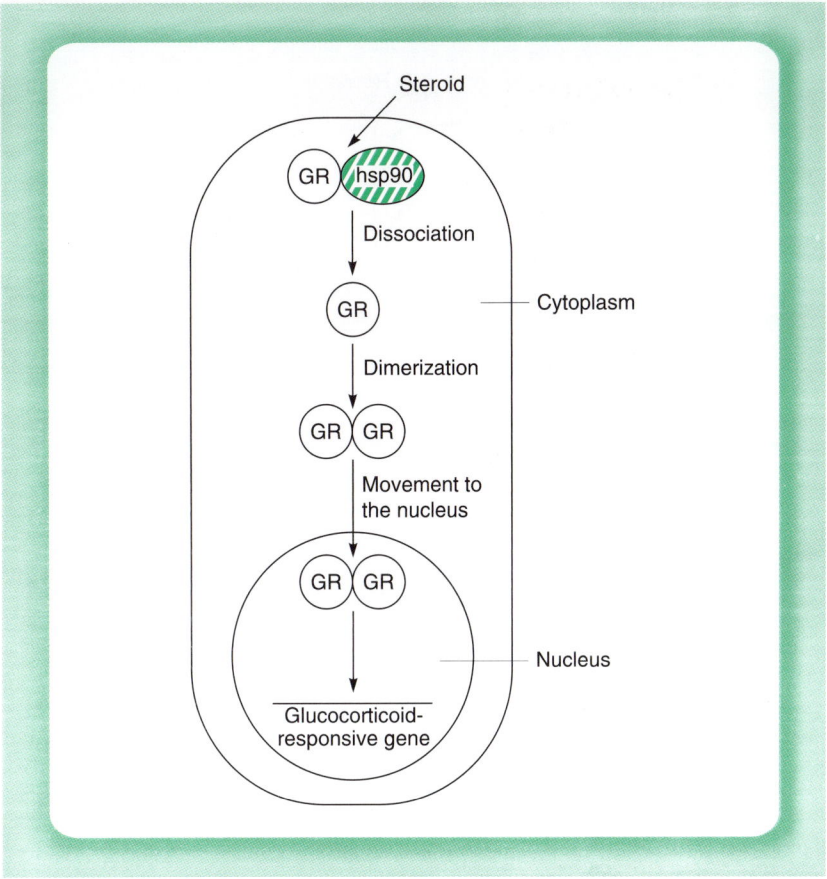

Figure 8.8

Activation of the glucocorticoid receptor (GR) by steroid involves dissociation of hsp90 allowing dimerization and movement to the nucleus.

domains and allowing DNA binding to occur (Fig. 8.9). Hence, activation of the steroid receptors involves a ligand-induced conformational change which results in the dissociation of an inhibitory protein.

In addition to the steroid-induced dissociation of the receptors from hsp90, it is clear that a second step following dissociation from hsp90 is also required for receptor activation. Thus, in a cell free system in which the progesterone receptor exists in a 4S form, free of bound hsp90, the addition of progesterone is still required for the activation of progesterone responsive genes. This indicates that the hormone has an additional effect on the receptor apart from dissociating it from hsp90. This effect involves the unmasking of a previously inactive transcriptional activation domain in the receptor allowing it to activate gene expression in a hormone-dependent manner following DNA binding. Thus, domain swapping experiments (see Chapter 2, section 2.4.1) have identified C-terminal

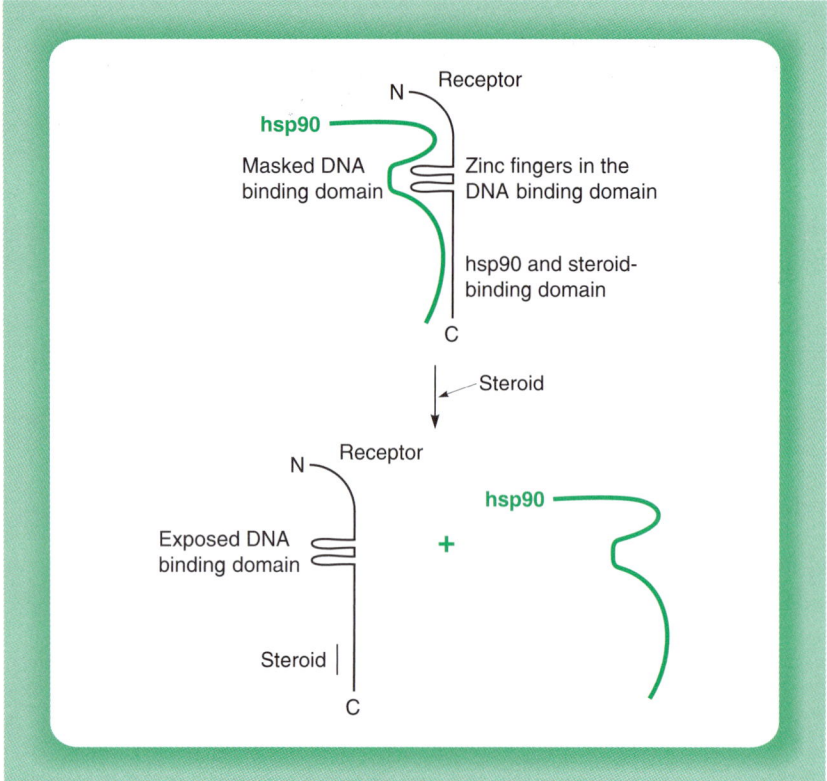

Figure 8.9

Interaction of hsp90 and the glucocorticoid receptor. Hsp90 binds to the receptor via the C-terminal region of the receptor which also binds steroid and may mask regions of the receptor necessary for dimerization or DNA binding. When steroid is added it binds to the receptor at the C-terminus displacing hsp90 and exposing the masked regions.

regions in both the glucocorticoid and estrogen receptors which when linked to the DNA binding domain of another factor, can activate transcription only following hormone addition (see Fig. 4.30). These regions hence constitute hormone-dependent activation domains.

Moreover, in the case of the estrogen receptor, it has been shown that the estrogen antagonist 4-hydroxytamoxifen induces the receptor to bind to DNA (presumably by promoting dissociation from hsp90 and dimerization) but does not induce gene activation, suggesting that it fails to activate the estrogen-responsive trans-activation domain. Hence, the mechanism by which the steroid receptors are activated is now thought to involve both dissociation from hsp90 and a change in their transcriptional activation ability (Fig. 8.10a). This second step is likely to involve a change in the activation domain which allows it to bind co-activator

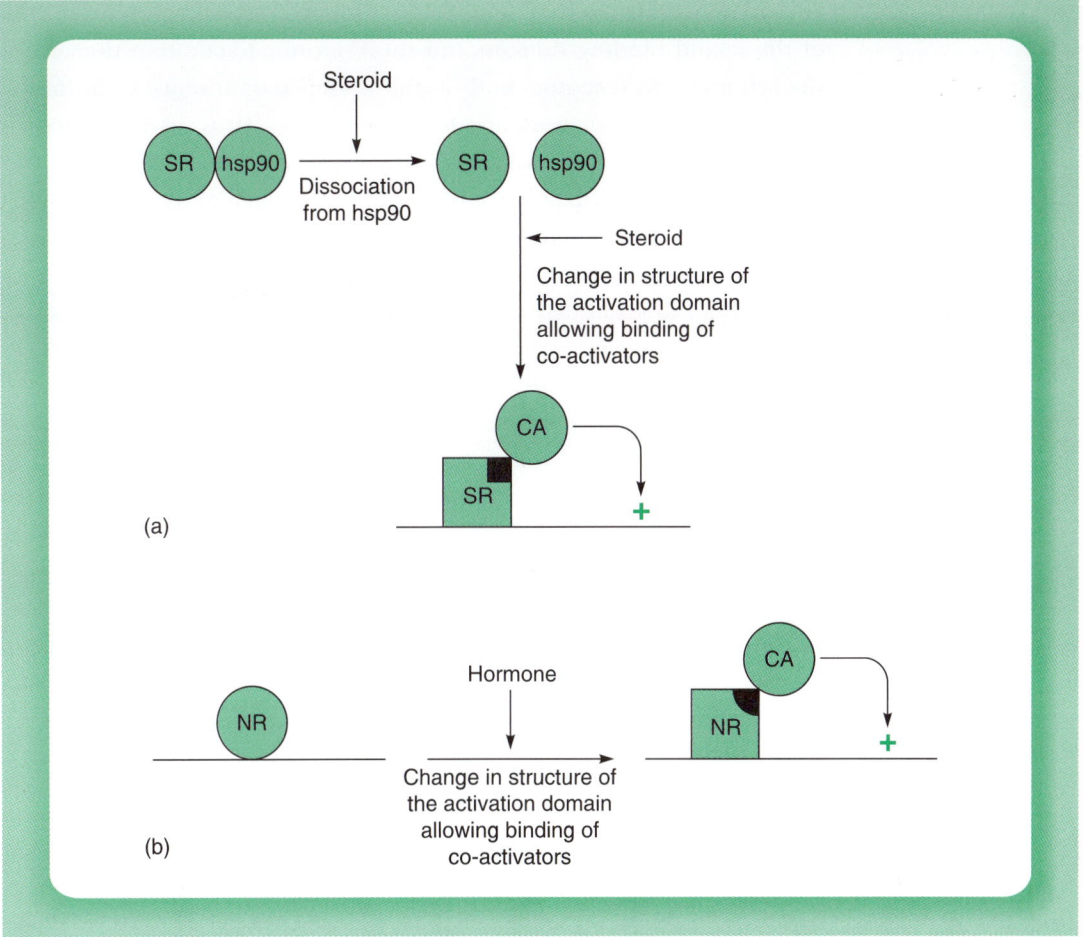

Figure 8.10

(a) Activation of the steroid receptors (SR) by treatment with steroid. As well as inducing dissociation of the receptor from hsp90, steroid treatment also increases the ability of the receptor to activate transcription following DNA binding by changing the structure of the activation domain (shaded) allowing it to bind co-activator proteins (CA) which stimulate transcription. (b) Activation of other members of the nuclear receptor family which bind non-steroids such as retinoic acid or thyroid hormone involves only the second of these stages.

proteins that are essential for transcriptional activation (see Chapter 5, section 5.4.3 for discussion of co-activator molecules).

Interestingly, other members of the nuclear receptor family which bind to substances that are related to steroids, such as retinoic acid or thyroid hormone, do not associate with hsp90 and are bound to DNA prior to exposure to ligand. Their activation by their appropriate ligand thus involves only the second stage discussed above, namely a ligand-induced structural change in their C-terminal activation domain (which is adjacent to the ligand binding domain) allowing it to bind co-activator molecules and activate transcription (Fig. 8.10b). Indeed, crystallographic studies of the ligand binding domain and the C-terminal activation domain of the retinoic acid receptors both in the presence or absence of hormone have provided direct evidence for this change. Thus, as illustrated in Plate 6, the activation domain is not closely associated with the ligand binding domain in the absence of ligand but is much more closely associated with it following ligand binding and forms a lid covering the ligand binding region (Renaud *et al.*, 1996).

Although first defined in the retinoic acid receptors, a similar structural change occurs upon ligand binding in other members of the nuclear receptor family including the glucocorticoid and estrogen receptors and the thyroid hormone receptor (Wurtz *et al.*, 1996). Indeed, it has been shown that whilst estrogen induces this realignment of the estrogen receptor activation domain, the estrogen antagonist raloxifene does not do so, thereby explaining its antagonistic action (Brzozosowski *et al.*, 1997) (Fig. 8.11). In turn, this ligand-induced structural change allows the activation domain to bind co-activator proteins, which bind to the receptors only after exposure to hormone and appear to play a key role in the ability of the receptors to activate transcription (Fig. 8.10) (see Chapter 5, section 5.4.3 for a discussion of co-activator molecules).

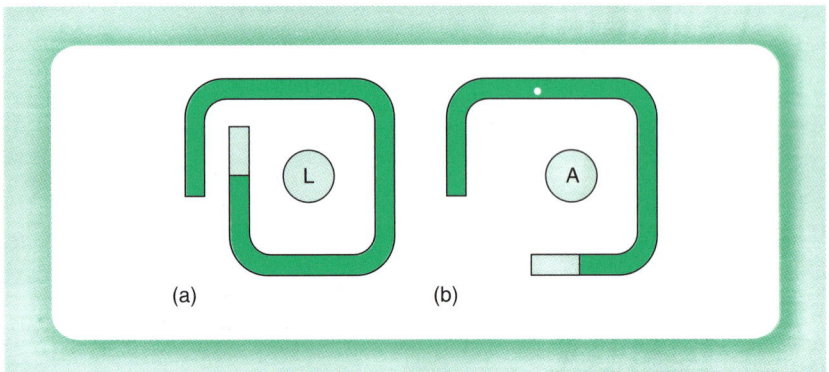

Figure 8.11

(a) The binding of the ligand (L) induces the realignment of the C-terminal activation domain of the nuclear receptors (light shading) so that it forms a lid over the ligand binding domain and the activation domain then stimulates transcription. (b) This realignment is not induced by binding of antagonists (A) which therefore do not stimulate transcriptional activation.

Interestingly, in the case of receptors such as the thyroid hormone receptor, where DNA binding is observed even prior to hormone treatment, the receptor actually represses transcription prior to thyroid hormone treatment. As discussed in Chapter 6 (section 6.3.2), this is because in the

absence of ligand, the receptor binds co-repressor molecules, which are dis-
placed by co-activators on hormone treatment. The importance of this conversion from repressor to activator is seen in the case of mutant forms of
the thyroid hormone receptor which cannot undergo this conformational
change because they do not bind thyroid hormone. This is observed not
only in the v-*erbA* oncogene as discussed in Chapter 9 (section 9.3.2) but
also in patients with generalized thyroid hormone resistance. Thus, these
patients have been shown to produce forms of the receptor which can
repress gene expression but which cannot activate genes in response to thyroid hormone. Most interestingly, the presence of these dominant negative
forms of the receptor results in impairment of physical and mental development which is much more severe than that observed if the receptor is
absent completely (Baniahmad *et al.*, 1992).

Hence, in all the nuclear receptors, activation by ligand involves a
structural change in the C-terminal activation domain which allows it to
bind co-activators. In the steroid hormone receptors, this is preceded by
an earlier step which involves the disruption of the receptor/hsp90 association. Activation of these steroid receptors, therefore, involves both the
ligand-induced conformational changes seen in ACE1 and DREAM, as
well as the dissociation of an inhibitor protein and thus combines the
mechanisms illustrated in Fig. 8.2a and b.

8.3 REGULATION BY PROTEIN–PROTEIN INTERACTIONS

8.3.1 INHIBITION OF TRANSCRIPTION FACTOR ACTIVITY BY PROTEIN–PROTEIN INTERACTION

As described above, the glucocorticoid receptor is regulated by its interaction with hsp90 which prevents it binding to DNA and activating transcription in the absence of steroid hormone. A similar mechanism is used
in the case of the NFκB factor which, as discussed above, only activates
transcription in mature B cells or in other cell types following treatment
with agents such as lipopolysaccharides or phorbol esters. In agreement
with this, no active form of NFκB capable of binding to DNA can be
detected in DNA mobility shift assays (see Chapter 2, section 2.2.1) using
either cytoplasmic or nuclear extracts prepared from pre-B cells or non-B
cell types. Interestingly, however, such activity can be detected in the cytoplasm but not the nucleus of such cells following denaturation and subsequent renaturation of the proteins in the extract. Hence, NFκB exists
in the cytoplasm of pre-B cells and other cell types in an inactive form
which is complexed with another protein, known as IκB, that inhibits

its activity (for reviews see Yamamoto and Gaynor, 2004; Hoffman and Baltimore, 2006; Hoffman *et al.*, 2006). The release of NFκB from IκB by the denaturation/renaturation treatment therefore results in the appearance of active NFκB, capable of binding to DNA (Fig. 8.12a).

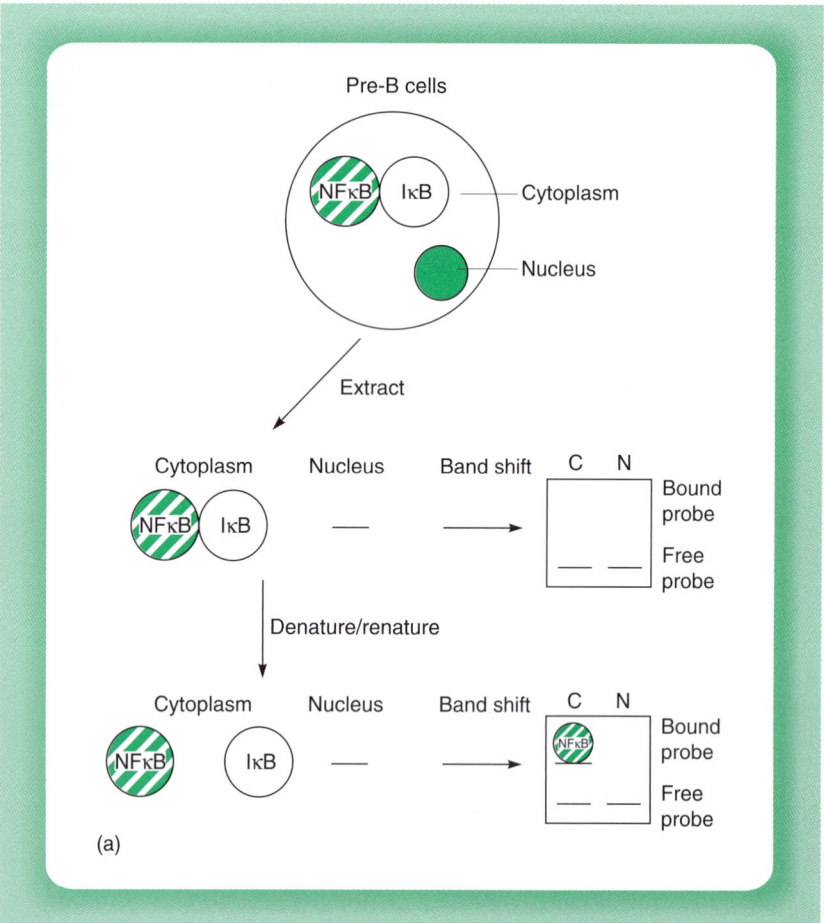

(a)

Figure 8.12(a)

Regulation of NFκB. Panel (a): In pre-B cells NFκB is located in the cytoplasm in an inactive form which is complexed to IκB. DNA mobility band shift assays do not therefore detect active NFκB. If a cytoplasmic extract is first denatured and renatured, however, active NFκB will be released from IκB and will be detected in a subsequent band shift assay.

These findings suggested therefore that treatments with substances such as lipopolysaccharides or phorbol esters do not activate NFκB by interacting directly with it in a manner analogous to the activation of the ACE1 factor by copper. Rather, they are likely to produce the dissociation of NFκB from IκB resulting in its activation. In agreement with this idea,

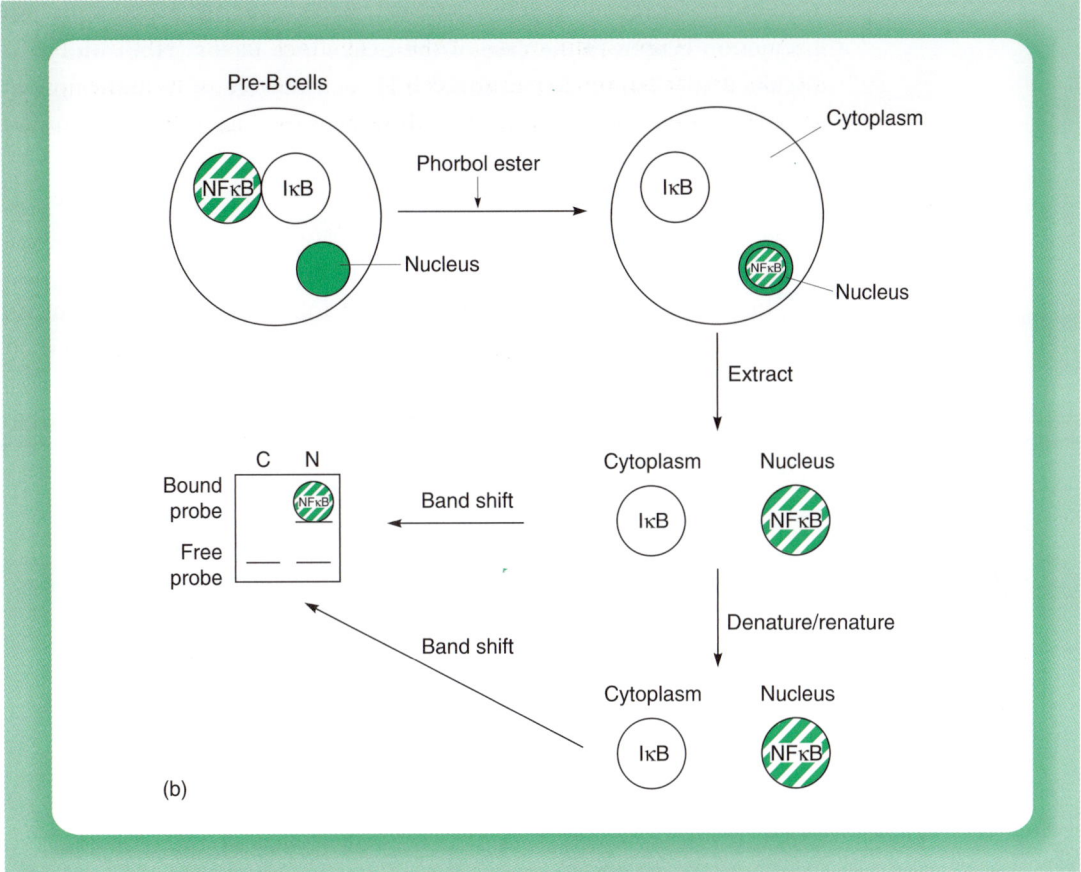

Figure 8.12(b)

Panel (b): In mature B cells, NFκB has been released from IκB and is present in the nucleus in an active DNA binding form. It can therefore be detected in a DNA mobility shift assay without a denaturation, renaturation step which has no effect on the binding activity.

phorbol ester treatment of cells prior to their fractionation eliminated the latent NFκB activity in the cytoplasm and resulted in the appearance of active NFκB in the nucleus (Fig. 8.12b). These substances act therefore by releasing NFκB from IκB, allowing NFκB to move to the nucleus where it can bind to DNA and activate gene expression. Hence, this constitutes an example of the activation of a factor by the dissociation of an inhibitory protein (Fig. 8.2b).

Such a mechanism is used to regulate the activity of many different transcription factors. Thus, apart from the NFκB/IκB and glucocorticoid receptor/hsp90 interactions, other examples of inhibitory interactions include those between DNA binding helix-loop-helix proteins and Id proteins (Chapter 6, section 6.2.2) and p53 and the MDM2 protein (Chapter 9, section 9.4.2). Hence, inhibitory interactions of this type are widely used to regulate the activity of specific transcription factors.

A highly complex example of such regulation by protein–protein interaction is seen in the case of the heat shock factor (HSF) which, as discussed in Chapter 5 (section 5.5.1), activates gene transcription in response to elevated temperature. HSF achieves this effect by binding to its binding site in target genes, which is known as the heat shock element (HSE) (see Chapter 1, section 1.3.3). The amount of HSF bound to the HSE increases with the time of exposure to elevated temperature and with the extent of temperature elevation. Moreover, increased protein binding to the HSE is also observed following exposure to other agents which also induce the transcription of the heat shock genes such as 2,4-dinitrophenol (Fig. 8.13). Thus, activation of the heat shock genes, mediated by the HSE, is accompanied by the binding of a specific transcription factor to this DNA sequence.

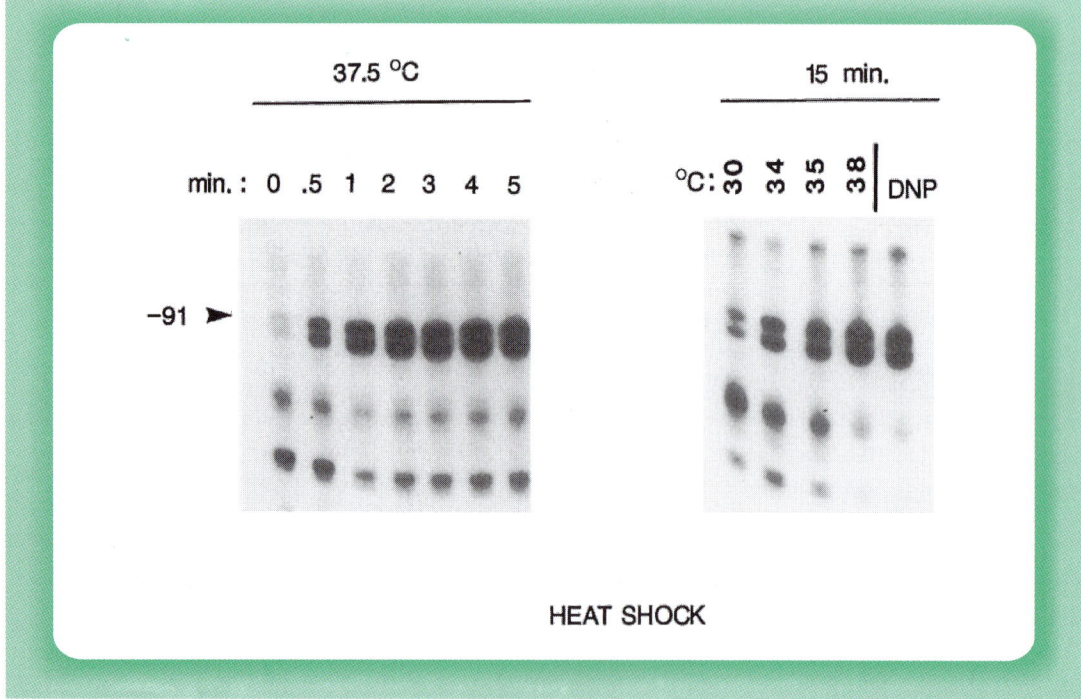

Figure 8.13

Detection of HSF binding to the HSE 91 bases upstream (−91) of the start site for transcription in the *Drosophila* hsp82 gene and protecting this region from digestion with exonuclease III. Note the increased binding of HSF with increasing time of exposure to heat shock or increased severity of heat shock. HSF binding is also induced by exposure to 2, 4-dinitrophenol (DNP) which is known to induce transcription of the heat shock genes.

As noted in section 8.1, this activation of HSF can occur in the absence of new HSF protein synthesis (for review see Morimoto, 1998). Thus, if cells are heat treated in the presence of cycloheximide, which is an inhibitor of protein synthesis, increased binding of HSF to the HSE is observed, exactly as in cells treated in the absence of the drug (Zimarino and Wu, 1987). This indicates that the observed binding of HSF following heat shock does not require *de novo* protein synthesis. Rather, this factor must pre-exist in non-heat treated cells in an inactive form whose ability to bind to the HSE sequence in DNA is activated post-translationally by heat. In agreement with this, activation of HSF can also be observed following heat treatment of cell extracts *in vitro* when new protein synthesis would not be possible (Larson *et al.*, 1988).

Analysis of the activation process using *in vitro* systems from human cells (Larson *et al.*, 1988) has indicated that it is a two-stage process (Fig. 8.14). In the first stage, the HSF is activated to a form which can bind to DNA by an ATP-independent mechanism, which is directly dependent on elevated temperature. Subsequently, this protein is further modified by phosphorylation, allowing it to activate transcription. Interestingly, the second of these two stages appears to be disrupted in murine erythroleukaemia (MEL) cells in which heat shock results in increased binding of HSF to DNA, but transcriptional activation of the heat shock genes is not observed (Hensold *et al.*, 1990).

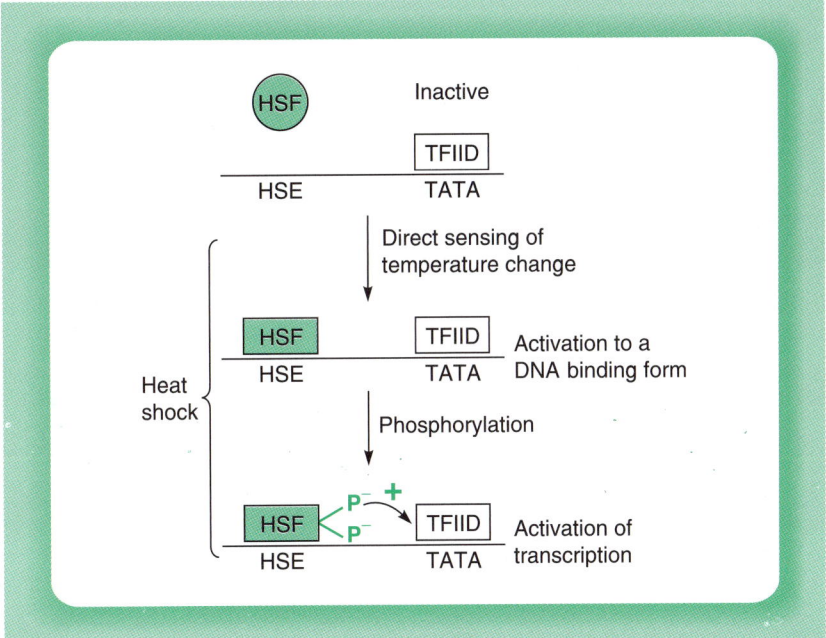

Figure 8.14

Stages in the activation of HSF in mammalian and *Drosophila* cells. Initial activation of HSF to a DNA binding form following elevated temperature is followed by its phosphorylation which converts it to a form capable of activating transcription.

The activation of HSF into a form capable of binding DNA involves its conversion from a monomeric to a trimeric form which can bind to the HSE (for review see Morimoto, 1998). The maintenance of the mono-meric form of HSF prior to heat shock is dependent on a region at the C-terminus of the molecule since when this region is deleted, HSF spon-taneously trimerizes and can bind to DNA even in the absence of heat shock. The C-terminal region contains a leucine zipper (see Chapter 4, section 4.5). As leucine zippers are known to be able to interact with one another, it is thought that this region acts by interaction with another leucine zipper, located adjacent to the N-terminal DNA binding domain, promoting intra-molecular folding which masks the DNA binding domain. Following heat shock, HSF unfolds, unmasking the DNA bind-ing domain and allowing a DNA binding trimer to form (Fig. 8.15).

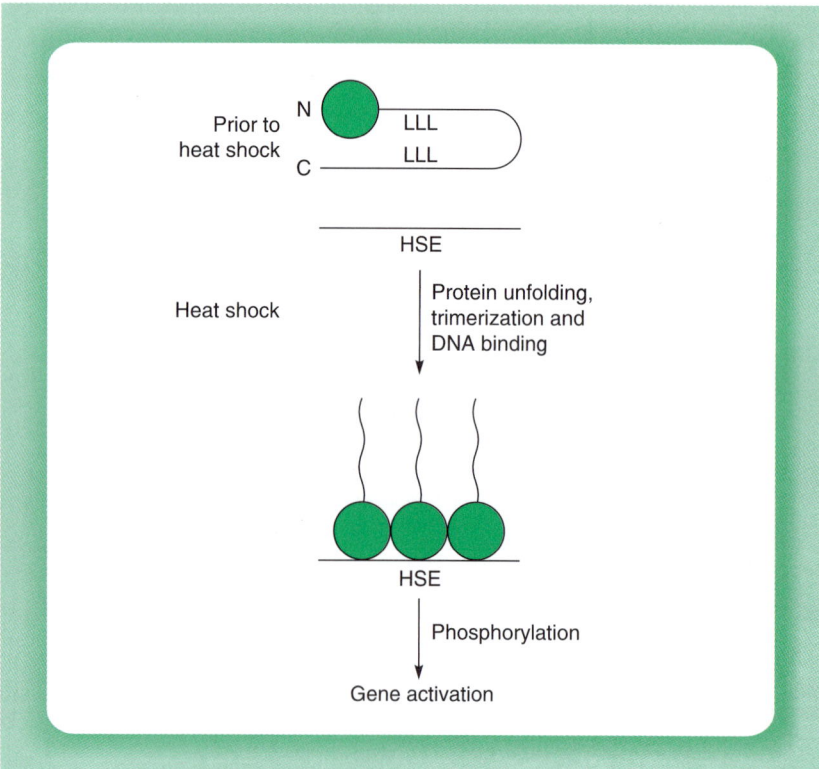

Figure 8.15

Prior to heat shock. HSF is present in a monomeric form in which the leucine zipper motifs (L) at the C-terminus and within the molecule promote intra-molecular folding which masks the N-terminal DNA binding domain (shaded) preventing binding to the HSE. Following heat shock, the protein unfolds and forms the DNA binding trimeric form. This form binds to the HSE and activates transcription following its subsequent phosphorylation.

Interestingly, as with the glucocorticoid receptor (see section 8.2.2), the conversion of HSF from a monomer to a DNA binding trimer involves the dissociation of hsp90 which binds to HSF in untreated cells and stabilizes it in the inactive form which cannot bind to DNA (Zou *et al.*, 1998). Interestingly, hsp90 acts as a so-called 'chaperone' protein, assisting the proper folding of other proteins. Evidently, following heat or other stress, the level of such unfolded proteins will increase. Hsp90 will therefore be 'called away' to deal with these unfolded proteins, leaving HSF free to trimerize and bind to DNA (Fig. 8.16).

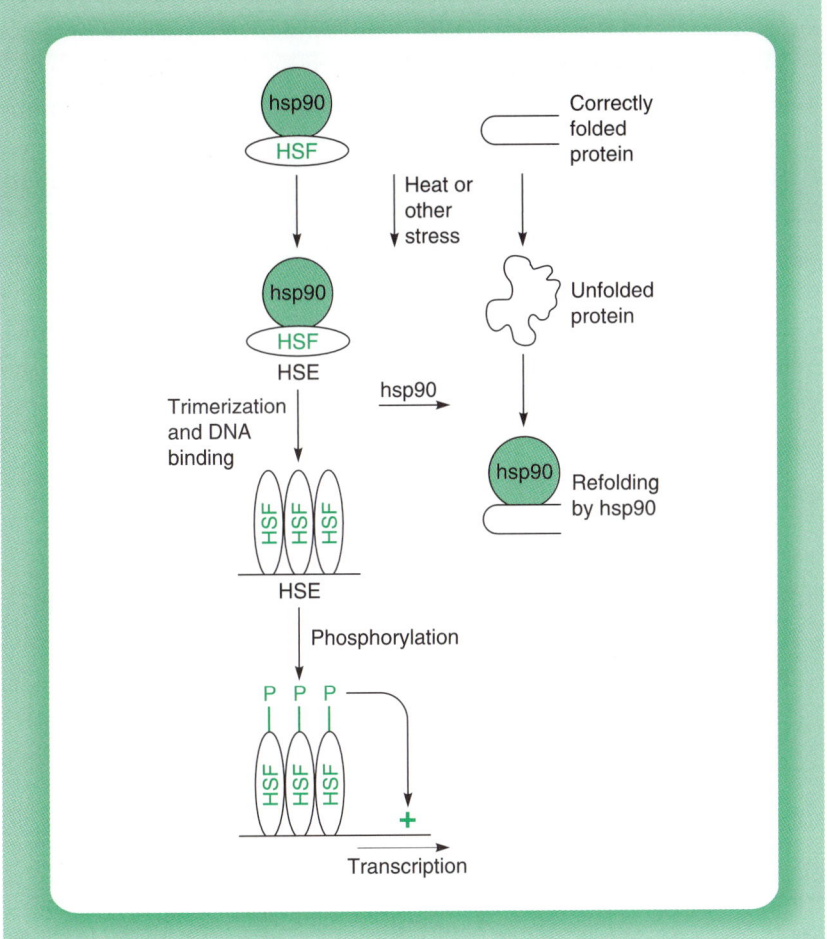

Figure 8.16

Prior to heat shock or other stress, HSF is bound to hsp90 which stabilizes its inactive monomeric form. Following heat shock, hsp90 dissociates from HSF to fulfil its function of refolding other proteins which have unfolded due to the elevated temperature. This allows HSF to trimerize and bind to DNA. However, transcriptional activation requires subsequent phosphorylation of HSF.

As well as the loss of hsp90, recent data have shown that the trimerization of HSF1 also requires at least two additional factors, which promote trimerization (Shamovsky *et al.*, 2006; for review see Kugel and Goodrich, 2006, Fig. 8.17). One of these is eEF1A, a factor which is normally involved in the translation of mRNAs into protein. As heat shock and other stresses shut down most cellular protein synthesis, this factor may be released from the translational apparatus, leaving it free to interact with HSF and stimulate its trimerization. The second factor which associates with HSF1 and stimulates its trimerization, is not a protein but an RNA of approximately 600 bases in size, known as HSR1. Thus, in contrast to the small RNAs, which inhibit gene expression, as described in Chapter 6 (section 6.4.2), HSR1 represents an example of an RNA which is involved in the activation of gene transcription (for reviews of the role of non-coding RNAs in the regulation of gene transcription see Goodrich and Kugel, 2006; Prasanth and Spector, 2007).

Hence, the response of HSF to stress involves both the loss of an inhibitory protein, the association of a stimulatory protein and an RNA, as well as changes in the HSF molecule itself (Fig. 8.17). Together these changes promote the transition from an HSF monomer to a DNA-binding trimer. However, this DNA binding by HSF is insufficient to produce transcriptional activation. This requires phosphorylation of HSF on serine 230 which allows the DNA-bound form of HSF to activate transcription (Figs. 8.15, 8.16) (Holmberg *et al.*, 2001).

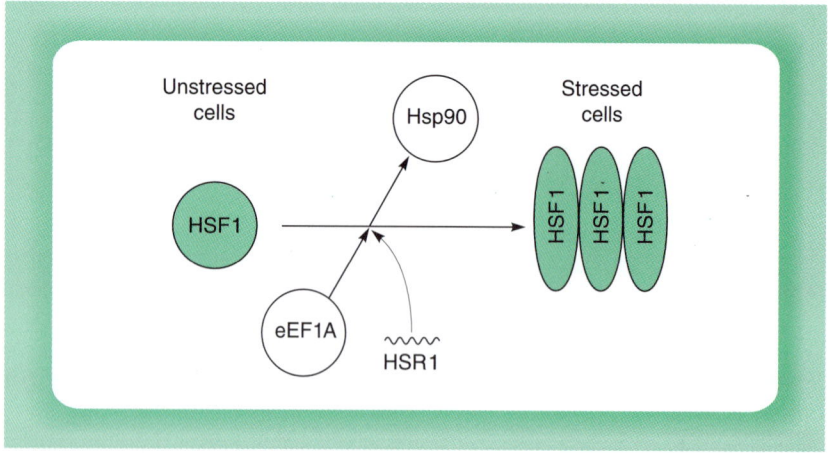

Figure 8.17

The transition of HSF1 from an inactive monomer to an active trimer involves the dissociation of the hsp90 protein from HSF1 and the association with it of the eEF1A protein and the HSR1 RNA.

This two-stage process, involving DNA binding induced by trimeriza-tion and transcriptional activation induced by serine phosphorylation, represents a common mechanism for the activation of HSF in higher eukaryotes such as *Drosophila* and mammals. In contrast, however, the *Saccharomyces cerevisiae* (budding yeast) HSF is activated by a much simpler mechanism. Thus, unlike *Drosophila* or mammalian HSF, the budding yeast protein lacks the C-terminal leucine zipper region which promotes monomer formation and therefore exists as a trimer prior to heat shock. As expected from this, HSF can be observed bound to the HSE even in non-heat shocked cells (Sorger *et al.*, 1987). HSF can activate transcrip-tion, however, only following heat treatment when the protein becomes phosphorylated. Interestingly, in *Schizosaccharomyces pombe* (fission yeast) HSF regulation follows the *Drosophila* and mammalian system, with HSF becoming bound to DNA only following heat shock (Gallo *et al.*, 1991).

Hence in mammals, *Drosophila* and fission yeast, activation of HSF is more complex than in budding yeast involving an initial stage activating the DNA binding ability of HSF in response to heat as well as the stage, common to all organisms, in which the ability to activate transcription is stimulated by phosphorylation (Fig. 8.18). It thus combines regulation by protein–protein interaction (between HSF itself and HSF/hsp90) as

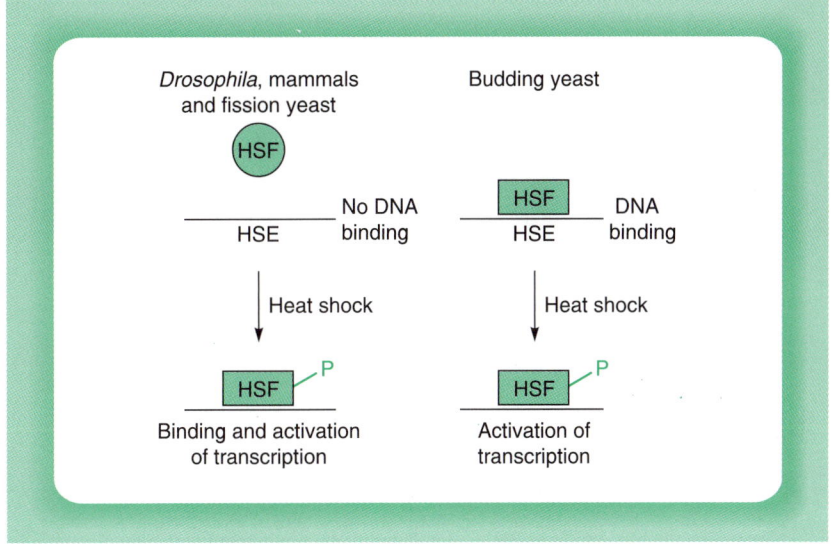

Figure 8.18

HSF activation in *Drosophila*, mammals and fission yeast compared to that in budding yeast. Note that in budding yeast HSF is already bound to DNA prior to heat shock and hence its activation by heat involves only the second of the two stages seen in other organisms, namely, its phosphorylation allowing it to activate transcription.

well as regulation by phosphorylation, which will be discussed more generally in section 8.4.

Interestingly, as well as being regulated by interacting with another transcription factor protein, it is also possible for a factor to be regulated by interaction with lipid within the cell. This is seen in the case of the Tubby factor, which regulates the expression of genes involved in fat metabolism. It has been shown that the Tubby protein is anchored at the plasma membrane by interaction with a phospholipid $PI(4,5)P_2$. However, following activation of specific G-protein coupled receptors in the plasma membrane, the enzyme phospholipase C is activated. This enzyme then cleaves $PI(4,5)P_2$, releasing Tubby and allowing it to move to the nucleus and activate transcription (Fig. 8.19) (for review see Cantley, 2001).

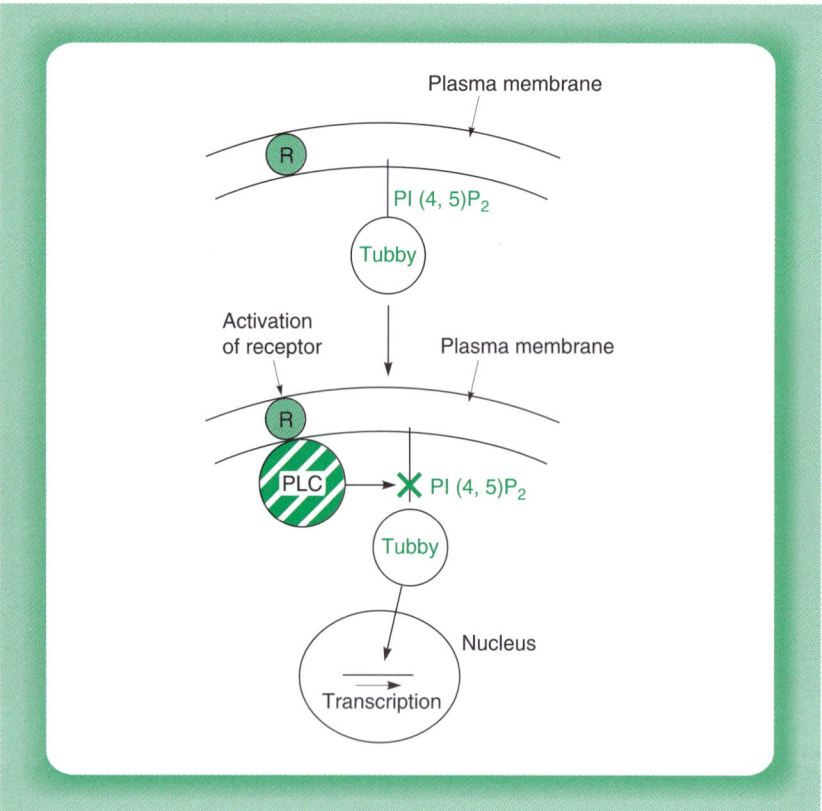

Figure 8.19

The Tubby transcription factor is anchored to the plasma membrane by binding to the phospholipid $PI(4,5)P_2$. Following activation of a membrane G-protein coupled receptor (R), phospholipase (PLC) is activated and cleaves $PI(4,5)P_2$. This releases Tubby, allowing it to move to the nucleus and activate gene expression.

This example is evidently similar to the glucocorticoid receptor/hsp90 and NFκB/IκB examples in that it involves the transcription factor moving from the cytoplasm to the nucleus but differs in that the activation process involves disruption of a protein–lipid interaction rather than a protein–protein interaction.

8.3.2 ACTIVATION OF TRANSCRIPTION FACTORS BY PROTEIN–PROTEIN INTERACTION

As described above, the activation of HSF involves both the disruption of an inhibitory interaction with hsp90 and a stimulatory interaction with eEF1A. Indeed, in many cases transcription factors may be inactive alone and may need to complex with a second factor in order to be active. This is seen in the case of the Fos protein, which cannot bind to DNA without first forming a heterodimer with the Jun protein (see Chapter 4, section 4.5). A similar mechanism also operates in the case of the Myc factor, which cannot bind to DNA except as a complex with the Max protein (see Chapter 9, section 9.3.3). Hence, protein–protein interactions between transcription factors can result in either inhibition or stimulation of their activity. The need for Fos and Myc to interact with another factor prior to DNA binding, arises from their inability to form a homodimer, coupled with the need for factors of this type to bind to DNA as dimers. Hence, they need to form heterodimers with another factor prior to DNA binding (see Chapter 4, section 4.5 for further discussion).

8.3.3 ALTERATION OF TRANSCRIPTION FACTOR FUNCTION BY PROTEIN–PROTEIN INTERACTION

Even in the case of factors such as Jun which can form DNA-binding homodimers, the formation of heterodimers with another factor offers the potential to produce a dimer with properties distinct from those of either homodimer. Thus, the Jun homodimer can bind strongly to AP1 sites but only weakly to the cyclic AMP response element (CRE). In contrast, a heterodimer of Jun and the CREB factor binds strongly to a CRE and more weakly to an AP1 site. Heterodimerization can therefore represent a means of producing multi-protein factors with unique properties different from that of either protein partner alone (for reviews see Jones, 1990; Lamb and McKnight, 1991).

Hence, as well as stimulating or inhibiting the activity of a particular factor, the interaction with another factor can also alter its properties, directing it to specific DNA binding sites to which it would not normally bind. Thus, as discussed in Chapter 4 (section 4.2.4) the *Drosophila* extradenticle protein changes the DNA binding specificity of the Ubx protein so that it binds to certain DNA binding sites with high affinity in the presence of extradenticle and with low affinity in its absence. Similarly, as described in Chapter 4 (section 4.2.4), the yeast α2 repressor factor forms heterodimers of different DNA binding specificities with the a1 or MCM1 transcription factors.

Although several examples of one transcription factor altering the DNA binding specificity of another have thus been defined, such protein–protein interactions can also change the specificity of a transcription factor in at least one other way. This is seen in the case of the *Drosophila* dorsal protein, which is related to the mammalian NFκB factors. Thus this factor is capable of both activating and repressing specific genes. Such an ability is not due, for example, to the production of different forms by alternative splicing, since both activation and repression take place in the same cell type. Rather, it appears to depend on the existence of a DNA sequence (the ventral repression element or VRE) adjacent to the dorsal binding site in genes such as zen which are repressed by dorsal, whereas the VRE sequence is absent in genes such as twist which are activated by dorsal.

It has been shown that DSP1 (dorsal switch protein), a member of the HMG family of transcription factors (see Chapter 4, section 4.6), binds to the VRE and interacts with the dorsal protein, changing it from an activator to a repressor. Hence, in genes such as twist where DSP1 cannot bind, dorsal activates expression whereas in genes such as zen which DSP1 can bind, dorsal represses expression (Fig. 8.20) (for review see Ip,

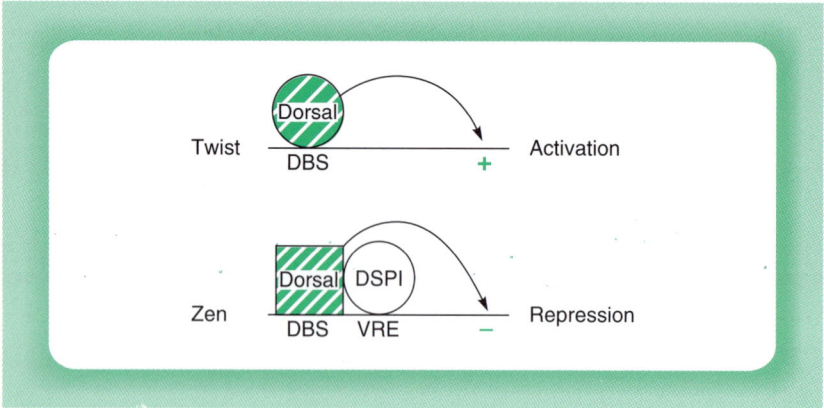

Figure 8.20

The interaction of DSPI bound at the ventral repression element (VRE) with the dorsal protein bound at its adjacent binding site (DBS) in the zen promoter results in dorsal acting as a repressor of transcription, whereas in the absence of binding sites for DSPI as in the twist promoter, it acts as an activator.

1995). It has been shown that DSP1 can interact with the basal transcriptional complex and disrupt the association of TFIIA with TBP (Kirov *et al.*, 1996). It therefore acts as an active transcriptional repressor interfering with the assembly of the basal transcriptional complex (see Chapter 6, section 6.3.2 for further discussion of this repression mechanism).

Protein–protein interactions between different factors can thus either stimulate or inhibit their activity or alter that activity, either in terms of DNA binding specificity or even from activator to repressor. It is likely that the wide variety of protein–protein interactions and their diverse effects allow the relatively small number of transcription factors which exist to produce the complex patterns of gene expression which are required in normal development and differentiation.

8.4 REGULATION BY PROTEIN MODIFICATION

8.4.1 TRANSCRIPTION FACTOR MODIFICATION

Many transcription factors are modified extensively following translation, for example, by phosphorylation, particularly on serine or threonine residues (for reviews see Hill and Treisman, 1995; Treisman, 1995) or via the modification of lysine residues either by acetylation or by addition of the small protein ubiquitin (for review see Freiman and Tjian, 2003). Such modifications represent obvious targets for agents that induce gene activation. Thus, such agents could act by altering the activity of a modifying enzyme, such as a kinase. In turn, this enzyme would modify the transcription factor, resulting in its activation and providing a simple and direct means of activating a particular factor in response to a specific signal (Fig. 8.2c). The various modifications which have been shown to affect transcription factor activity will be discussed in turn.

8.4.2 PHOSPHORYLATION

Many cellular signalling pathways involve the activation of cascades of kinase enzymes, which ultimately lead to the phosphorylation of specific transcription factors. The most direct example of such an effect of a signalling pathway on a transcription factor is seen in the case of gene activation by the interferons α and γ. Thus, these molecules bind to cell surface receptors, which are associated with factors having tyrosine kinase activity. The binding of interferon to the receptor stimulates the kinase activity and results in the phosphorylation of transcription factors known as STATs (Signal Transducers and Activators of Transcription). In turn, this results in the dimerization of the STAT proteins, allowing them to move to the nucleus where they bind to DNA and activate interferon-responsive genes (Fig. 8.21) (for reviews see Horvath, 2000; Ihle, 2001).

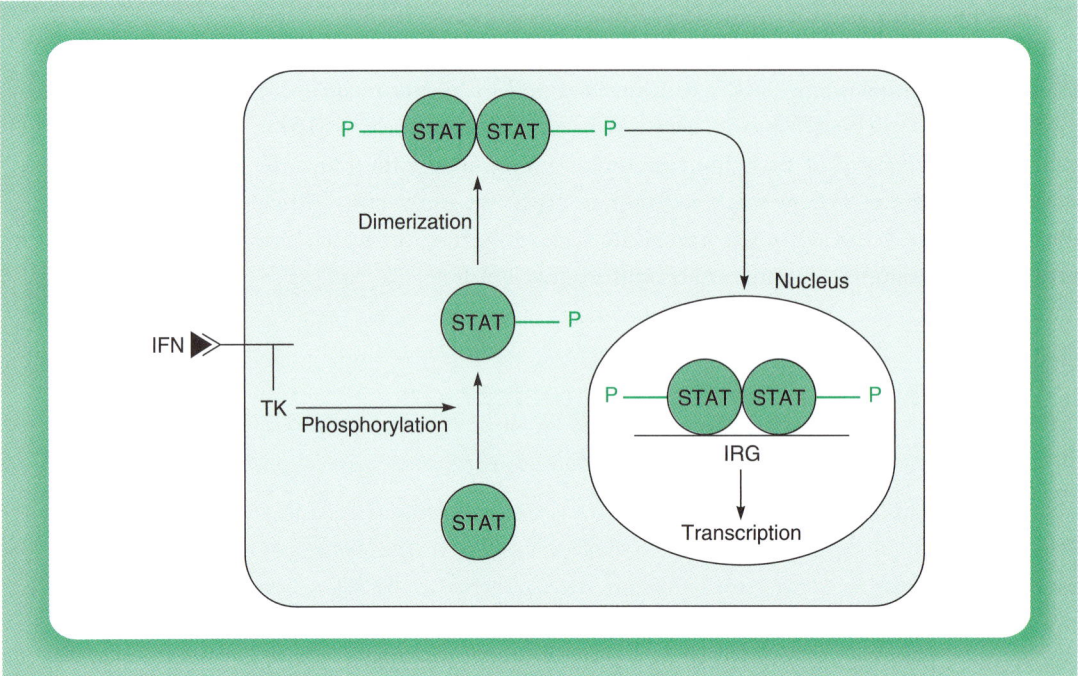

Figure 8.21

Binding of interferon (IFN) to its receptor results in activation of an associated tyrosine kinase (TK) activity leading to phosphorylation of a STAT transcription factor, allowing it to dimerize and move to the nucleus and stimulate interferon responsive genes (IRG).

Another example of this type is provided by the CREB factor which mediates the induction of specific genes in response to cyclic AMP treatment. As discussed in Chapter 5 (section 5.4.3), CREB binds to DNA in its non-phosphorylated form but only activates transcription following phosphorylation by the protein kinase A enzyme, which is activated by cyclic AMP. Hence, in this case, the activation of a specific enzyme by the inducing agent allows the transcription factor to activate transcription and hence results in the activation of cyclic AMP-inducible genes.

Similarly, the phosphorylation of the heat shock factor (HSF) following exposure of cells to elevated temperature increases the activity of its activation domain, leading to increased transcription of heat-inducible genes (see section 8.3.1), whilst the ability of the retinoic acid receptor to stimulate transcription is enhanced by phosphorylation of its activation domain by the basal transcription factor TFIIH (see Chapter 3, section 3.5.1).

In contrast to these effects on transcriptional activation ability, phosphorylation of the serum response factor (SRF), which mediates the induction of several mammalian genes in response to growth factors or serum addition, increases its ability to bind to DNA rather than directly increasing the activity of its activation domain. Interestingly, SRF normally binds to DNA in association with an accessory protein p62^{TCF}. The ability of p62^{TCF} to associate with SRF is itself stimulated by phosphorylation.

Similarly, as discussed in Chapter 5 (section 5.4.3), phosphorylation of CREB on serine 133 by protein kinase A allows it to stimulate transcription because it allows it to associate with the CBP co-activator. Protein kinase A can also phosphorylate the equivalent serine residue in the CREM transcription factor, which is closely related to CREB (see Chapter 7, section 7.3.2). As well as allowing it to activate its target genes, this phosphorylation also enhances the ability of CREM to bind to the DREAM repressor protein, discussed in section 8.2.1. As binding to CREM removes DREAM from its binding site in the dynorphin promoter, it provides an alternate means of activating this promoter, apart from direct calcium binding to DREAM (for review see Costigan and Woolf, 2002) (Fig. 8.22). Hence, the phosphorylation state of a transcription

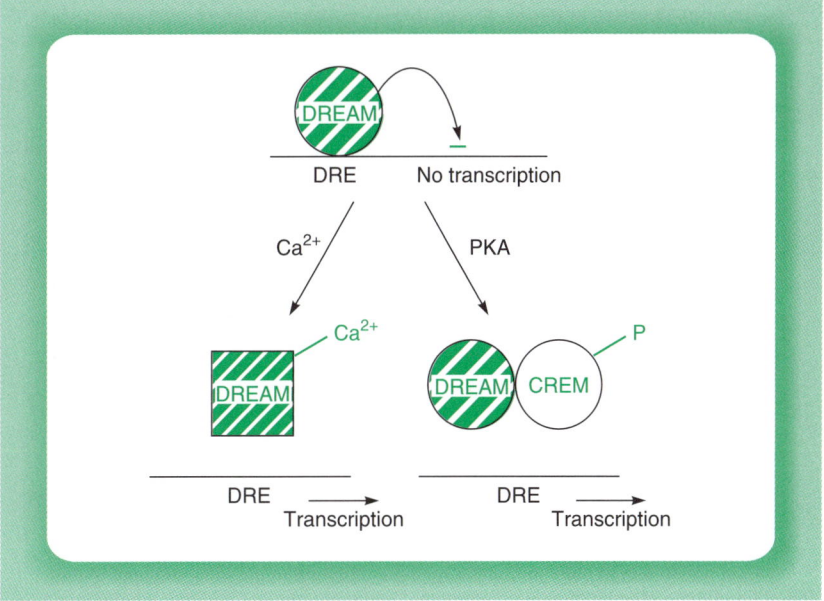

Figure 8.22

The DREAM repressor can be removed from its binding site (DRE) in the dynorphin promoter either by direct binding of calcium (compare Fig. 8.5) or by binding to DREAM of the CREM transcription factor following its phosphorylation by protein kinase A.

factor can control its ability to associate with other factors and regulate their activity as well as its ability to enter the nucleus, bind to DNA or stimulate transcription.

The effect of phosphorylation on protein–protein interactions is also involved in the dissociation of NFκB and its associated inhibitory protein IκB, which was discussed above (section 8.3.1). In this case, however, the target for phosphorylation is the inhibitory protein IκB, rather than the potentially active transcription factor itself. Thus, following treatment with phorbol esters or other stimuli such as tumour necrosis factor or interleukin 1, IκB becomes phosphorylated. Such phosphorylation results in the dissociation of the NFκB/IκB complex and targets IκB for rapid degradation. This breakdown of the complex results in NFκB being free to move to the nucleus and activate transcription (Fig. 8.23)

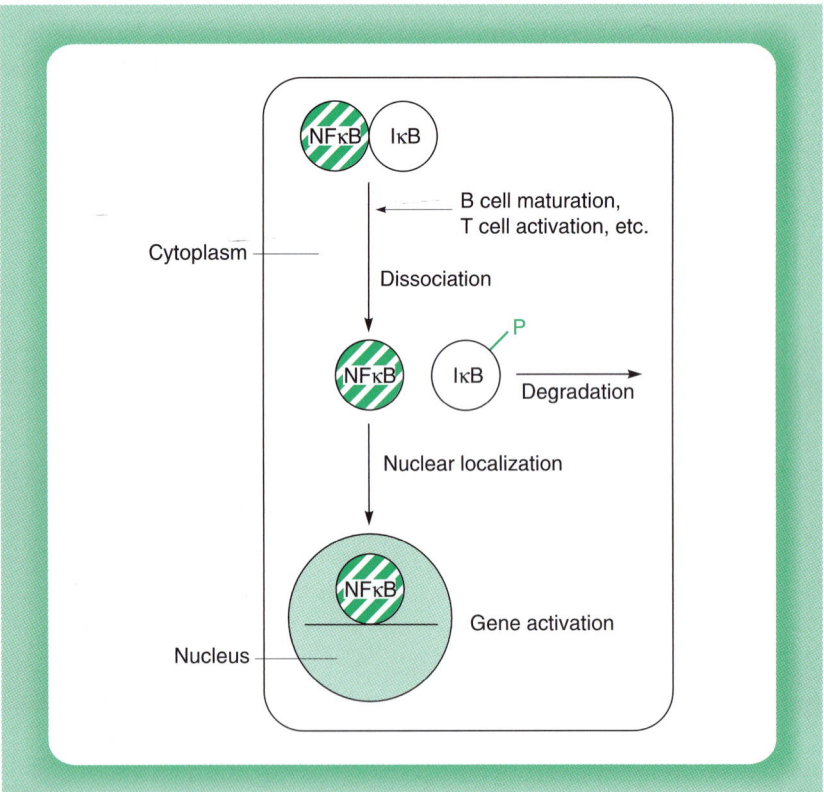

Figure 8.23

Activation of NFκB by dissociation of the inhibitory protein IκB, allowing NFκB to move to the nucleus and switch on gene expression. Note that dissociation of IκB from NFκB is caused by its phosphorylation (P) and degradation. NFκB is shown as a single factor for simplicity, although it normally exists as a heterodimer of two subunits p50 and p65.

(for review see Hayden and Ghosh, 2004; Yamamoto and Gaynor, 2004; Hoffman and Baltimore, 2006). Hence, in this case, as before, the inducing agent has a direct effect on the activity of a kinase enzyme but the resulting phosphorylation inactivates the IκB inhibitory transcription factor rather than stimulating an activating factor.

This example therefore involves a combination of two of the post-translational activation mechanisms we have discussed, namely, protein modification (Fig. 8.2c) and dissociation of an inhibitory protein (Fig. 8.2b). Moreover, as with the glucocorticoid receptor and its disassociation from hsp90 or the release of Tubby from PI(4.5)P$_2$, discussed in section 8.3.1, the net effect of the activation process is the movement of the activating factor from the cytoplasm to the nucleus, where it can bind to DNA. Thus, regulatory processes can activate a transcription factor by changing its localization in the cell as well as by altering its inherent ability to bind to DNA or to activate transcription (for review see Vandromme *et al.*, 1996).

Clearly, a key role in the regulation of the NFκB pathway will therefore be played by the enzymes which actually phosphorylate IκB in response to specific stimuli. Several IκB kinases have been identified and shown to be activated following treatment with substances which stimulate NFκB activity (for reviews see Yamamoto and Gaynor, 2004; Scheidereit, 2006). Hence, such stimuli act by activating the IκB kinase, resulting in phosphorylation of IκB leading to its degradation and thus activation of NFκB (Fig. 8.24a).

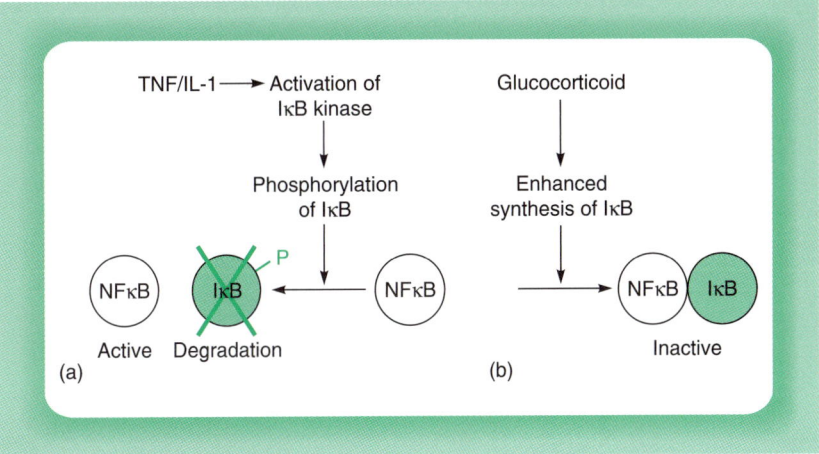

Figure 8.24

Regulation of NFκB activity by IκB can be modulated by stimuli which result in its phosphorylation and degradation leading to activation of NFκB (a) or by stimuli which enhance its synthesis thereby inactivating NFκB (b).

In contrast, other stimuli such as glucocorticoid hormone treatment can inhibit NFκB activity. This appears to involve at least two distinct effects of the glucocorticoid receptor. Thus, as described in Chapter 6 (section 6.5), the glucocorticoid receptor can repress transcriptional elongation by blocking the recruitment of the P-TEFb kinase, which phosphorylates RNA polymerase II on serine 2 of its C-terminal domain. This effect has been shown to repress NF-κB-mediated activation of some, but not all, of its target promoters, such as that of the interleukin-8 gene (Luecke and Yamamoto, 2005).

In addition, however, glucocorticoid has been shown to induce enhanced IκB synthesis, resulting in inhibition of NFκB (for review see Marx, 1995) (Fig. 8.24b). Hence, the inhibitory effect of the glucocorticoid receptor on NFκB involves both its ability to activate gene expression (of the IκB gene) and its ability to repress gene expression (of NFκB-dependent genes, such as interleukin-8). Moreover, these findings indicate that the ability of IκB to interfere with NFκB is modulated both by processes which alter the activity of IκB by phosphorylating it (Fig. 8.24a) and by altering its rate of synthesis (Fig. 8.24b).

Interestingly, one form of IκB is actually induced by activated NFκB. Hence, following activation of NFκB, new IκB is synthesized and binds to NFκB. As this binding inhibits NFκB, a feedback loop is created, which limits the effects of activating the NFκB pathway (Fig. 8.25) (for review see Ting and Endy, 2002).

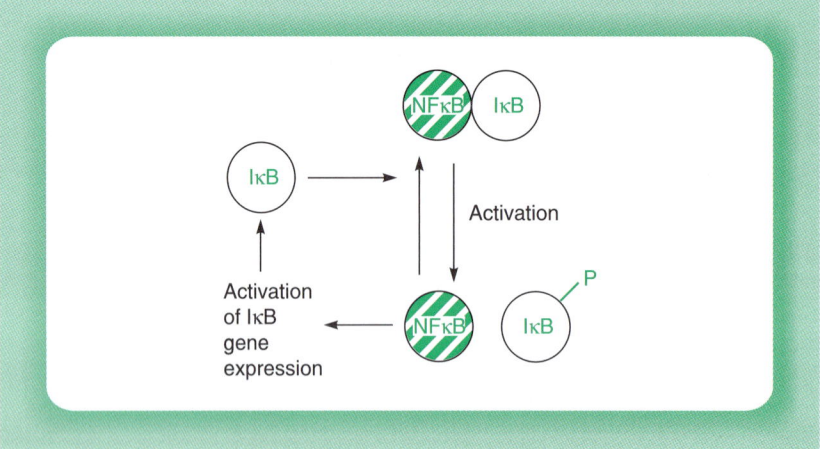

Figure 8.25

Following the release of active NFκB from the inhibitory IκB protein, it can activate the gene encoding one form of IκB. This newly synthesized IκB can bind to active NFκB and inactivate it, thereby creating a negative feedback loop which limits NFκB activity.

In addition to its activation of NFκB, treatment with phorbol esters also results in the increased expression of several cellular genes which contain specific binding sites for the transcription factor AP-1. As discussed in Chapter 9 (section 9.3.1), this transcription factor in fact consists of a complex mixture of proteins including the proto-oncogene products Fos and Jun. Following treatment of cells with phorbol esters, the ability of Jun to bind to AP-1 sites in DNA is stimulated. This effect, together with the increased levels of Fos and Jun produced by phorbol ester treatment, results in the increased transcription of phorbol ester-inducible genes. As with the activation of NFκB, phorbol esters appear to increase DNA binding of Jun by activating protein kinase C. Paradoxically, however, it has been shown (Boyle *et al.*, 1991) that the increased DNA binding ability of Jun following phorbol ester treatment is mediated by its dephosphorylation at three specific sites located immediately adjacent to the basic DNA binding domain, indicating that protein kinase C acts by stimulating a phosphatase enzyme which in turn dephosphorylates Jun (Fig. 8.26).

Such an inhibitory effect of phosphorylation on the activity of a transcription factor is not unique to the Jun protein. Indeed, in the case of the Ets-1 factor (see Chapter 9, section 9.3.4), multiple phosphorylation events have an additive effect, inhibiting DNA binding activity. Thus, when Ets-1 is unphosphorylated it has a high DNA binding activity which progressively decreases as it is phosphorylated on one, two or three of its phosphorylation sites (Pufall *et al.*, 2005) (Fig. 8.27). Hence, as well as acting as an on/off switch, controlling whether a transcription factor is active or not, phosphorylation can also quantitatively regulate the activity of a transcription factor.

Moreover, DNA binding ability is not the only target for such inhibitory effects of phosphorylation. Thus, phosphorylation of the bicoid protein reduces its ability to activate transcription without affecting its DNA binding activity, presumably by inhibiting the activity of its activation domain (Ronchi *et al.*, 1993). Similarly, phosphorylation of the Rb-1 anti-oncogene protein inhibits its ability to bind to the E2F transcription factor and inhibit its activity (see Chapter 9, section 9.4.3, for discussion of the Rb-1/E2F interaction).

As well as targeting transcription factors themselves, phosphorylation has also been shown to affect the activity of co-activators (see Chapter 5, section 5.4.3). For example, it has recently been shown that the Polo kinase enzyme phosphorylates the Ndd1p co-activator and thereby stimulates the transcription of yeast cell cycle genes (Darieva *et al.*, 2006).

Similarly, phosphorylation can also modulate the activity of histone modifying enzymes, which in turn regulate chromatin structure. Thus, in the absence of calcium stimulation, the MEF2 transcription factor is bound to the promoters of muscle-specific genes. However, gene activation

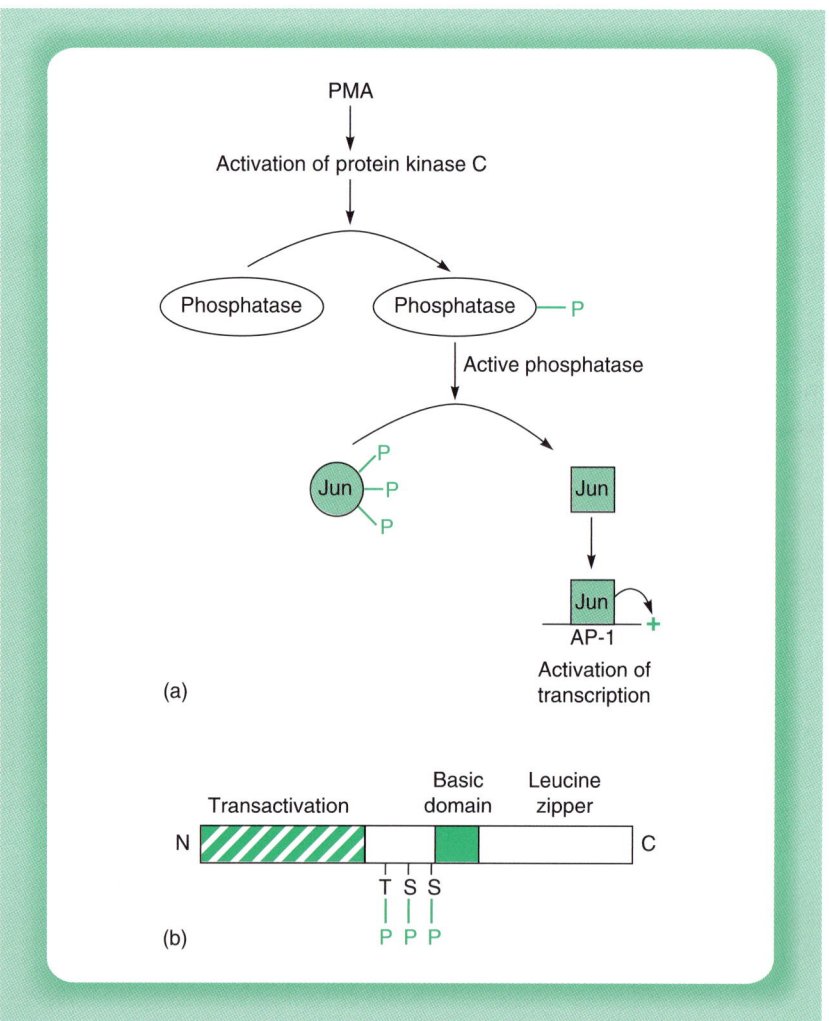

Figure 8.26

(a) Activation of Jun binding to DNA by dephosphorylation. The dephosphorylation of Jun protein following PMA treatment increases its ability to bind to AP-1 sites and activate PMA-responsive genes. This is likely to be mediated via the PMA-dependent activation of protein kinase C which in turn phosphorylates a phosphatase enzyme allowing it to dephosphorylate Jun. (b) Position in the Jun protein of the two serine (S) and one threonine (T) residues which are dephosphorylated in response to PMA. Note the close proximity to the basic domain (shaded) which mediates DNA binding. The positions of the transactivation domain and leucine zipper are also indicated.

does not occur since histone deacetylase enzymes are bound to MEF-2 and, as discussed in Chapter 1 (section 1.2.3), a lack of acetylated histones produces a tightly packed chromatin structure incompatible with transcription. However, in response to calcium, kinase enzymes are activated and phosphorylate the histone deacetylases. This phosphorylation

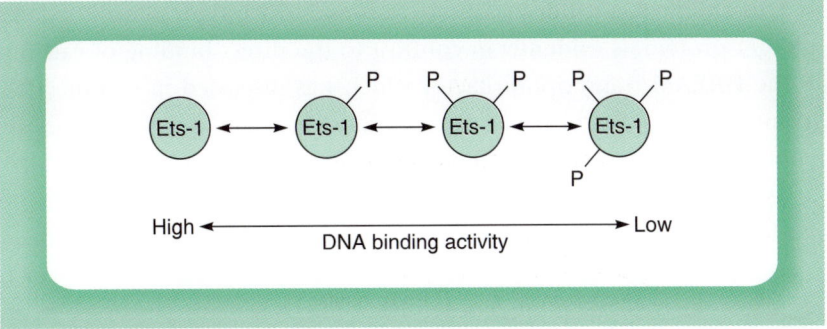

Figure 8.27

The DNA binding activity of the Ets-1 transcription factor is progressively reduced as it is phosphorylated at multiple sites.

results in the histone deacetylase enzymes being exported from the nucleus, allowing MEF-2 to fulfil its function and activate muscle-specific gene expression (Fig. 8.28) (for reviews see Stewart and Crabtree, 2000; McKinsey *et al.*, 2002).

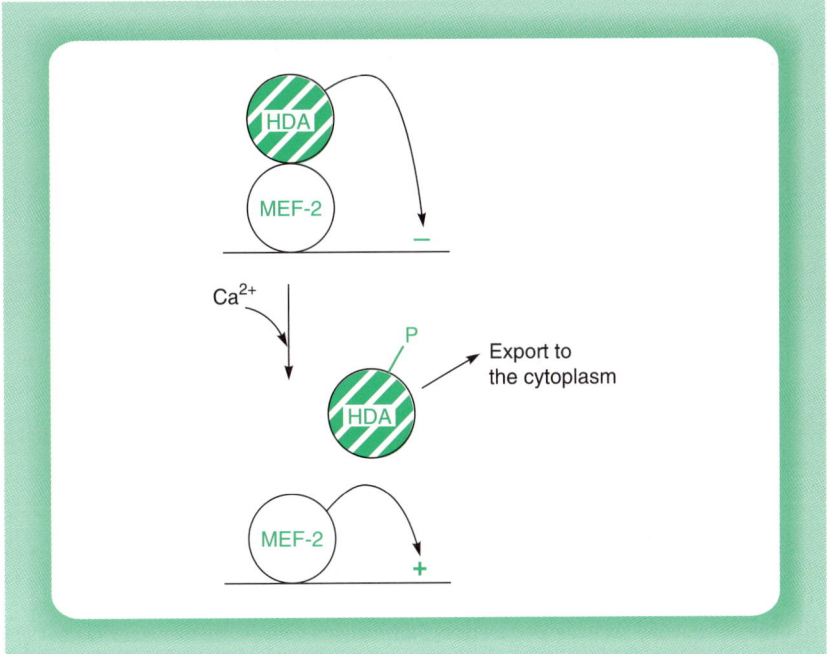

Figure 8.28

The ability of MEF-2 transcription factor to activate gene transcription can be blocked by histone deacetylase enzymes (HAD) which deacetylate histones and thereby block transcription. Following calcium treatment, the histone deacetylases are phosphorylated and exported from the nucleus, allowing MEF-2 to activate gene transcription.

This ability of calcium to activate a kinase, which then phosphorylates a target protein, is evidently in contrast to the direct binding of calcium to the DREAM transcription factor which was discussed in section 8.2.1 (compare Fig. 8.5 and Fig. 8.28) and illustrates the fact that a specific stimulus can use multiple mechanisms to activate transcription.

As well as affecting factors involved in transcriptional initiation or the modification of chromatin structure, phosphorylation can also affect transcriptional elongation. Thus, the yeast Hog1 kinase enzyme, which regulates the expression of genes in response to osmotic stress, has been shown to stimulate transcriptional elongation (Proft *et al.*, 2006). Indeed, this kinase has an effect on transcriptional initiation and chromatin structure, as well as transcriptional elongation. Thus, ChIP analysis (see Chapter 2, section 2.4.3) has shown that Hog1 forms part of the transcriptional complex, at genes induced by osmotic stress (for review see Edmunds and Mahadevan, 2006). As well as phosphorylating transcription factors, it also stimulates the recruitment of RNA polymerase II and of a histone deacetylase complex, which promotes a more open chromatin structure (see Chapter 5, section 5.5.2). Hog1 therefore stimulates transcription, via multiple mechanisms (Fig. 8.29).

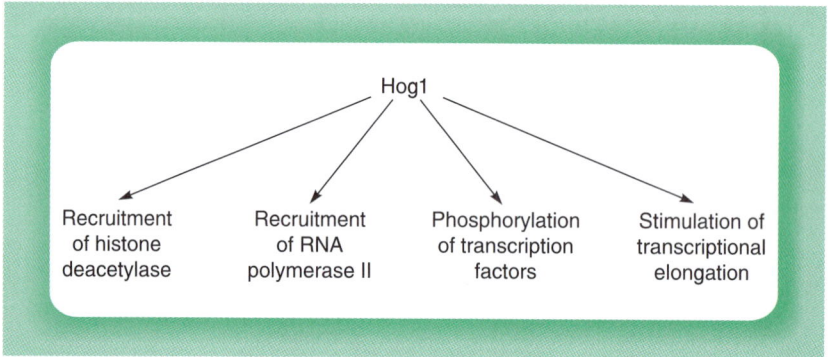

Figure 8.29
The Hog1 kinase uses multiple mechanisms to promote the transcription of genes activated by osmotic stress.

Hence, protein modification by phosphorylation can have a wide variety of effects on transcription factors, either stimulating or inhibiting their activity and acting via a direct effect on the ability of the factor to enter the nucleus, bind to DNA, associate with another protein or activate transcription or by an indirect effect affecting the activity of an inhibitory protein, a co-activator, or a histone-modifying enzyme. Moreover, such phosphorylation can affect both transcriptional initiation and transcriptional elongation. The directness and rapidity of this means of transcription factor activation evidently renders it of particular importance in the response to cellular signalling pathways.

8.4.3 ACETYLATION

In view of the directness and rapidity of using post-translational modification as a means of modulating the activity of transcription factors, it is not surprising that other transcription factor modifications apart from phosphorylation, are used in this way.

In particular, acetylation of transcription factors, particularly on lysine residues, has now been defined as an important means of regulating their activity. Thus, although acetylation was initially defined as a modification able to modulate histone activity (see Chapter 1, section 1.2.3), it has now been shown also to occur for transcription factors themselves (for review see Freiman and Tjian, 2003).

Thus, the addition of acetyl residues to the C-terminal domain of the p53 protein (see Chapter 9, section 9.4.2) increases the DNA binding activity of p53. This acetylation of p53 is carried out by the p300 co-activator molecule, which as described in Chapter 5 (section 5.4.3) associates with p53 as well as with a wide variety of other transcription factors. This finding indicates that as well as acetylating histones and thereby modifying chromatin structure (see Chapter 1, section 1.2.3), p300 and the related CBP co-activator also use their acetyltransferase activity to acetylate specific transcription factors and thereby modify their activity (Fig. 8.30).

Hence, acetylation can modulate the activity of p53 by targeting its C-terminus. However, the N-terminus of p53 can be modified by phosphorylation and this reduces its ability to bind to the MDM2 inhibitory protein (see Chapter 9, section 9.4.2), thereby enhancing the stability of p53. Therefore, the activity of p53 can be modified by phosphorylation and by acetylation, indicating that different post-translational modification can target the same transcription factor molecule.

Acetylation also occurs in the NFκB/IκB system, which also involves regulated phosphorylation, as discussed above (section 8.4.2). Thus, it has been shown that NFκB is acetylated and that this inhibits its interaction with IκB. Hence, interaction of NFκB with IκB requires both deacetylated NFκB and dephosphorylated IκB (Fig. 8.31) (for review see Perkins, 2006).

As well as targeting the same transcription factor (as in the case of p53) or two interacting transcription factors (as in the case of NFκB/IκB), there is evidence that the phosphorylation and acetylation systems can interact with one another. Thus, for example, the ATF-2 transcription

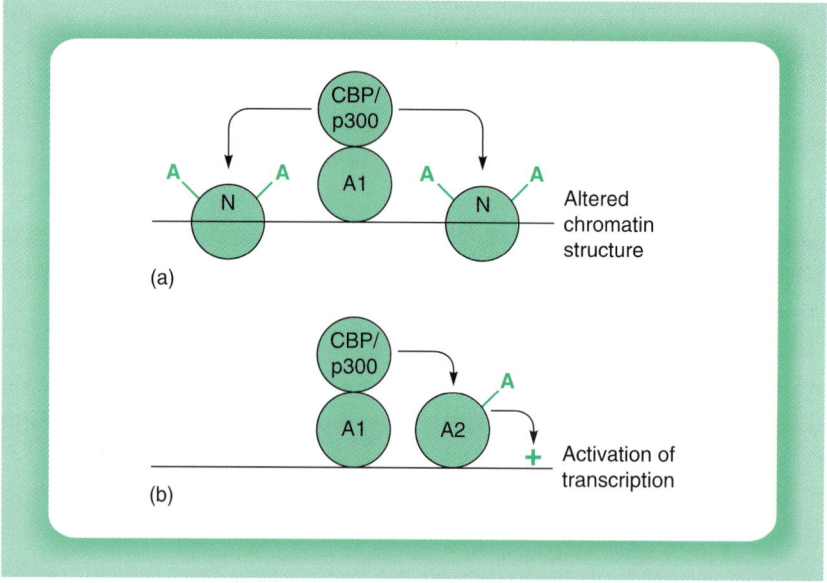

Figure 8.30
Possible mechanisms of action of CBP/p300. Following recruitment to DNA by an activating molecule (A1) the acetyltransferase activity of CBP/p300 may either (a) acetylate histones producing a more open chromatin structure or (b) acetylate another activating transcription factor (A2) allowing it to stimulate transcription.

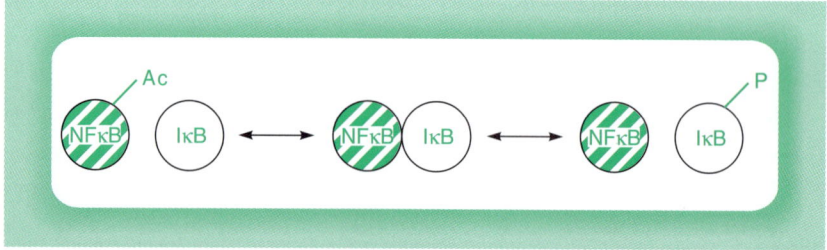

Figure 8.31
Either acetylation of NFκB or phosphorylation of IκB can inhibit the NFκB/IκB interaction.

factor has been shown to have histone acetyltransferase activity and this activity is stimulated by ATF-2 phosphorylation (Kawasaki *et al.*, 2000).

8.4.4 METHYLATION

As with acetylation, methylation has been shown to play an important role in the modification of histones (see Chapter 1, section 1.2.3) and as described in Chapter 6 (section 6.4.1), the polycomb repressor complex contains an activity capable of methylating histones. However, as with acetylation, methylation has also been shown to occur for factors other than histones.

Thus, methylation can affect co-activators, such as CBP and the related p300 factor, discussed in Chapter 5 (section 5.4.3). Both these factors are modified by methylation on specific arginine residues (for review see Gamble and Freedman, 2002). Most interestingly, such methylation affects the ability of CBP/p300 to bind to the various transcription factors with which they interact. Thus, methylation abolishes the ability of CBP/p300 to bind to the CREB factor but has no effect on its ability to bind to nuclear receptors, such as the steroid receptors. Hence, the competition between different transcription factors for binding to CBP/p300 (see Chapter 6, section 6.6) can be altered by modification of the co-activator, resulting in a different balance between the different factors under different conditions (Fig. 8.32).

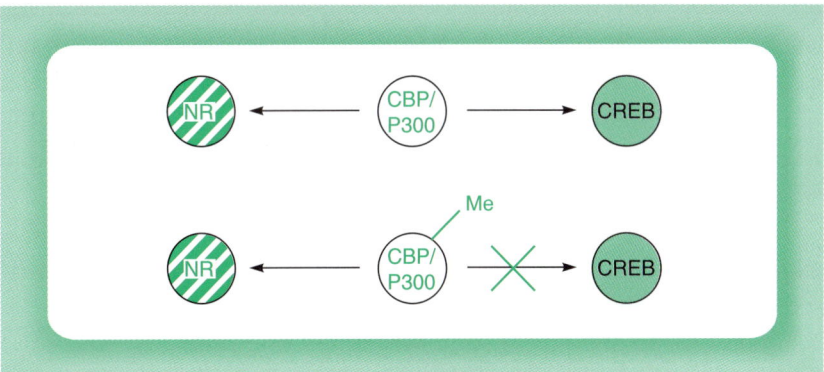

Figure 8.32
The ability of CBP/p300 to bind to different transcription factors is affected by the methylation of CBP/p300 which blocks binding to the CREB factor whilst not affecting binding to the nuclear receptors (NR).

As well as being modified by methylation, CBP is also modified by phosphorylation. Thus, phosphorylation of CBP on serine 436 enhances its ability to interact with the AP-1 (see Chapter 9, section 9.3.1) and Pit-1 (see Chapter 4, section 4.2.7) transcription factors (for review see Gamble and Freedman, 2002).

The post-translational modification of co-activators can therefore modulate their interaction with different activating molecules, allowing them to preferentially activate different pathways under different conditions. This effect evidently parallels the phosphorylation of transcription factors such as CREB which affects their ability to interact with CBP/p300 and thus produce transcriptional activation (see Chapter 5, section 5.4.3 and section 8.4.2 of this chapter).

8.4.5 UBIQUITINATION AND SUMOYLATION

Although phosphorylation, acetylation and methylation all involve the addition of relatively small chemical groups to the transcription factor molecule, it is possible for a much larger entity to be added. Thus, many proteins in the cell, including transcription factors, become modified by the addition of ubiquitin, which is itself a 76 amino acid protein. This small protein is linked to the transcription factor by a covalent bond between the C-terminal of ubiquitin and an internal lysine residue of the transcription factor (for reviews see Freiman and Tjian, 2003; Welchman *et al.*, 2005).

In many cases, this ubiquitination serves to target the molecule for degradation, since it is recognized by the proteolytic machinery of the cell as marking the protein for destruction. Indeed, in the NFκB/IκB case discussed above (section 8.4.2), phosphorylation of IκB leads in turn to its ubiquitination and hence targets it for destruction, releasing NFκB to activate gene expression (Fig. 8.33) (for review see Karin and Ben-Neriah, 2000; Perkins, 2006).

An interesting example of such ubiquitin-mediated control of gene expression is provided by the hypoxia-inducible factor, HIF-1 (for review see Bruick and McKnight, 2001; Kaelin, 2002). This factor consists of two subunits, HIF-1α and HIF-1β, and is activated when cells are exposed to low oxygen. It then activates the expression of genes which are required in this situation. This activation of HIF-1 is controlled at the level of protein degradation. In the presence of oxygen, the HIF-1α subunit is rapidly ubiquitinated and degraded. When oxygen levels fall, HIF-1α is no longer ubiquitinated and can therefore associate with HIF-1β and activate gene transcription (Fig. 8.34).

This obviously leads to the question of how the ubiquitination of HIF-1α is regulated by oxygen. It has been shown that in the presence of oxygen, HIF-1α is modified by the addition of a hydroxyl (OH) group to a proline amino acid by a proline hydroxylase enzyme. This novel transcription factor modification allows the HIF-1α to be recognized by the von Hippel–Lindau anti-oncogene product (VHL) (see Chapter 9, section 9.4.4), which is part of a multi-protein complex necessary for the addition of ubiquitin. Following a fall in oxygen levels, the proline hydroxylation of HIF-1α does not occur since the activity of the proline hydroxylase enzyme is directly regulated by oxygen. Hence, the VHL

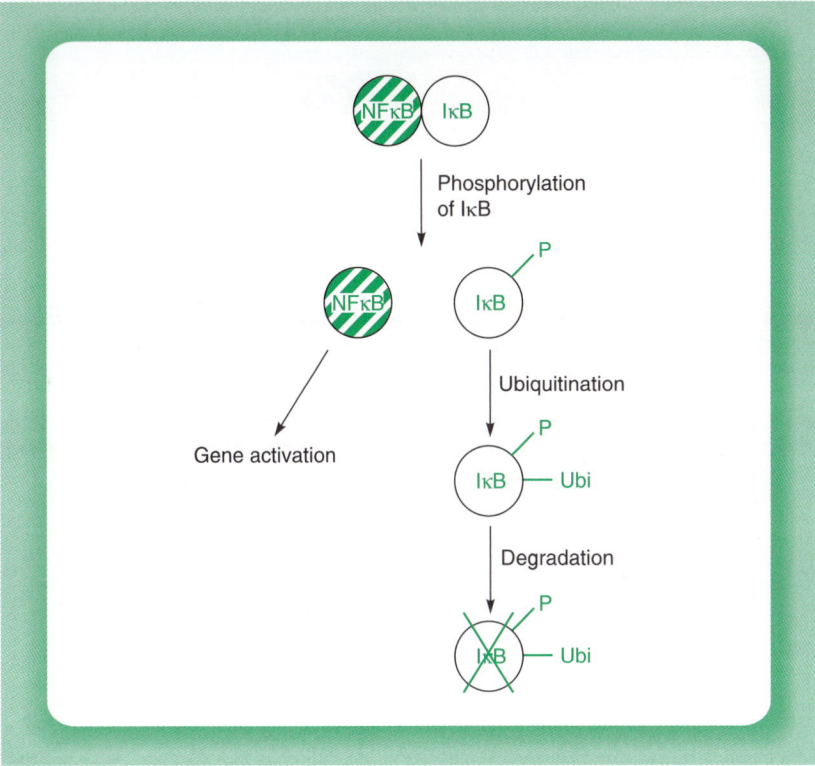

Figure 8.33

Phosphorylation of IκB is followed by its ubiquitination which targets it for destruction.

product cannot bind and HIF-1α is stabilized (for review see Semenza, 2001; Zhu and Bunn, 2001) (Fig. 8.35).

In this case, therefore, a novel transcription factor modification is recognized by the VHL protein and leads to further modification by ubiquitination. This is evidently analogous to the phosphorylation of IκB discussed above, which is necessary for its subsequent ubiquitination. The structural basis for the role of hydroxyproline in regulating the interaction of HIF-1α and VHL has been defined. Thus, the hydroxyproline residue on HIF-1α inserts into a pocket in VHL, allowing only hydroxyproline-modified HIF-1α to bind to VHL (Hon *et al.*, 2002; Min *et al.*, 2002) (Fig. 8.36).

Interestingly, the pocket in VHL which binds the hydroxyproline has been shown to be a hot spot for mutations which inactivate VHL and result in cancer. Hence, the anti-oncogenic function of VHL appears to involve its ability to bind to proteins such as HIF-1α via hydroxyproline residues. Indeed, patients with cancer caused by mutation of VHL show

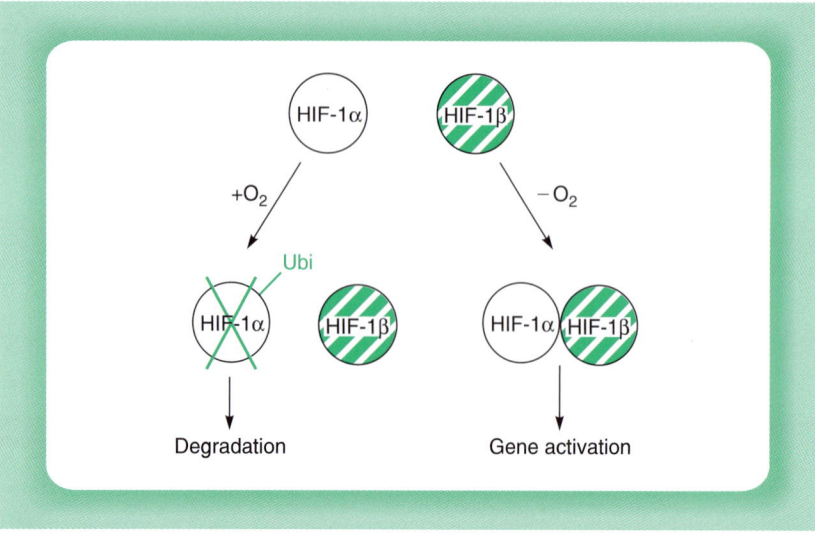

Figure 8.34

In the presence of oxygen, the HIF-1α factor is modified by addition of ubiquitin (Ubi) and then degraded. In the absence of oxygen this addition of ubiquitin does not occur and the HIF-1α is stabilized, allowing it to dimerize with HIF-1β and activate transcription.

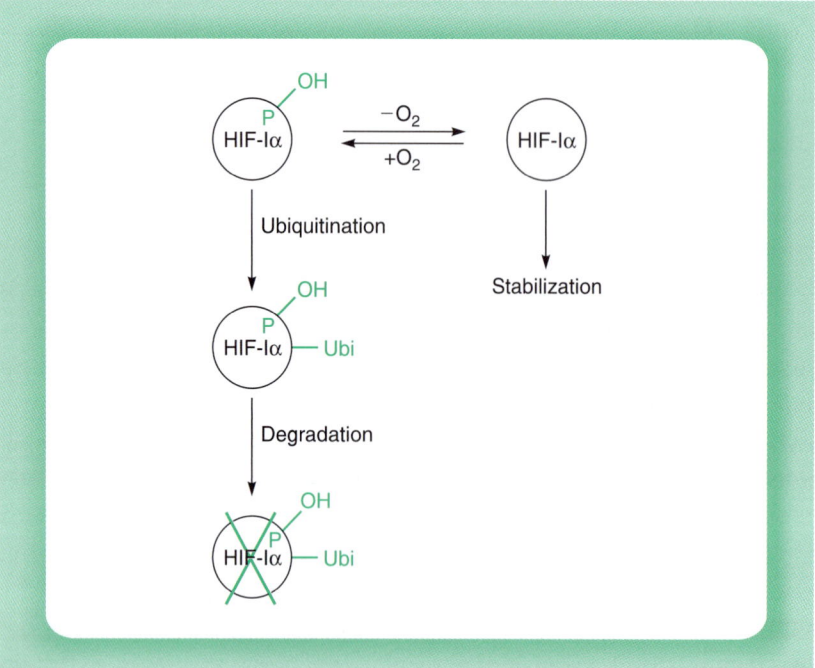

Figure 8.35

Oxygen induces the modification of HIF-1α by addition of a hydroxyl group (OH) on a proline (P) amino acid which results in its ubiquitination and degradation.

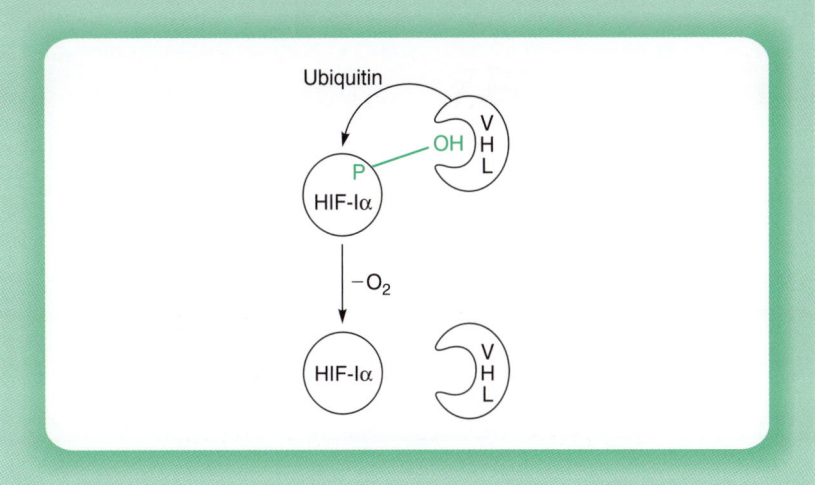

Figure 8.36

Proline hydroxylation allows recognition of HIF-1α by the von Hippel–Lindau (VHL) protein which catalyses the addition of ubiquitin.

expression of HIF-1-activated genes even in the presence of oxygen (see Chapter 9, section 9.4.4 for further discussion).

In the case of HIF-1α therefore, a novel modification involving the hydroxylation of proline residues stimulates ubiquitination and consequent degradation. In addition, however, HIF-1α is also modified by a further novel modification, involving addition of a hydroxyl group to an asparagine amino acid. Like hydroxylation of proline, this modification is also inhibited by reduced oxygen levels. However, rather than controlling protein stability, the loss of the hydroxyl group on asparagine facilitates the binding of the p300 transcriptional co-activator (for review see Bruick and McKnight, 2002; Kaelin, 2002). This binding of p300 enhances the ability of HIF-1α to activate transcription (see Chapter 5, section 5.4.3 for discussion of CBP/p300). Hence, reduced oxygen levels stabilize the HIF-1α protein by inhibiting hydroxylation of proline and enhance the ability of the stabilized protein to activate transcription by inhibiting hydroxylation of asparagine (Fig. 8.37).

As discussed in Chapter 6 (section 6.5), HIF-1α is not the only target for ubiquitination by the VHL complex. Thus, the phosphorylated form of the large subunit of RNA polymerase II is also ubiquitinated by the VHL complex resulting in its degradation. This specifically blocks the elongation step of transcription, since this phosphorylated form of RNA polymerase II is specifically required for transcriptional elongation (see Chapter 6, section 6.5). As with HIF-1α, the ubiquitination of the large subunit of RNA polymerase II also requires prior proline hydroxylation

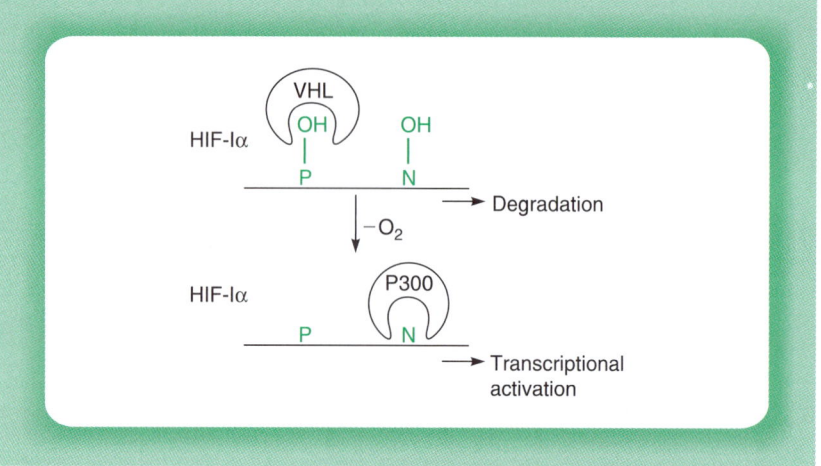

Figure 8.37

In the HIF-1α factor, removal of oxygen not only blocks the addition of hydroxyl residues to proline (P) preventing VHL binding but also blocks addition of hydroxyl residues to asparagine (N) promoting the binding of the p300 co-activator molecule. This allows the stabilized protein to stimulate transcription.

of the polymerase subunit (Kuznetsova *et al.*, 2003) suggesting that this may be a general mechanism for targeting of proteins by VHL, accounting for its importance and its anti-oncogenic function (see above).

The use of ubiquitination to target proteins such as NFκB, HIF-1α or the large subunit of RNA polymerase II for degradation is not unique to transcription factors but is widely used in the turnover of a variety of different proteins. However, a further role of ubiquitin exists, which is specific to transcription factors. Thus, it has been shown that modification by ubiquitination may be necessary for activation domains to stimulate transcription (see Chapter 5, section 5.2 for a discussion of activation domains). In experiments in yeast, the VP16 acidic activation domain could not activate transcription in a yeast strain which could not add ubiquitin to the VP16 protein. However, if a modified VP16 was prepared in which ubiquitin had already been added to the activation domain, then transcription was activated (Salghetti *et al.*, 2001; Fig. 8.38). This indicates that modification of the VP16 activation domain by ubiquitination is necessary for it to activate transcription.

Although activation of transcription by ubiquitination of a transcription factor appears to be entirely distinct from the promotion of protein degradation by ubiquitination, this may not always be the case. Thus, in the case of the yeast transcriptional activator GCN4, it has been shown that ubiquitination and enhanced degradation can actually stimulate gene expression (Lipford *et al.*, 2005; for review see Arndt and Winston,

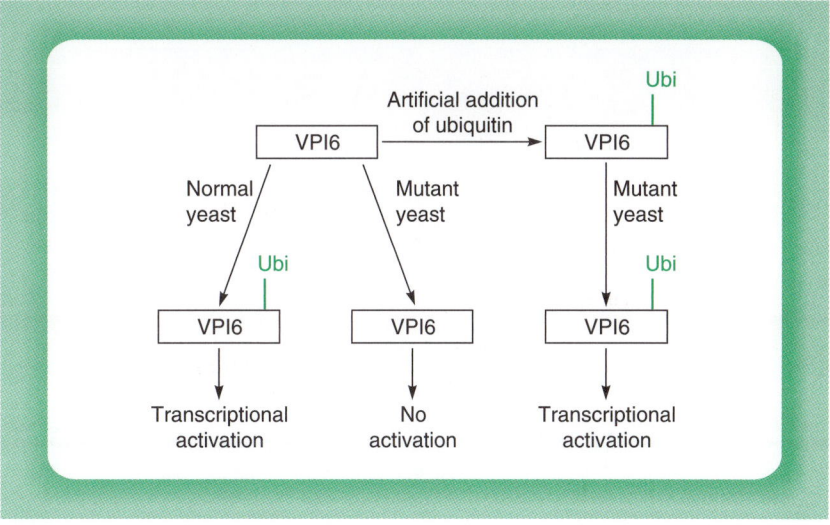

Figure 8.38

The transcriptional activator VP16 can activate transcription in normal yeast which can modify it by the addition of ubiquitin (Ubi), but not in mutant yeast which cannot carry out this modification. However, if the VP16 is modified by artificial addition of ubiquitin, then it can activate transcription even in the mutant yeast.

2005). One explanation of this paradox is that after a round of transcription has occurred, ubiquitination and degradation removes 'spent' inactive GCN4 from the promoter, allowing new highly active molecules of the factor to bind (Fig. 8.39).

Hence, in this case, activator molecules are rapidly turned over and replaced, allowing continuing stimulation of gene expression. It should be noted, however, that in other cases, such as that of HSF (see section 8.3.1), the factor remains stably bound to the promoter and continues to activate transcription for as long as the inducing stimulus is present (see for example Nalley *et al.*, 2006; Yao *et al.*, 2006).

As well as modification by addition of ubiquitin, transcription factors can also be modified by addition of the small ubiquitin-related protein, SUMO (for reviews see Hay, 2005; Marx, 2005). For example, modification of a specific lysine residue in IκB by addition of SUMO-1 has been shown to prevent the addition of ubiquitin and thereby protect IκB from degradation (Desterro *et al.*, 1998) (Fig. 8.40). Hence, different modifications of the same residue may produce opposite effects on transcription factor activity, providing a further mechanism for regulating such activity. As lysine residues are the target for acetylation (section 8.4.3) as well as for addition of ubiquitin or SUMO-1, several different modification enzymes may compete to modify a specific lysine amino acid in a

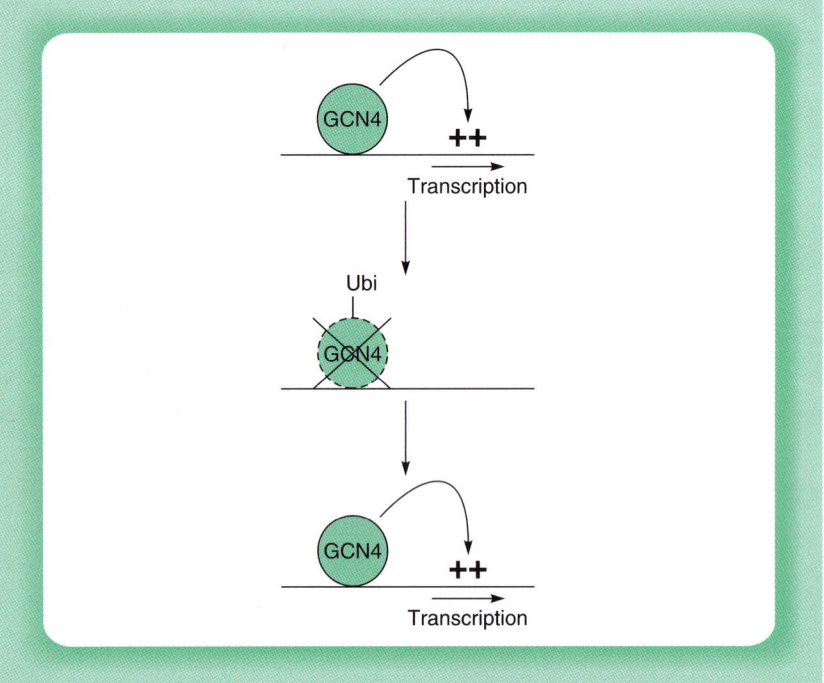

Figure 8.39
Following a round of transcription mediated by the GCN4 transcription factor, less active 'spent' GCN4 is removed by ubiquitination and degradation, allowing new active GCN4 molecules to bind and stimulate transcription.

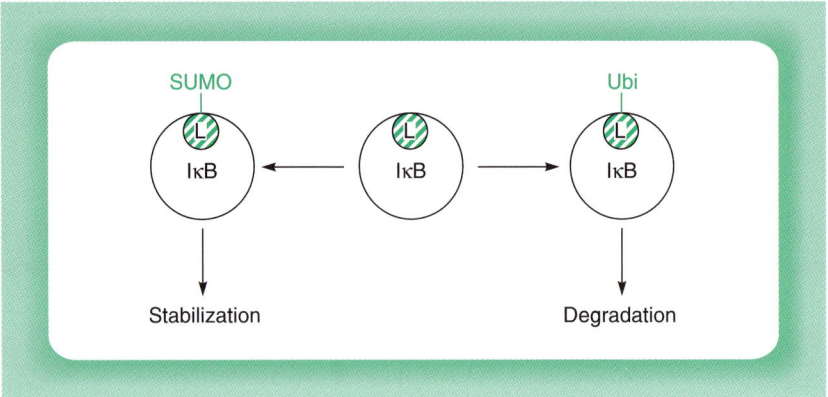

Figure 8.40
Modification of a specific lysine residue (L) in IκB by addition of ubiquitin promotes degradation of the protein. In contrast, addition of the ubiquitin-like protein SUMO-1 to the same lysine residue blocks addition of ubiquitin and hence stabilizes IκB.

transcription factor with different consequences for its functional activity (for review see Freiman and Tjian, 2003).

Indeed, such a complex interaction of multiple modifications is seen in the case of the MEF-2A factor, which was discussed in section 8.4.3. As well as regulating muscle differentiation, as previously discussed, this factor also regulates gene expression in neuronal cells, where it can act as either a transcriptional activator or a repressor. This switch is controlled by post-translational modification of MEF-2A, with acetylated MEF-2A acting as a transcriptional activator, whereas when the factor is modified by sumoylation and phosphorylation it acts as a transcriptional repressor (Fig. 8.41) (for review see Beg and Scheiffele, 2006).

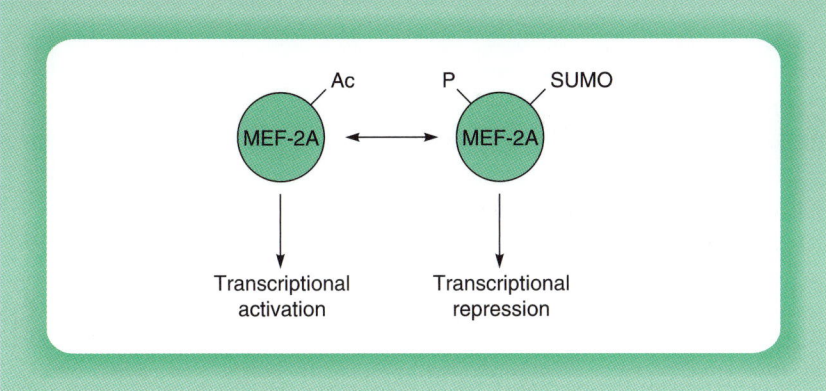

Figure 8.41
When acetylated, the MEF-2A transcription factor acts as an activator of transcription. However, when modified by phosphorylation and sumoylation, it acts as a transcriptional repressor.

Interestingly, sumoylation and phosphorylation of MEF2 takes place in the sequence motif Lysine-X-GlutamicAcid-X-X-Serine (where X is any amino acid), which is also found in several other transcription factors such as HSF (for review see Yang and Gregoire, 2006). In several such cases it has been shown that phosphorylation of the serine amino acid in this motif stimulates subsequent sumoylation of the lysine amino acid producing a transcriptional repressor (Fig. 8.42). Hence, this appears to represent a general mechanism which occurs in several different transcription factors and which links together two different post-translational modifications.

The modification of transcription factors by phosphorylation, acetylation, methylation, ubiquitination and sumoylation thus provides a means of controlling transcription factor activity both by the effects of individual modifications and the interactions of different modifications. It is of

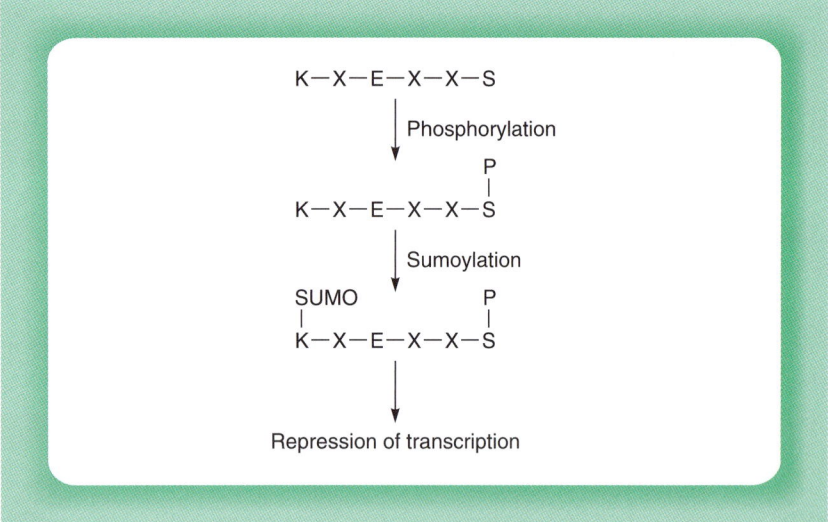

Figure 8.42
The amino acid sequence Lysine-X-GlutamicAcid-X-X-Serine (K-X-E-X-X-S in the one letter code for amino acids where X is any amino acid) is found in several different transcription factors. Its modification by phosphorylation promotes its subsequent sumoylation and results in transcriptional repression.

critical importance, particularly in allowing gene expression to be modulated by specific signalling pathways.

8.5 REGULATION BY PROTEIN DEGRADATION AND PROCESSING

Evidently, a number of the cases discussed in the previous section involve regulating the degradation of a specific factor such as HIF-1α or IκB to ensure that it is stable and can fulfil its function in one situation but is rapidly degraded in another situation and so cannot fulfil its function. A number of other cases of this type have been described (see Fig. 8.2d), including that of the ID2 factor (whose degradation relieves its inhibitory effect on the E12/E47 transcription factors, see Chapter 6, section 6.2.2) and allows neuronal differentiation to occur (for reviews see Jackson, 2006; Qiu, 2006).

Interestingly, such control of transcription factor degradation is combined with the regulation of transcription factor synthesis in the case of another regulator of neuronal differentiation. Thus, as discussed in Chapter 6 (section 6.1), the REST factor is synthesized in non-neuronal cells and represses the expression of neuronal-specific genes. As pluripotent stem cells differentiate into neuronal precursors, REST is degraded at an enhanced rate but it continues to be synthesized. This results in REST protein being present at reduced levels, which are still sufficient to repress neuronal-specific genes, but leave them poised for expression. As the neuronal progenitors differentiate into neurones, REST synthesis is switched off and neuronal-specific genes are activated (Ballas *et al.*, 2005; for review see Lunyak and Rosenfeld, 2005) (Fig. 8.43).

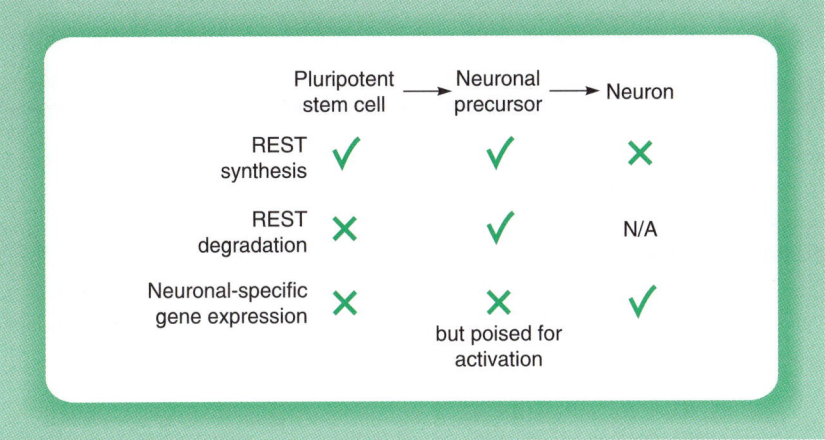

Figure 8.43

In pluripotent stem cells, the REST factor is synthesized and represses the expression of neuronal-specific genes. In neuronal progenitors, REST continues to be synthesized but it is degraded, so that its level falls to a point where neuronal-specific genes are still repressed but are poised for activation. This activation occurs subsequently when REST synthesis is switched off and neuronal differentiation occurs. N/A = not applicable since REST is not being synthesized.

It is clear therefore that regulating the stability of a transcription factor so that it is different in different situations is an important means of regulating transcription factor activity (Fig. 8.44a).

In addition, however, proteolysis can also be used to activate a transcription factor. This can be achieved by cleaving an inactive precursor to produce an active form of the transcription factor (Fig. 8.44b). This form of regulation is also seen in the NFκB family. Thus, an NFκB-related protein p105 is synthesized as a single molecule in which the NFκB portion is linked to an IκB-like region that inhibits its activity, resulting in an inactive precursor protein. Following exposure to an activating stimulus, the IκB-like portion is phosphorylated by the same IκB kinases which phosphorylate IκB. The phosphorylated protein is then cleaved

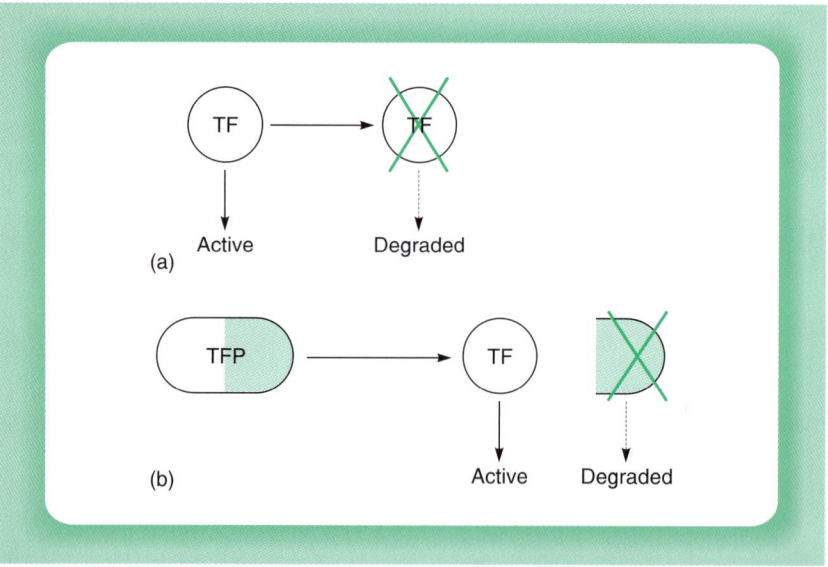

Figure 8.44

Proteolytic cleavage of a transcription factor can be used either (a) to degrade the factor so preventing it from acting or (b) to cleave an inactive precursor molecule to produce an active factor.

to release active NFκB (Fig. 8.45) (for reviews see Schmitz *et al.*, 2001; Pomerantz and Baltimore, 2002). This mechanism evidently resembles the regulation of NFκB by IκB described above (section 8.4.2), except that in this case the NFκB and IκB-like activities are contained in the same molecule rather than in different molecules.

This regulatory mechanism is also seen in the case of the SREBP transcription factors, which activate gene expression in response to removal of cholesterol (for review see Brown and Goldstein, 1997). In the presence of cholesterol, these factors are anchored in the endoplasmic reticulum by a specific region of the protein. When cells are deprived of cholesterol, this region of the protein is cleaved off, allowing the protein to move to the nucleus and switch on genes whose protein products are required for cholesterol biosynthesis (Fig. 8.46). Interestingly, following DNA binding and activation of transcription, SREBP is rapidly ubiquitinated, leading to its degradation (Punga *et al.*, 2006). Hence, SREBP is one of the group of factors which are rapidly degraded once they have stimulated transcription (see section 8.4.5) and its activity is regulated both by cleavage of an inactive precursor and degradation of the active protein.

Regulation by proteolytic cleavage of an inactive precursor thus represents another means of regulating transcription factor activity. Both the

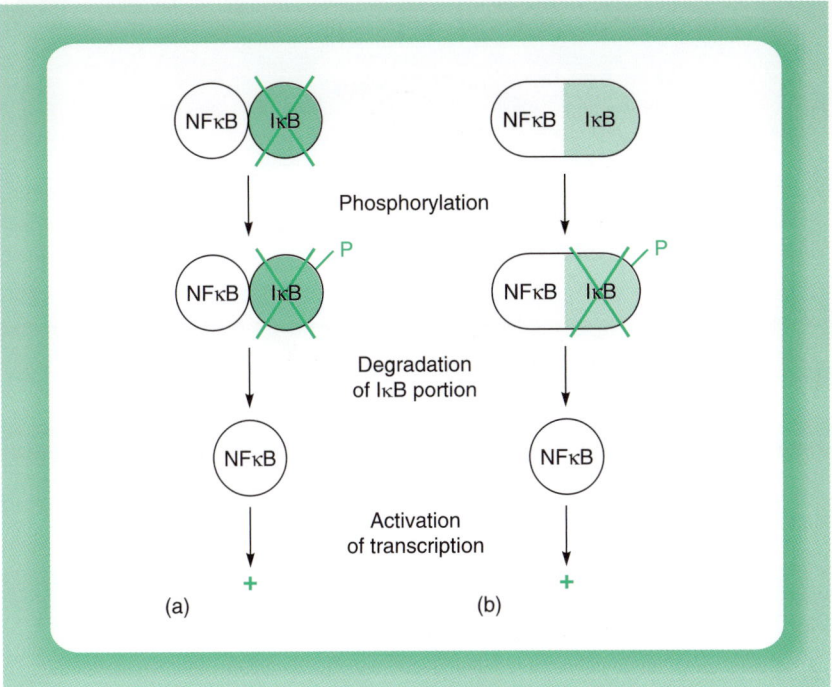

Figure 8.45

In the NFκB family, activation of NFκB can be achieved either (a) by phosphorylation and degradation of an associated IκB protein or (b) by phosphorylation of the IκB-like portion of a large precursor protein (p105) resulting in its proteolytic processing to release active NFκB.

cases of regulation by proteolytic cleavage we have described result in a change in localization of the transcription factor with the NFκB portion of p105 moving from the cytoplasm to the nucleus and the activated SREBP factor moving from the endoplasmic reticulum membrane to the nucleus. This further underlies the importance of changes in transcription factor localization brought about by regulatory processes.

Another aspect of this process is illustrated by the recent description of a transcription factor, which is produced by cleavage of a calcium channel protein (Gomez-Ospina *et al.*, 2006). Thus, in this case the precursor protein is located in the cell membrane and functions as a channel to allow calcium to enter the cell. However, it can be cleaved to produce a C-terminal fragment which then moves to the nucleus and acts as a transcription factor binding to the promoters of specific genes and regulating their transcription. Hence, in this case both precursor and processed proteins are functional but have different functions in different cellular compartments (Fig. 8.47).

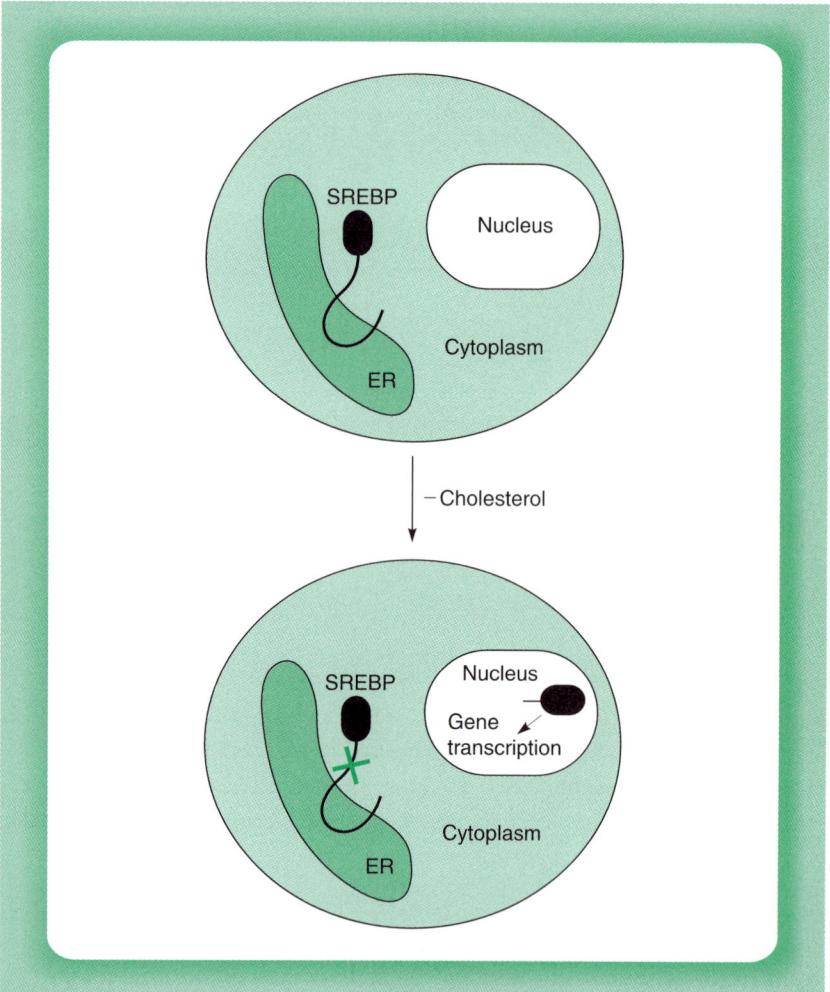

Figure 8.46

In the presence of cholesterol, the SREBP factor is anchored in the membrane of the endoplasmic reticulum and hence cannot enter the nucleus. On removal of cholesterol, the SREBP precursor is cleaved, releasing the active form of the protein which can move to the nucleus and activate the expression of genes involved in cholesterol biosynthesis.

8.6 ROLE OF REGULATED ACTIVITY

In addition to its ability to produce a very rapid activation of gene expression, modification of the activity of a pre-existing protein also allows specific targets for modification to be used in different cases. Thus, the various regulatory processes discussed above affect the activity of transcription factors at a wide variety of different stages. For example, in the case of phosphorylation (section 8.4.2) we have seen how in different

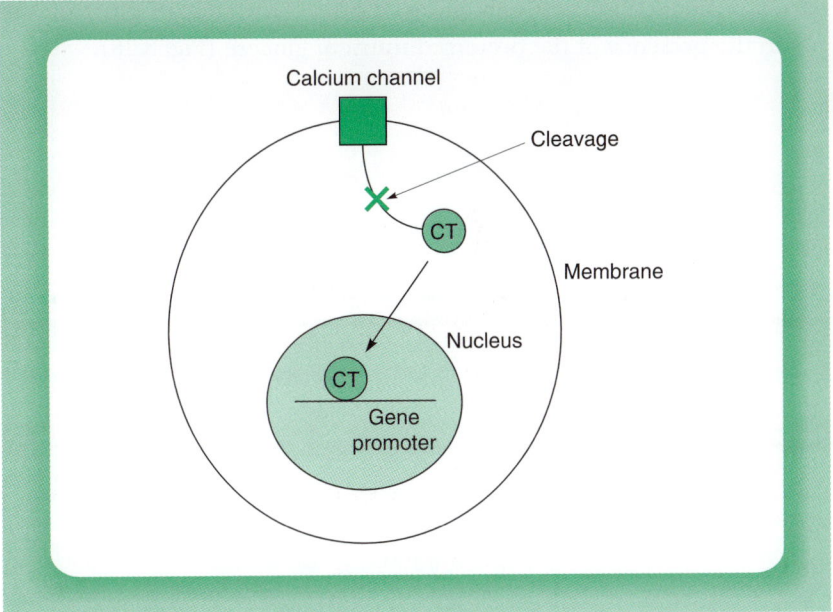

Figure 8.47

The C-terminal domain (CT) of a cell membrane-localized calcium channel can be released by proteolytic cleavage. The C-terminal domain can then move to the nucleus and act as a transcription factor to regulate gene expression.

cases a single process can alter the DNA binding ability of a factor, its localization within the cell, its trans-activation ability, its ability to associate with another protein, or its degradation.

Clearly, therefore, post-translational mechanisms for activating pre-existing protein could be used independently to stimulate either the DNA binding or the transcriptional activation activities of a single factor in different situations within a complex regulatory pathway. Indeed, such a combination of mechanisms is actually used to regulate the activity of the yeast GAL4 transcription factor. Thus, in the absence of galactose, the GAL4 transcription factor is bound to DNA but its activity is inhibited by the inhibitory GAL80 protein, so that GAL4 only activates transcription when galactose is present (for review see Sellick and Reece, 2005). Interestingly, however, this effect only occurs when the cells are grown in the presence of glycerol as the main carbon source. By contrast however, in the presence of glucose, GAL4 does not bind to DNA and the addition of galactose has no effect. Hence, by having a system in which glucose modulates the DNA binding of the factor and galactose modulates the activation of bound factor, it is possible for glucose to inhibit the stimulatory effect of galactose. This ensures that the enzymes required for galactose metabolism are only induced in the presence of glycerol and not in the presence of the preferred nutrient glucose (Fig. 8.48).

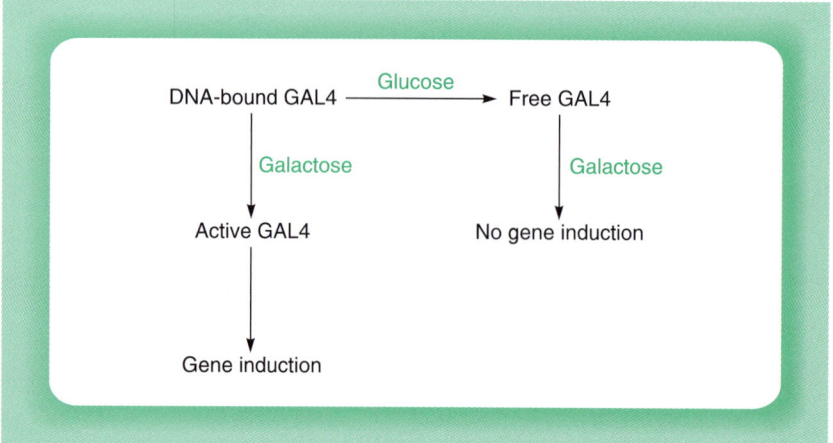

Figure 8.48

Effects of glucose and galactose on GAL4 activity. Note that whilst galactose stimulates the ability of DNA-bound GAL4 to activate transcription, this effect does not occur in the presence of glucose which results in the release of GAL4 from DNA.

Such a system in which two different activities of a single factor are independently modulated could clearly not be achieved by stimulating the *de novo* synthesis of the factor, which would simply result in more of it being present. Hence, in addition to its rapidity, the activation of pre-existing factor has the advantage of flexibility in potentially being able to generate different forms of the factor with different activities. It should be noted, however, that this effect can also be achieved, for example, by alternative splicing of the RNA encoding the factor (section 7.3.2) which can, for example, generate forms of the protein with and without the DNA binding domain, as in the case of the Era-1 factor, with and without the activation domain, as in the case of CREM or Oct-2 or with and without the ligand binding domain, as in the case of the thyroid hormone receptor.

8.7 CONCLUSIONS

In this chapter and the previous one, we have discussed how the regulation of gene expression by transcription factors is achieved both by the regulated synthesis or by the regulated activity of these factors. Although there are exceptions, the regulation of synthesis of a particular factor is used primarily in cases of factors which mediate tissue specific or developmentally regulated gene expression, where a factor is only required in a small proportion of cell types and is never required in most cell types. In contrast, however, the rapid induction of transcription in response to inducers of gene expression is primarily achieved by the activation of pre-existing inactive forms of transcription factors that are present in most cell types since this process, although more metabolically expensive, provides the required rapidity in response.

Although these two processes have been discussed separately, it should not be thought that a given factor can only be regulated either at the level of synthesis or at the level of activity. In fact, in many cases of inducible gene expression, which involve activation of pre-existing factors, such activation is supplemented by the slower process of synthesizing new factor in response to the inducing agent. Thus, in the case of the stimulation of genes containing AP-1 sites by phorbol esters discussed above (section 8.4.2), the phorbol ester-induced increase in the DNA binding of pre-existing Jun protein is supplemented by increased synthesis of both Fos and Jun following phorbol ester treatment. Such newly synthesized Fos and Jun will clearly eventually become a major part of the increased AP-1 activity observed following phorbol ester treatment (see Chapter 9, section 9.3.1). Similarly, the activation of NFκB by dissociation from IκB following treatment with substances such as phorbol esters which activate T cells (see section 8.3.1 and 8.4.2) has been shown to be supplemented by increased synthesis of NFκB and its corresponding mRNA following T cell activation, whilst increased synthesis of IκB itself occurs in response to glucocorticoid (section 8.4.2).

Hence, in many cases the rapid effects of post-translational processes in activating gene expression are supplemented by *de novo* synthesis of the factor which, although slower, will enhance and maintain the effect. Interestingly, the same factor can be regulated by enhanced synthesis or enhanced activity in different situations. Thus, as described in section 8.4.5, the activity of the HIF-1α factor is enhanced by hypoxia, by means of post-translational modifications which enhance its stability and its association with the p300 co-activator. However, treatment with angiotensin II, enhances the synthesis of HIF-1α by enhancing the transcription of the gene encoding it and the translation of its mRNA (Page *et al.*, 2002).

This combination of regulated synthesis and regulated activity is also seen in the case of factors which mediate tissue specific gene expression and which are synthesized in only a few cell types. Thus, in the case of the MyoD factor which regulates muscle-specific genes, the factor and its corresponding mRNA are synthesized only in cells of the muscle lineage (see Chapter 7, section 7.2.1). The activation of MyoD-dependent genes which occurs when myoblast cells within this lineage differentiate into myotubes is not, however, mediated by new synthesis of MyoD, which is present at equal levels in both cell types. Rather, it occurs due to the decline in the level of the inhibitory protein Id, resulting in the post-transcriptional activation of pre-existing MyoD and the transcription of MyoD-dependent genes (see Chapter 4, section 4.5.3). Hence, in this case regulation of synthesis is used to avoid the wasteful production of MyoD in cells of non-muscle lineage, whilst the activation of pre-existing MyoD ensures a rapid response to agents which induce differentiation within cells of the muscle lineage. Thus, in a number of cases a combination of both regulated synthesis and regulated activity allows the precise requirements of a particular response to be fulfilled rapidly but with minimum unnecessary wastage of energy.

In summary, therefore, the different properties of regulated synthesis and regulated activity allow these two processes, both independently and in combination, to efficiently regulate the complex processes of inducible, tissue specific and developmentally regulated gene expression.

REFERENCES

Abu-Amer, Y. and Tondravi, M.M. (1997) NFκB and bone – the breaking point. Nature Medicine 3, 1189–1190.

Arndt, K. and Winston, F. (2005) An unexpected role for ubiquitylation of a transcriptional activator. Cell 120, 733–734.

Baeuerle, P.A. (1995) Enter a polypeptide messenger. Nature 373, 661–662.

Baeuerle, P.A. and Baltimore, D. (1996) NF-κB: Ten years after. Cell 87, 13–20.

Ballas, N., Grunseich, C., Lu, D.D., Speh, J.C. and Mandel, G. (2005) REST and its corepressors mediate plasticity of neuronal gene chromatin throughout neurogenesis. Cell 121, 645–657.

Baniahmad, A., Tsai, S.Y., O'Malley, B.W. and Tsai, M-J. (1992) Kindred S thyroid hormone receptor is an active and constitutive silencer and a repressor for thyroid hormone and retinoic acid responses. Proceedings of the National Academy of Sciences, USA 89, 10633–10637.

Barolo, S. and Posakony, J.W. (2002) Three habits of highly effective signalling pathways: principles of transcriptional control by developmental cell signalling. Genes and Development 16, 1167–1181.

Becker, P.B., Gloss, B., Schmid, W., Strahle, U. and Schutz, G. (1986) In vivo protein-DNA interactions in a glucocorticoid response element require the presence of the hormone. Nature 324, 686–688.

Beg, A.A. and Scheiffele, P. (2006) SUMO wrestles the synapse. Science 311, 962–963.

Boyle, W.J., Smeal, T., Defize, L.H.K., Angel, P., Woodgett, J.R., Karin, M. and Hunter, T. (1991) Activation of protein kinase C decreases phosphorylation of c Jun at sites that negatively regulate its DNA binding activity. Cell 64, 573–584.

Brivanlou, A.H. and Darnell, J.E. (2002) Signal transduction and the control of gene expression. Science 295, 813–818.

Brown, M.S. and Goldstein, J.L. (1997) The SREBP pathway: regulation of cholesterol metabolism by proteolysis of a membrane-bound transcription factor. Cell 89, 331–340.

Bruick, R.K. and McKnight, S.L. (2001) Building better vasculature. Genes and Development 15, 2497–2502.

Bruick, R.K. and McKnight, S.L. (2002) Oxygen sensing gets a second wind. Science 295, 807–808.

Brzozosowski, A.M., Pike, A.C.W., Dauter, Z., Hubbard, R.E., Bonn, T., Engstrom, O., Ohman, L., Greene, G.L., Gustafsson, J.A. and Carlquist, M. (1997) Molecular basis of agonism and antagonism in the oestrogen receptor. Nature 389, 753–758.

Cantley, L.C. (2001) Translocating Tubby. Science 292, 2019–2021.

Costigan, M. and Woolf, C.J. (2002) No DREAM, no pain: closing the spinal gate. Cell 108, 297–300.

Darieva, Z., Bulmer, R., Pic-Taylor, A., Doris, K.S., Geymonat, M., Sedgwick, S.G., Morgan, B.A. and Sharrocks, A.D. (2006) Polo kinase controls cell-cycle-dependent transcription by targeting a coactivator protein. Nature 444, 494–498.

Desterro, J.M., Rodriguez, M.S. and Hay, R.T. (1998) SUMO-1 modification of IκBα inhibits NF-κB activation. Molecular Cell 2, 233–239.

Dixit, V. and Mak, T.W. (2002) NF-κB signalling. Many roads lead to Madrid. Cell 111, 615–619.

Edmunds, J.W. and Mahadevan, L.C. (2006) Protein kinases seek close encounters with active genes. Science 313, 449–451.

Freiman, R.N. and Tjian, R. (2003) Regulating the regulators: lysine modifications make their mark. Cell 112, 11–17.

Gallo, G.J., Schuetz, T.J. and Kingston, R.E. (1991) Regulation of heat shock factor in *Schizosaccharomyces pombe* more closely resembles regulation in mammals than in Saccharomyces cerevisiae. Molecular and Cellular Biology 11, 281–288.

Gamble, M.J. and Freedman, L.P. (2002) A coactivator code for transcription. Trends in Biochemical Sciences 27, 165–167.

Gomez-Ospina, N., Tsuruta, F., Barreto-Chang, O., Hu, L. and Dolmetsch, R. (2006) The C terminus of the L-type voltage-gated calcium channel ca(v)1.2 encodes a transcription factor. Cell 127, 591–606.

Goodrich, J.A. and Kugel, J.F. (2006) Non-coding-RNA regulators of RNA polymerase II transcription. Nature Reviews Molecular Cell Biology 7, 612–616.

Hay, R.T. (2005) SUMO: a history of modification. Molecular Cell 18, 1–12.

Hayden, M.S. and Ghosh, S. (2004) Signalling to NF-κB. Genes and Development 18, 2195–2224.

Hayden, M.S., West, A.P. and Ghosh, S. (2006) NF-κB and the immune response. Oncogene 25, 6758–6780.

He, J. and Furmanski, P. (1995) Sequence specificity and transcriptional activation in the binding of lactoferrin to DNA. Nature 373, 721–724.

Hensold, J.O., Hunt, C.R., Calderwood, S.K., Housman, D.E. and Kingston, R.E. (1990) DNA binding of heat shock factor to the heat shock element is insufficient for transcriptional activation in murine erythroleukaemia cells. Molecular and Cellular Biology 10, 1600–1608.

Hill, C.S. and Treisman, R. (1995) Transcriptional regulation by extracellular signals: mechanisms and specificity. Cell 80, 199–211.

Hoffmann, A. and Baltimore, D. (2006) Circuitry of nuclear factor κB signalling. Immunological Reviews 210, 171–186.

Hoffmann, A., Natoli, G. and Ghosh, G. (2006) Transcriptional regulation via the NF-κB signalling module. Oncogene 25, 6706–6716.

Holmberg, C.I., Hietakangas, V., Mikhailov, A., Rantanen, J.O., Kallio, M., Meinander, A., Hellman, J., Morrice, N., MacKintosh, C., Morimoto, R.I., Eriksson, J.E. and Sistonen, L. (2001) Phosphorylation of serine 230 promotes inducible transcriptional activity of heat shock factor 1. EMBO Journal 20, 3800–3810.

Hon, W-C., Wilson, J.I., Harlos, K., Claridge, T.D.W., Schofield, C.J., Pugh, C.W., Maxwell, P.H., Ratcliffe, P.J., Stuart, D.I. and Jones, E.Y. (2002) Structural basis for the recognition of hydroxyproline in HIF-1α by pVHL. Nature 417, 975–978.

Horvath, C.M. (2000) STAT proteins and transcriptional responses to extracellular signals. Trends in Biochemical Sciences 25, 496–502.

Ihle, J.N. (2001) The Stat family in cytokine signalling. Current Opinion in Cell Biology 13, 211–217.

Ip, Y.T. (1995) Converting an activator into a repressor. Current Biology 5, 1–3.

Jackson, P.K. (2006) A destructive switch for neurons. Nature 442, 365–366.

Jones, N. (1990) Transcriptional regulation by dimerization: two sides to an incestuous relationship. Cell 61, 9–11.

Kaelin, W.G. (2002) How oxygen makes its presence felt. Genes and Development 16, 1441–1445.

Karin, M. and Ben-Neriah, Y. (2000) Phosphorylation meets ubiquitination: the control of NF-κB activity. Annual Reviews of Immunology 18, 621–663.

Kawasaki, H., Schiltz, L., Chiu, R., Itakura, K., Taira, K., Nakatani, Y. and Yokoyama, K.K. (2000) ATF-2 has intrinsic histone acetyltransferase activity which is modulated by phosphorylation. Nature 405, 195–200.

Kirov, N.C., Lieberman, P.M. and Rushlow, C. (1996) The transcriptional co-repressor DSP1 inhibits activated transcription by disrupting TFIIA-TBP complex formation. EMBO Journal 15, 7079–7087.

Kugel, J.F. and Goodrich, J.A. (2006) Beating the heat: a translation factor and an RNA mobilize the heat shock transcription factor HSF1. Molecular Cell 22, 153–154.

Kuznetsova, A.V., Meller, J., Schnell, P.O., Nash, J.A., Ignacak, M.L., Sanchez, Y., Conaway, J.W., Conaway, R.C. and Czyzyk-Krzeska, M.F. (2003) von Hippel–Lindau protein binds hyperphosphorylated large subunit of RNA polymerase II through a proline hydroxylation motif and targets it for ubiquitination. Proceedings of the National Academy of Sciences, USA 100, 2706–2711.

Lamb, P. and McKnight, S.L. (1991) Diversity and specificity in transcriptional regulation: the benefits of heterotypic dimerization. Trends in Biochemical Sciences 16, 417–422.

Larson, J.S., Schuetz, T.J. and Kingston, R.E. (1988) Activation in vitro of sequence-specific DNA binding by a human regulatory factor. Nature 335, 372–375.

Lipford, J.R., Smith, G.T., Chi, Y. and Deshaies, R.J. (2005) A putative stimulatory role for activator turnover in gene expression. Nature 438, 113–116.

Luecke, H.F. and Yamamoto, K.R. (2005) The glucocorticoid receptor blocks P-TEFb recruitment by NFκB to effect promoter-specific transcriptional repression. Genes and Development 19, 1116–1127.

Lunyak, V.V. and Rosenfeld, M.G. (2005) No rest for REST: REST/NRSF regulation of neurogenesis. Cell 121, 499–501.

Mandel, G. and Goodman, R.H. (1999) DREAM on without calcium. Nature 398, 29–30.

Marx, J. (1995) How the glucocorticoids suppress immunity. Science 270, 232–233.

Marx, J. (2005) SUMO wrestles its way to prominence in the cell. Science 307, 836–839.

McKinsey, T.A., Zhang, C.L. and Olson, E.N. (2002) MEF2: a calcium-dependent regulator of cell division, differentiation and death. Trends in Biochemical Sciences 27, 40–47.

Min, J.-H., Yang, H., Ivan, M., Gertler, F., Kaelin, W.G. and Pavletich, N.P. (2002) Structure of an HIF-1α-pVHL complex: hydroxyproline recognition in signalling. Science 296, 1886–1889.

Morimoto, R.I. (1998) Regulation of the heat shock transcriptional response: cross talk between a family of heat shock factors, molecular chaperones and negative regulators. Genes and Development 12, 3788–3796.

Nalley, K., Johnston, S.A. and Kodadek, T. (2006) Proteolytic turnover of the Gal4 transcription factor is not required for function in vivo. Nature 442, 1054–1057.

Pagé, E.L., Robitaille, G.A., Pouysségur, J. and Richard, D.E. (2002) Induction of hypoxia-inducible factor-1α by transcriptional and translational mechanisms. Journal of Biological Chemistry 277, 48403–48409.

Perkins, N.D. (2006) Post-translational modifications regulating the activity and function of the nuclear factor kappa B pathway. Oncogene 25, 6717–6730.

Pomerantz, J.L. and Baltimore, D. (2002) Two pathways to NF-κB. Molecular Cell 10, 693–695.

Prasanth, K.V. and Spector, D.L. (2007) Eukaryotic regulatory RNAs: an answer to the 'genome complexity' conundrum. Genes and Development 21, 11–42.

Pratt, W.B. (1997) The role of the hsp90-based chaperone system in signal transduction by nuclear receptors and receptor signalling via MAP kinase. Annual Review of Pharmacology and Toxicology 37, 297–326.

Proft, M., Mas, G., de Nadal, E., Vendrell, A., Noriega, N., Struhl, K. and Posas, F. (2006) The stress-activated hog1 kinase is a selective transcriptional elongation factor for genes responding to osmotic stress. Molecular Cell 23, 241–250.

Pufall, M.A., Lee, G.M., Nelson, M.L., Kang, H.S., Velyvis, A., Kay, L.E., McIntosh, L.P. and Graves, B.J. (2005) Variable control of Ets-1 DNA binding by multiple phosphates in an unstructured region. Science 309, 142–145.

Punga, T., Bengoechea-Alonso, M.T. and Ericsson, J. (2006) Phosphorylation and ubiquitination of the transcription factor sterol regulatory element-binding protein-1 in response to DNA binding. Journal of Biological Chemistry 281, 25278–25286.

Qiu, J. (2006) Degrading id. Nature Reviews Neuroscience 7, 592.

Renaud, J-P., Rochel, N., Ruff, M., Vivat, V., Chambon, P., Gronemeyer, H. and Moras, D. (1990) Crystal structure of the RAR-γ ligand binding domain bound to all trans-retinoic acid. Nature 378, 681–689.

Ronchi, E., Treisman, J., Dostatri, N., Struhl, G. and Desplan, C. (1993) Down-regulation of the Drosophila morphogen bicoid by the torso receptor mediated signal transduction cascade. Cell 74, 347–355.

Salghetti, S.E., Caudy, A.A., Chenoweth, J.G. and Tansey, W.P. (2002) Regulation of transcriptional activation domain function by ubiquitin. Science 293, 1651–1653.

Scheidereit, C. (2006) IκB kinase complexes: gateways to NF-κB activation and transcription. Oncogene 25, 6685–6705.

Schmitz, M.L., Bacher, S. and Kracht, M. (2001) IκB-independent control of NF-κB activity by modulatory phosphorylations. Trends in Biochemical Sciences 26, 186–190.

Sellick, C.A. and Reece, R.J. (2005) Eukaryotic transcription factors as direct nutrient sensors. Trends in Biochemical Sciences 30, 405–412.

Semenza, G.L. (2001) HIF-1, O_2, and the 3 PHDs: how animal cells signal hypoxia to the nucleus. Cell 107, 1–3.

Shamovsky, I., Ivannikov, M., Kandel, E.S., Gershon, D. and Nudler, E. (2006) RNA-mediated response to heat shock in mammalian cells. Nature 440, 556–560.

Sorger, P.K., Lewis, M.J. and Pelham, H.R.B. (1987) Heat shock factor is regulated differently in yeast and HeLa cell. Nature 329, 81–84.

Stewart, S. and Crabtree, G.R. (2000) Regulation of the regulators. Nature 408, 46–47.

Tergaonkar, V. (2006) NFκB pathway: a good signalling paradigm and therapeutic target. International Journal of Biochemistry and Cell Biology 38, 1647–1653.

Thiele, D.J. (1992) Metal regulated transcription in eukaryotes. Nucleic Acids Research 20, 1183–1191.

Ting, A.Y. and Endy, D. (2002) Decoding NF-κB signalling. Science 298, 1189–1190.

Treisman, R. (1996) Regulation of transcription by MAP kinase cascades. Current Opinion in Cell Biology 8, 205–215.

Tsukiyama, T., Becker, P.B. and Wu, C. (1994) ATP-dependant nucleosome disruption at a heat-shock promoter mediated by DNA binding of GAGA transcription factor. Nature 367, 525–532.

Vandromme, M., Gauthier-Rouviere, C., Lamb, N. and Fernadez, A. (1996) Regulation of transcription factor localisation: fine tuning of gene expression. Trends in Biochemical Sciences 21, 59–64.

Weatherman, R.V., Fletterick, R.J. and Scanlan, T.S. (1999) Nuclear-receptor ligands and ligand-binding domains. Annual Reviews of Biochemistry 68, 559–581.

Welchman, R L., Gordon, C. and Mayer, R.J. (2005) Ubiquitin and ubiquitin-like proteins as multifunctional signals. Nature Reviews Molecular Cell Biology 6, 599–609.

Wilmann, T. and Beato, M. (1986) Steroid-free glucocorticoid receptor binds specifically to mouse mammary tumour DNA. Nature 324, 688–691.

Wood, M.J., Storz, G. and Tjandra, N. (2004) Structural basis for redox regulation of Yap1 transcription factor localization. Nature 430, 917–921.

Wurtz, J-M., Bourget, W., Renaud, J-P., Vivat, V., Chambon, P., Moras, D. and Gronemeyer, H. (1996) A canonical structure for the ligand binding domain of nuclear receptors. Nature Structural Biology 3, 87–94.

Yamamoto, Y. and Gaynor, R.B. (2004) IκB kinases: key regulators of the NF-κB pathway. Trends in Biochemical Sciences 29, 72–79.

Yang, X.J. and Gregoire, S. (2006) A recurrent phospho-sumoyl switch in transcriptional repression and beyond. Molecular Cell 23, 779–786.

Yao, J., Munson, K.M., Webb, W.W. and Lis, J.T. (2006) Dynamics of heat shock factor association with native gene loci in living cells. Nature 442, 1050–1053.

Zhu, H. and Bunn, H.F. (2001) How do cells sense oxygen? Science 292, 449–451.

Zimarino, V. and Wu, C. (1987) Induction of sequence-specific binding of *Drosophila* heat-shock activator protein without protein synthesis. Nature 327, 727–730.

Zou, J., Guo, Y., Guettouche, T., Smith, D.F. and Voellmy, R. (1998) Repression of heat shock transcription factor HSF1 activation by HSP90 (HSP90 complex) that forms a stress-sensitive complex with HSF1. Cell 94, 471–480.

CHAPTER 9

TRANSCRIPTION FACTORS AND HUMAN DISEASE

9.1 DISEASES CAUSED BY TRANSCRIPTION FACTOR MUTATIONS

In previous chapters we have discussed a number of examples of the involvement of transcription factors in normal cellular regulatory processes, for example constitutive, inducible, cell type specific or developmentally regulated transcription. It is not surprising that aspects of this complex process can go wrong and that the resulting defects in transcription factors can result in disease (for reviews see Engelkamp and van Heyningen, 1996; Latchman, 1996).

For example, mutations in several classes of transcription factor which control gene expression during development have been shown to result in human developmental disorders. Thus, mutations in the gene encoding the POU family transcription factor Pit-1 (see Chapter 4, section 4.2.6), result in a failure of pituitary gland development leading to congenital dwarfism, whilst mutations in the genes encoding the Pax family transcription factors Pax3 and Pax6 (see Chapter 4, section 4.2.7) result in eye defects. Similarly, mutations in genes encoding homeobox proteins (see Chapter 4, section 4.2) result in a variety of congenital abnormalities (for review see Boncinelli, 1997).

Interestingly, the mutations in Pit-1 and Pax6 discussed above are both dominant, with one single copy of the mutant gene being sufficient to produce the disease even in the presence of a functional copy. However, this dominance arises for different reasons (for review see Latchman, 1996). In the case of Pit-1, the mutant Pit-1 can bind to its DNA binding site but cannot activate gene expression. It therefore not only fails to stimulate transcription of its target genes but can also act as a dominant negative factor inhibiting gene activation by preventing the wild type protein from binding to DNA (Fig. 9.1a). This mechanism is similar to one mode of action of transcriptional repressors, which act by preventing an activator from binding to its DNA target site (see Chapter 6, section 6.2.1). In contrast, the dominant nature of the Pax6 mutation does not reflect any dominant negative action of the mutant protein since such mutations often involve complete deletion of the gene. Rather it reflects a phenomenon known as haploid insufficiency in which the amount of protein produced by a single functional copy of the gene is not enough to allow it to activate its target genes effectively (Fig. 9.1b).

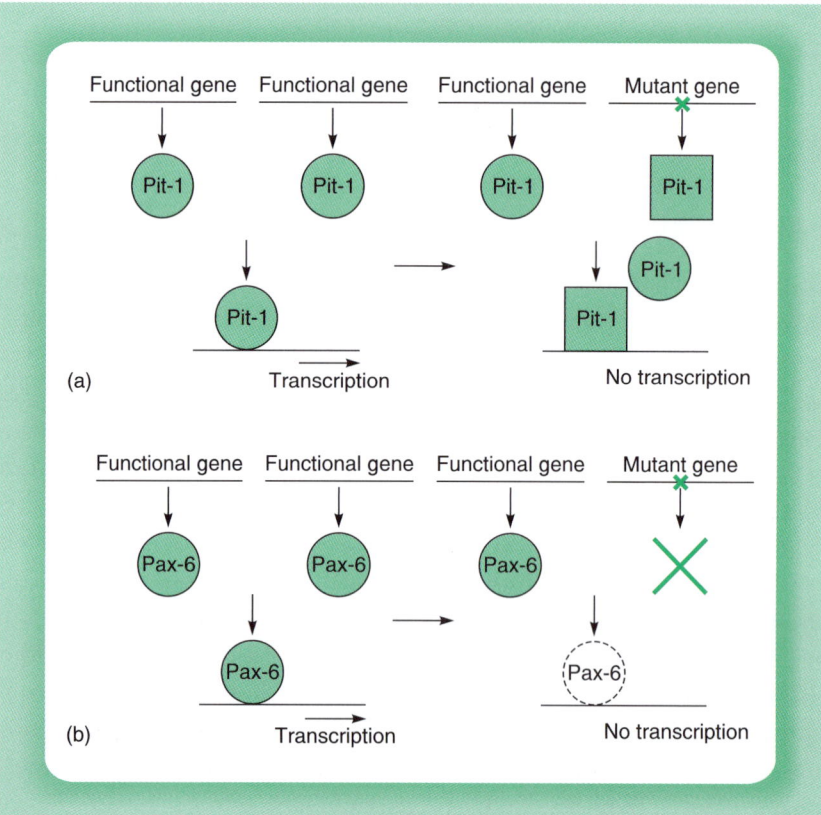

Figure 9.1

Different mechanisms by which mutations in genes encoding transcription factors can be dominant producing disease in the presence of a functional copy of the gene. (a) In the case of Pit-1, the mutation produces a dominant negative form of the factor (square) which binds to the appropriate binding site and not only fails to activate transcription but also prevents binding and activation by the functional protein (circle). (b) In the case of Pax-6, one functional gene cannot produce enough functional protein (circle) to activate its target genes (dotted circle).

As well as resulting from mutations in the genes encoding DNA binding factors, developmental disorders can also result from mutations in genes encoding other types of transcription factors, such as components of the basal transcriptional complex, co-activators or factors which alter chromatin structure. Thus, for example, mutation in the gene encoding the SNF2 factor which is part of the SWI/SNF chromatin remodelling complex (see Chapter 1, section 1.2.2) results in a lack of α-globin gene expression and a variety of other symptoms such as mental retardation. This indicates that this factor is necessary for opening the chromatin structure of the α-globin genes and a number of other genes so preventing their transcription when it is absent (Gibbons *et al.*, 1995) (Fig. 9.2).

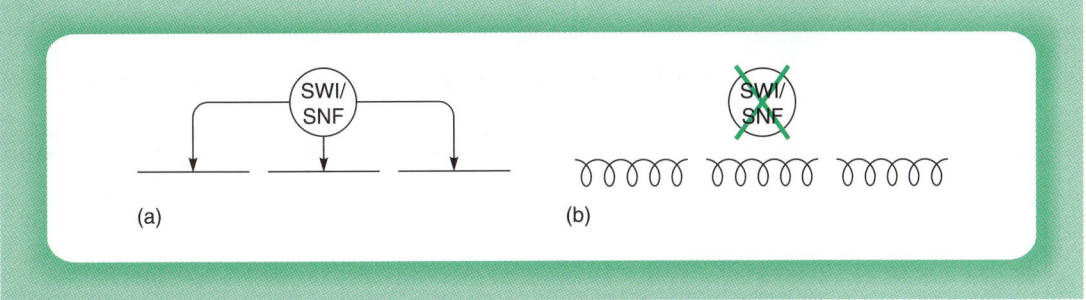

(a)   (b)

Figure 9.2

(a) Functional SWI/SNF alters the chromatin structure of its target genes to a more open configuration (solid line). (b) If SWI/SNF is inactive these genes remain in a closed chromatin structure (wavy line) and are thus not transcribed.

Similarly, the Williams syndrome transcription factor (WSTF) which is mutated in the human disease Williams syndrome, associates with the SWI/SNF complex (Poot *et al.*, 2004). This chromatin remodelling complex then interacts with the vitamin D receptor, a member of the nuclear receptor family (discussed in Chapter 4, section 4.4). Chromatin remodelling by the WSTF-containing complex is essential for recruitment of the vitamin D receptor to its target genes, and the failure of this process in Williams syndrome results in mental retardation and growth deficiency (for review see Belandia and Parker, 2003).

Hence, a number of different DNA binding factors and chromatin remodelling factors can be affected in human disease. Although it might be thought that the critical role of the RNA polymerase II basal transcriptional complex (see Chapter 3, section 3.5) would result in any mutations in its components being lethal, this is not always the case. Thus, mutations in specific subunits of the basal transcription factor TFIIH (see Chapter 3, section 3.5) result in the skin disease xeroderma pigmentosum. Interestingly, the mutant TFIIH proteins found in these patients show defects in their ability to respond to the transcriptional activator FBP and the transcriptional repressor FIR (Fig. 9.3) (Liu *et al.*, 2001). Hence, in this case, disease results from an inability of a component of the basal transcriptional complex to respond appropriately to activating or repressing signals.

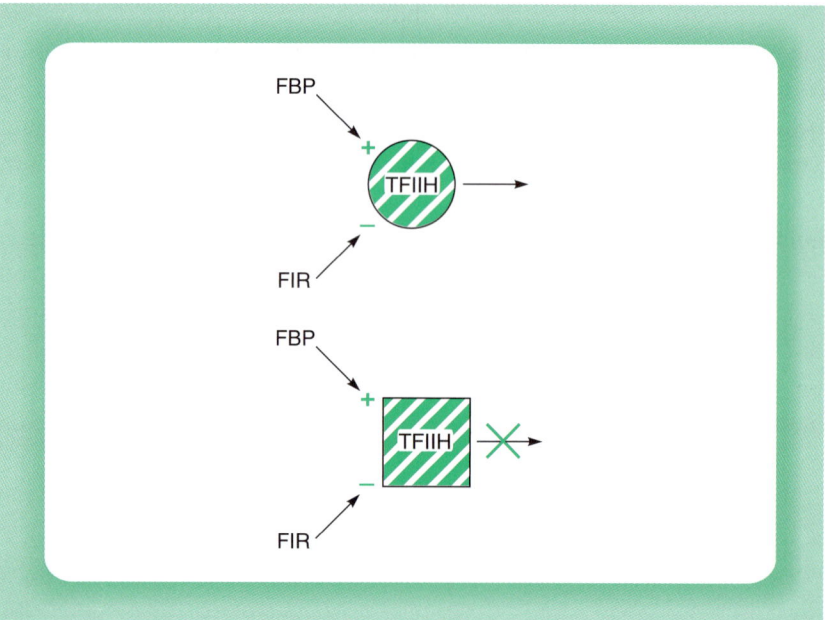

Figure 9.3

In the human disease xeroderma pigmentosum, the mutant TFIIH (square) has lost the ability of the wild type protein (circle) to respond appropriately to the activating factor FBP and the inhibitory factor FIR.

In addition to affecting DNA binding factors, chromatin remodelling factors and components of the basal transcriptional complex, mutation can also affect co-activator molecules which transmit the signal between DNA binding factors and the basal complex. Thus, mutation in the gene encoding the CBP factor which acts as a co-activator for a variety of other transcription factors (see Chapter 5, section 5.4.3) results in the severe developmental disorder known as Rubinstein–Taybi syndrome, which is characterized by mental retardation and physical abnormalities (for review see D'Arcangelo and Curran, 1995). This indicates that CBP is an important co-activator for developmentally regulated as well as inducible gene expression. Interestingly, no individuals with mutations inactivating both copies of the CBP gene have ever been identified and it is likely that a lack of functional CBP is incompatible with life. Individuals with Rubinstein–Taybi syndrome have a single functional CBP gene and a single mutant gene indicating that the mutation is dominant. As with Pax6, however, this dominance apparently reflects a haploid insufficiency in which a single copy of the CBP gene cannot produce enough functional protein. This is not surprising since, as discussed in Chapter 6 (section 6.6), the amount of CBP in the cell is limited and different transcription factors compete for it.

As well as being the cause of Rubinstein–Taybi syndrome, CBP is also involved in the neurodegenerative disease Huntington's chorea. However, in this case, the CBP protein is entirely normal and the disease is caused by mutations in a protein known as Huntingtin. The mutant Huntingtin gene contains abnormally increased numbers of the triplet DNA sequence CAG. This results in an abnormal protein containing a run of additional glutamine amino acids, resulting in Huntington's disease being referred to as a triplet repeat or polyglutamine disease (for review of transcription and polyglutamine diseases see Riley and Orr, 2006). Although Huntingtin is not a transcription factor, the mutant protein can bind to CBP and sequester it into protein aggregations. Since the amounts of CBP in the cell are limiting, this prevents it binding to transcriptional activators and therefore causes disease.

Hence, inactivation of CBP can occur by mutation or by its binding of another protein and consequent inactivation (Fig. 9.4). Interestingly, it has been shown that as well as targeting CBP, the mutant Huntingtin can also disrupt the function of DNA binding factors such as Sp1 and components of the basal transcriptional complex such as TFIID and TFIIF, indicating that it can target DNA binding factors and the basal transcriptional complex as well as co-activators (Zhai *et al.*, 2005).

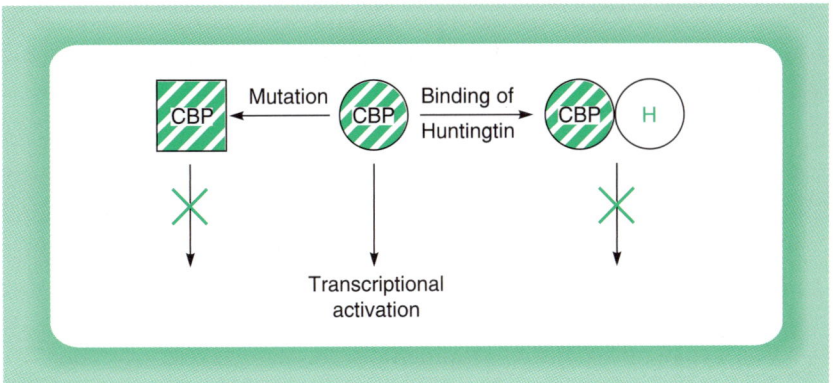

Figure 9.4
CBP can be inactivated by mutation producing Rubinstein–Taybi syndrome or by binding to mutant Huntingtin protein in Huntington's disease.

Obviously, the ability of mutant Huntingtin to interfere with the function of specific transcription factors will produce abnormalities in the expression of target genes regulated by such factors. One such target gene whose expression is affected by mutant Huntingtin in this way has recently been identified. Thus, it has been shown that mutant

Huntingtin interferes with activation of the PGC-1α gene by the CREB transcription factor and its co-activator CBP (Cui *et al.*, 2006; for review see Ross and Thompson, 2006). Interestingly, the PGC-1α gene actually encodes a transcriptional co-activator, which is involved in activation by several members of the nuclear receptor transcription factor family (see Chapter 4, section 4.4 for discussion of these factors). Hence, transcriptional repression of the PGC-1α gene by mutant Huntingtin will in turn affect the expression of a number of other genes, whose activation requires PGC-1α.

Interestingly, Huntington's disease is not the only triplet repeat disease to involve transcriptional abnormalities (for reviews see Helmlinger *et al.*, 2006; Riley and Orr, 2006). For example, ataxin, the gene which is mutated in human SCA7 disease, normally encodes a histone acetyltransferase, a type of protein which is involved in the remodelling of chromatin structure to allow gene activation to occur (see Chapter 5, section 5.5). The mutant form of ataxin found in SCA7 acts as a dominant negative factor, inhibiting the wild type protein and therefore preventing chromatin remodelling and activation of target genes (Fig. 9.5) (for review see Helmlinger *et al.*, 2006).

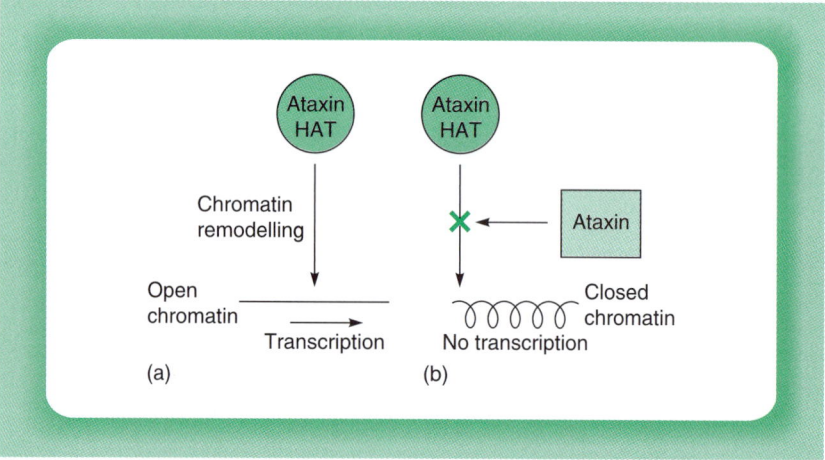

Figure 9.5
The ataxin protein (circle) has histone acetyltransferase (HAT) activity and can remodel chromatin resulting in activation of target genes (panel a). However, in human SCA7 disease, an abnormal dominant negative form of the protein (square) interferes with this activity. Hence chromatin remodelling and activation of target genes do not occur (panel b).

Hence, developmental disorders can arise from mutations in genes encoding DNA binding activator proteins such as Pit-1 and Pax-6, components of the basal transcriptional complex such as TFIIH, co-activators such as CBP or components of chromatin modulating complexes such as SNF2 or the Williams syndrome transcription factor (WSTF) (Fig. 9.6).

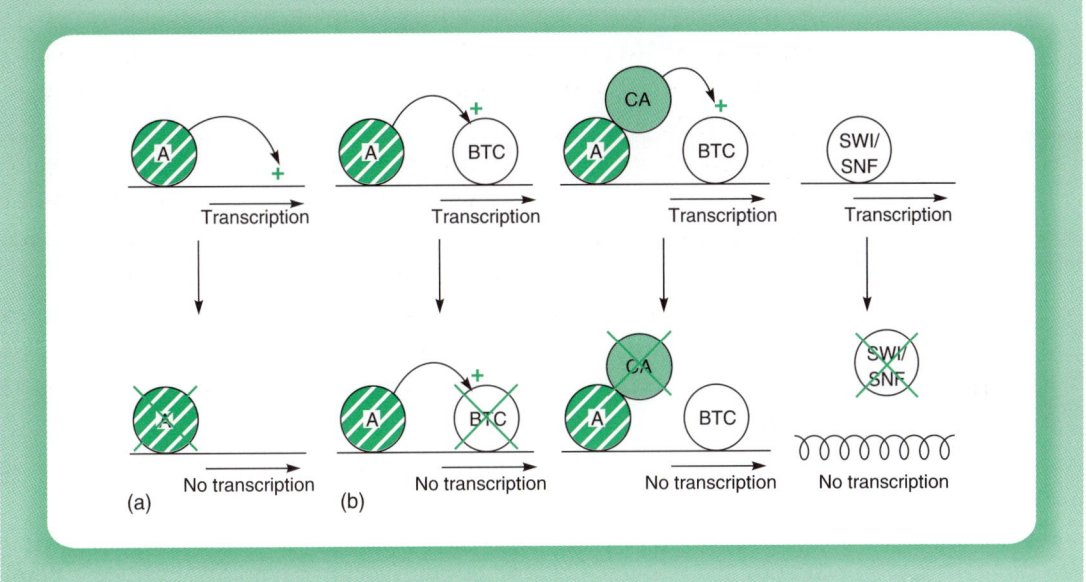

Figure 9.6

Mutations can occur in genes encoding (a) DNA binding activators (A), (b) components of the basal transcriptional complex (BTC), (c) co-activators (CA) or (d) factors which alter chromatin structure (SWI/SNF).

As well as such developmental defects, mutations in genes encoding the nuclear receptor transcription factor family (see Chapter 4, section 4.4) can produce a failure to respond to the hormone which normally binds to the receptor and regulates transcription. Such mutations have been reported, for example, in the receptors for glucocorticoid, thyroid hormone and vitamin D (for review see Latchman, 1996). Similarly, mutations in the gene encoding the peroxisome proliferator-activated receptor γ (PPARγ), another member of the nuclear receptor family, have been identified in patients who show resistance to insulin, diabetes and high blood pressure. Although such mutations are rare, they indicate that PPARγ is involved in insulin responses and may therefore be a valuable therapeutic target to enhance these responses in the much more numerous cases of diabetes which have a normal PPARγ protein (for reviews see Kersten *et al.*, 2000; Rosen and Spiegelman 2001; Evans *et al.*, 2004) (see section 9.5).

Thus, human disease-causing mutations in proteins regulating transcription are not solely involved in developmental disorders. Indeed, mutations in the MEF2 transcription factor, discussed in Chapter 8 (section 8.4.2), have been shown to produce coronary artery disease in middle-aged patients (Wang *et al.*, 2003).

Hence, human diseases can involve a variety of different factors, which play a role in RNA polymerase II-mediated transcription and can produce both developmental disorders and adult-onset diseases. Interestingly, however, other human diseases can involve factors associated with other RNA polymerases. Thus, the abnormal craniofacial development, which is observed in Treacher Collins syndrome, is due to abnormalities in a protein, which interacts with the RNA polymerase I basal transcription factor, UBF (Valdez *et al.*, 2004) (see Chapter 3, section 3.3) whilst mutations in the RNA polymerase I-associated protein, CBF, cause Cockayne's syndrome, a human disease involving abnormalities in the nervous system and skeleton (Bradsher *et al.*, 2002).

9.2 CANCER

Despite the existence of transcription factor mutations producing developmental defects or non-responsiveness to hormone, a special place in the human diseases which can involve alterations in transcription factors is occupied by cancer. Thus, because this disease results from growth in an inappropriate place or at an inappropriate time, it can be caused not only by deficiencies in particular genes but also by the enhanced expression or activation of specific cellular genes involved in growth regulatory processes, which are normally only expressed at low levels or very transiently.

Interestingly, many cancer-causing genes of this type, known as oncogenes (for general reviews see Broach and Levine, 1997; Hunter, 1997), were originally identified within cancer-causing retroviruses which had picked them up from the cellular genome. Within the virus, the oncogene has become activated either by over-expression or by mutation and is therefore responsible for the ability of the virus to transform cells to a cancerous phenotype. In contrast, the homologous gene within the cellular genome is clearly not always cancer-causing since all cells are not cancerous. It can be activated, however, into a cancer-causing form either by over-expression or by mutation and hence these genes can play an important role in the generation of human cancer (Fig. 9.7). The forms of the oncogene isolated from the retrovirus and from the normal cellular genome are distinguished by the prefixes v and c respectively, as in v-onc and c-onc.

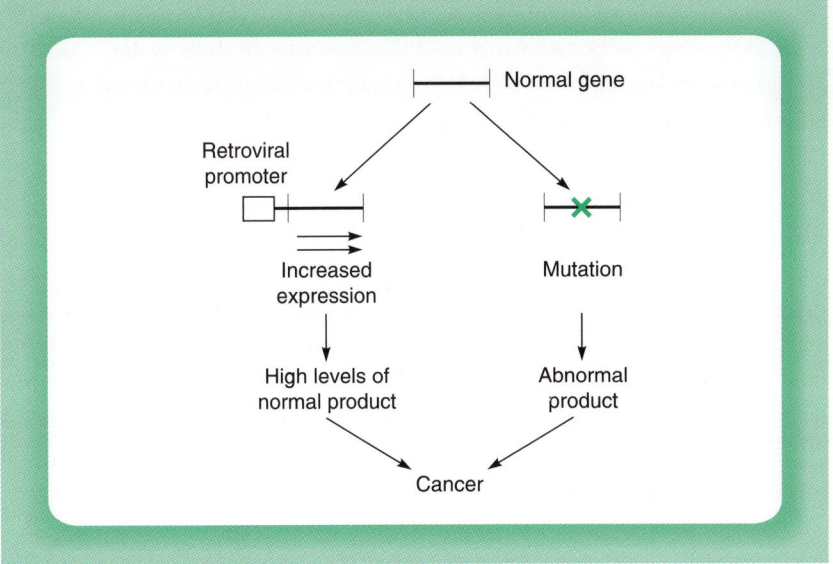

Figure 9.7

A cellular proto-oncogene can be converted into a cancer-causing oncogene by increased expression or by mutation.

Despite this potential to cause cancer, the c-onc genes are highly conserved in evolution, being found not only in the species from which the original virus was isolated but in a wide range of other eukaryotes. This indicates that the products of these oncogenes play a critical role in the regulation of normal cellular growth processes, their malregulation or mutation resulting therefore in abnormal growth and cancer. In agreement with this idea, oncogenes identified in this way include genes encoding many different types of protein involved in growth control such as growth factors, growth factor receptors and G proteins. They also include, however, several genes encoding cellular transcription factors which normally regulate specific sets of target genes. Similarly, a number of other genes encoding transcription factors have been identified at the break points of the chromosomal translocations characteristic of human leukaemias with their activation being involved in the resulting cancer. Section 9.3 of this chapter therefore discusses several cases of this type and the insights they have provided into the processes regulating gene expression in normal cells and their malregulation in cancer (for reviews see Rabbits, 1994; Latchman, 1996).

Following the discovery of cellular oncogenes, it subsequently became clear that another class of genes existed whose protein products appeared to restrain cellular growth. The deletion or mutational inactivation of these so-called anti-oncogenes therefore results in the abnormal unregulated growth characteristic of cancer cells (for reviews see Lowe *et al.*, 2006; Vogelstein and Kinzler, 2006). As some of these anti-oncogenes also encode transcription factors, they are discussed in section 9.4 of this chapter.

9.3 CELLULAR ONCOGENES AND CANCER

9.3.1 FOS, JUN AND AP1

The AP1 binding site is a DNA sequence that renders genes which contain it inducible by treatment with phorbol esters such as TPA. The activity binding to this site is referred to as AP1 (activator protein 1). It is clear, however, that preparations of AP1 purified by affinity chromatography on an AP1 binding site contain several different proteins (for reviews of AP1 see Kerppola and Curran, 1995; Karin *et al.*, 1997; Shaulian and Karin, 2002; Ozanne *et al.*, 2007).

A possible clue as to the identity of one of these AP1-binding proteins was provided by the finding that the yeast protein GCN4, which induces transcription of several yeast genes involved in amino acid biosynthesis, does so by binding to a site very similar to the AP1 site (Fig. 9.8). In turn, the DNA binding region of GCN4 shows strong homology at the amino acid level to v-*jun*, the oncogene of avian sarcoma virus ASV17 (Fig. 9.9). This suggested therefore that the protein encoded by the cellular homologue of this gene, c-*jun*, which was known to be a nuclearly located DNA-binding protein, might be one of the proteins which bind to the AP1 site.

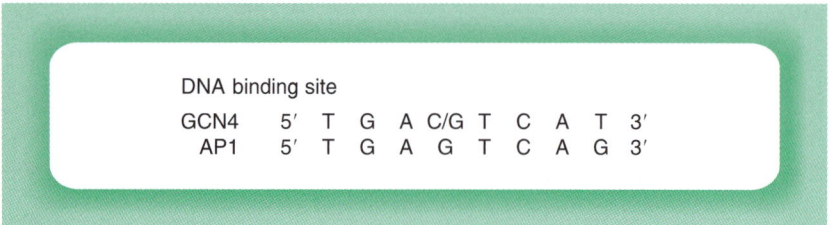

Figure 9.8

Relationship of the DNA binding sites for the yeast transcription factor GCN4 and the mammalian transcription factor AP1.

In agreement with this, antibodies against the Jun protein react with purified AP1-binding proteins, whilst Jun protein expressed in bacteria can bind to AP1 binding sites. Hence, the Jun protein is capable of binding to the AP1 binding site and constitutes one component of purified AP1 preparations which also contain other Jun-related proteins such as

Figure 9.9

Comparison of the carboxyl-terminal amino acid sequences of the chicken Jun protein and the yeast transcription factor GCN4. Boxes indicate identical residues.

Jun B (Fig. 9.10). Moreover, co-transfection of a vector expressing the Jun protein with a target promoter resulted in increased transcription if the target gene contained AP1 binding sites but not if it lacked them, indicating that Jun was capable of stimulating transcription via the AP1 site (Fig. 9.11). Hence, the Jun oncogene product is a sequence specific transcription factor capable of stimulating transcription of genes containing its binding site.

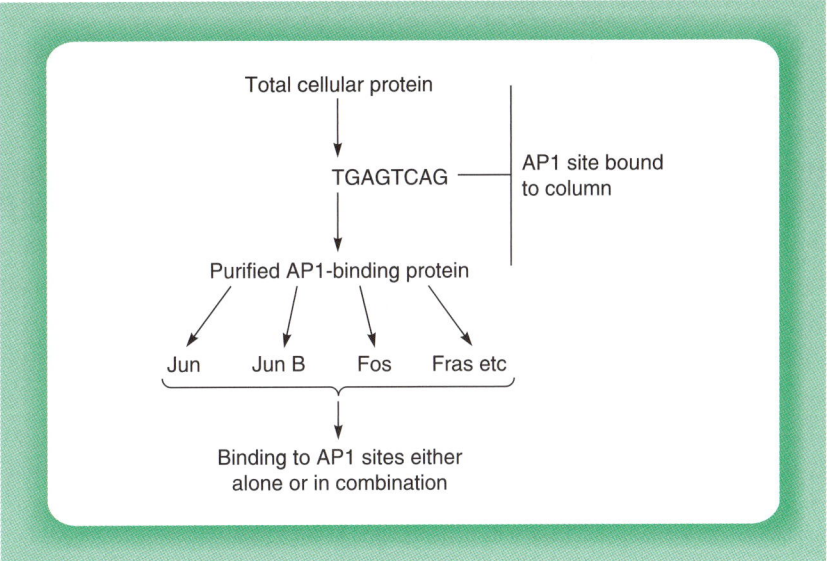

Figure 9.10

Passage of total cellular proteins through a column containing an AP1 site results in the purification of several cellular proteins including Jun, Jun B, Fos and Fos-related antigens (Fras) which are capable of binding to the AP1 site either alone or in combination.

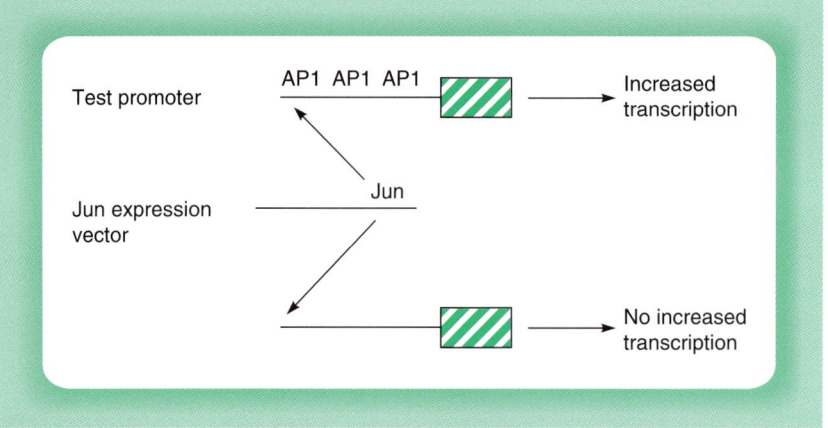

Figure 9.11

Artificial expression of the Jun protein in an expression vector results in the activation of a target promoter containing several AP1 binding sites but has no effect on a similar promoter lacking these sites, indicating that Jun can specifically activate gene expression via AP1 binding sites.

In addition to Jun and Jun-related proteins, purified AP1 preparations also contain the product of another oncogene c-*fos*, as well as several Fos-related proteins known as the Fras (Fos-related antigens; see Fig. 9.10). Unlike Jun, however, Fos cannot bind to the AP1 site alone but can do so only in the presence of another protein p39, which is identical to Jun (see Chapter 4, section 4.5). Hence, in addition to its ability to bind to AP1 sites alone, Jun can also mediate binding to this site by the Fos protein. Such DNA binding by Fos and Jun is dependent on the formation of a dimeric molecule. Although Jun can form a DNA binding homodimer, Fos cannot do so. Hence DNA binding by Fos is dependent upon the formation of a heterodimer between Fos and Jun, which binds to the AP1 site with approximately thirty-fold greater affinity than the Jun homodimer (Fig. 9.12).

It is clear therefore that both Fos and Jun, which were originally isolated in oncogenic retroviruses, are also cellular transcription factors that play an important role in activating specific cellular genes following phorbol ester treatment. Increased levels of Fos and Jun occur in cells following treatment with phorbol esters, indicating that these substances act, at least in part, by increasing the levels of Fos and Jun, which in turn bind to the AP1 sites in phorbol ester-responsive genes and activate their expression.

Similar increases in the levels of Fos, and Jun, as well as Jun-B and the Fos-related protein Fra-1, are also observed when quiescent cells are stimulated to grow by treatment with growth factors or serum. This indicates that these substances act, at least in part, by increasing the levels

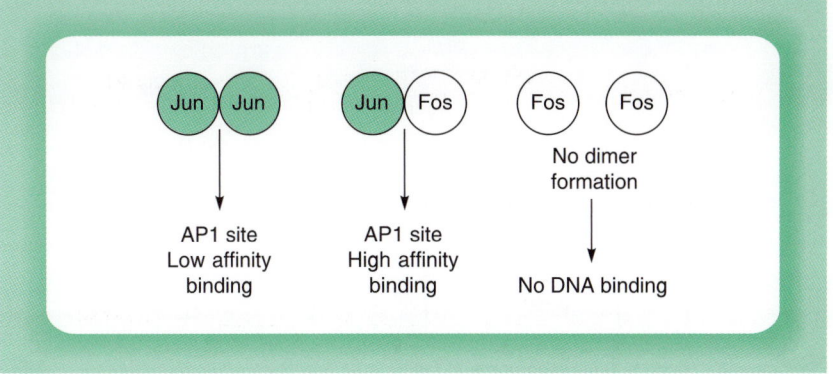

Figure 9.12

Heterodimer formation between Fos and Jun results in a complex capable of binding to an AP1 site with approximately thirty-fold greater affinity than a Jun homodimer, whilst a Fos homodimer cannot bind to the AP1 site.

of Fos and Jun, which in turn will switch on genes whose products are necessary for growth itself (Fig. 9.13). In agreement with this idea, cells derived from mice in which the c-*jun* gene has been inactivated grow very slowly in culture and the mice themselves die early in embryonic development. Hence, Fos and Jun play a critical role in normal cells, as transcription factors inducing phorbol ester or growth-dependent genes.

Normally, levels of Fos and Jun increase only transiently following growth factor treatment resulting in a period of brief controlled growth. Clearly, continuous elevation of these proteins, such as would occur when cells become infected with a retrovirus expressing one of them, would result in cells which exhibited continuous uncontrolled growth and were not subject to normal growth regulatory signals. Since such uncontrolled growth is one of the characteristics of cancer cells, it is relatively easy to link the role of Fos and Jun in inducing genes required for growth with their ability to cause cancer. Normally, however, the transformation of a cell to a transformed cancerous phenotype requires more than simply its conversion to a continuously growing immortal cell (for review see Land *et al.*, 1983). Since repeated treatments with phorbol esters can promote tumour formation in immortalized cells, the prolonged induction of phorbol ester responsive genes by elevated levels of Fos and Jun may therefore result in the conversion of already continuously growing cells into the tumourigenic phenotype characteristic of cancer cells (Fig. 9.14).

Hence, the ability of Fos and Jun to cause cancer represents an aspect of their ability to induce transcription of specific cellular genes. In agreement with this idea, mutations in Fos which abolish its ability to dimerize with Jun and hence prevent it from binding to AP1 sites also

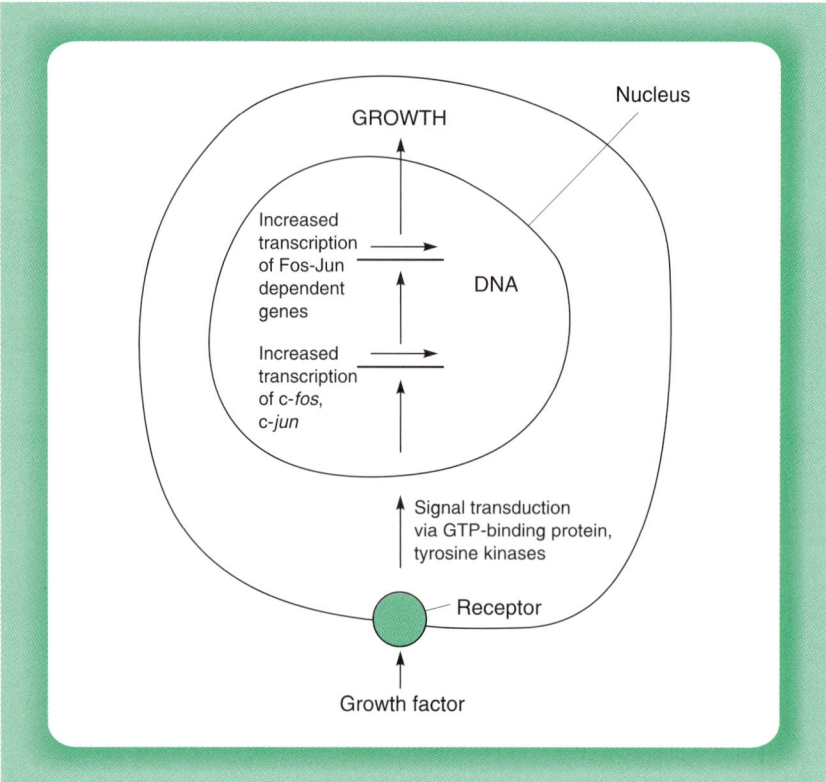

Figure 9.13

Growth factor stimulation of cells results in increased transcription of the c-*fos* and c-*jun* genes which in turn stimulates transcription of genes which are activated by the Fos–Jun complex.

abolish its ability to transform cells to a cancerous phenotype. It should be noted, however, that in addition to their over-expression within a retrovirus, there is also evidence that mutational changes render the viral proteins more potent transcriptional activators than the equivalent cellular proteins. Thus the v-Jun protein appears to activate transcription more efficiently than c-Jun due to a loss of a phosphorylation site which is involved in targeting the c-Jun protein for degradation and the loss of another region of the protein (the delta domain) which mediates its interaction with a histone deacetylase enzyme that reduces its ability to stimulate transcription (Fig. 9.15) (Vogt and Bader, 2005).

Interestingly, in addition to its central role in the growth response, the Fos, Jun, AP1 system also appears to represent a target for other oncogenes. Thus, for example, the *ets* oncogene which, like *fos* and *jun* encodes a cellular transcription factor, acts via a DNA binding site known as PEA3 which is located adjacent to the AP1 site in a number of TPA-responsive genes, such as collagenase and stromelysin. Moreover, the Ets

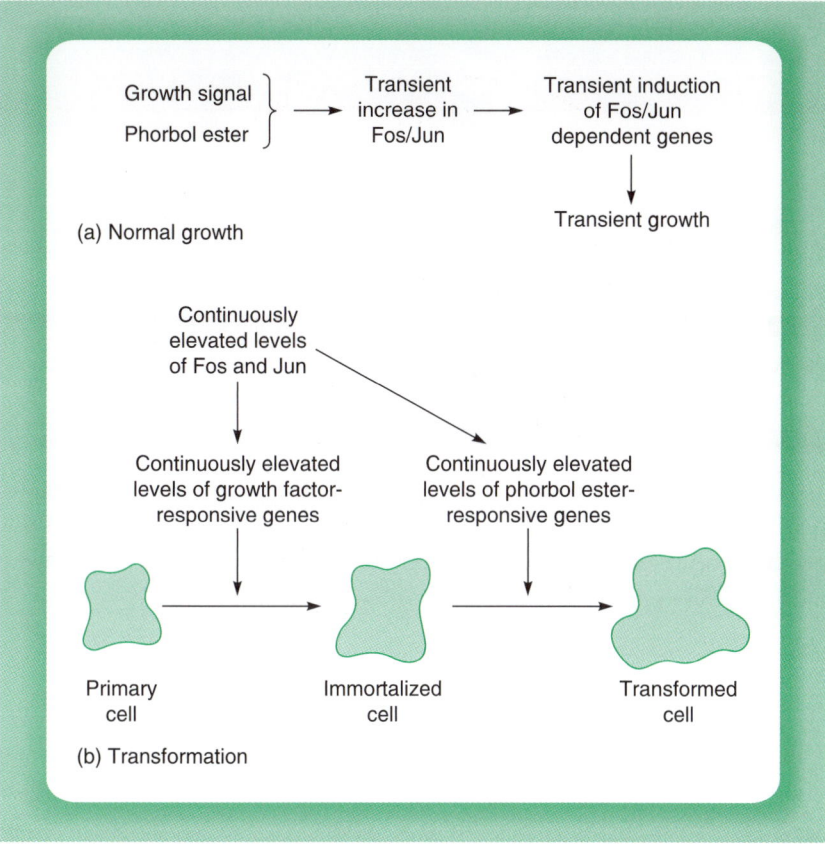

Figure 9.14

Effects of Fos and Jun on cellular growth. In normal cells (a) a brief exposure to a growth signal or phorbol ester will lead to a brief period of growth via the transient induction of Fos and Jun and hence of Fos/Jun-dependent genes. In contrast, the continuous elevation of Fos and Jun produced for example by infection with a retrovirus expressing Fos or Jun results in continuous unlimited growth and cellular transformation (b).

protein co-operates with Fos and Jun to produce high level activation of these promoters.

In addition to interacting positively with other factors, the Fos/Jun complex can also inhibit the action of other transcription factors. Thus, as described in Chapter 6 (section 6.6), the Fos/Jun complex requires the CBP co-activator in order to activate transcription. It therefore competes with the activated glucocorticoid receptor for CBP, hence preventing the receptor from activating transcription. Similarly, both Fos and Jun can inhibit the activation of muscle-specific promoters by the MyoD transcription factor (see Chapter 7, section 7.2.1) thereby preventing cells from differentiating into non-dividing muscle cells and allowing cellular proliferation to continue (Li *et al.*, 1992).

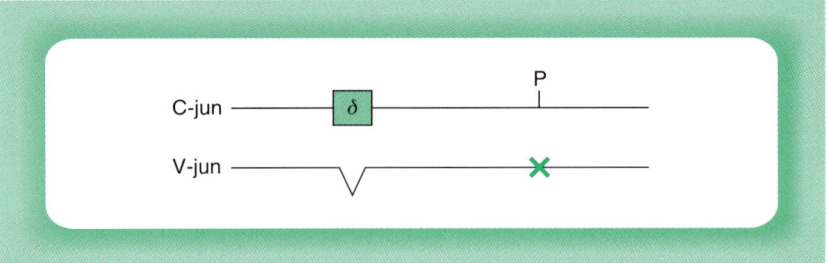

Figure 9.15

In addition to being over-expressed, the v-Jun protein is a more potent transcriptional activator than c-Jun since it (1) lacks the δ domain, which mediates the interaction of c-Jun with a histone deacetylase enzyme, and (2) is not phosphorylated at a site which targets c-Jun for degradation.

Hence, the Fos and Jun oncogene products play a critical role in the regulation of specific cellular genes in normal cells, interacting with the products of other transcription factors to produce the controlled activity of their target genes necessary for normal controlled growth.

9.3.2 V-ERBA AND THE THYROID HORMONE RECEPTOR

The v-*erbA* oncogene is one of two oncogenes carried by avian erythroblastosis virus (AEV). The cellular equivalent of this oncogene c-*erbA*, has been shown to encode the cellular receptor for thyroid hormone (Sap *et al.*, 1986; Weinberger *et al.*, 1986) which is a member of the nuclear receptor super family discussed in Chapter 4 (section 4.4). Following the binding of thyroid hormone, the receptor/hormone complex binds to its appropriate recognition site in the DNA of thyroid hormone responsive genes and activates their transcription (Fig. 9.16).

Hence, the protein encoded by the c-*erbA* gene represents a bona fide cellular transcription factor involved in the activation of thyroid hormone responsive genes. Unlike the case of the *fos* and *jun* gene products which regulate genes involved in growth, it is not immediately obvious how the form of thyroid hormone receptor encoded by the viral v-*erbA* gene can transform cells to a cancerous phenotype.

The solution to this problem is provided by a comparison of the cellular ErbA protein, which is a functional thyroid hormone receptor, and the viral ErbA protein encoded by AEV. Thus, in addition to being fused to the retroviral gag protein at its N-terminus, the viral ErbA protein contains several mutations in the regions of the receptor responsible for binding to DNA and for binding thyroid hormone as well as a small deletion in the hormone binding domain (Fig. 9.17).

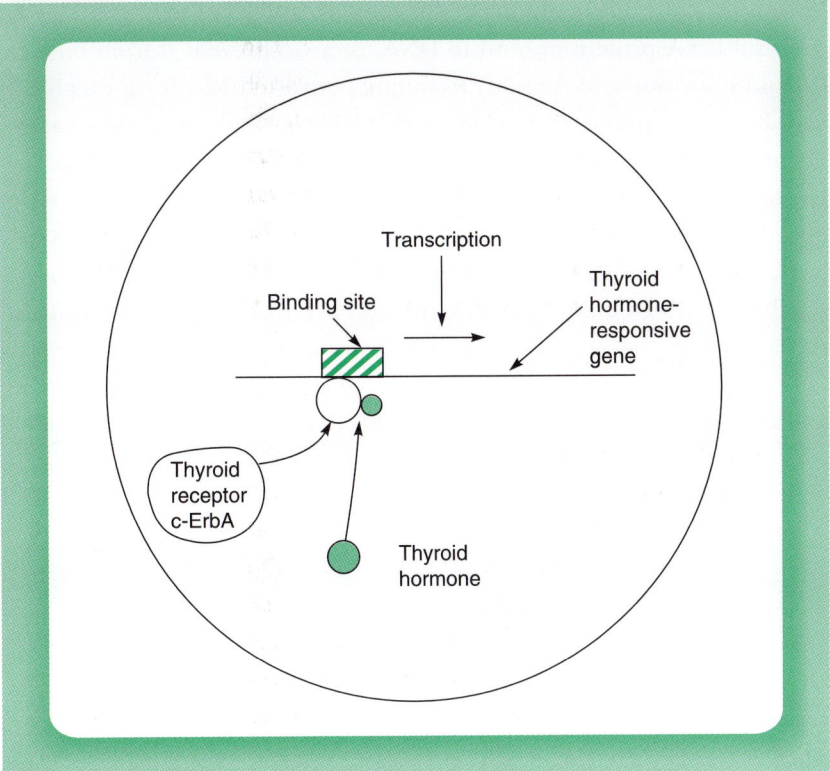

Figure 9.16
The c-*erb*A gene encodes the thyroid hormone receptor and activates transcription
in response to thyroid hormone.

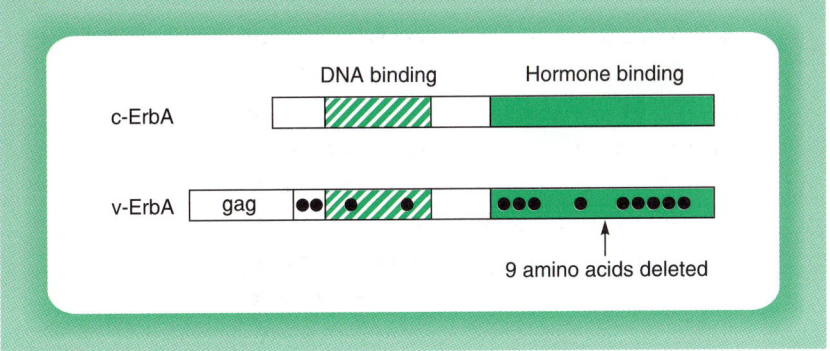

Figure 9.17
Relationship of the cellular ErbA protein and the viral protein. The black dots
indicate single amino acid differences between the two proteins while the arrow
indicates the region where nine amino acids are deleted in the viral protein.

Interestingly, although these changes do not abolish the ability of the viral ErbA protein to bind to DNA, they do prevent it from binding thyroid hormone and thereby becoming converted to a form which can activate transcription (Sap *et al.*, 1986, 1989). However, these changes do not affect the inhibitory domain, which, as discussed in Chapter 6 (section 6.3.2), allows the thyroid hormone receptor to repress transcription. Hence, the viral v-ErbA protein can inhibit the induction of thyroid hormone responsive genes when cells are treated with thyroid hormone. It achieves this by binding to the thyroid hormone response elements in the gene promoters and dominantly repressing their transcription, as well as preventing binding of the activating complex of thyroid hormone and the cellular ErbA protein (Fig. 9.18). In agreement with this critical role for repression in producing transformation by v-ErbA, a mutation in

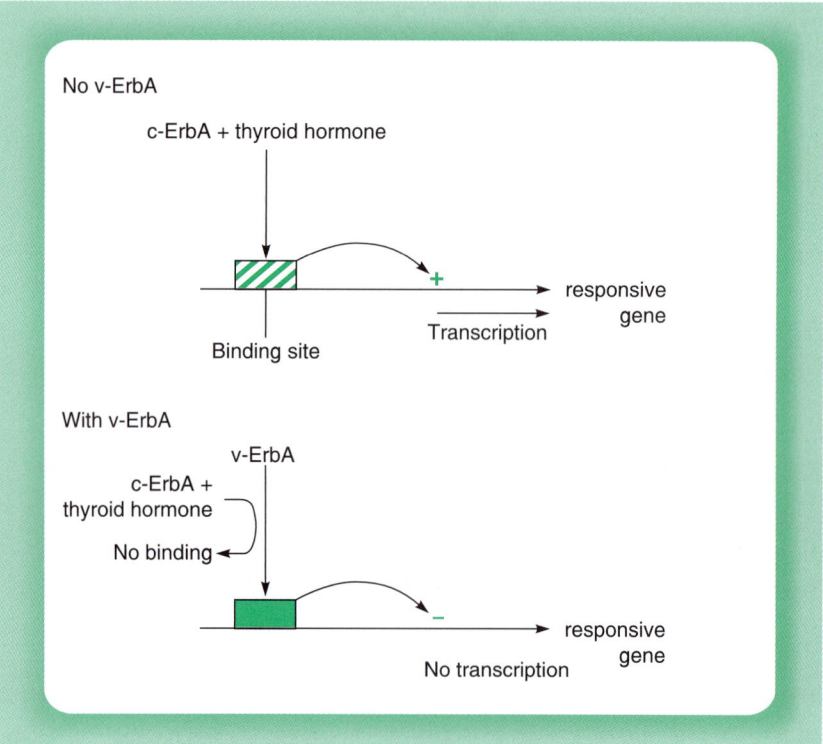

Figure 9.18

Inhibitory effect of the viral ErbA protein on gene activation by the cellular protein, in response to thyroid hormone. The viral protein both inhibits binding by the activated c-ErbA protein and also dominantly represses transcription by means of its inhibitory domain. Note the similarity to the action of the alpha-2 form of the c-ErbA protein, illustrated in Figure 6.14.

v-ErbA which abolishes its ability to repress transcription by preventing it binding its co-repressor (see Chapter 6, section 6.3.2) also abolishes its ability to transform cells (Perlmann and Vennstrom, 1995).

Hence, the viral ErbA protein acts as a dominant repressor of thyroid hormone responsive genes being both incapable of activating transcription itself and able to prevent activation by intact receptor. This mechanism of action is clearly similar to the repression of thyroid hormone responsive genes by the naturally occurring alternatively spliced form of the thyroid hormone receptor which, as discussed in Chapter 6 (section 6.3.2), lacks the hormone binding domain and therefore cannot bind hormone. Thus the same mechanism of gene repression by a non-hormone binding receptor is used naturally in the cell and by an oncogenic virus.

One of the targets for repression by the viral ErbA protein is the erythrocyte anion transporter gene, which is one of the genes normally induced when avian erythroblasts differentiate into erythrocytes. This differentiation process has been known for some time to be inhibited by the ErbA protein and it is now clear that it achieves this effect by blocking the induction of the genes needed for differentiation. In turn, such inhibition will allow continued proliferation of these cells, rendering them susceptible to transformation into a tumour cell type by the product of the other AEV oncogene v-*erbB*, which encodes a truncated form of the epidermal growth factor receptor and therefore renders cell growth independent of external growth factors (Fig. 9.19).

The two cases of Fos/Jun and ErbA therefore represent contrasting examples of the involvement of transcription factors in oncogenesis both in terms of the mechanism of transformation and the manner in which the cellular form of the oncogene becomes an active transforming gene. Thus, in the case of Fos and Jun, transformation is achieved by the continuous activation of genes necessary for growth in normal cell types. Moreover, it occurs, at least in part, via the natural activity of the cellular oncogene in inducing these genes being enhanced by their over-expression such that it occurs at an inappropriate time or place (Fig. 9.20a). In contrast, in the ErbA case, transformation is achieved by inhibiting the expression of genes whose products are required for the differentiation of a particular cell type therefore allowing growth to continue. Moreover, this occurs via the activity of a mutated form of the transcription factor which rather than carrying out its normal function more efficiently, actually interferes with the normal role of the thyroid hormone receptor in inducing thyroid hormone responsive genes required for differentiation (Fig. 9.20b).

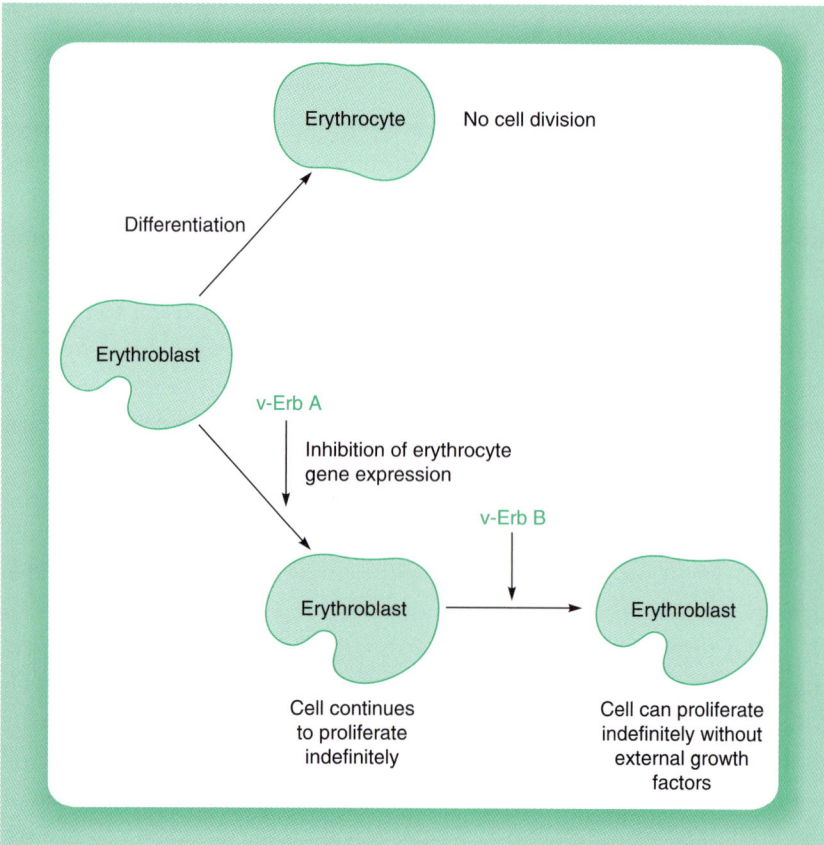

Figure 9.19

Inhibition of erythrocyte-specific gene expression by the v-ErbA protein prevents erythrocyte differentiation and allows transformation by the v-ErbB protein.

9.3.3 THE Myc ONCOGENE

Interestingly, for a considerable period, the techniques of molecular biology failed in the case of the c-*myc* oncogene, which was one of the earliest cellular oncogenes to be identified, with its expression being dramatically increased in a wide variety of transformed cells (for review see Levens, 2002; Nilsson and Cleveland, 2003; Adhikary and Eilers, 2005). Thus, the Myc protein has a number of properties suggesting that it is a transcription factor, notably nuclear localization, the possession of several motifs characteristic of transcription factors such as the helix-loop-helix and leucine zipper elements (see Chapter 4, section 4.5) and the ability to activate target promoters in co-transfection assays. Despite exhaustive efforts, however, no DNA sequence to which the Myc protein binds could be defined, rendering its mechanism of action uncertain.

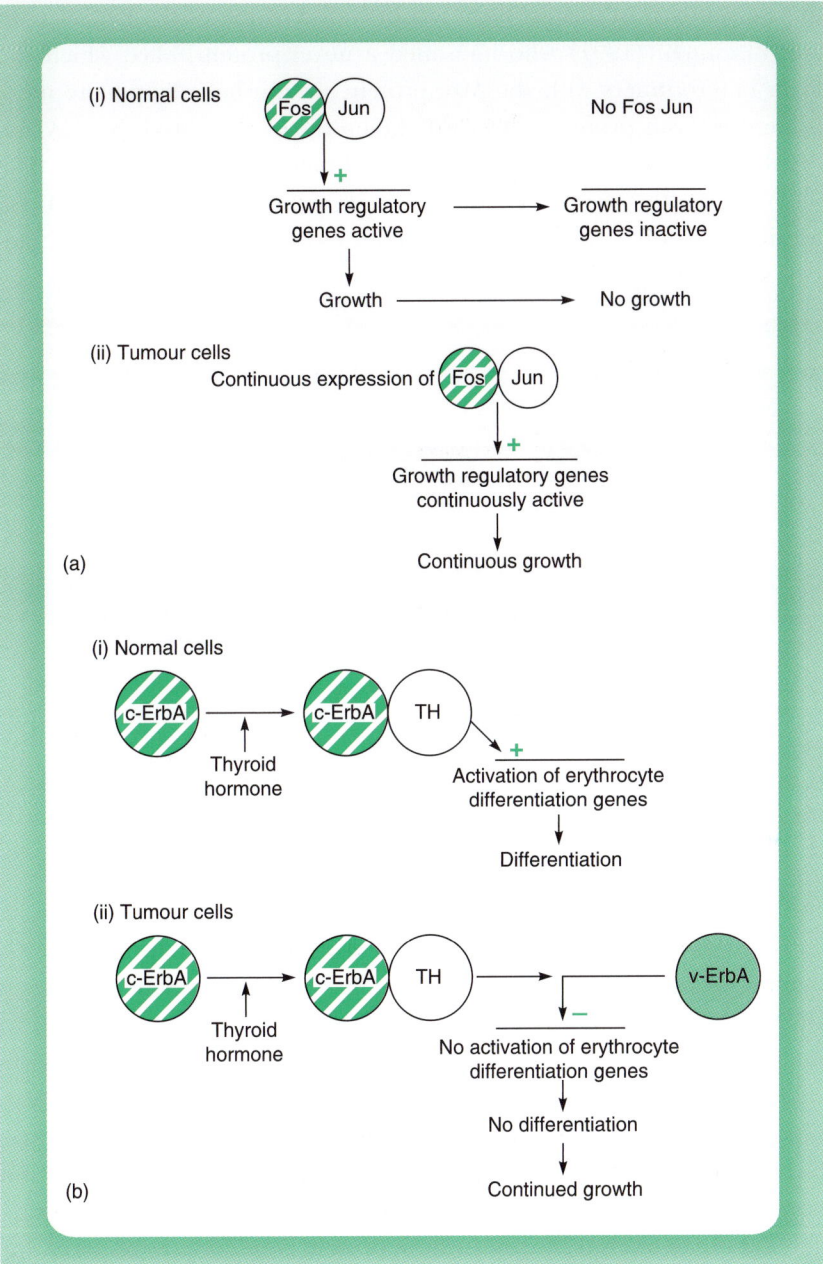

Figure 9.20

Transformation mechanisms of Fos/Jun (panel a) and ErbA (panel b). Note that Fos/Jun-induced transformation occurs because the proteins induce the continual activation of growth regulatory genes which are normally expressed only transiently whilst v-ErbA-induced transformation occurs because the protein interferes with the action of its cellular homologue and hence inhibits the induction of genes involved in erythrocyte differentiation.

The solution to this problem was provided by the work of Blackwood and Eisenman (1991) who identified a novel protein, Max, which can form heterodimers with the Myc protein via the helix-loop-helix motif present in both proteins. Myc/Max heterodimers can bind to DNA and regulate transcription, whereas Myc/Myc homodimers cannot do so (Fig. 9.21). This effect evidently parallels the requirement of the Fos protein for dimerization with Jun in order to bind with high affinity to AP1 sites (see section 9.3.1).

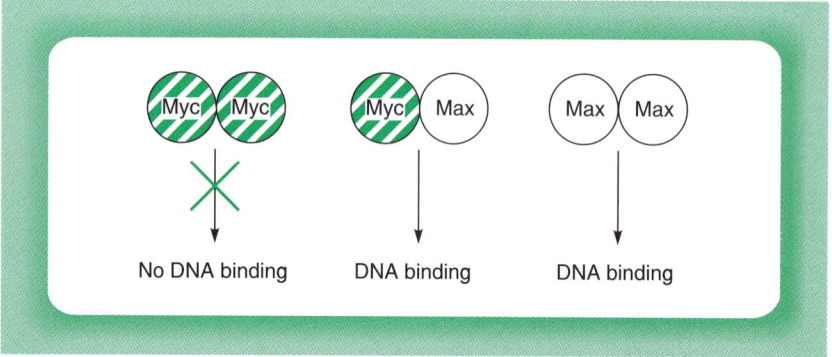

Figure 9.21

Both Myc/Max heterodimers and Max/Max homodimers can bind to DNA whereas Myc/Myc homodimers cannot.

The Max protein therefore plays a critical role in allowing the DNA binding of Myc, and the structure of a Myc/Max heterodimer bound to DNA has been defined (Nair and Burley, 2003). Moreover, the ability to interact with Max, bind to DNA and modulate gene expression is critical for the ability of the Myc protein to transform since mutations in Myc which abolish its ability to heterodimerize with Max also abolish its transforming ability. Hence, as was previously speculated, the Myc protein is a transcription factor whose over-expression causes transformation presumably via the activation of genes whose protein products are required for cellular growth (for reviews see Levens, 2002; Adhikary and Eilers, 2005). Indeed, a recent global analysis, using ChIP methods (Chapter 2, section 2.4.3), identified 668 genes regulated by Myc, including 48 encoding other transcription factors (Zeller *et al.*, 2006). This indicates that Myc is a major regulator of genes involved in cellular growth, including many which act as transcriptional regulators of other, downstream genes.

Interestingly, as well as modulating protein-coding genes, Myc can also induce a number of different small RNAs, which, as discussed in

Chapter 6 (section 6.4.2), act to inhibit the expression of target genes. Hence, Myc acts both directly and indirectly to regulate gene expression (Fig. 9.22) (for reviews see Esquela-Kerscher and Slack, 2006). One of the target genes inhibited by a Myc-induced small RNA in this way is that encoding the E2F transcription factor. As discussed in section 9.4.3, E2F induces the expression of the c-*myc* gene. Hence, the inhibition of E2F by a Myc-induced small RNA constitutes a negative feedback loop in which E2F induces Myc expression and E2F expression is then in turn inhibited by the Myc-induced small RNA (Fig. 9.23) (for reviews see Meltzer, 2005; Esquela-Kerscher and Slack, 2006).

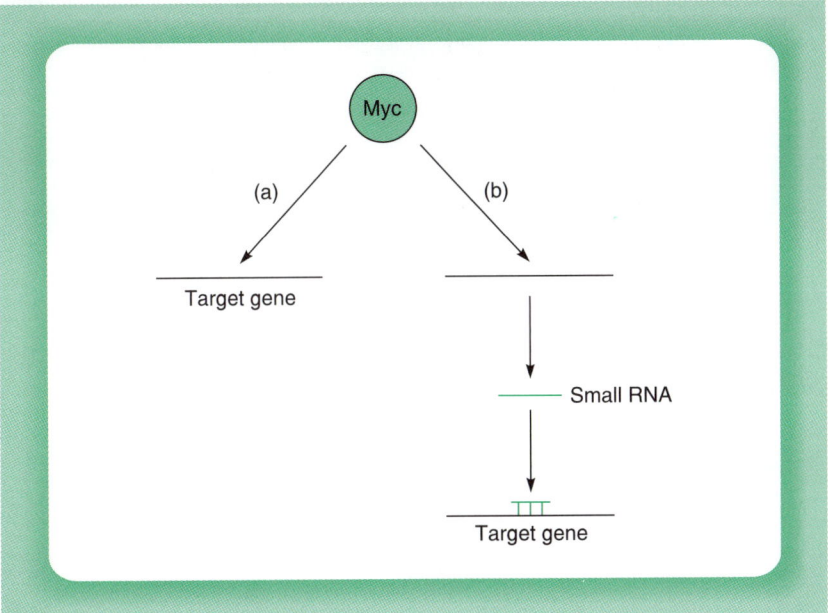

Figure 9.22

The Myc oncogene can regulate gene expression directly (a) but can also act indirectly (b) by inducing the expression of small RNAs, which inhibit the expression of specific genes.

Interestingly, the Max protein does not appear to represent a passive partner which merely serves to deliver Myc to the DNA of target genes. Rather, it plays a key role in regulating the activity of target genes containing the appropriate binding site. Thus, it has been shown that whereas Myc/Max heterodimers can activate transcription, Max/Max homodimers can bind to the same site and weakly repress transcription. Moreover, Max can also heterodimerize with another member of the helix-loop-helix family, known as Mad, to form a strong repressor of transcription (Fig. 9.24).

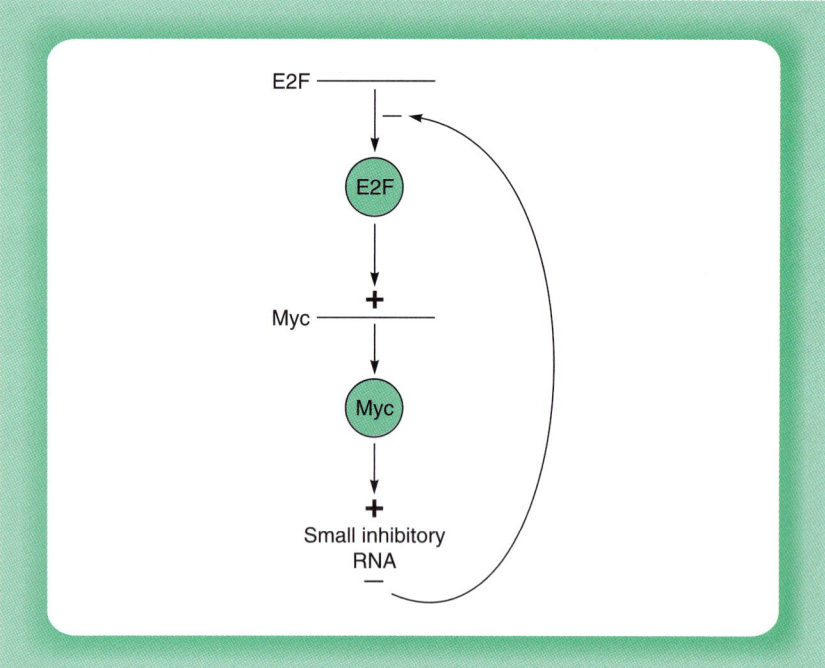

Figure 9.23
The E2F transcription factor induces expression of the c-*myc* gene. In turn the Myc transcription factor induces a small inhibitory RNA, which inhibits E2F gene expression, producing a negative feedback loop.

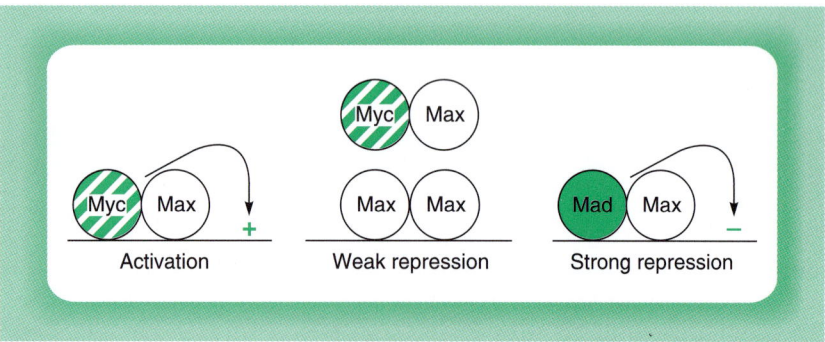

Figure 9.24
Functional effects of Max/Max homodimers and of Myc/Max or Mad/Max heterodimers. Note that Max/Max homodimers repress transcription only weakly by passively blocking activator binding whereas Mad/Max heterodimers actively repress transcription and therefore have a much stronger effect.

The Max/Max homodimer appears to act as a weak repressor simply by preventing the Myc/Max activator from binding to its appropriate binding sites and thereby preventing it from activating transcription. In contrast, the Mad/Max heterodimer appears to act as an active repressor, which is capable of reducing transcription below that which would be observed in the absence of any activator binding (see Chapter 6, for discussion of the mechanisms of transcriptional repression). Thus, it has been shown that the Mad protein can bind the same co-repressor complex of N-CoR, mSIN-3 and mRPD3, which mediates active repression by nuclear receptors such as the thyroid hormone receptor in the absence of hormone (see Chapter 6, section 6.3.2) (for review see Wolffe, 1997). As this complex includes the mRPD3 protein which has histone deacetylase activity, it is possible that the Mad/Max heterodimer may repress transcription, at least in part, by recruiting a complex which deacetylates histones thereby organizing a more tightly packed chromatin structure (Fig. 9.25).

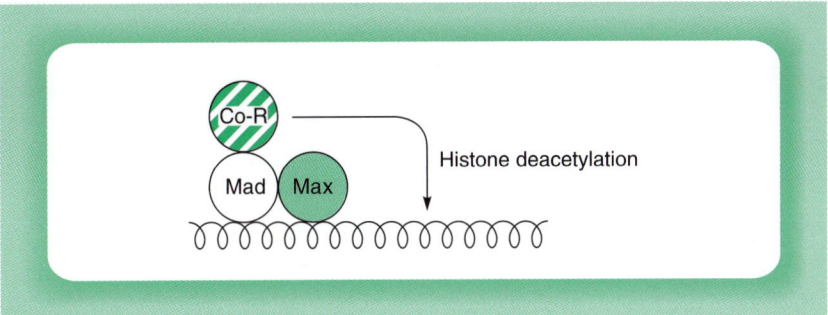

Figure 9.25

The Max/Mad heterodimer can recruit a co-repressor complex (Co-R) with histone deacetylase activity which can produce a more tightly packed chromatin structure (compare with Fig. 6.30).

In the case of the nuclear receptors, the switch from the repressed state of target genes to their activation is mediated by the addition of hormone. In contrast, however, in the case of the Myc family it is mediated by signals which produce a rise in Myc expression and a corresponding fall in the expression of Mad. Thus, Myc is expressed at very low levels in resting cells and its expression is induced when cells begin to grow, whereas Max is expressed at similar high levels in both resting and proliferating cells and Mad is expressed at high levels only in resting cells and not in proliferating cells. Hence, in resting cells Mad and not Myc will be expressed and the expression of Myc-dependent genes will be repressed by Mad/Max homodimers. In contrast, expression will be activated by Myc/Max heterodimers as the cells receive signals to proliferate, resulting in increased Myc expression and decreased Mad expression (Fig. 9.26). Clearly, the over-expression of the Myc gene which is observed in many cancer cells would result in a similar production of activating Myc/Max heterodimers leading to gene activation. Hence, as in the case of the Fos/Jun system, transformation by the Myc oncogene appears to depend primarily on its over-expression resulting in the activation of genes required for cellular growth.

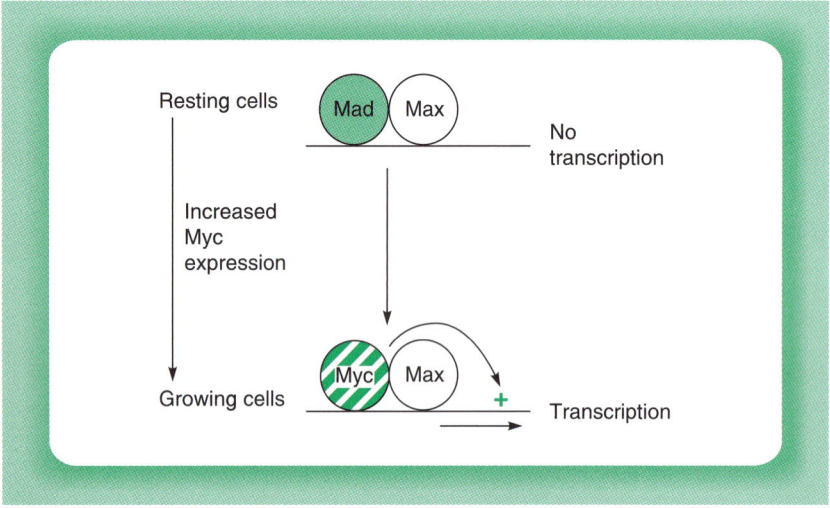

Figure 9.26
In resting cells, Myc-dependent genes will be repressed by a Mad/Max heterodimer. As cells begin to grow, the expression of Myc increases, resulting in the formation of Myc/Max heterodimers which activate transcription.

Interestingly, it has been shown that Myc can also interact with another transcription factor Miz-1 (Myc interacting zinc finger protein-1), which is a zinc finger protein (see Chapter 4, section 4.3 for discussion of this type of transcription factor). Unlike the situation with Max, however, in the absence of Myc, Miz-1 recruits the p300 co-activator (see Chapter 5, section 5.4.3) and therefore acts as an activator of genes promoting growth arrest. In the presence of Myc, however, this recruitment of p300 is blocked and instead the complex recruits a DNA methyltransferase enzyme, which methylates the DNA and therefore inhibits transcription (Fig. 9.27) (for reviews see Adhikary and Eilers, 2005; Gartel, 2006) (for discussion of DNA methylation as a means of inhibiting transcription see Chapter 6, section 6.4.2 and Latchman, 2005). Hence the rise in Myc levels in transformed cells stimulates the activity of growth promoting genes via Myc/Max-mediated gene activation and represses growth inhibitory genes via a repression of Miz-1 activity.

As well as regulating growth by altering the transcription of specific protein-coding genes by RNA polymerase II, another means by which Myc can alter growth has been demonstrated. Thus, it has been shown

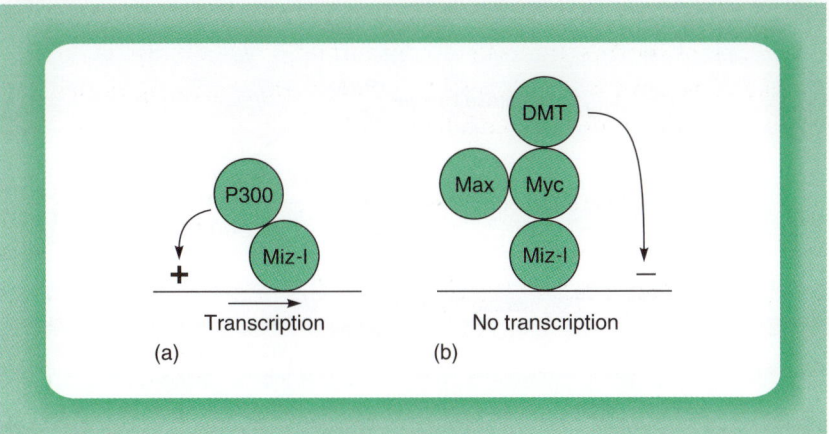

Figure 9.27

The Miz-1 transcription factor binds the p300 co-activator and therefore activates transcription (panel a). However, in the presence of the Myc/Max complex, p300 is not recruited and instead the complex recruits a DNA methyltransferase (DMT) which methylates the DNA and blocks transcription (panel b).

that Myc interacts with the TFIIIB transcription factor which is essential for transcription by RNA polymerase III (see Chapter 3, section 3.4) and stimulates the transcription of the genes encoding tRNA and 5S ribosomal RNA. Similarly, Myc can also enhance the transcription of the ribosomal RNA genes by RNA polymerase I (for review see Oskarsson and Trump, 2005). Hence, Myc can influence transcription by all three RNA polymerases. Moreover, the RNA polymerase II-transcribed genes induced by Myc include those encoding ribosomal proteins. Therefore, in the case of each of the three RNA polymerases, Myc acts to induce proteins and RNAs which are essential for protein synthesis by the ribosome and hence for cellular growth, and this provides a further means by which transcriptional regulation by Myc can regulate cellular growth (Fig. 9.28).

9.3.4 OTHER ONCOGENIC TRANSCRIPTION FACTORS

In view of the likely need for multiple different trancription factors to regulate genes involved in cellular growth processes, it is not surprising that several other genes encoding transcription factors have also been identified as oncogenes, as well as playing a key role in gene expression in specific cell types. Thus, for example, the *myb* oncogene and the *maf* oncogene, both of which were originally isolated from avian retroviruses, play key roles in gene regulation in monocytes and erythroid cells respectively (for reviews see Blank and Andrews, 1997; Motohashi *et al.*, 1997).

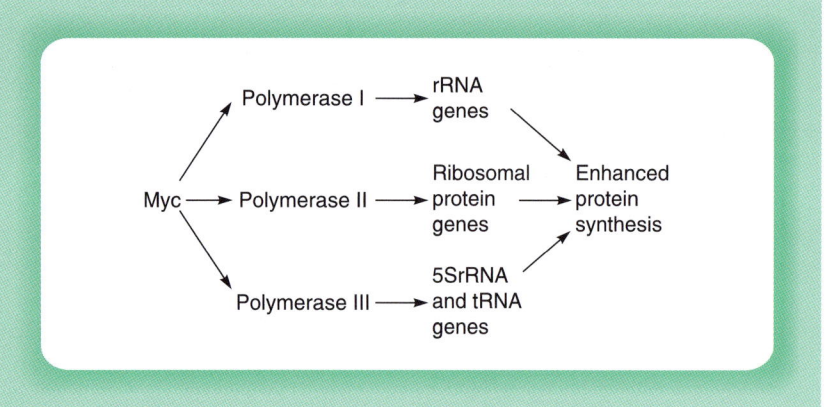

Figure 9.28

Myc stimulates total protein synthesis and therefore enhances cellular growth by stimulating the transcription of the genes encoding (a) the major ribosomal RNAs (by RNA polymerase I), (b) the genes encoding ribosomal proteins (by RNA polymerase II) and (c) the genes encoding tRNAs and the 5S ribosomal RNA (by RNA polymerase III).

Similarly, the rel oncogene of the avian retrovirus Rev-T is a member of the NFκB family of transcription factors discussed in Chapter 8 whilst the Bcl-3 oncogene is a member of the IκB family which interacts with the NFκB proteins (for reviews of the involvement of NFκB family members in cancer see Viatour *et al.*, 2005; Basseres and Baldwin, 2006; Courtois and Gilmore, 2006; Karin, 2006).

As well as these oncogenic transcription factors, cancer can also be caused by mutations in factors which regulate chromatin structure (for review see Gregory and Shiekhattar, 2004). Thus, for example, GASC, a gene which is over-expressed in squamous cell carcinoma, demethylates histone H3 on lysine 9 (Cloos *et al.*, 2006), thereby promoting a more open chromatin structure (see Chapter 6, section 6.4.1).

Interestingly, the Bcl-3 factor illustrates another facet of the mechanisms by which transcription factor genes become oncogenic. Thus this factor was not identified as a retroviral oncogene but on the basis that it was located at the break point of chromosomal rearrangements, which resulted in its translocation to a position adjacent to the immunoglobulin gene in some B cell chronic leukaemias. A number of other transcription factors have also been shown to be capable of causing cancer when translocated in this way. This can occur because their expression is increased due to their being translocated to a highly expressed locus, such as the immunoglobulin gene loci in B cells or the T cell receptor gene loci in T cells (Fig. 9.29a). Alternatively, it can occur because the translocation results in the production of a novel form of the transcription factor due to its truncation or its linkage to another gene (encoding either another transcription factor or another class of protein) following the translocation (Fig. 9.29b).

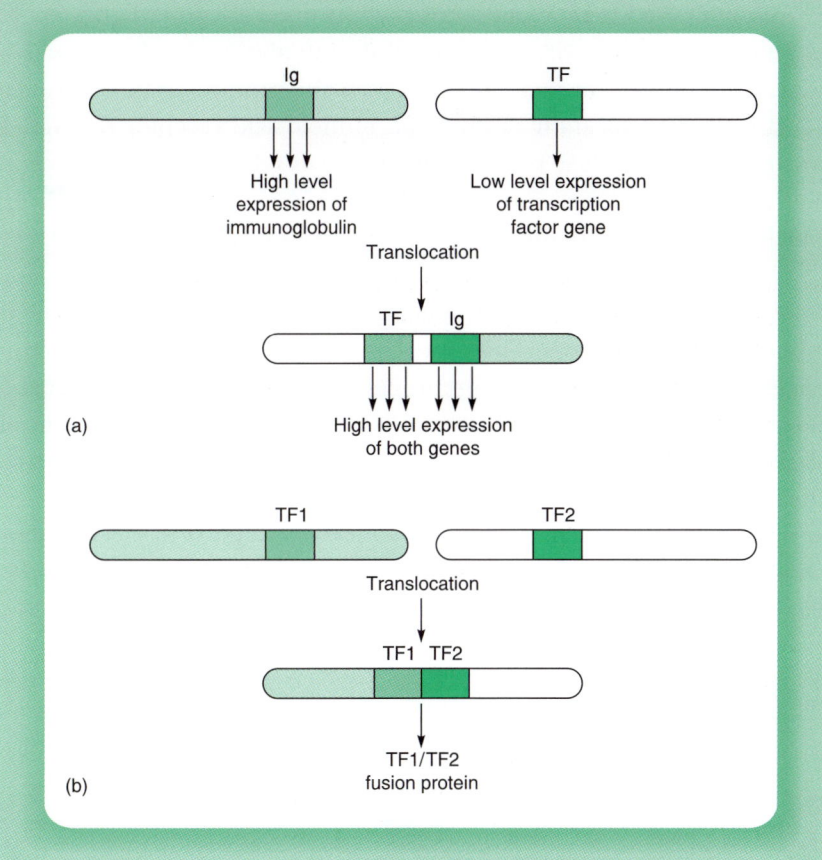

Figure 9.29

Chromosomal translocations can result in cancer when (a) the gene encoding a transcription factor is translocated next to a highly transcribed locus such as the immunoglobulin gene (Ig) and is therefore expressed at a high level, or (b) the translocation results in the fusion of the genes encoding two different transcription factors, resulting in a fusion protein with oncogenic properties.

Factors translocated in these ways include both factors which were originally identified in oncogenic retroviruses and others which had not previously been shown to have oncogenic potential (for reviews see Rabbits, 1994; Latchman, 1996; Look, 1997). Thus, for example, expression of the c-*myc* oncogene (section 9.3.3) is dramatically increased by its translocation into the immunoglobulin heavy chain locus which occurs in the human B cell malignancy known as Burkitts lymphoma (for review see Spencer and Groudine, 1991) whilst the gene encoding the Ets transcription factor discussed above (section 9.3.1) is fused to the gene for the platelet derived growth factor receptor to create a novel oncogenic fusion protein in patients with chronic myelomonocytic leukaemia (for review see Sawyers and Denny, 1994).

Although the Ets gene was originally identified due to its being found in an oncogenic retrovirus, translocations can also involve factors which were not identified in this way. Thus, in acute promyelocytic leukaemia (PML), the translocation involves the retinoic acid receptor α gene (a member of the nuclear receptor family discussed in Chapter 4, section 4.4) and the PML transcription factor, a zinc finger transcription factor, which was originally identified on the basis of its involvement in PML.

The RAR–PML fusion protein produced by this translocation, acts as a transcriptional repressor, even though the retinoic acid receptor alone is a transcriptional activator. This involves the ability of RAR–PML to recruit histone deacetylases which can produce an inactive chromatin structure (for review see Lin *et al.*, 2001) (Fig. 9.30). Indeed, such enhanced deacetylation is also induced by other oncogenic fusion proteins, such as the AML–ETO fusion protein found in 15% of acute myeloid leukaemias, suggesting that the altered chromatin structure produced in this way plays an important role in human leukaemias (for reviews see Minucci *et al.*, 2001; Hake *et al.*, 2004) and may be a target for therapy (see section 9.5).

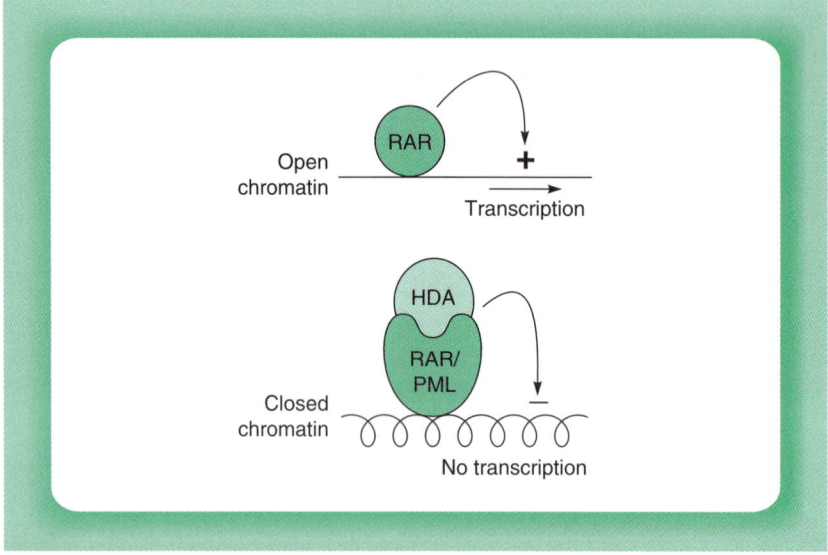

Figure 9.30

The retinoic acid receptor (RAR) is an activator of transcription (panel a). However, the RAR/PML fusion protein recruits a histone deacetylase (HAD), which produces an inactive chromatin and is therefore a repressor of transcription (panel b).

These findings provide further evidence that transcription factor genes are not only rendered oncogenic by transfer into a retrovirus but are also involved in the causation of human cancers playing a key role, for example, in the oncogenic effects of the chromosome translocations which are characteristic of specific cancers. Indeed, whilst such translocations are most widespread in leukaemias, they have also been detected in human solid tumours such as prostate cancer (for review see Marx, 2005).

9.4 ANTI-ONCOGENES AND CANCER

9.4.1 NATURE OF ANTI-ONCOGENES

As noted in section 9.2, a number of genes exist whose normal function is to encode proteins that function in an opposite manner to those of onco-genes, acting to restrain cellular growth. The deletion or mutational inactivation of these anti-oncogenes (also known as tumour suppressor genes) therefore results in cancer (for reviews see Hunter, 1997; Lowe *et al.*, 2006; Vogelstein and Kinzler, 2006) (Fig. 9.31). This effect evidently parallels the production of cancer by the over-expression or mutational activation of cellular proto-oncogenes (compare Figs. 9.7 and 9.31).

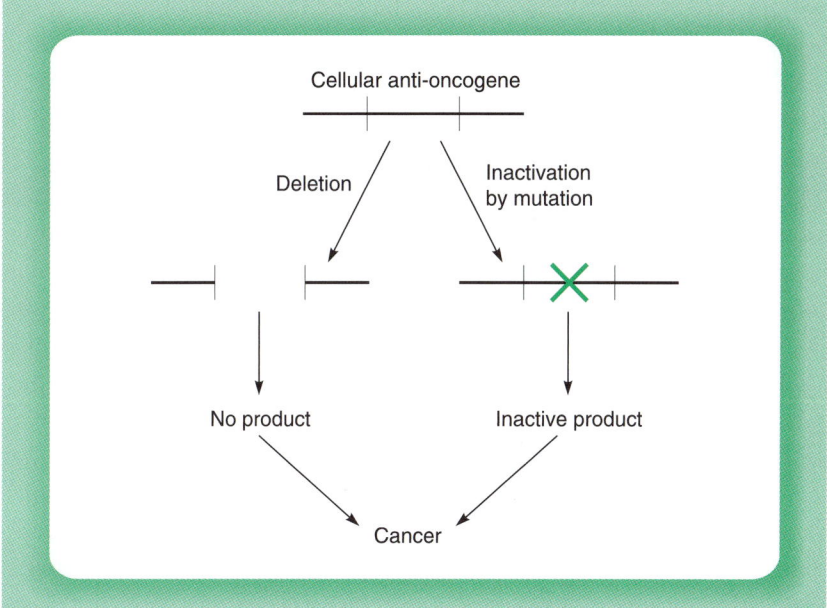

Figure 9.31

Cancer can result from the deletion of specific anti-oncogenes or their inactivation by mutation.

A number of anti-oncogenes of this type have been defined and several encode transcription factors. The two best characterized of these act by different mechanisms. Thus, p53 acts by binding to the DNA of its target genes and regulating their expression whereas the retinoblastoma gene product (Rb-1) acts primarily via protein–protein interactions with other DNA binding transcription factors. The p53 and Rb-1 proteins are therefore discussed in sections 9.4.2 and 9.4.3 as examples of these two mechanisms of action. Other anti-oncogenes encoding transcription factors are discussed in section 9.4.4.

9.4.2 *P53*

The gene encoding the 53 kilo-dalton protein known as p53 is mutated in a very wide variety of human tumours, especially carcinomas (for review see Vogelstein *et al.*, 2000; Haupt *et al.*, 2002; Sharpless and DePinho, 2002; Mills, 2005). In normal cells, expression of this protein is induced by agents which cause DNA damage and its over-expression results in growth arrest of cells containing such damage or their death by the process of programmed cell death (apoptosis). Hence, p53 has been called the 'guardian of the genome' (Lane, 1992), which allows cells to proliferate only if they have intact undamaged DNA. This would prevent the development of tumours containing cells with mutations in their DNA and the inactivation of the p53 gene by mutation would therefore result in an enhanced rate of tumour formation. In agreement with this idea, mice in which the p53 gene has been inactivated do not show any gross abnormalities in normal development but do exhibit a very high rate of tumour formation leading to early death (for review see Berns, 1994).

The molecular analysis of the p53 gene product showed that it contains a DNA binding domain and a region capable of activating transcription. The majority of the mutations in p53 which occur in human tumours are located in the DNA binding domain. These mutations result in a failure of the mutant p53 protein to bind to DNA indicating that this ability is crucial for the ability of the normal p53 protein to regulate cellular growth and suppress cancer.

The p53 protein therefore functions, at least in part, by activating the expression of genes whose proteins products act to inhibit cellular growth (Fig. 9.32a). The absence of functional p53 either due to gene deletion (Fig. 9.32b) or to its inactivation by mutation (Fig. 9.32c) results in a failure to express these genes leading to uncontrolled growth.

Interestingly, some p53 mutations have a dominant negative effect, interfering with the activity of the wild type protein. Unlike inactivating mutations, these dominant negative mutations can therefore produce

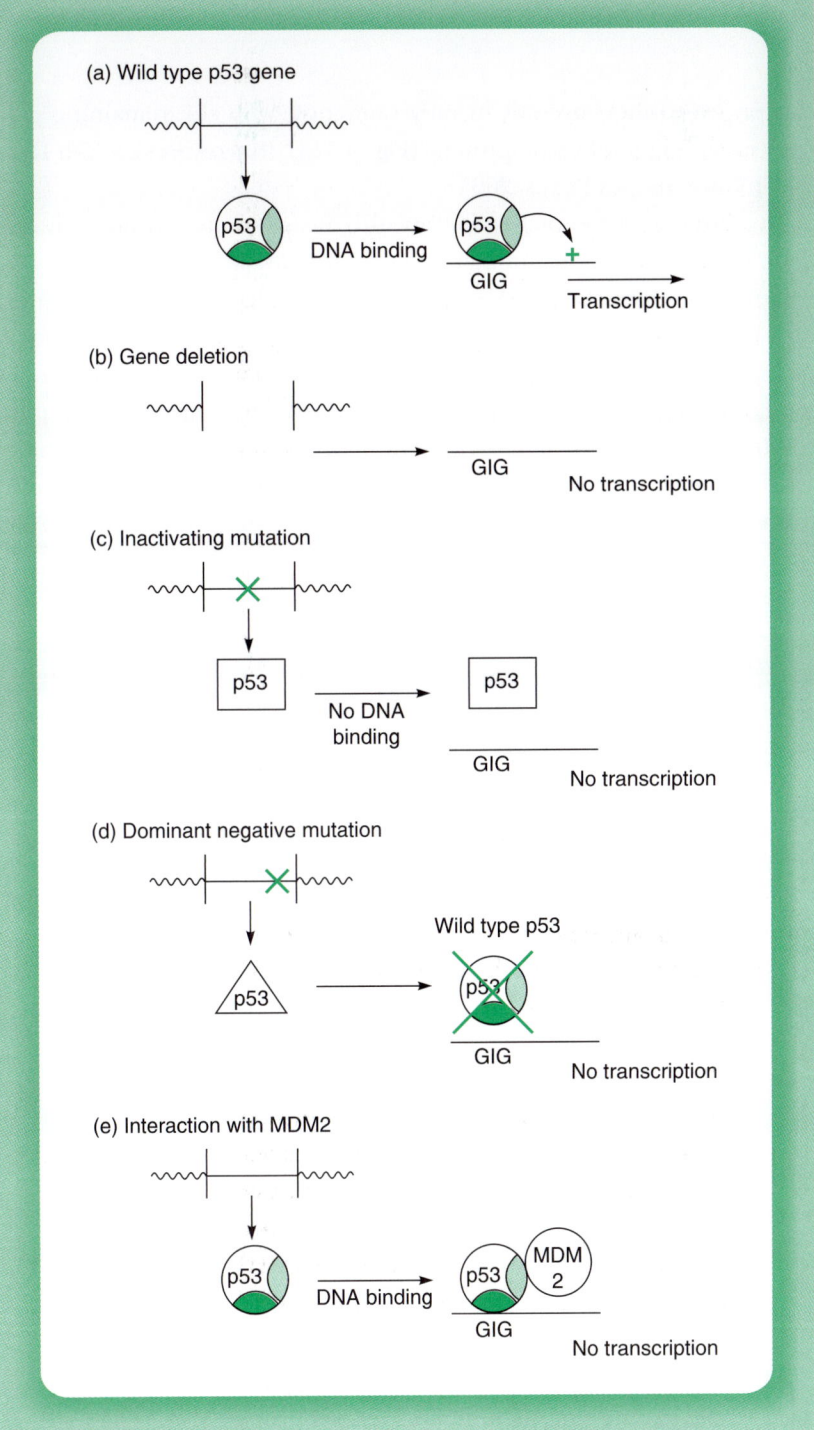

Figure 9.32

The ability of wild type p53 to activate genes encoding growth inhibiting proteins (GIG) (panel a) can be abolished by deletion of the p53 gene (panel b), inactivating mutations in the DNA binding domain (solid) which prevent it binding to DNA (panel c) dominant negative mutations in p53, which inhibit the wild type p53 protein (panel d) or by the interaction of functional p53 with the MDM2 protein which prevents it from activating transcription (panel e).

cancer, even when present in only one copy, with the remaining gene copy encoding a wild type protein (Fig. 9.32d) (for reviews see van Dyke, 2005; Vousden and Prives, 2005).

In addition, functional p53 can also be prevented from activating gene transcription by interaction with the MDM2 oncoprotein (Fig. 9.32e). The interaction of MDM2 with p53 results in the rapid degradation of p53. Thus, MDM2 causes the addition of ubiquitin residues to p53, thereby promoting its degradation (Haupt *et al.*, 1997; for review see Lane and Hall, 1997). Interestingly, whilst addition of two ubiquitin residues to p53 results in its degradation, the addition of a single ubiquitin residue results in its export from the nucleus, with either mechanism preventing p53 from activating its target genes (Fig. 9.33) (Li *et al.*, 2005; for review see Yang *et al.*, 2004).

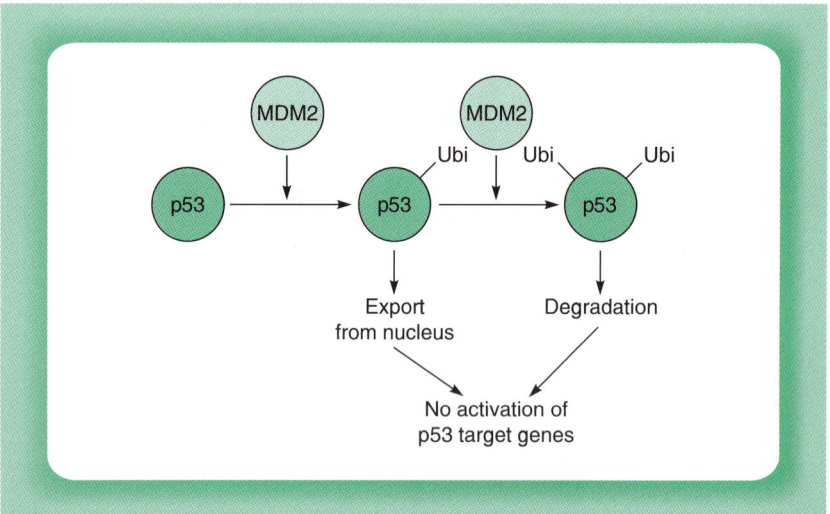

Figure 9.33
Addition of a single ubiquitin residue to p53 results in nuclear export of p53 whilst the addition of two ubiquitin residues results in degradation of p53. In either case p53 can no longer activate its target genes.

The addition of ubiquitin to p53, targeting it for degradation, is paralleled by the addition of the ubiquitin-related protein SUMO-1 to MDM2. This modification of MDM2 paradoxically enhances its ability to add ubiquitin to p53 and thereby induce p53 degradation (Buschmann *et al.*, 2000) (see Chapter 8, section 8.4.5 for discussion of the regulation of transcription factors by the addition of ubiquitin or SUMO-1).

The inhibitory effect of MDM2 on p53 is of particular importance in many human soft tissue sarcomas, where the p53 gene is intact and encodes wild type p53 but the protein is functionally inactivated due to the high levels of MDM2 resulting from amplification of the *mdm2* gene encoding it. Indeed the major function of MDM2 even in normal cells may be to inhibit the action of p53 by interacting with it. Thus, mice in which the gene encoding MDM2 is inactivated are non-viable but can be rendered viable by the additional inactivation of the p53 gene (de OcaLuna *et al.*, 1995). Interestingly, different human individuals have different sequences in the promoter driving MDM2 gene expression. This results in a different affinity for the transcription factor Sp1 and hence different levels of MDM2 gene expression. In turn, individuals with higher levels of MDM2 gene expression show an enhanced risk of cancer, further emphasizing the importance of the p53/MDM2 interaction (for review see Vousden and Prives, 2005).

Both partners in the p53/MDM2 interaction are subject to modification by phosphorylation and these modifications affect their interaction with one another (for reviews see Prives, 1998; Mayo and Donner, 2002). Thus, following exposure to DNA damage/stress, p53 is phosphorylated. This enhances its retention in the nucleus and inhibits its interaction with MDM2, so allowing it to activate transcription of its target genes. Conversely, stimuli which inhibit apoptosis, lead to phosphorylation of MDM2. This promotes its movement from the cytoplasm to the nucleus and its association with p53 and hence allows it to inhibit p53 and its pro-apoptotic effect (Fig. 9.34) (for review see Gottifredi and Prives, 2001).

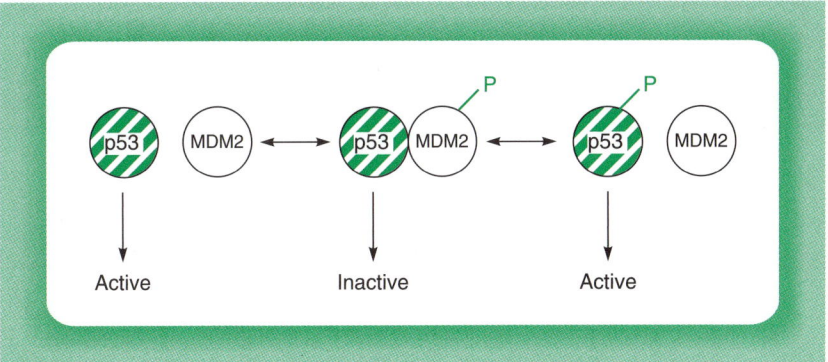

Figure 9.34

The interaction of p53 and MDM2, which inactivates p53, is promoted by phosphorylation of MDM2 but inhibited by phosphorylation of p53.

As well as affecting binding to MDM2, phosphorylation also enhances the binding of p53 to the Pin1 protein (Zheng *et al.*, 2002). Pin1 is a member of the class of proteins known as peptidyl prolyl isomerases, which have the ability to change the structure of the peptide bond between proline residues and adjacent amino acids in a process known as cis-trans isomerization. In the case of p53, interaction with Pin1 and the consequent isomerization of peptide bonds within the p53 protein, stimulates the DNA binding and transactivation ability of p53. This therefore provides a second mechanism for phosphorylation to stimulate the activity of p53 (Fig. 9.35) (for further details of the effect of phosphorylation on transcription factors see Chapter 8, section 8.4.2).

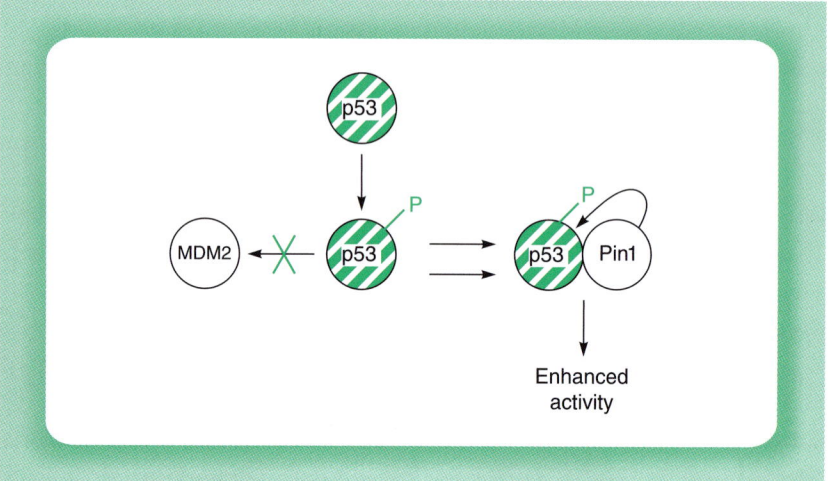

Figure 9.35
Phosphorylation of p53 blocks its interaction with MDM2, so stabilizing the protein, and also enhances its interaction with the peptidyl prolyl isomerase Pin1, which stimulates the activity of p53.

It should be noted that this effect of a prolyl isomerase enzyme on a transcription factor is not unique to p53. Thus, DNA binding activity of the c-*myb* proto-oncogene protein (see section 9.3.4) has been shown to be negatively regulated by its interaction with the peptidyl prolyl isomerase, Cyp40 (for review see Hunter, 1997). Similarly, proline isomerization of histone H3 has been shown to inhibit H3 methylation on lysine 36 and thereby modulates chromatin structure (Nelson *et al.*, 2006; for review see Amoils, 2006).

Hence, signals such as DNA damage/stress can activate p53 by inducing its phosphorylation. As noted in Chapter 8 (section 8.4.3), p53 is also subject to acetylation, which stimulates its DNA binding activity (for review see Prives and Manley, 2001; Mayo and Donner, 2002). It has been demonstrated that histone deacetylase enzymes such as Sir2 can specifically deacetylate p53, thereby reducing its ability to activate transcription (Luo *et al.*, 2001; Vaziri *et al.*, 2001). Indeed, it appears that MDM2 exists in a complex with a histone deacetylase enzyme and that deacetylation of p53 actually enhances its degradation by MDM2 (Ito *et al.*, 2002; Li *et al.*, 2002). Hence, this inhibitory complex can deacetylate p53, reducing its activity and targeting it for degradation by MDM2 (Fig. 9.36). A similar deacetylation of p53 is also produced by the RAR–PML fusion protein discussed in section 9.3.4, providing an example of an anti-oncogene protein being targeted by an oncogenic fusion protein (Insinga *et al.*, 2004).

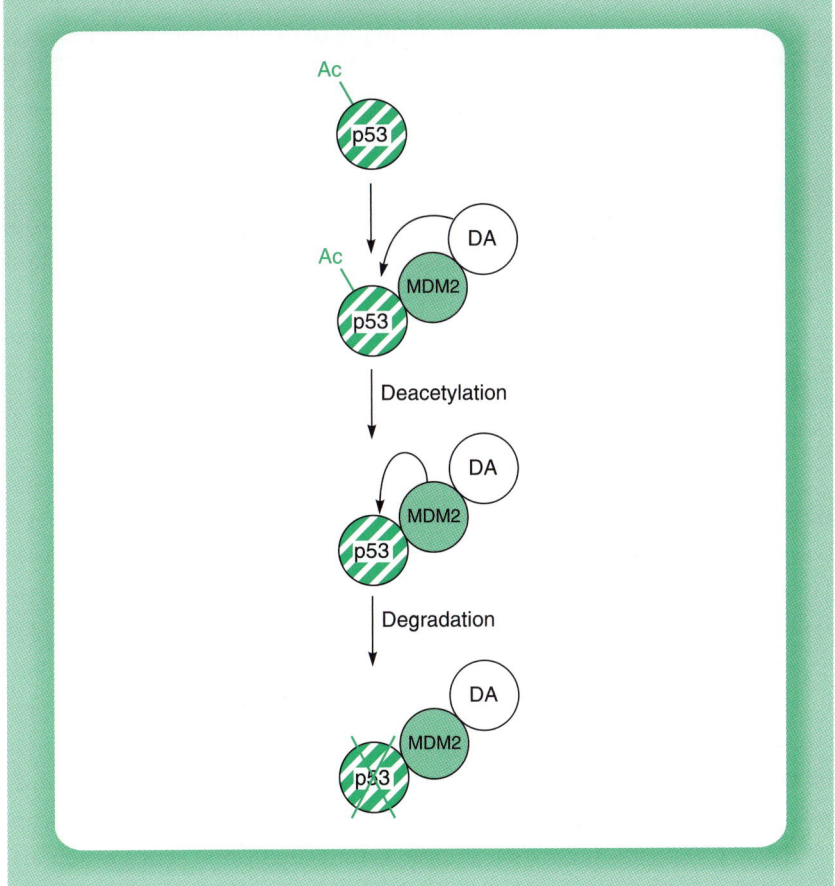

Figure 9.36

Binding to p53 of a complex of MDM2 and a histone deacetylase enzyme (DA) results in deacetylation of p53 which promotes its subsequent degradation by MDM2.

The activity of p53 is thus regulated, in part, by the balance between its acetylation by molecules such as the p300 co-activator (as discussed in Chapter 8, section 8.4.3) and its deacetylation by molecules such as Sir2 and RAR/PML. In addition, p53 is also modified by methylation on specific lysine residues. As with the histones (see Chapter 1, section 1.2.3), methylation of p53 can have opposite effects depending on the specific amino acid involved. Thus, methylation of lysine 370 represses p53 activity (Huang *et al.*, 2006) whilst methylation of lysine 372 stimulates its nuclear localization and enhances its stability, thereby increasing its activity (Chuikov *et al.*, 2004). Similarly, p53 can be modified by glycosylation, which enhances its stability by reducing its ubiquitination (Yang *et al.*, 2006).

Hence, multiple modifications are used to alter the activity of the p53/MDM2 system and these include many of the modifications that can affect transcription factors, such as phosphorylation, acetylation, methylation and modification by addition of ubiquitin or SUMO-1 (see Chapter 8, section 8.4 for a discussion of the modulation of transcription factor activity by post-translational modifications).

The interaction of p53 with the MDM2 oncogenic protein is paralleled by its interaction with another inhibitory cellular protein, MDM4 (also known as MDMX). As discussed in Chapter 6 (section 6.2.3), this interaction masks the activation domain of p53 and therefore prevents it activating transcription. It therefore reduces the activity of any p53 which is present, thereby complementing the enhanced degradation of p53 produced by MDM2 (for review see Toledo and Wahl, 2006).

As well as interacting with cellular proteins, p53 also interacts with the transforming proteins of several DNA viruses. Indeed, p53 was originally discovered as a protein which interacted with the large T oncoprotein of the DNA tumour virus SV40 (for review see Liu and Marmorstein, 2006). The functional inactivation of p53 produced by this interaction appears to play a critical role in the ability of these DNA viruses to transform cells, paralleling the similar action of MDM2. These interactions suggest that functional antagonism between oncogene and anti-oncogene products is likely to be critical for the control of cellular growth with changes in this balance, which activate oncogenes or inactivate anti-oncogenes, resulting in cancer.

These considerations evidently focus attention on the genes which are activated by p53. One such gene is that encoding a 21 kilo-dalton protein (p21), which acts as an inhibitor of cyclin-dependent kinases (for review of p53-dependent genes see Vogelstein *et al.*, 2000). As the cyclin-dependent kinases are enzymes that stimulate cells to enter cell division, the finding that p53 stimulates the expression of an inhibitor of these enzymes is entirely consistent with its role in restraining growth, since the inhibition of the cyclin-dependent kinases will prevent cells replicating their DNA and undergoing cell division (Fig. 9.37).

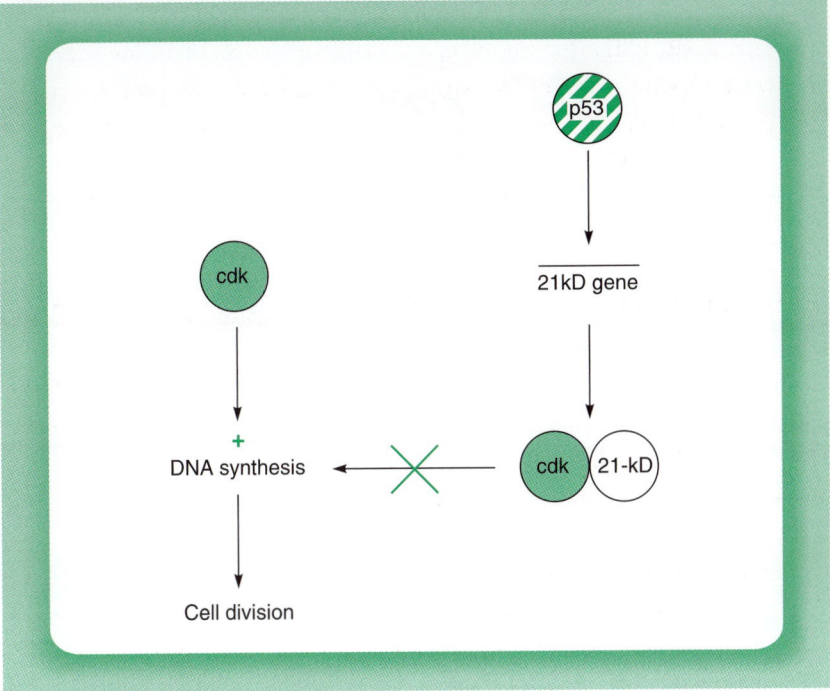

Figure 9.37

p53 activates the gene for the 21-kD inhibitor of cyclic dependent kinases (cdk). This inhibitor then prevents the cyclin-dependent kinases from stimulating DNA synthesis and consequent cell division.

As well as enhancing p21 gene expression, p53 also stimulates expression of the *mdm2* gene whose protein product interferes with the activity of p53, as described above. This effect is likely to be part of a negative feedback loop in which p53, having fulfilled its function, activates *mdm2* expression resulting in p53 inactivation (Fig. 9.38). This would allow, for example, cells which had repaired the damage to their DNA to inactivate p53 and resume cell division. Similarly, p53 also stimulates the expression of the bax gene whose protein product stimulates programmed cell death or apoptosis, allowing p53 to promote the death of cells whose damaged DNA is irreparable.

Interestingly, acetylation of p53 at a specific site (lysine 120) has recently been shown to have a differential effect on its ability to stimulate different target genes. Thus, such acetylation enhances the ability of p53 to stimulate genes encoding pro-apoptotic proteins such as Bax but does not affect its ability to stimulate the expression of genes encoding growth arrest proteins such as p21 (for review see Tyteca *et al.*, 2006). Hence, although acetylation at some sites within p53 can affect its overall

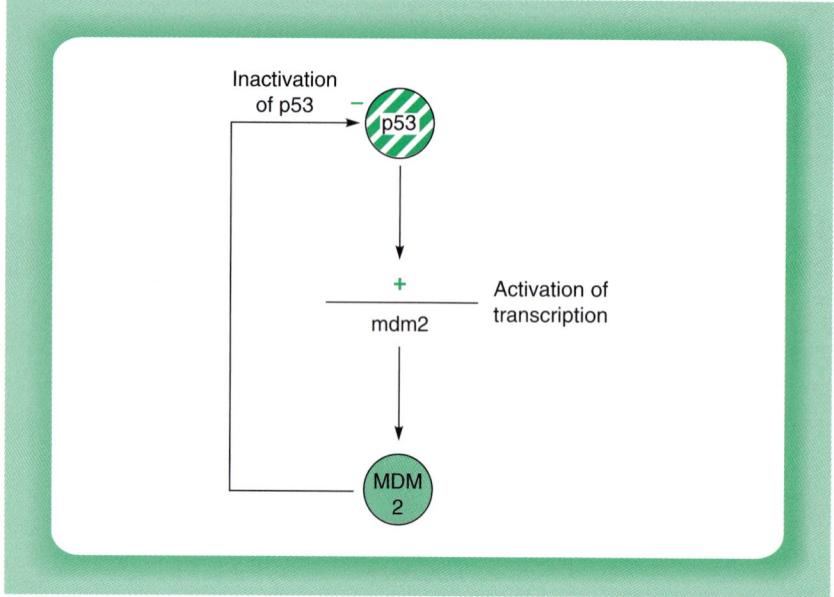

Figure 9.38

p53 activates the gene for the MDM2 protein which acts in a negative feedback loop to inactivate p53.

activity (see above), acetylation at lysine 120 appears to play a critical role in deciding whether p53 expression causes growth arrest while cells repair their damaged DNA or undergo cell death when the damage is irreparable.

As with AP1 (section 9.3.1), transcriptional activation by p53 requires the CBP co-activator or the closely related p300 protein (Avantaggiati *et al.*, 1997). Hence, as with AP1 and the steroid receptors (see Chapter 6, section 6.6), AP1 and p53 can compete for CBP/p300, resulting in antagonism between the oncogenic activity of AP1 and the anti-oncogenic activity of p53.

As well as regulating genes involved in cell growth and apoptosis, p53 also activates the promoter of the Huntingtin gene, which was discussed in section 9.1 (Feng *et al.*, 2006). Moreover, there is evidence that p53 is involved in the human disease, Huntington's chorea, which is caused by a mutant Huntingtin gene. Thus, as well as up regulating Huntingtin gene expression, p53 also interacts with the mutant Huntingtin protein, producing abnormal gene expression (for review see La Spada and Morrison, 2005). Hence, p53 joins CBP, Sp1, TFIID and TFIIE as targets for the abnormal Huntingtin protein.

The p53 gene product therefore plays a key role in regulating cellular growth by binding to DNA and activating the expression of specific genes (for review see Almog and Rotter, 1997). Its inactivation by mutation or by interaction with oncogene products is likely to play a critical role in most human cancers. Interestingly, two novel p53-related proteins which encode transcription factors, known as p73 and p63, have been described (for review see Yang *et al.*, 2002). It is currently unclear whether either p63 or p73 play a role as anti-oncogenes whose inactivation results in human cancers. However, inactivation of p63 or p73 in knock out mice results in gross developmental abnormalities, whereas this is not the case for p53 knock out mice (see above). Similarly, inactivation of p63 is the cause of EEC syndrome (ectrodactyly, ectodermal dysplasia and cleft lip) in humans, in which patients have limb defects and facial clefts. These findings further emphasize the importance of p53 and the proteins related to it in the regulation of normal embryonic development and cellular proliferation/survival and in the development of cancer.

9.4.3 *THE RETINOBLASTOMA PROTEIN*

The Retinoblastoma gene (Rb-1) was the first anti-oncogene to be defined and is so named because its inactivation in humans results in the formation of eye tumours, known as retinoblastomas (for reviews see Du and Pogoriler, 2006; Macaluso *et al.*, 2006). Like p53, the Rb-1 gene product is a transcription factor which exerts its anti-oncogenic effect by modulating the expression of specific target genes. In contrast to p53, however, it exerts this effect via protein–protein interactions with other transcription factors rather than by direct DNA binding.

One of the major targets for Rb-1 is the transcription factor E2F, which plays a critical role in stimulating the expression of genes encoding growth promoting proteins such as Myc (section 9.3.3), DNA polymerase α and thymidine kinase (for reviews see Attwooll *et al.*, 2004; Dimova and Dyson, 2005; Du and Pogoriler, 2006) and the structure of Rb-1 bound to E2F has been defined (for review see Münger, 2003). The association of Rb-1 and E2F does not inhibit the DNA binding of E2F, but prevents it from stimulating the transcription of these growth promoting genes and hence inducing growth arrest (Fig. 9.39a).

It appears that Rb-1 exerts its inhibiting effect on transcription in two distinct ways. First, it acts as an indirect repressor by blocking the ability of DNA-bound E2F to activate transcription. This is achieved by binding of Rb-1, resulting in the masking of several key residues in the activation domain of E2F, thereby preventing transcriptional activation (Lee *et al.*, 2002) (see Chapter 6, section 6.2.3 for a discussion of this quenching mechanism of transcriptional repression).

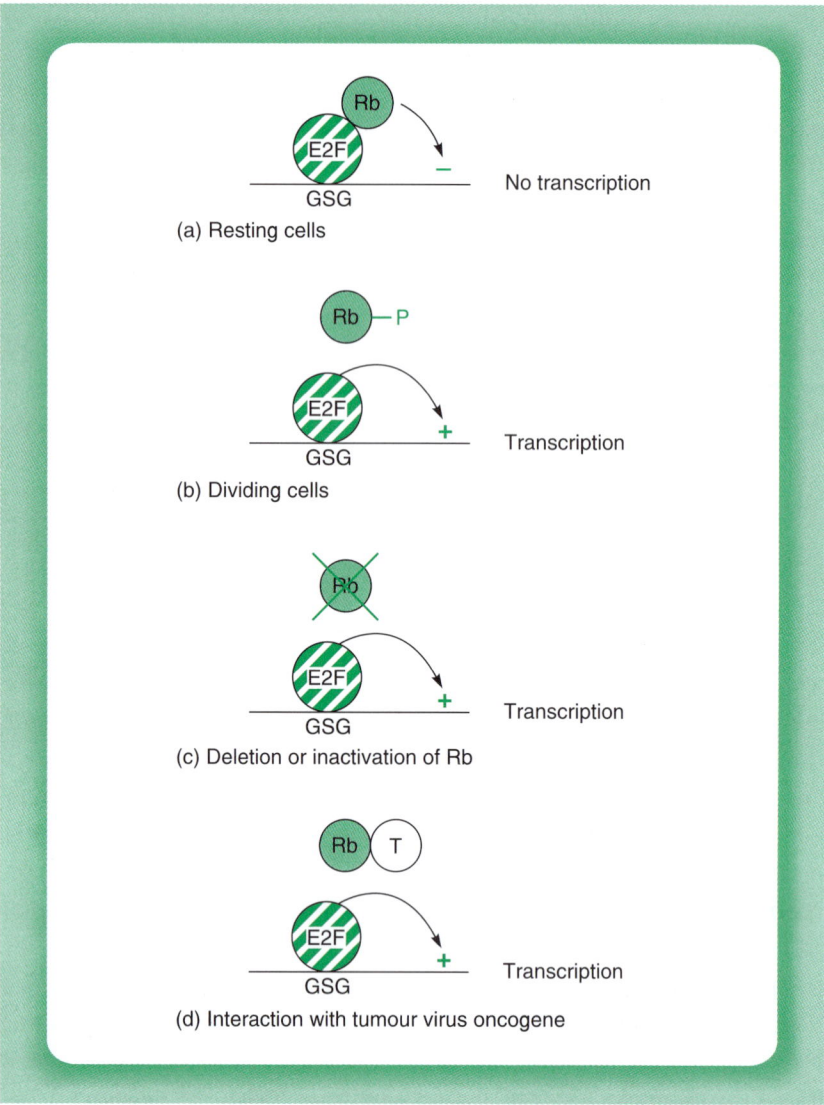

Figure 9.39

In resting cells, the Rb-1 protein binds to E2F and prevents it activating the transcription of genes encoding growth stimulating proteins (GSG) as well as directly inhibiting transcription of these genes (panel a). In normal dividing cells, the Rb-1 protein is phosphorylated at the G1/S transition in the cell cycle which prevents it from interacting with E2F and hence allows E2F to activate transcription (panel b). This release of E2F can also occur in tumour cells where the Rb-1 gene is deleted or inactivated by mutation (panel c) or following the interaction of Rb-1 with tumour virus oncogenes (T) (panel d).

Secondly, the Rb/E2F complex acts directly to inhibit transcription, by organizing a tightly packed chromatin structure incompatible with transcription. This involves the ability of Rb-1 to recruit histone deacetylases and methyltransferases which, as discussed in Chapter 1 (section 1.2.3), promote a more tightly packed chromatin structure (Macaluso *et al.*, 2006; Kotake *et al.*, 2007) (Fig. 9.40). It appears that this second effect, involving chromatin structure, may be of greater importance since it has been shown to be essential for the growth-arresting effect of Rb-1 (Zhang *et al.*, 1999).

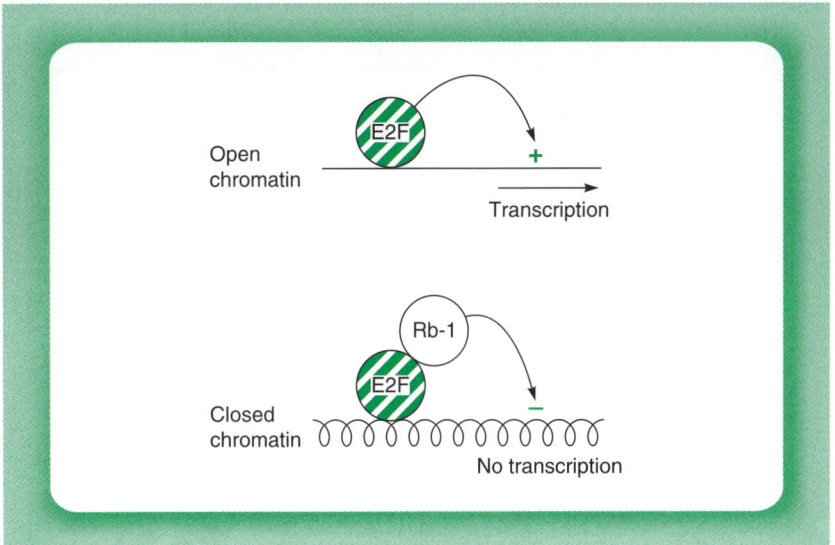

Figure 9.40
Binding of Rb-1 to E2F represses transcription both indirectly by inhibiting activation by E2F and directly by organizing a closed chromatin structure incompatible with transcription.

Hence, Rb-1 exerts its anti-oncogenic effect by inhibiting the transcription of growth promoting genes, using both indirect and direct inhibiting mechanisms (see Chapter 6) rather than, as with p53, promoting the transcription of growth inhibiting genes. In normal dividing cells, this interaction of Rb-1 and E2F is inhibited as cells move from G to S phase in the cell cycle. This effect is dependent on the phosphorylation of Rb-1 which prevents it interacting with E2F (Fig. 9.39b). Hence, the controlled growth of normal cells can be controlled by the regulated phosphorylation of Rb-1 which in turn regulates its ability to interact with E2F and modulate its activity (for review of Rb-1 and the cell cycle see Genovese *et al.*, 2006; Giacinti and Giordano *et al.*, 2006).

Interestingly, the phosphorylation of Rb-1 in the cell cycle is carried out by the cyclin-dependent kinases. This provides a link between p53 and the regulation of Rb-1 activity since, as noted above (section 9.4.2) p53 activates the gene encoding the p21 protein, which inhibits cyclin-dependent kinases and would thus prevent the phosphorylation of Rb-1 and cell cycle progression (Fig. 9.41). To add to the complexity still further, it appears that the activity of both p53 itself and E2F is also altered following phosphorylation by cyclin-dependent kinases, indicating that a complex network of interacting transcription factors, kinases and their inhibitors regulates cellular growth.

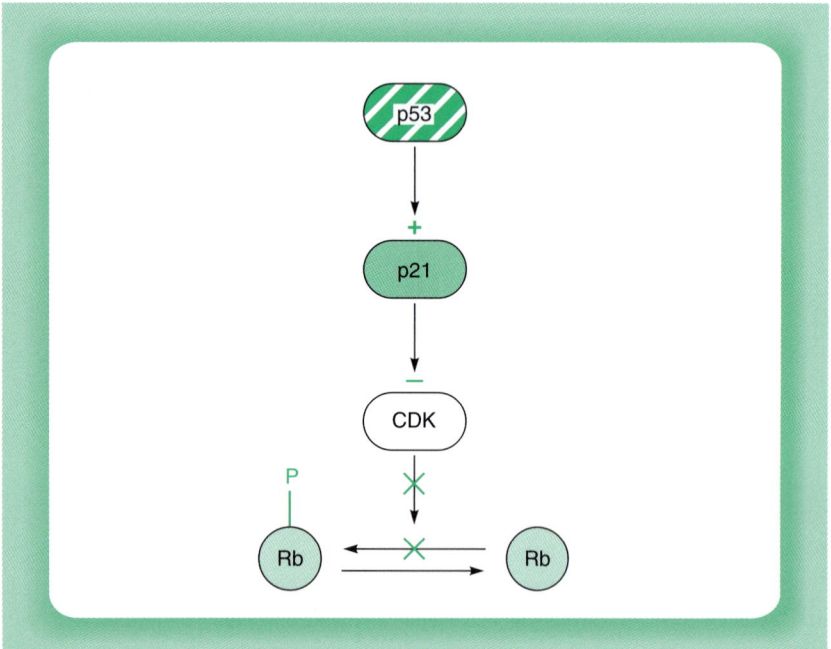

Figure 9.41

By activating the p21 gene whose protein product inhibits cyclin-dependent kinases (CDK), p53 produces a fall in CDK activity which results in more Rb being in the growth inhibitory unphosphorylated form.

Further links exist between p53 and Rb-1. Thus, like p53 (see section 9.4.2), Rb-1 is ubiquitinated by MDM2, which targets it for degradation (Sdek *et al.*, 2005; Uchida *et al.*, 2005). Hence, MDM2 can induce the degradation of two key anti-oncogene-encoded proteins (Fig. 9.42). Interestingly, as well as being modified by phosphorylation and ubiquitination, Rb-1 can also be modified by acetylation, which hinders its phosphorylation (Chan *et al.*, 2001). As with p53, Rb-1 is therefore modified by multiple post-translational modifications which interact with one another (Fig. 9.43).

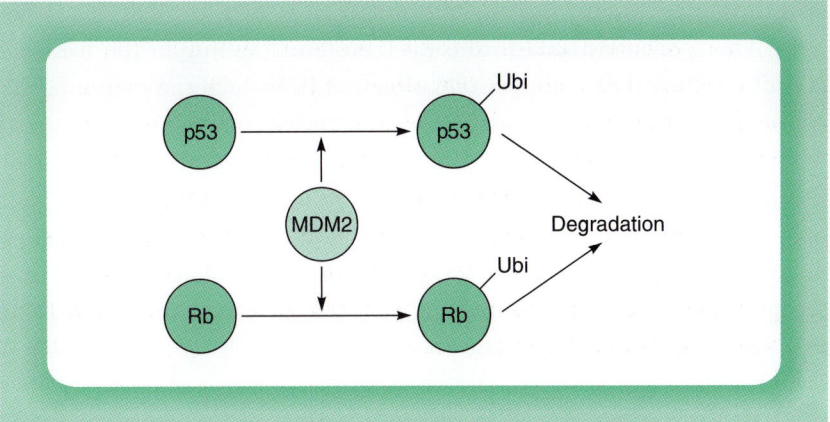

Figure 9.42
MDM2 can ubiquitinate both p53 and Rb, thereby promoting their degradation.

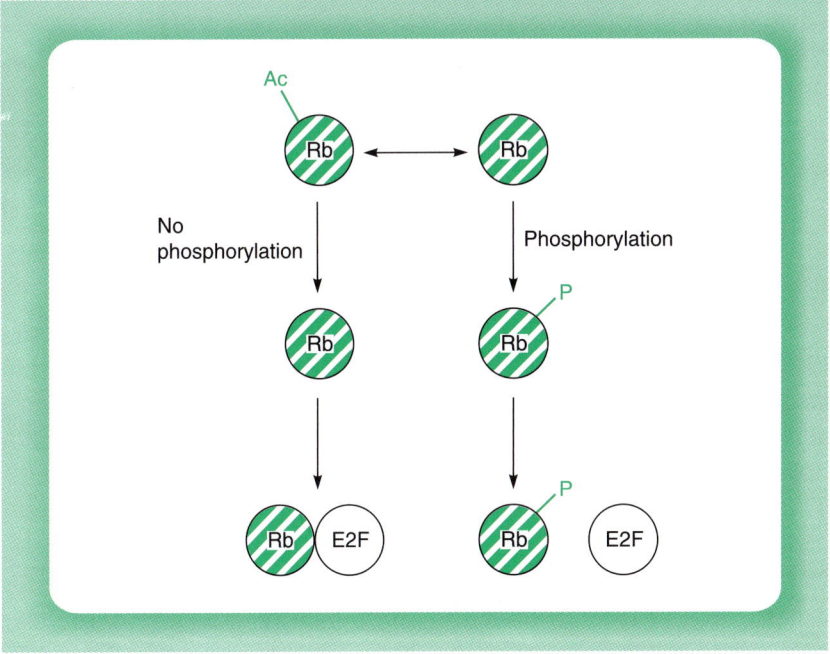

Figure 9.43
Acetylation of Rb-1 inhibits its phosphorylation and thereby promotes the
formation of the Rb/E2F complex.

Clearly, abolishing the activity of Rb-1, either by deletion of its gene
or by mutation will result in the unregulated activity of E2F, leading
to the uncontrolled growth which is characteristic of cancer cells (Fig.
9.39c). Interestingly, the inactivation of Rb-1 can also be achieved by the

transforming proteins of DNA tumour viruses, such as SV40 or adenovirus. These proteins bind to the Rb-1 protein resulting in the dissociation of the Rb-1/E2F complex releasing free E2F which can activate gene expression (Fig. 9.39d) (for review see Liu and Marmorstein, 2006).

One potential problem in this simple model is that inactivation of Rb-1 still leaves the p53 pathway intact. Hence, one might expect that p53 would induce growth arrest or programmed cell death of the Rb-1 mutated cells and so prevent tumour formation. This problem has recently been solved by the finding that the gene encoding MDM2 or that encoding MDM4 (also known as MDMX) are frequently amplified in retinoblastoma tumours. The enhanced expression of either MDM2 or MDM4 this produces will therefore inactivate p53 (see section 9.4.2) and allow the mutation in Rb-1 to produce uncontrolled growth (for reviews see Wallace, 2006; Sage, 2007).

Although E2F is a major target of Rb-1, there are also other factors with which Rb-1 interacts. Thus Rb-1 has been shown to inactivate the UBF factor which plays a critical role in transcription of the ribosomal RNA genes by RNA polymerase I (see Chapter 3, section 3.3). Due to the need for these ribosomal RNAs for the effective functioning of the ribosomes, the inactivation of UBF by Rb-1 will lead to a decrease in the levels of total protein synthesis, which would in turn lead to the arrest of cell growth.

Obviously, the 5S ribosomal RNA and the transfer RNAs which are produced by RNA polymerase III are also necessary for ribosomal function and protein synthesis. Indeed, it has been shown that Rb can also inhibit RNA polymerase III transcription by interacting directly with the polymerase III transcription factor TFIIIB (see Chapter 3, section 3.4) and inhibiting its activity (Crighton et al., 2003). This is evidently the opposite effect to that produced by interaction of the Myc protein with TFIIIB, which stimulates transcription of the genes encoding tRNA and 5S RNA (see section 9.3.3).

Hence, Rb-1 can directly inhibit transcription of genes involved in cellular growth, both by inhibiting the transcription of E2F-dependent genes by RNA polymerase II and the transcription of all the genes transcribed by RNA polymerases I and III. It therefore has a remarkable ability to modulate transcription by all three RNA polymerases (for review see White, 2004) and is likely to play a critical role not only in preventing cancer but also in normal cells by promoting the growth arrest which is necessary for terminal differentiation (Fig. 9.44) (for reviews of Rb-1 and differentiation see De Falco et al., 2006; Khidr and Chen, 2006).

In agreement with this idea, mice in which the Rb-1 gene has been inactivated die before birth and show gross defects in cellular differentiation

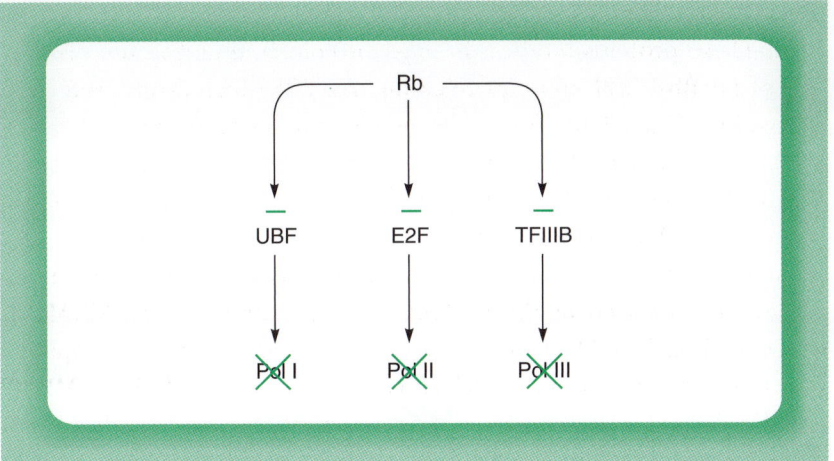

Figure 9.44

Rb can inhibit all transcription by RNA polymerases I and III by inhibiting the activity of UBF and TFIIIB as well as inhibiting the activity of E2F and hence inhibiting the ability of RNA polymerase II to transcribe genes whose protein products stimulate growth.

(for review see Dyson, 2003). This indicates that Rb-1 plays a key role in normal development, as well as acting as an anti-oncogene and contrasts with the viability of mice in which the p53 gene has been inactivated (see section 9.4.2). Interestingly, many of the developmental defects observed in mice lacking Rb-1 can be rescued by also inactivating the gene encoding Id2, which is an inhibitory transcription factor having a helix-loop-helix motif but lacking a DNA binding domain (see Chapter 4, section 4.5 and Chapter 6, section 6.2.2). This indicates that during normal development, Id2 and Rb-1 antagonize one another so that the effects of inactivating both are less severe than inactivating Rb-1 alone (Lasorella *et al.*, 2000). In agreement with this, Id2 has been shown to interact directly with the non-phosphorylated form of Rb-1 via a protein–protein interaction and inactivate it.

Hence, the correct balance between the antagonistic factors Id2 and Rb-1 is essential for normal development. It has been shown that in cells over-expressing the Myc oncogene protein (see section 9.3.3), the expression of Id2 is transcriptionally activated by Myc. The excess Id2 then inactivates Rb-1, thereby promoting tumour formation (Fig. 9.45) (for review of Id factors and cancer see Perk *et al.*, 2005).

Hence, the Rb protein plays a key role in regulating cellular growth and differentiation, by interacting with transcription factors involved in transcription by RNA polymerases I, II and III. Its inactivation, either

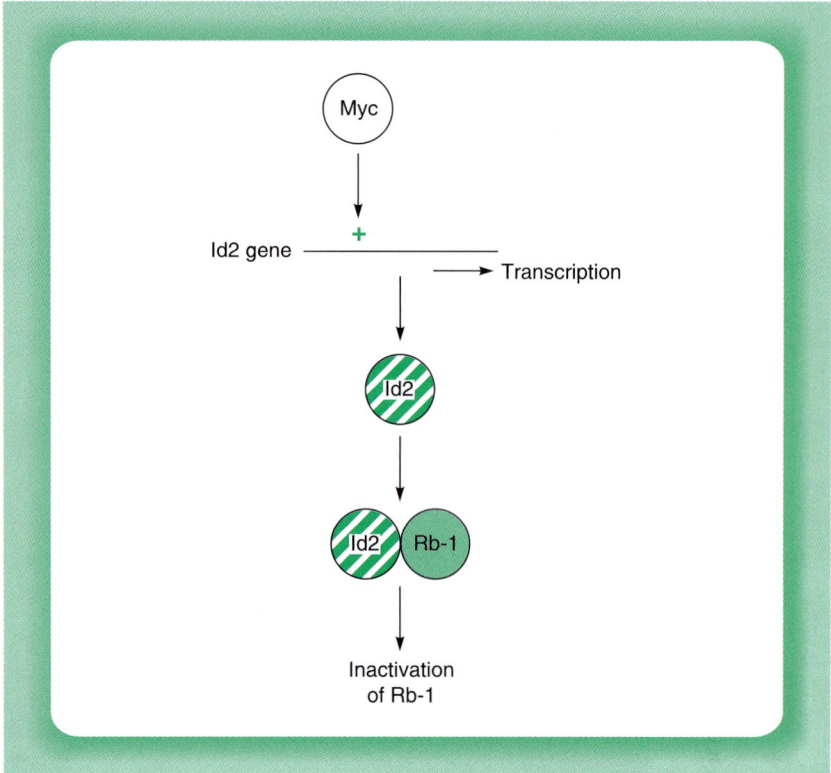

Figure 9.45

The Myc oncogene product can transcriptionally activate the gene encoding the Id2 transcription factor. Id2 then binds to Rb-1 and inhibits its tumour suppressor function.

by mutation or by specific oncogenes, therefore results in uncontrolled proliferation and cancer. When taken together with the similar role of p53 in growth regulation and as a target for oncogenes, this suggests that anti-oncogenes are likely to play a key role in regulating cellular growth, which is likely to be controlled by the balance between the antagonistic effects of oncogene and anti-oncogene products.

9.4.4 OTHER ANTI-ONCOGENIC TRANSCRIPTION FACTORS

Normally, anti-oncogenes are identified on the basis of their inactivation in specific human cancers and their functional role subsequently characterized. For some time, only three anti-oncogene products were known to be transcription factors, namely the p53 and Rb-1 proteins discussed in previous sections and the Wilm's tumour gene product (for review see Hastie, 2001).

More recently, however, other anti-oncogene products have also been implicated in transcriptional control. For example, the BRCA-1 and BRCA-2 anti-oncogenes, which are mutated in many cases of familial breast cancer, appear to function primarily in controlling the repair of damaged DNA, but there is also evidence that they influence transcription. Thus, BRCA-1 can interact with both p53 and Rb-1 to modulate their activity (for review see Mullan *et al.*, 2006). Moreover, BRCA-1 can interact with the C-terminal domain of RNA polymerase II and inhibit its phosphorylation, thereby inhibiting transcriptional elongation (Moisan *et al.*, 2004) (for discussion of the C-terminal domain of RNA polymerase II and its role in transcriptional elongation see Chapter 3, section 3.7).

In contrast to these features suggesting that BRCA-1 can influence transcription rates within the nucleus, the adenomatous polyposis coli (APC) anti-oncogene which is mutated in most human colon tumours (for review see Nelson and Nusse, 2004) appears to influence transcription indirectly. Thus, APC acts by interacting with a protein known as β-catenin which is involved both in cell adhesion and also acts as a transcription factor. This interaction between APC and β-catenin results in the export of β-catenin to the cytoplasm and its rapid degradation (Fig. 9.46a) (Rosin-Arbesfeld *et al.*, 2000).

In normal cells, specific secreted proteins known as WNT proteins (or wingless proteins, after the first member of the family which was discovered in *Drosophila*) activate a kinase enzyme, glycogen synthase kinase, and this kinase phosphorylates and thereby stabilizes β-catenin preventing it from being degraded (for review see Taipale and Beachy, 2001; Reya and Clevers, 2005). The β-catenin then moves to the nucleus and interacts with the LEF-1 transcription factor discussed in Chapter 1 (section 1.3.6) and stimulates its ability to activate transcription (Fig. 9.46b). One of the genes activated by the LEF-1/β-catenin complex is that encoding the Pitx2 transcription factor, which in turn activates the cyclin D2 gene, thereby stimulating cellular proliferation (Kioussi *et al.*, 2002). Another β-catenin target gene is the c-*jun* oncogene, discussed in section 9.3.1, and such activation of c-*jun* by β-catenin has been shown to be important in the development of intestinal tumours (Nateri *et al.*, 2005).

Interestingly, as well as promoting the nuclear export of β-catenin, APC can also enter the nucleus and inhibit β-catenin activity (for review see Xiong and Kotake, 2006). Thus, APC can bind to β-catenin, which is bound to LEF-1, and recruit a histone deacetylase. As discussed in Chapter 6 (section 6.4.1), histone deacetylation produces a tightly packed chromatin structure and hence represses target gene activation. APC can therefore repress β-catenin activity by two distinct mechanisms (Fig. 9.47).

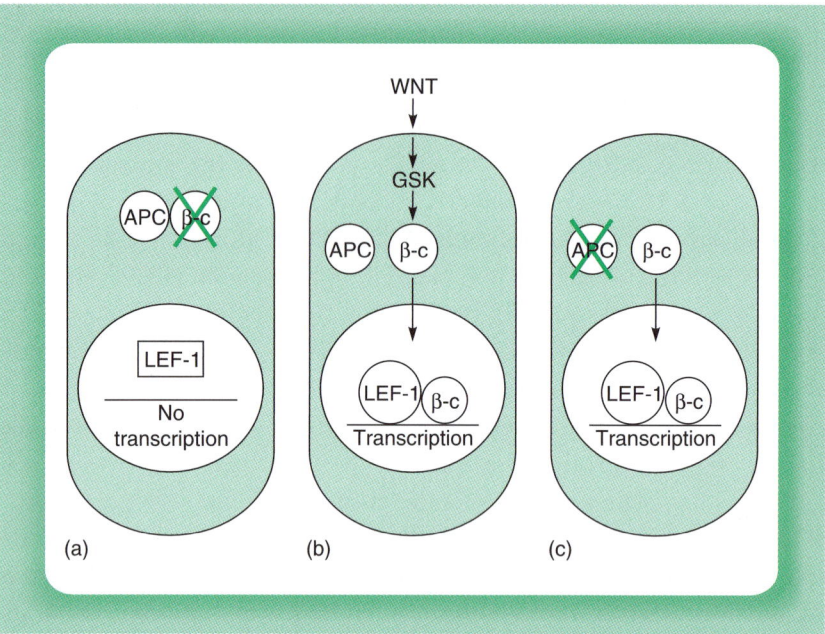

Figure 9.46

(a) Interaction of the anti-oncogenic protein APC and the oncogenic protein β-catenin resulting in degradation of β-catenin. (b) Following activation of glycogen synthase kinase (GSK) by WNT proteins, β-catenin is stabilized. It then moves to the nucleus and interacts with the LEF-1 transcription factor promoting its ability to stimulate transcription. (c) In cancer, the APC factor is inactivated, resulting in the constitutive activation of β-catenin.

In a normal situation, therefore, this ability of β-catenin to stimulate the activity of LEF-1 is tightly regulated by the presence or absence of WNT proteins, so ensuring appropriate control of cellular growth. Any change which causes this pathway to become constitutively active results in cancer. For example, if the APC gene is mutated so that APC cannot inactivate β-catenin, cancer will result from the constitutive activation of β-catenin (Fig. 9.46c). Hence APC acts as an anti-oncogene whose inactivation by mutation causes cancer.

As well as illustrating how an anti-oncogene can act indirectly to influence transcription, this example also illustrates how oncogene products interact with one another. Clearly, mutations in the β-catenin gene which enhance β-catenin stability or mutations in the WNT genes which result in their over-expression, will also cause cancer and hence the genes encoding β-catenin or the WNT proteins are oncogenes whose products act in the same pathway as the APC anti-oncogene product.

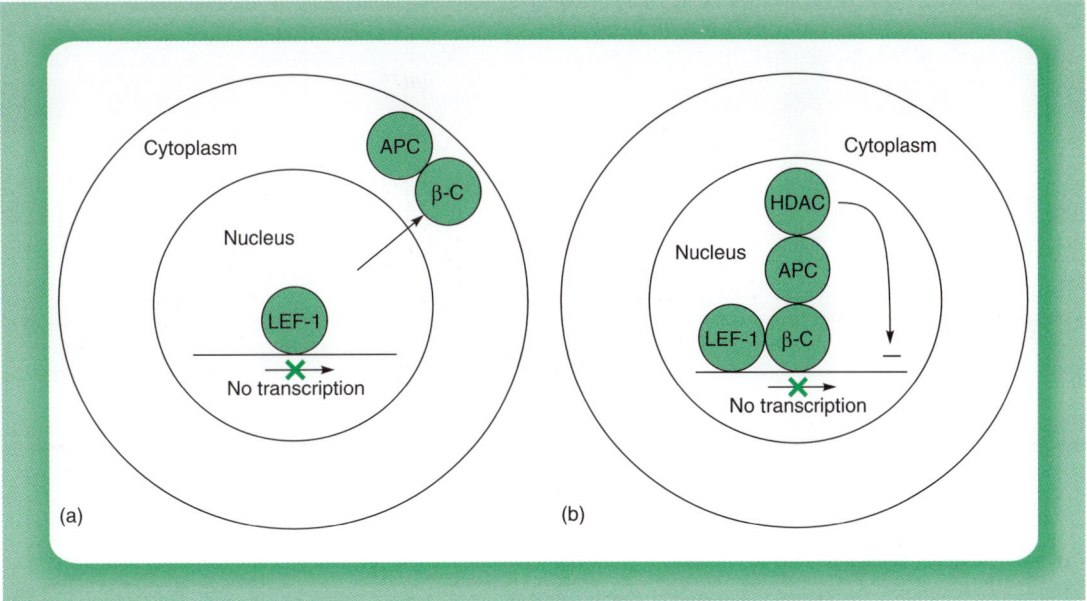

Figure 9.47

APC can inhibit β-catenin (β-c) both by promoting its export to the cytoplasm (panel a) and by binding to it in the nucleus and recruiting a histone deacetylase (HDAC) which inhibits transcription.

In addition to being stabilized by phosphorylation (see above), β-catenin is also acetylated by the CBP co-activator (see Chapter 5, section 5.4.3). This acetylation apparently reduces the ability of β-catenin to activate one of its target genes, the c-*myc* proto-oncogene (see section 9.3.3), providing an example of CBP acting to inhibit transcription rather than its normal role as a co-activator (Wolf *et al.*, 2002). Interestingly, the lysine residue in β-catenin, which is a target for acetylation, is often found mutated to a non-acetylatable amino acid in human cancers and, as expected, from its non-acetylatability, this mutant protein is a strong activator of c-*myc* oncogene expression (Fig. 9.48).

Hence, β-catenin offers an example of a proto-oncogene whose activity is regulated both by phosphorylation and acetylation, paralleling the similar regulation of the anti-oncogene proteins p53 (section 9.4.2) and Rb-1 (section 9.4.3) by multiple post-translational modifications.

Like the majority of transcription factors, the anti-oncogenic proteins discussed so far act by directly or indirectly altering the rate at which transcription is initiated by RNA polymerase. This is apparently not the case, however, for the von Hippel–Lindau (VHL) anti-oncogene protein which is mutated in multiple forms of cancer. Thus, as discussed in Chapter 6 (section 6.5), this factor acts to inhibit transcriptional elongation by promoting the degradation of the large subunit of RNA polymerase II.

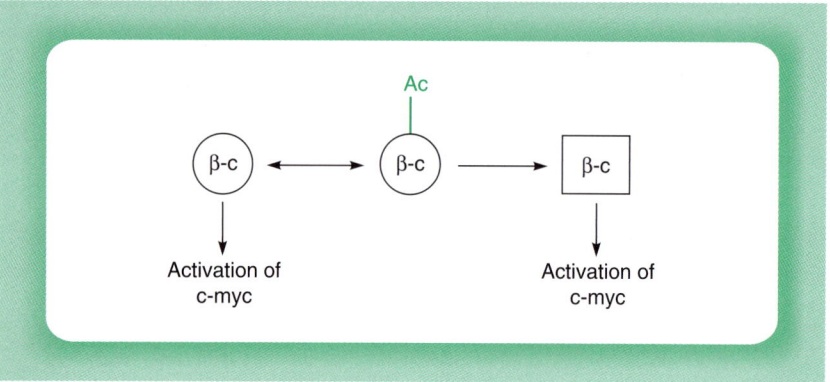

Figure 9.48

The ability of β-catenin to activate the c-*myc* gene is reduced by its acetylation but enhanced by its mutation to a non-acetylatable form.

Interestingly, the mutant forms of the VHL protein found in human tumours do not inhibit transcriptional elongation, indicating that the anti-oncogenic action of the protein is mediated, at least in part, by its effect on transcriptional elongation. This is likely to be because several oncogenes, such as c-*fos* and c-*myc*, are regulated at the stage of transcriptional elongation with many of the RNA transcripts which are initiated, not being elongated to produce a full length functional mRNA (see Chapter 5, section 5.6). It is possible therefore that in the absence of the VHL protein, too much full length mRNA is produced resulting in over-production of the corresponding oncogenic proteins.

However, as discussed in Chapter 8 (section 8.4.5), the VHL protein also acts to promote the degradation of the hypoxia-inducible factor, HIF-1α, thereby ensuring that it activates its target genes only in response to lowered oxygen levels (for review of HIF-1 see Dery *et al.*, 2005; Pouyssegur *et al.*, 2006). The mutations of the VHL protein, which occur in cancer, also block its interaction with HIF-1α and in these tumour cells, HIF-dependent genes are expressed at high levels even in the presence of oxygen. As one of the roles of HIF-1 is to activate genes involved in blood vessel formation in response to falling oxygen levels in tissues, the inappropriate activation of HIF-1 and its target genes in tumours may enhance the blood supply to the tumour, allowing it to grow more rapidly.

Hence, the VHL protein may target multiple pathways promoting the degradation of proteins which are important in regulating normal growth, including transcriptional activation by HIF-1α and transcriptional elongation. Interestingly, the gene encoding another transcriptional elongation factor ELL is found at the break point of chromosomal translocations in several leukaemias (see section 9.3.4) indicating that it can be oncogenic under certain circumstances (for review see Conaway and Conaway, 1999). The existence of both oncogenic and anti-oncogenic transcription factors which modulate transcriptional elongation indicates that this is an important target for processes that regulate normal cellular growth and hence for malregulation in cancer (for review see Li and Green, 1996).

More generally, the examples of BRCA-1, APC and VHL, given in this section, add considerable variety to the three 'classical' anti-oncogenes encoding transcription factors (p53, Rb-1 and the Wilm's tumour gene) and indicate the key role of such gene products in different forms of transcriptional regulation in normal cells and in cancer. Interestingly, as well as its other actions, VHL enhances p53 activity, acting in two ways (for review see Semenza, 2006). Thus, it suppresses MDM2-mediated ubiquitination of p53, thereby inhibiting its degradation. In addition, it promotes the interaction of p53 and the p300 transcriptional co-activator. This enhances the acetylation of p53 and thereby increases its ability to activate transcription (Fig. 9.49) (for discussion of p53 see section 9.4.2). Hence, different anti-oncogene products can act together to produce a strong anti-oncogenic effect.

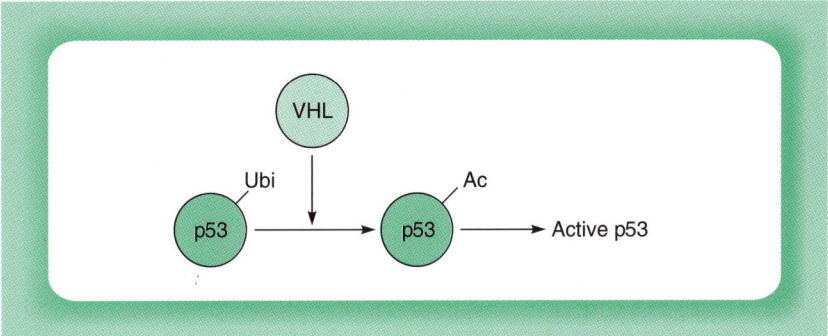

Figure 9.49
The von Hippel–Landau anti-oncogenic factor stimulates the acetylation of p53 and inhibits its ubiquitination and thereby enhances its activity.

9.5 TRANSCRIPTION FACTORS AND TREATMENT OF HUMAN DISEASE

The clear importance of transcription factors in many biological processes and their aberrant activity in human diseases, indicates their potential as therapeutic targets to treat human patients. This could involve artificially altering the expression of the target transcription factor or its activity, paralleling the manner in which transcription factors are normally controlled within the cell by regulating their synthesis or their activity (see Chapters 7 and 8 respectively).

Depending on the role of a particular factor, it may be necessary to increase or decrease its expression in order to achieve a therapeutic benefit in a specific disease. The over-expression of the gene encoding a transcription factor could be achieved by delivering the gene to cells in a gene therapy procedure in the same manner as genes encoding any other protein (for review of gene therapy see French Anderson, 1998, 2000) (Fig. 9.50a).

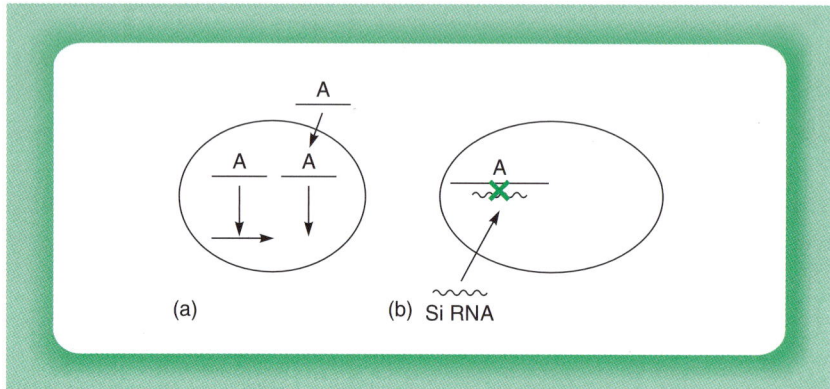

Figure 9.50

The expression of a gene encoding a transcription factor (A) can be enhanced by delivering exogenous copies of the gene into the cell using a gene therapy procedure (panel a). Similarly, small inhibitory RNAs (si RNA) can be targeted to the endogenous gene encoding the transcription factor in order to inhibit its expression (panel b).

Similarly, procedures to inhibit the expression of genes encoding transcription factors have been developed based on the small inhibitory RNAs discussed in Chapter 6 (section 6.4.2). Thus, small inhibitory RNAs can be designed to bind to the gene encoding a transcription factor and inhibit gene transcription or destabilize its mRNA and/or block its translation (Fig. 9.50b) (for reviews of the therapeutic potential of small inhibitory RNAs see Hannon and Rossi, 2004; Rossi, 2004).

Although these methods have potential, they suffer from two limitations. First, they are not specific to transcription factors but could potentially be used to enhance or inhibit the expression of genes encoding any type of protein. Secondly, and more importantly, the delivery of genes or small inhibitory RNAs to human patients is difficult and methods for doing this require considerable improvement before they can be used routinely. For this reason, considerable attention has focused on developing small molecule drugs which can be delivered easily to patients and which will specifically target a particular transcription factor.

Interestingly, a number of therapeutic drugs which are used to treat patients with a variety of diseases have been shown to target transcription factors (for review see Latchman, 2000). In a number of cases, these drugs were introduced on the basis of their efficacy in the treatment of a particular disease and it was only many years later that they were shown to target transcription factors.

Thus, for example, one of the most commonly used drugs, salicylate (aspirin), has been used therapeutically for many years and was subsequently shown to inhibit the phosphorylation of IκB, thereby blocking the dissociation of IκB and NFκB (Pierce et al., 1996). As discussed in Chapter 8 (section 8.4.2), this will prevent the activation of NFκB and will produce an anti-inflammatory effect, since NFκB activation plays a key role in inflammation. Similarly, the commonly used anti-inflammatory drugs, cyclosporin and FK506 (tacrolimus) have been shown to block the dephosphorylation of the NF-AT transcription factor, which is required for its activation and an effective immune response.

These examples illustrate that the post-translational modifications of transcription factors represent an obvious therapeutic target for the development of a new generation of therapeutic drugs, which are specifically designed to affect such modification of transcription factors.

Indeed, different drugs can be designed to target the various different post-translational modifications of transcription factors which were discussed in Chapter 8 (section 8.4). Thus, as discussed in Chapter 8 (section 8.4.5), the HIF-1 transcription factor is modified by proline-hydroxylation so allowing it to be recognized by the von Hippel–Lindau anti-oncogene product and be rapidly degraded. HIF-1 is an attractive therapeutic target since it promotes the development of blood vessels, which could be of therapeutic benefit in individuals suffering from cardiovascular disease. Conversely, the inhibition of this process may be of benefit in cancer patients by starving the tumour of oxygen by restricting its blood supply (for reviews of the therapeutic potential of HIF-1 see Dery et al., 2005; Pouyssegur et al., 2006). For these reasons, small molecule drugs have been developed which can alter the activity of the enzyme which hydroxylates HIF-1, prolyl hydroxylase, and that could ultimately be used clinically to enhance or reduce HIF-1 hydroxylation (Asikainen et al., 2005).

Similarly, as described earlier in this chapter (section 9.4.2), the ubiquitination of p53 by MDM2 produces its rapid degradation. It has recently been shown that p53 is activated by drugs which block the ubiquitination activity of MDM2 and can therefore exert its anti-oncogenic effect more effectively (for review see Poyurovsky and Prives, 2006). Hence, the modulation of transcription factor activity for therapeutic means can in principle be achieved by targeting various different post-translational modifications that are essential for transcription factor activity (Fig. 9.51).

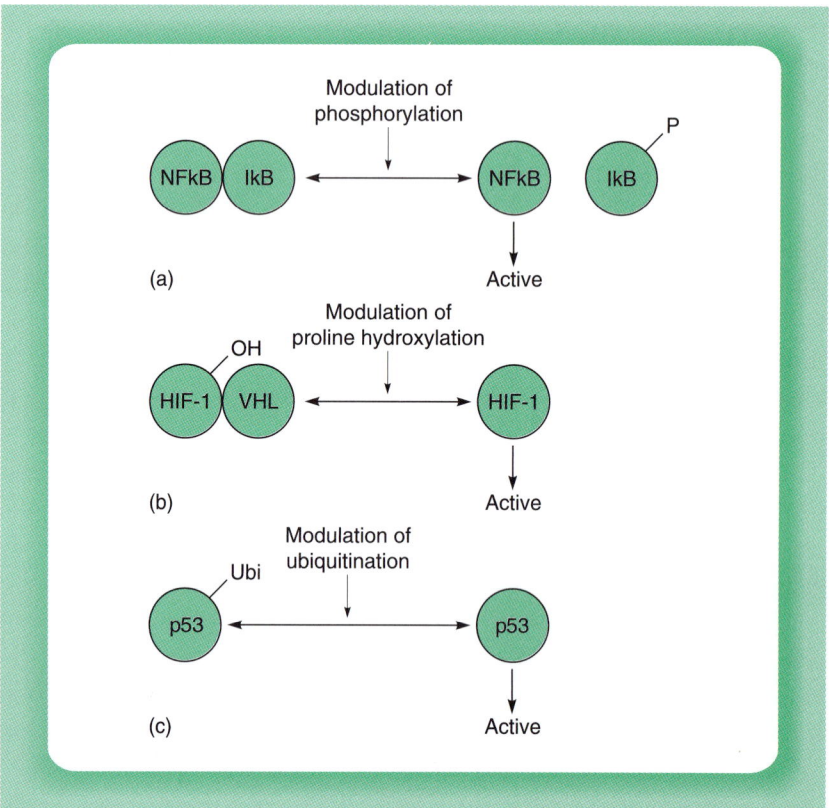

Figure 9.51

Therapeutic drugs can target transcription factors by modulating their post-translational modification, for example by targeting their phosphorylation (panel a), hydroxylation (panel b), or ubiquitination (panel c).

Evidently, as well as targeting transcription factor modification, therapeutic drugs could target any of the potential mechanisms by which transcription factors are activated and which were discussed in Chapter 8. Thus, the ubiquitination of p53 by MDM2 requires a protein–protein interaction between the two proteins (see section 9.4.2) and is an example of transcription factor activity being modulated by protein–protein inter-action (see Chapter 8, section 8.3).

Hence, the protein–protein interaction of p53 and MDM2 repre-sents another target for therapeutic drugs, paralleling the development of drugs which inhibit the ubiquitination activity of MDM2, as discussed above. It has been demonstrated using structural studies that p53 binds to MDM2 by inserting itself into a deep pocket in the MDM2 molecule. In one of the most successful examples of drug design based on struc-tural studies, a small molecule (nutlin-2) was synthesized which was designed to fill this pocket and thereby block the p53/MDM2 interaction (Fig. 9.52) (Plate 7). This drug successfully activated p53 and reduced tumour growth both in culture and the intact animal, indicating that it has significant therapeutic potential (Vassilev *et al.*, 2004; for reviews see Lane and Fischer, 2004; Poyurovsky and Driver, 2006).

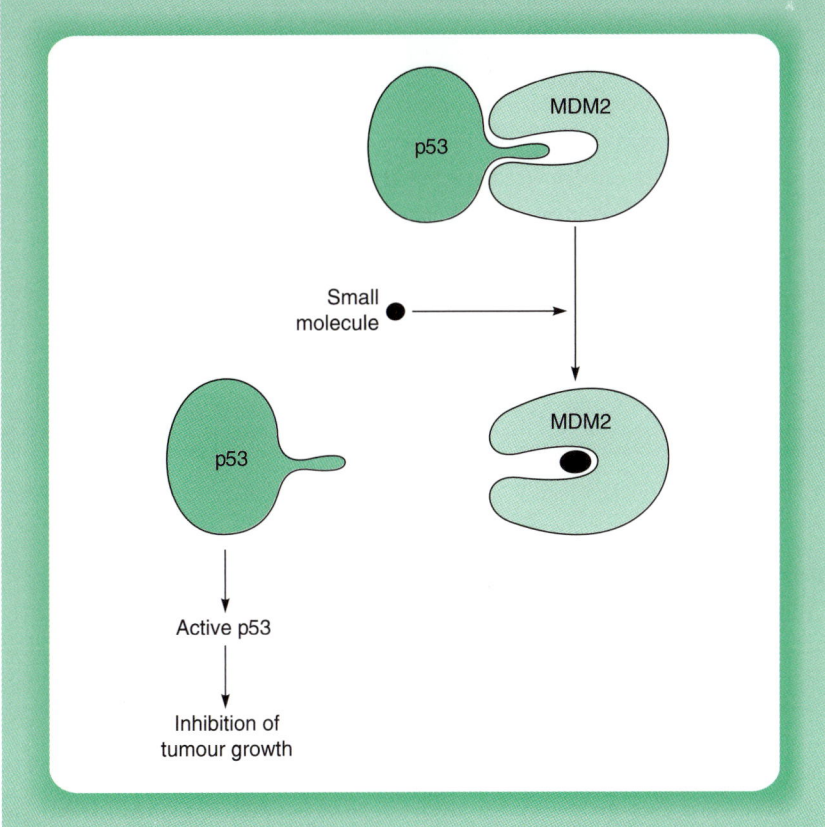

Figure 9.52

A small chemical molecule (solid) designed to fill the pocket in MDM2 where p53 normally binds, can prevent p53/MDM2 binding. This activates p53 allowing it to inhibit tumour growth.

Obviously, this type of approach targeting protein–protein interaction, need not be confined to p53. Thus, for example, small molecules have been developed which inhibit the interaction between the CREB transcription factor and the CBP co-activator (Best *et al.*, 2004). This interaction is essential for transcriptional activation by CREB (see Chapter 5, section 5.4.3) and its inhibition may be of therapeutic importance, since enhanced CREB activity has been demonstrated in acute myeloid leukaemia (for review see Conkright and Montminy, 2005).

Hence, drugs designed to modulate transcription factor modification or the protein–protein interactions of transcription factors are likely to be of therapeutic benefit in the future, supplementing the use of well-established drugs which were identified by other means and subsequently shown to modulate transcription factor activation.

As described in Chapter 8 (section 8.2), transcription factors can also be activated by the binding of a specific ligand. Indeed, drugs aimed at modulating this form of activation, are already used clinically, being primarily directed against members of the nuclear receptor family of transcription factors which are activated by ligand binding (see Chapter 4, section 4.4). Thus, one could envisage inhibiting the ligand-mediated activation of a member of this family, by using an antagonist which competes with the ligand for binding but does not activate the receptor. This effect has been used clinically in the case of anti-estrogen drugs such as tamoxifen, which are used to inhibit the growth of estrogen-dependent breast cancer cells. Thus, tamoxifen competes with estrogen for binding to the estrogen receptor but does not produce receptor activation following binding (Fig. 9.53) (for review see Jordan, 2007).

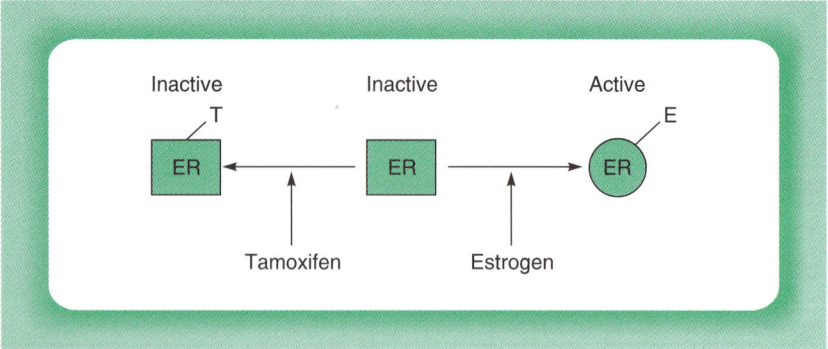

Figure 9.53

Tamoxifen (T) competes with estrogen (E) for binding to the estrogen receptor (ER) but does not activate the receptor. It therefore antagonizes estrogen-mediated activation of the receptor.

As well as using drugs which inhibit nuclear receptor family members for therapeutic benefit, it is also possible to achieve a therapeutic benefit in other situations by activating a member of this family. Thus, as described above (section 9.1), the PPARγ member of the nuclear receptor family is mutated in a few rare cases of patients with diabetes, indicating that its activity is important for preventing the disease process. The vast majority of patients with diabetes have a normal receptor and their disease is due to other abnormalities. However, in these patients, the disease can be treated by stimulating the activity of the receptor by using synthetic drugs, known as thiazolidinediones, which bind to the receptor and stimulate its activity (for reviews see Rosen and Spiegelman, 2001; Evans *et al.*, 2004).

Such activation of a transcription factor/receptor can also be used therapeutically in the case of the abnormal RAR–PML fusion protein, which is involved in cases of promyelocytic leukaemia. Thus, as described in section 9.3.4, this fusion protein causes cancer because it represses transcription, unlike the normal retinoic acid receptor which activates transcription. It is possible, therefore, to treat this form of leukaemia by administering retinoic acid and stimulating the activity of the retinoic acid receptor portion of the fusion protein, so as to overcome the transcriptional inhibition, which is normally produced by the fusion protein (Fig. 9.54a).

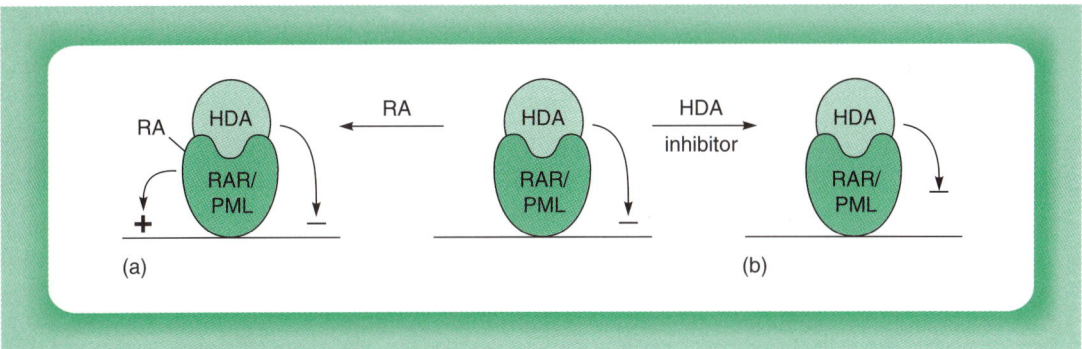

(a)   (b)

Figure 9.54

The inhibitory effect of the oncogenic RAR/PML fusion protein can be overcome by (a) treating with retinoic acid (RA) to stimulate gene activation by the RAR component or (b) by using histone deacetylase inhibitors to block the inhibitory effect of histone deacetylases (HAD) which are recruited by the fusion protein.

Although this treatment is often effective in the short term, it is not usually an effective long term treatment. This has led to the development of alternative therapies, which rely on the fact that the inhibitory effect of the RAR–PML fusion protein is due to its ability to recruit histone deacetylases, which organize an inactive chromatin structure (see section 9.3.4). Thus, considerable effort has gone into the development of histone deacetylase inhibitors, which could have a role in the treatment of promyelocytic leukaemia by blocking the inhibitory effect of the fusion protein (Fig. 9.54b). Indeed, the production of an inactive chromatin structure by histone deacetylases, appears to occur in other forms of leukaemia (such as those caused by the AML–ETO oncogenic fusion protein: see section 9.3.4) as well as in other cancers. For this reason, histone deacetylase inhibitors are currently in clinical trials in a variety of leukaemias and other cancers (for review see Minucci and Pelicci, 2006).

Such treatments may be applicable to a wide range of human diseases where alterations in chromatin structure are involved in the disease (for review see Egger *et al.*, 2004). One potential group of diseases which could be treated in this manner are the polyglutamine diseases, discussed in section 9.1 (for review see Butler and Bates, 2006). Thus, as discussed in section 9.1, in the disease SCA7, the ataxin gene is mutated so that the mutant form inhibits the histone acetylation activity of the wild type ataxin protein. Similarly, the CBP co-activator, which is inhibited in Huntington's disease, also has histone acetyltransferase activity (see Chapter 5, section 5.4.3). Diseases of this type could therefore be treated by using histone deacetylase inhibitors to restore the altered balance between histone acetylation and deacetylation.

Hence, a number of drugs which are being used clinically target processes involved in gene transcription and a number of other approaches are being developed and have considerable therapeutic potential. These methods evidently target a specific gene or genes which are naturally regulated by a particular transcription factor or by modulation of chromatin structure in a particular situation.

However, it has also proved possible to develop potential therapies which take advantage of a specific property of a group of transcription factors to develop engineered transcription factors that can potentially target any gene in the genome, whose expression needs to be manipulated for therapeutic benefit. Thus, as described in Chapter 4 (section 4.3.2), the two cysteine, two histidine zinc finger DNA binding domain has an α-helical region which contacts the DNA, with the precise sequence at the N-terminus of this α-helical region determining the exact DNA sequence to which the zinc finger binds. Moreover, it is now possible to predict from the amino acid sequence at the N-terminus, which DNA sequence will be bound by the factor.

This therefore allows artificial zinc finger-containing proteins to be synthesized, which contain a particular sequence at the N-terminus of the α-helical region and which will therefore bind a specific DNA sequence. Thus, it is possible to introduce an artificial zinc finger into cells which will bind specifically to any gene carrying the target sequence of the zinc finger. Hence, artificial zinc fingers can be produced which will target any particular gene on the basis of a specific DNA sequence that it contains (for reviews see Klug, 2005a,b).

If this designer zinc finger is then linked to a domain which inhibits transcription (see Chapter 6, section 6.3.2), the zinc finger will deliver the inhibitory domain to the target gene and inhibit its transcription (Fig. 9.55a). This method has been used to specifically inhibit the expression of the CHK2 gene, which is involved in cell proliferation and cancer, without inhibiting the expression of any other cellular gene (Tan *et al.*, 2003). Similarly, it has been used to inhibit the expression of particular viral genes in infected cells and thereby inhibit infection of cultured cells with viruses causing human disease, such as herpes simplex virus and human immunodeficiency virus (Papworth *et al.*, 2003; Reynolds *et al.*, 2003).

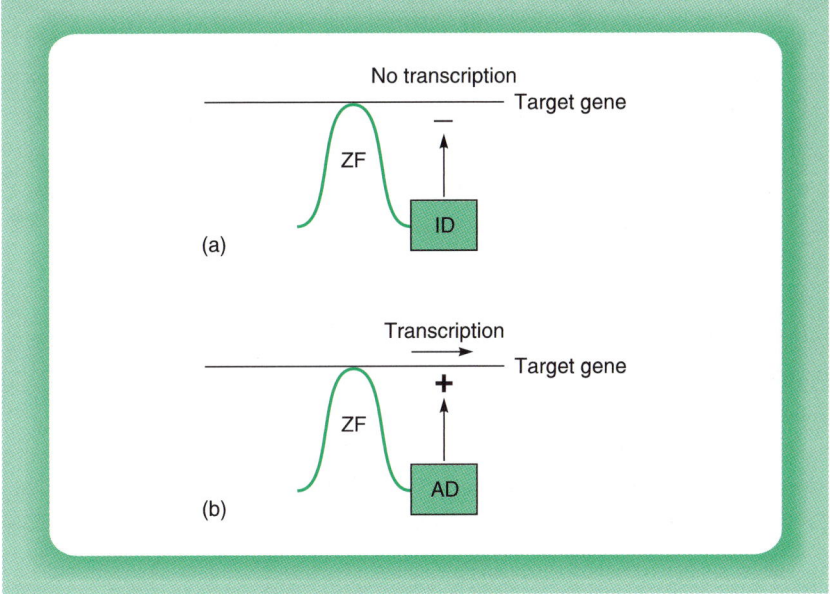

Figure 9.55

A zinc finger designed to bind specifically to a DNA sequence in a target gene can be used to deliver either a transcriptional inhibitory domain, thereby repressing transcription (panel a), or a transcriptional activation domain, thereby activating transcription (panel b).

Similarly, it is possible to link the designer finger to a transcriptional activation domain (see Chapter 5, section 5.2) so that it activates transcription of the target gene bound by the zinc finger (Fig. 9.55b). This approach has been used, for example, to activate the gene encoding the VEGF growth factor in the intact animal *in vivo*. This resulted in enhanced levels of VEGF, which were functional in inducing the formation of new blood vessels and could therefore be of therapeutic value in diseases where patients suffer from poor blood supply (for review see Pasqualini *et al.*, 2002) (Fig. 9.56).

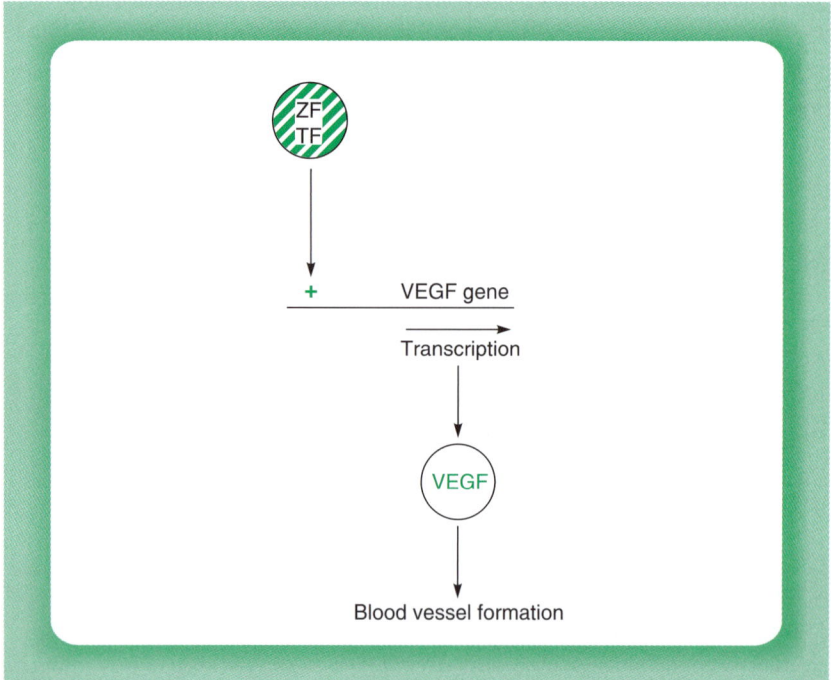

Figure 9.56
The synthesis of a zinc finger transcription factor (ZFTF) with a novel DNA binding specificity that allows it to bind to the VEGF gene results in VEGF gene transcription. The resulting VEGF protein then induces blood vessel formation.

Obviously, this method is not limited to delivering domains which modulate transcription but can also be used to deliver a variety of different protein domains. Another potential application of this technique therefore involves using a designer zinc finger to specifically deliver an endonuclease to a particular gene. The endonuclease cuts the DNA of the target gene. In turn, this facilitates homologous recombination in which a mutant gene sequence involved in a particular disease is replaced by the corresponding wild type sequence (for reviews see Kaiser, 2005; Klug, 2005b). This example does not involve the manipulation of gene expression but rather exploits a particular property of zinc finger transcription factors to specifically deliver a protein which cuts DNA.

The use of designer zinc fingers thus offers an attractive therapeutic method both for manipulating gene expression and for other purposes. The use of this method in patients, however, will obviously require the development of gene therapy procedures which are able to deliver the designer zinc finger and its associated domain, both safely and effectively. Currently, it therefore suffers from the same limitations as the methods for delivering genes encoding transcription factors or for inhibiting their expression with RNAi, which were discussed above. Nonetheless, when taken together with the available and developing drugs which manipulate transcription factor activity, it is clear that there are a number of actual and potential therapeutic approaches which involve targeting transcription factors or taking advantage of their particular properties. It is likely that the number of treatments involving transcription factors in these ways will increase dramatically in the future.

9.6 CONCLUSIONS

The ability to affect cellular transcriptional regulatory processes is crucial to the ability of many different viruses to transform cells. Thus, for example, the large T oncogenes of the small DNA tumour viruses SV40 and polyoma and the ElA protein of adenovirus can all affect cellular gene expression and this ability is essential for the transforming ability of these viruses (for review see Moran, 1993).

In this chapter we have seen that several RNA viruses also have this ability, containing transcription factors which can act as oncogenes, either by promoting the expression of genes required for growth or by inhibiting the expression of genes required for the production of non-proliferating differentiated cells.

Although the oncogenes of both DNA and some RNA tumour viruses can therefore affect transcription, their origins are completely different. Thus, whilst the oncogenes of the DNA viruses do not have equivalents in cellular DNA and appear to have evolved within the viral genome, the oncogenes of retroviruses have, as we have seen, been picked up from the cellular genome. The fact that, despite their diverse origins, both types of oncogenes can affect transcription, indicates therefore that the modulation of transcription represents an effective mechanism for the transformation of cells.

In addition, however, the origin of retroviral oncogenes from the cellular genome allows several other features of transcription to be studied.

Thus, for example, the conversion of a normal cellular transcription factor into a cancer-causing viral oncogene allows insights to be obtained into the processes whereby oncogenes become activated.

In general, such oncogenes, whether they encode growth factors, growth factor receptors or other types of protein, can be activated within a virus either by over-expression driven by a strong retroviral promoter or by mutation. The transcription factors we have discussed in this chapter illustrate both these processes. Thus, the Fos, Jun and Myc oncogenes for example become cancer-causing both by continuous expression of proteins which are normally made only transiently, leading to constitutive stimulation of genes required for growth, as well as in some cases by mutations in the viral forms of the protein which render them more potent transcriptional activators. Similarly, the ErbA oncogene is activated by deletion of a part of the protein coding region, leading to a protein with different or enhanced properties. Although such effects of mutation or over-expression have initially been defined in tumourigenic retroviruses, it is clear that such changes can also occur within the cellular genome, over-expression of the c-*myc* oncogene for example being characteristic of many different human tumours (for review see Spencer and Groudine, 1991) whilst several other transcription factor genes are activated by the translocations characteristic of particular human leukaemias (see section 9.3.4).

In addition, since cellular oncogenes clearly also play an important role in the regulation of normal cellular growth and differentiation, their identification via tumourigenic retroviruses has, paradoxically, greatly aided the study of normal cellular growth regulatory processes. Thus, for example, the prior isolation of the c-*fos* and c-*jun* genes greatly aided the characterization of the AP1 binding activity and of its role in stimulating genes involved in cellular growth.

A similar boost to our understanding of growth regulation in normal cells has also emerged from studies of the anti-oncogene proteins. Thus, studies on the Rb-1 gene, which was originally identified on the basis of its inactivation in retinoblastomas, have led to an understanding of its key role in regulating the balance between cellular growth and differentiation. Similarly, work on p53, which was originally identified as a protein interacting with the product of the SV40 large T oncogene, has led to the identification of its key role as the so-called 'guardian of the genome'.

The interaction of p53 and SV40 large T indicates another aspect of anti-oncogenes, namely their antagonistic interaction with oncogene products. Thus both p53 and Rb-1 have been shown to bind cellular and viral oncogene proteins, with the activity of the anti-oncogene product being inhibited by this interaction. Such interactions are not confined to the oncogenes and anti-oncogenes, which encode factors directly regulating transcription. Thus, as discussed in section 9.4.4, the APC anti-oncogene protein does not act directly as a transcription factor. Rather it interacts with the β-catenin oncogene product to promote its degradation and inhibit its activity. Hence, cancer can result from mutations in the β-catenin oncogene, which enhance the stability of its protein product or directly enhance its ability to stimulate transcription, or from mutations in the APC protein, which inactivate it and prevent it interfering with the function of β-catenin. Similarly, it has been shown that the Jun oncoprotein can promote tumour formation by antagonizing the pro-apoptotic effect of p53 (Eferl *et al.*, 2003).

Hence, the interaction between oncogene and anti-oncogene products is likely to play a key role in regulating cellular growth and survival. The uncontrolled growth characteristic of cancer cells therefore results from changes in this balance due either to over-expression or mutational activation of oncogenes or to deletion or mutational inactivation of anti-oncogenes.

It is clear therefore that, as with other oncogenes and anti-oncogenes, the study of the oncogenes and anti-oncogenes which encode transcription factors can provide considerable information on both the processes regulating normal growth and differentiation and on how these processes are altered in cancer. When taken together with the involvement of transcription factor mutations in disorders of development or hormone responses, discussed in section 9.1, they illustrate the key role played by transcription factors, the manner in which alterations in their activity can result in disease and their potential as therapeutic targets.

REFERENCES

Adhikary, S. and Eilers, M. (2005) Transcriptional regulation and transformation by Myc proteins. Nature Reviews Molecular Cell Biology 6, 635–645.

Almog, N. and Rotter, V. (1997) Involvement of p53 in differentiation and development. Biochemica et Biophysica Acta 1333, F1–F27.

Amoils, S. (2006) A twist in the tail. Nature Reviews Molecular Cell Biology 7, 796–797.

Anderson, M.E. and Tegtmeyer, P. (1995) Giant leap for p53, small step for drug design. Bioessays 17, 3–7.

Asikainen, T.M., Schneider, B.K., Waleh, N.S., Clyman, R.I., Ho, W.B., Flippin, L.A., Gunzler, V. and White, C.W. (2005) Activation of hypoxia-inducible factors in hyperoxia through prolyl 4-hydroxylase blockade in cells and explants of primate lung. Proceedings of the National Academy of Sciences, USA 102, 10212–10217.

Attwooll, C., Denchi, E.L. and Helin, K. (2004) The E2F family: specific functions and overlapping interests. EMBO Journal 23, 4709–4716.

Avantaggiati, M.L., Ogryzko, V., Gardner, K., Giordano, A., Levine, A.S. and Kelly, K. (1997) Recruitment of p300/CBP in p53 dependent signalling pathway. Cell 89, 1175–1184.

Basseres, D.S. and Baldwin, A.S. (2006) Nuclear factor-κB and inhibitor of κB kinase pathways in oncogenic initiation and progression. Oncogene 25, 6817–6830.

Baudino, T.A. and Cleveland, J.L. (2001) The Max network gone mad. Molecular and Cellular Biology 21, 691–702.

Belandia, B. and Parker, M.G. (2003) Nuclear receptors: a rendezvous for chromatin remodeling factors. Cell 114, 277–280.

Berns, A. (1994) Is p53 the only real tumour suppressor gene? Current Biology 4, 137–139.

Best, J.L., Amezcua, C.A., Mayr, B., Flechner, L., Murawsky, C.M., Emerson, B., Zor, T., Gardner, K.H. and Montminy, M. (2004) Identification of small-molecule antagonists that inhibit an activator:coactivator interaction. Proceedings of the National Academy of Sciences, USA 101, 17622–17627.

Blackwood, E.M. and Eisenman, R.N. (1991) Max: a helix-loop-helix zipper protein that forms a sequence – specific – DNA binding complex with myc. Science 251, 1211–1217.

Blank, V. and Andrews, N.C. (1997) The Maf transcription factors: regulators of differentiation. Trends in Biochemical Sciences 22, 437–441.

Boncinelli, E. (1997) Homeobox genes and disease. Current Opinion in Genetics and Development 7, 331–337.

Bradsher, J., Auriol, J., Proietti, de Santis, Iben, S., Vonesch, J.L., Grummt, I. and Egly, J.M. (2002) CSB is a component of RNA pol I transcription. Molecular Cell 10, 819–829.

Broach, J.R. and Levine, A.J. (1997) Oncogenes and cell proliferation. Current Opinion in Genetics and Development 7, 1–6.

Brown, T.R.P., Scott, P.H., Stein, T., Winter, A.G. and White, R.J. (2000) RNA Polymerase III transcription: its control by tumour suppressors and its deregulation by transforming agents. Gene Expression 9, 15–28.

Buschmann, T., Fuchs, S.Y., Lee, C-G., Pan, Z-Q. and Ronai, Z. (2000) SUMO-1 modification of Mdm2 prevents its self-ubiquitination and increases Mdm2 ability to ubiquitinate p53. Cell 101, 753–762.

Butler, R. and Bates, G.P. (2006) Histone deacetylase inhibitors as therapeutics for polyglutamine disorders. Nature Reviews Neuroscience 7, 784–796.

Chan, H.M., Krstic-Demonacos, M., Smith, L., Demonacos, C. and La Thangue, N.B. (2001) Acetylation control of the retinoblastoma tumour-suppressor protein. Nature Cell Biology 3, 667–674.

Chuikov, S., Kurash, J.K., Wilson, J.R., Xiao, B., Justin, N., Ivanov, G.S., McKinney, K., Tempst, P., Prives, C., Gamblin, S.J., Barlev, N.A. and Reinberg, D. (2004) Regulation of p53 activity through lysine methylation. Nature 432, 353–360.

Cloos, P.A., Christensen, J., Agger, K., Maiolica, A., Rappsilber, J., Antal, T., Hansen, K.H. and Helin, K. (2006) The putative oncogene GASC1 demethylates tri- and dimethylated lysine 9 on histone H3. Nature 442, 307–311.

Conaway, J.W. and Conaway, R.C. (1999) Transcription elongation and human disease. Annual Reviews of Biochemistry 68, 301–319.

Conkright, M.D. and Montminy, M. (2005) CREB: the unindicted cancer co-conspirator. Trends in Cell Biology 15, 457–459.

Courtois, G. and Gilmore, T.D. (2006) Mutations in the NF-κB signalling pathway: implications for human disease. Oncogene 25, 6831–6843.

Crighton, D., Woiwode, A., Zhang, C., Mandavia, N., Morton, J.P., Warnock, L.J., Milner, J., White, R.J. and Johnson, D.L. (2003) p53 represses RNA polymerase III transcription by targetting TBP and inhibiting promoter occupancy by TFIIIB. EMBO Journal 22, 2810–2820.

Cui, L., Jeong, H., Borovecki, F., Parkhurst, C.N., Tanese, N. and Krainc, D. (2006) Transcriptional repression of PGC-1α by mutant huntingtin leads to mitochondrial dysfunction and neurodegeneration. Cell 127, 59–69.

D'Arcangelo, G. and Curran, T. (1995) Smart transcription factors. Nature 389, 149–152.

De Falco, G., Comes, F. and Simone, C. (2006) pRb: master of differentiation. Coupling irreversible cell cycle withdrawal with induction of muscle-specific transcription. Oncogene 25, 5244–5249.

de OcaLuna, R.M., Wagner, D.S. and Lozano, G. (1995) Rescue of early embryonic lethality in mdm2-deficent mice by deletion of p53. Nature 378, 203–206.

Dery, M-A.C., Michaud, M.D. and Richard, D.E. (2005) Hypoxia-inducible factor 1: regulation by hypoxic and non-hypoxic activators. International Journal of Biochemistry and Cell Biology 37, 535–540.

Dimova, D.K. and Dyson, N.J. (2005) The E2F transcriptional network: old acquaintances with new faces. Oncogene 24, 2810–2826.

Du, W. and Pogoriler, J. (2006) Retinoblastoma family genes. Oncogene 25, 5190–5200.

Dyson, N. (2003) A twist in a mouse tale. Nature 421, 903–904.

Eferl, R., Ricci, R., Kenner, L., Zenz, R., David, J-P., Rath, M. and Wagner, E.F. (2003) Liver tumour development: c-Jun antagonises the proapoptotic activity of p53. Cell 112, 181–192.

Egger, G., Liang, G., Aparicio, A. and Jones, P.A. (2004) Epigenetics in human disease and prospects for epigenetic therapy. Nature 429, 457–463.

Engelkamp, D. and van Heyningen, V. (1996) Transcription factors in disease. Current Opinion in Genetics and Development 6, 334–342.

Esquela-Kerscher, A. and Slack, F.J. (2006) Oncomirs – microRNAs with a role in cancer. Nature Reviews Cancer 6, 259–269.

Evans, R.M., Barish, G.D. and Wang, Y.X. (2004) PPARs and the complex journey to obesity. Nature Medicine 10, 355–361.

Feng, Z., Jin, S., Zupnick, A., Hoh, J., de Stanchina, E., Lowe, S., Prives, C. and Levine, A.J. (2006) p53 tumor suppressor protein regulates the levels of huntingtin gene expression. Oncogene 25, 1–7.

Freiman, R.N. and Tjian, R. (2002) A glutamine-rich trail leads to transcription factors. Science 296, 2149–2150.

French Anderson, W. (1998) Human gene therapy. Nature 392, Supplement, 25–30.

French Anderson, W. (2000) Gene therapy scores against cancer. Nature Medicine 6, 862–863.

Friend, S. (1994) p53: a glimpse at the puppet behind the shadow play. Science 265, 334–335.

Gartel, A.L. (2006) A new mode of transcriptional repression by c-myc: methylation. Oncogene 25, 1989–1990.

Genovese, C., Trani, D., Caputi, M. and Claudio, P.P. (2006) Cell cycle control and beyond: emerging roles for the retinoblastoma gene family. Oncogene 25, 5201–5209.

Giacinti, C. and Giordano, A. (2006) RB and cell cycle progression. Oncogene 25, 5220–5227.

Gibbons, R.J., Picketss, D.S., Villard, L. and Higgs, D.R. (1995) Mutations in a putative global regulator cause X-linked mental retardation with α-Thalassaemia (ATR-X) syndrome. Cell 80, 837–845.

Gottifredi, V. and Prives, C. (2001) Getting p53 out of the nucleus. Science 292, 1851–1852.

Grandori, C., Cowley, S.M., James, L.P. and Eisenman, R.N. (2000) The Myc/Max/Mad network and the transcriptional control of cell behaviour. Annual Review of Cell and Developmental Biology 16, 653–699.

Gregory, R.I. and Shiekhattar, R. (2004) Chromatin modifiers and carcinogenesis. Trends in Cell Biology 14, 695–702.

Hake, S.B., Xiao, A. and Allis, C.D. (2004) Linking the epigenetic 'language' of covalent histone modifications to cancer. British Journal of Cancer 90, 761–769.

Hannon, G.J. and Rossi, J.J. (2004) Unlocking the potential of the human genome with RNA interference. Nature 431, 371–378.

Hastie, N.D. (2001) Life, sex, and WT1 isoforms – three amino acids can make all the difference. Cell 106, 391–394.

Haupt, Y., Maya, R., Kazaz, A. and Oren, M. (1997) Mdm2 promotes rapid degradation of p53. Nature 387, 296–299.

Haupt, Y., Robles, A.I., Prives, C. and Rotter, V. (2002) Deconstruction of p53 functions and regulation. Oncogene 21, 8223–8231.

Helmlinger, D., Tora, L. and Devys, D. (2006) Transcriptional alterations and chromatin remodeling in polyglutamine diseases. Trends in Genetics 22, 562–570.

Huang, J., Perez-Burgos, L., Placek, B.J., Sengupta, R., Richter, M., Dorsey, J.A., Kubicek, S., Opravil, S., Jenuwein, T. and Berger, S.L. (2006) Repression of p53 activity by Smyd2-mediated methylation. Nature 444, 629–632.

Hunter, T. (1998) Prolyl isomerases and nuclear function. Cell 92, 141–143.

Insinga, A., Monestiroli, S., Ronzoni, S., Carbone, R., Pearson, M., Pruneri, G., Viale, G., Appella, E., Pelicci, P.G. and Minucci, S. (2004) Impairment of p53 acetylation, stability and function by an oncogenic transcription factor. EMBO Journal 23, 1144–1154.

Ito, A., Kawaguchi, Y., Lai, C-H., Kovacs, J.J., Higashimoto, Y., Appella, E. and Yao, T-P. (2002) MDM2-HDAC1-mediated deacetylation of p53 is required for its degradation. EMBO Journal 21, 6236–6245.

Jordan, V.C. (2007) Chemoprevention of breast cancer with selective oestrogen-receptor modulators. Nature Reviews Cancer 7, 46–53.

Kaiser, J. (2005) Putting the fingers on gene repair. Science 310, 1894–1896.

Karin, M. (2006) Nuclear factor-kappaB in cancer development and progression. Nature 441, 431–436.

Karin, M., Liu, Z-G. and Zandi, E. (1997) AP-1 function and regulation. Current Opinion in Cell Biology 9, 240–246.

Kerppola, T. and Curran, T. (1995) Zen and the art of Fos and Jun. Nature 373, 199–200.

Kersten, S., Desvergne, B. and Wahli, W. (2000) Roles of PPARs in health and disease. Nature 405, 421–424.

Khidr, L. and Chen, P.L. (2006) RB, the conductor that orchestrates life, death and differentiation. Oncogene 25, 5210–5219.

Kioussi, C., Briata, P., Baek, S.H., Rose, D.W., Hamblet, N.S., Herman, T., Ohgi, K.A., Lin, C., Gleiberman, A., Wang, J., Brault, V., Ruiz-Lozano, P., Nguyen, H.D., Kemler, R., Glass, C.K., Wynshaw-Boris, A. and Rosenfeld, M.G. (2002) Identification of a Wnt/Dvl/b-Catenin → Pitx2 pathway mediating cell-type-specific proliferation during development. Cell 111, 673–685.

Klug, A. (2005a) Towards therapeutic applications of engineered zinc finger proteins. FEBS Letters 579, 892–894.

Klug, A. (2005b) The discovery of zinc fingers and their development for practical applications in gene regulation. Proceedings of the Japan Academy Series B: Physical and Biological Sciences 81, Ser. B, 87–102.

Kotake, Y., Cao, R., Viatour, P., Sage, J., Zhang, Y. and Xiong, Y. (2007) pRB family proteins are required for H3K27 trimethylation and Polycomb repression complexes binding to and silencing p16INK4a tumor suppressor gene. Genes and Development 21, 49–54.

La Spada, A.R. and Morrison, R.S. (2005) The power of the dark side: Huntington's disease protein and p53 form a deadly alliance. Neuron 47, 1–3.

Land, H., Parada, L.F. and Weinberg, R. (1983) Cellular oncogenes and multistep carcinogenesis. Science 222, 771–778.

Lane, D.P. (1992) p53, guardian of the genome. Nature 358, 15–16.

Lane, D.P. and Fischer, P.M. (2004) Turning the key on p53. Nature 427, 789–790.

Lane, D.P. and Hall, P.A. (1997) MDM2 – arbiter of p53's destruction. Trends in Biochemical Sciences 22, 373–374.

Lasorella, A., Noseda, M., Beyna, M. and Lavarone, A. (2000) Id2 is a retinoblastoma protein target and mediates signalling by Myc oncoproteins. Nature 407, 592–598.

Latchman, D.S. (1996) Transcription factor mutations and human diseases. New England Journal of Medicine 334, 28–33.

Latchman, D.S. (2000) Transcription factors as potential targets for therapeutic drugs. Current Pharmaceutical Biotechnology 1, 57–61.

Latchman, D.S. (2005) Gene regulation – a eukaryotic perspective. Fifth edition. Taylor and Francis, Oxford and New York.

Lee, C., Chang, J.H., Lee, H.S. and Cho, Y. (2002) Structural basis for the recognition of the E2F transactivation domain by the retinoblastoma tumour suppressor. Genes and Development 16, 3199–3212.

Levens, D. (2002) Disentangling the MYC web. Proceedings of the National Academy of Sciences, USA 99, 5757–5759.

Li, B., Pattenden, S.G., Lee, D., Gutierrez, J., Chen, J., Seidel, C., Gerton, J. and Workman, J.L. (2005) Preferential occupancy of histone variant H2AZ at inactive promoters influences local histone modifications and chromatin remodeling. Proceedings of the National Academy of Sciences, USA 102, 18385–18390.

Li, L., Chambard, J.C., Karin, M. and Olson, E.M. (1992) Fos and Jun repress transcriptional activation by myogenin and MyoD: the amino terminus of Jun can mediate repression. Genes and Development 6, 676–689.

Li, M., Brooks, C.L., Wu-Baer, F., Chen, D., Baer, R. and Gu, W. (2003) Mono- versus polyubiquitination: differential control of p53 fate by Mdm2. Science 302, 1972–1975.

Li, M., Luo, J., Brooks, C.L. and Gu, W. (2002) Acetylation of p53 inhibits its ubiquitination by Mdm2. Journal of Biological Chemistry 52, 50607–50611.

Li, X-Y. and Green, M.R. (1996) Transcriptional elongation and cancer. Current Biology 6, 943–944.

Lin, R.J., Sternsdorf, T., Tini, M. and Evans, R.M. (2001) Transcriptional regulation in acute promyelocytic leukemia. Oncogene 20, 7204–7215.

Liu, J., Akoulitchev, S., Weber, A., Ge, H., Chuikov, S., Libutti, D., Wang, X.W., Conaway, J.W., Harris, C.C., Conaway, R.C., Reinberg, D. and Levens, D. (2001) Defective interplay of activators and repressors with TFIIH in xeroderma pigmentosum. Cell 104, 353–363.

Liu, X. and Marmorstein, R. (2006) When viral oncoprotein meets tumor suppressor: a structural view. Genes and Development 20, 2332–2337.

Look, A.T. (1997) Oncogenic transcription factors in the human acute leukaemias. Science 278, 1059–1064.

Lowe, S.W., Cepero, E. and Evan, G. (2006) Intrinsic tumour suppression. Nature Milestones Cancer Supplement, S25–S33.

Luo, J., Nikolaev, A.Y., Imai, S-I., Chen, D., Su, F., Shiloh, A., Guarente, L. and Gu, W. (2001) Negative control of p53 by Sir2a promotes cell survival under stress. Cell 107, 137–148.

Macaluso, M., Montanari, M. and Giordano, A. (2006) Rb family proteins as modulators of gene expression and new aspects regarding the interaction with chromatin remodeling enzymes. Oncogene 25, 5263–5267.

Marx, J. (1997) Possible function found for breast cancer genes. Science 276, 531–532.

Marx, J. (2005) Fused genes may help explain the origins of prostate cancer. Science 310, 603.

Mayo, L.D. and Donner, D.B. (2002) The PTEN, Mdm2, p53 tumour suppressor-oncoprotein network. Trends in Biochemical Sciences 27, 462–467.

Mills, A.A. (2005) p53: link to the past, bridge to the future. Genes and Development 19, 2091–2099.

Minucci, S. and Pelicci, P.G. (2006) Histone deacetylase inhibitors and the promise of epigenetic (and more) treatments for cancer. Nature Reviews Cancer 6, 38–51.

Minucci, S., Nervi, C., Lo, Coco F. and Pelicci, P.G. (2001) Histone deacetylases: a common molecular target for differentiation treatment of acute myeloid leukemias? Oncogene 20, 3110–3115.

Moisan, A., Larochelle, C., Guillemette, B. and Gaudreau, L. (2004) BRCA1 Can Modulate RNA Polymerase II Carboxy-Terminal Domain Phosphorylation Levels. Molecular and Cellular Biology 24, 6947–6956.

Moran, E. (1993) DNA tumour virus transforming proteins and the cell cycle. Current Opinion in Genetics and Development 3, 63–70.

Motohashi, H., Shavit, J.A., Igarashi, K., Yamamoto, K. and Engel, J.D. (1997) The world according to MAF. Nucleic Acids Research 25, 2953–2959.

Mullan, P.B., Quinn, J.E. and Harkin, D.P. (2006) The role of BRCA1 in transcriptional regulation and cell cycle control. Oncogene 25, 5854–5863.

Müller, H. and Helin, K. (2000) The E2F transcription factors: key regulators of cell proliferation. Biochimica et Biophysica Acta 1470, M1–M12.

Münger, K. (2003) Clefts, grooves, and (small) pockets: the structure of the retinoblastoma tumour suppressor in complex with its cellular target E2F unveiled. Proceedings of the National Academy of Sciences, USA 100, 2165–2167.

Nair, S.K. and Burley, S.K. (2003) X-ray structures of Myc-Max and Mad-Max recognising DNA: molecular bases of regulation by proto-oncogenic transcription factors. Cell 112, 193–205.

Nateri, A.S., Spencer-Dene, B. and Behrens, A. (2005) Interaction of phosphorylated c-Jun with TCF4 regulates intestinal cancer development. Nature 437, 281–285.

Nelson, C.J., Santos-Rosa, H. and Kouzarides, T. (2006) Proline isomerization of histone h3 regulates lysine methylation and gene expression. Cell 126, 905–916.

Nelson, W.J. and Nusse, R. (2004) Convergence of Wnt, beta-catenin, and cadherin pathways. Science 303, 1483–1487.

Nilsson, J.A. and Cleveland, J.L. (2003) Myc pathways provoking cell suicide and cancer. Oncogene 22, 9007–9021.

Oskarsson, T. and Trumpp, A. (2005) The Myc trilogy: lord of RNA polymerases. Nature Cell Biology 7, 215–217.

Ozanne, B.W., Spence, H.J., McGarry, L.C. and Hennigan, R.F. (2007) Transcription factors control invasion: AP-1 the first among equals. Oncogene 26, 1–10.

Papworth, M., Moore, M., Isalan, M., Minczuk, M., Choo, Y. and Klug, A. (2003) Inhibition of herpes simplex virus 1 gene expression by designer zinc-finger transcription factors. Proceedings of the National Academy of Sciences, USA 100, 1621–1626.

Pasqualini, R., Barbas, C.F. and Arap, W. (2002) Vessel maneuvers: Zinc fingers promote angiogenesis. Nature Medicine 8, 1353–1354.

Peifer, M. (1997) β-caterin as oncogene: the smoking gun. Science 275, 1752–1753.

Perk, J., Iavarone, A. and Benezra, R. (2005) Id family of helix-loop-helix proteins in cancer. Nature Reviews Cancer 5, 603–614.

Perlmann, T. and Vennstrom, B. (1995) Nuclear receptors: the sound of silence. Nature 377, 387–388.

Peukert, K., Staller, P., Schneider, A., Carmichael, G., Hanel, F. and Eilers, M. (1997) An alternative pathway for gene regulation by Myc. EMBO Journal 16, 5672–5686.

Pierce, J.W., Read, M.A., Ding, H., Luscinskas, F.W. and Collins, T. (1996) Salicylates inhibit I κ B-α phosphorylation, endothelial-leukocyte adhesion molecule expression, and neutrophil transmigration. Journal of Immunology 156, 3961–3969.

Poot, R.A., Bozhenok, L., van den Berg, D.L., Steffensen, S., Ferreira, F., Grimaldi, M., Gilbert, N., Ferreira, J. and Varga-Weisz, P.D. (2004) The Williams syndrome transcription factor interacts with PCNA to target chromatin remodelling by ISWI to replication foci. Nature Cell Biology 6, 1236–1244.

Pouyssegur, J., Dayan, F. and Mazure, N.M. (2006) Hypoxia signalling in cancer and approaches to enforce tumour regression. Nature 441, 437–443.

Poyurovsky, M.V. and Prives, C. (2006) Unleashing the power of p53: lessons from mice and men. Genes and Development 20, 125–131.

Prives, C. (1998) Signalling to p53: breaking the MDM2-p53 circuit. Cell 95, 5–8.

Prives, C. and Manley, J.L. (2001) Why is p53 acetylated? Cell 107, 815–818.

Rabbits, T.H. (1994) Chromosomal translocations in human cancer. Nature 372, 143–149.

Reya, T. and Clevers, H. (2005) Wnt signalling in stem cells and cancer. Nature 434, 843–850.

Reynolds, L., Ullman, C., Moore, M., Isalan, M., West, M.J., Clapham, P., Klug, A. and Choo, Y. (2003) Repression of the HIV-1 5' LTR promoter and inhibition of HIV-1 replication by using engineered zinc-finger transcription factors. Proceedings of the National Academy of Sciences, USA 100, 1615–1620.

Riley, B.E. and Orr, H.T. (2006) Polyglutamine neurodegenerative diseases and regulation of transcription: assembling the puzzle. Genes and Development 20, 2183–2192.

Rosen, E.D. and Spiegelman, B.M. (2001) PPARgamma: a nuclear regulator of metabolism, differentiation and cell growth. Journal of Biological Chemistry 276, 37731–37734.

Rosin-Arbesfeld, R., Townsley, F. and Bienz, M. (2000) The APC tumour suppressor has a nuclear export function. Nature 406, 1009–1012.

Ross, C.A. and Thompson, L.M. (2006) Transcription meets metabolism in neurodegeneration. Nature Medicine 12, 1239–1241.

Rossi, J.J. (2004) A cholesterol connection in RNAi. Nature 432, 155–156.

Sage, J. (2007) Hope in sight for retinoblastoma. Nature Medicine 13, 30–31.

Sap, J., Munoz, A., Damm, K., Goldberg, Y., Ghysdael, J., Leutz, A., Beug, H. and Vennstrom, B. (1986) The c-erb-A protein is a high-affinity receptor for thyroid hormone. Nature 324, 635–640.

Sap, J., Munoz, A., Schmitt, A., Stunnenberg, H. and Vennstrom, B. (1989) Repression of transcription at a thyroid hormone response element by the v-*erbA* oncogene product. Nature 340, 242–244.

Sawyers, C.L. and Denny, C.T. (1994) Chronic myelomonocytic leukemia. Tel-a-kinase what ets all about. Cell 77, 171–173.

Schwartz, M.W. and Kahn, S.E. (1999) Insulin resistance and obesity. Nature 402, 860–861.

Scully, R., Anderson, S.F., Chao, D.M., Wei, W., Ye, L., Young, R.A., Livingston, D.M. and Parvin, J.D. (1997) BRCA1 is a component of the RNA polymerase II holoenzyme. Proceedings of the National Academy of Sciences, USA 94, 5605–5610.

Sdek, P., Ying, H., Chang, D.L., Qiu, W., Zheng, H., Touitou, R., Allday, M.J. and Jim Xiao, Z.X. (2005) MDM2 promotes proteasome-dependent ubiquitin-independent degradation of retinoblastoma protein. Molecular Cell 20, 699–708.

Semenza, G.L. (2006) VHL and p53: tumor suppressors team up to prevent cancer. Molecular Cell 22, 437–439.

Sharpless, N.E. and DePinho, R.A. (2002) p53: Good cop/bad cop. Cell 110, 9–12.

Shaulian, E. and Karin, M. (2002) AP-1 as a regulator of cell life and death. Nature Cell Biology 4, E131–E136.

Spencer, C.A. and Groudine, M. (1991) Control of c-*myc* regulation in normal and neoplastic cells. Advances in Cancer Research 56, 1–48.

Taipale, J. and Beachy, P.A. (2001) The Hedgehog and Wnt signalling pathways in cancer. Nature 411, 349–354.

Tan, S., Guschin, D., Davalos, A., Lee, Y.L., Snowden, A.W., Jouvenot, Y., Zhang, H.S., Howes, K., McNamara, A.R., Lai, A., Ullman, C., Reynolds, L., Moore, M., Isalan, M., Berg, L.P., Campos, B., Qi, H., Spratt, S.K., Case, C.C., Pabo, C.O., Campisi, J. and Gregory, P.D. (2003) Zinc-finger protein targeted gene regulation: genomewide single-gene specificity. Proceedings of the National Academy of Sciences, USA 100, 11997–12002.

Toledo, F. and Wahl, G.M. (2006) Regulating the p53 pathway: in vitro hypotheses, in vivo veritas. Nature Reviews Cancer 6, 909–923.

Tyteca, S., Legube, G. and Trouche, D. (2006) To die or not to die: a HAT trick. Molecular Cell 24, 807–808.

Uchida, C., Miwa, S., Kitagawa, K., Hattori, T., Isobe, T., Otani, S., Oda, T., Sugimura, H., Kamijo, T., Ookawa, K., Yasuda, H. and Kitagawa, M. (2005) Enhanced Mdm2 activity inhibits pRB function via ubiquitin-dependent degradation. EMBO Journal 24, 160–169.

Valdez, B.C., Henning, D., So, R.B., Dixon, J. and Dixon, M.J. (2004) The Treacher Collins syndrome (TCOF1) gene product is involved in ribosomal DNA gene transcription by interacting with upstream binding factor. Proceedings of the National Academy of Sciences, USA 101, 10709–10714.

Van Dyke, T. (2005) Sense out of missense. Nature 434, 287–288.

Vassilev, L.T., Vu, B.T., Graves, B., Carvajal, D., Podlaski, F., Filipovic, Z., Kong, N., Kammlott, U., Lukacs, C., Klein, C., Fotouhi, N. and Liu, E.A. (2004) In vivo activation of the p53 pathway by small-molecule antagonists of MDM2. Science 303, 844–848.

Vaziri, H., Dessain, S.K., Eaton, E.N., Imai, S-I., Frye, R.A., Pandita, T.K., Guarente, L. and Weinberg, R.A. (2001) hSIR2^{SIRT1} functions as an NAD-dependent p53 deacetylase. Cell 107, 149–159.

Viatour, P., Merville, M-P., Bours, V. and Chariot, A. (2005) Phosphorylation of NF-κB and IκB proteins: implications in cancer and inflammation. Trends in Biochemical Sciences 30, 43–52.

Vogelstein, B. and Kinzler, K.W. (2006) Cancer genes and the pathways they control. Nature Milestones Cancer Supplement, S33–S42.

Vogelstein, B., Lane, D. and Levine, A.J. (2000) Surfing the p53 network. Nature 408, 307–310.

Vogt, P.K. and Bader, A.G. (2005) Jun: stealth, stability, and transformation. Molecular Cell 19, 432–433.

Vousden, K.H. and Prives, C. (2005) p53 and prognosis: new insights and further complexity. Cell 120, 7–10.

Wallace, V.A. (2006) Second step to retinal tumours. Nature 442, 45–46.

Wang, L., Fan, C., Topol, S.E., Topol, E.J. and Wang, Q. (2003) Mutation of MEF2A in an inherited disorder with features of coronary artery disease. Science 302, 1578–1581.

Weinberger, C., Thompson, C.C., Ong, E.S., Lebo, R., Gruol, D.J. and Evans, R.M. (1986) The c-erb-A gene encodes a thyroid hormone receptor. Nature 324, 641–646.

White, R.J. (2004) RNA polymerase III transcription and cancer. Oncogene 23, 3208–3216.

Wolf, D., Rodova, M., Miska, E.A., Calvet, J.P. and Kouzarides, T. (2002) Acetylation of β-Catenin by CREB-binding protein (CBP). Journal of Biological Chemistry 277, 25562–25567.

Wolffe, A. (1997) Sinful repression. Nature 387, 16–17.

Wyllie, A. (1997) Clues in the p53 murder mystery. Nature 389, 237–238.

Xiong, Y. and Kotake, Y. (2006) No exit strategy? No problem: APC inhibits beta-catenin inside the nucleus. Genes and Development 20, 637–642.

Yang, A., Kaghad, M., Caput, D. and Mckeon, F. (2002) On the shoulders of giants: p63, p73 and the rise of p53. Trends in Genetics 18, 90–95.

Yang, W.H., Kim, J.E., Nam, H.W., Ju, J.W., Kim, H.S., Kim, Y.S. and Cho, J.W. (2006) Modification of p53 with O-linked N-acetylglucosamine regulates p53 activity and stability. Nature Cell Biology 8, 1074–1083.

Yang, Y., Li, C-C.H. and Weissman, A.M. (2004) Regulating the p53 system through ubiquitination. Oncogene 23, 2096–2106.

Zeller, K.I., Zhao, X., Lee, C.W., Chiu, K.P., Yao, F., Yustein, J.T., Ooi, H.S., Orlov, Y.L., Shahab, A., Yong, H.C., Fu, Y., Weng, Z., Kuznetsov, V.A., Sung, W.K., Ruan, Y., Dang, C.V. and Wei, C.L. (2006) Global mapping of c-Myc binding sites and target gene networks in human B cells. Proceedings of the National Academy of Sciences, USA 103, 17834–17839.

Zhai, W., Jeong, H., Cui, L., Krainc, D. and Tjian, R. (2005) In vitro analysis of huntingtin-mediated transcriptional repression reveals multiple transcription factor targets. Cell 123, 1241–1253.

Zhang, H.S., Postigo, A.A. and Dean, D.C. (1999) Active transcriptional repression by the Rb-E2F complex mediates G1 arrest triggered by p16^{INK4a}, TGFβ, and contact inhibition. Cell 97, 53–61.

Zheng, H., You, H., Zhou, X.Z., Murray, S.A., Uchida, T., Wulf, G., Gu, L., Tang, X., Lu, K.P. and Xiao, Z-X.J. (2002) The prolyl isomerase Pin1 is a regulator of p53 in genotoxic response. Nature 419, 849–853.

CHAPTER 10

CONCLUSIONS AND FUTURE PROSPECTS

At the time the first edition of this book was published (1991), enormous progress had been made in understanding the nature and role of transcription factors. Thus, the roles of specific factors in processes such as constitutive, inducible, tissue specific and developmentally regulated gene expression had been defined, as had their involvement in diseases such as cancer. Moreover, by studying these factors in detail, it proved possible to analyse how they fulfil their function in these processes, by binding to specific sites in the DNA of regulated genes and activating or repressing transcription, as well as the regulatory processes which result in their doing so only at the appropriate time and place. Moreover, the regions of individual factors which mediate these effects and the critical amino acids within them which are of importance had been identified in a number of cases.

In the intervening years, up to the current publication of the fifth edition, much further progress has been made in these areas. In addition, the ability to prepare 'knock out' mice, in which the gene encoding an individual factor has been inactivated, has allowed the *in vivo* functional role of many factors to be directly assessed, whilst numerous studies have elucidated the structure of specific factors either in isolation or bound to DNA, as illustrated in the colour plate section. It has become increasingly clear, however, that the activity of a particular factor cannot be considered in isolation. Thus, very often the activity of a factor can be stimulated either positively or negatively by its interaction with another factor. For example, the Fos protein needs to interact with the Jun protein to form a DNA binding complex (see Chapter 4, section 4.5 and Chapter 9, section 9.3.1). Conversely, the DNA binding ability of the glucocorticoid receptor is inhibited by its association with hsp90 (see Chapter 8, section 8.2.2) whilst that of the MyoD factor is inhibited by its association with Id (Chapter 4, section 4.5.3 and Chapter 6, section 6.2.2).

Additionally, however, it has become clear that, as well as stimulating or inhibiting factor activity, such protein–protein interactions can also alter the specificity of a factor. Thus, differences in the ability to interact with other proteins can affect the DNA binding specificity of particular factors and hence the target genes to which they bind. This can result in factors with identical DNA binding specificities having entirely different functional effects, as in the case of Ubx and Antp (see Chapter 4, section 4.2.4). Alternatively, it may completely change the factor from activator to repressor, as in the case of the dorsal/DSP1 interaction (see Chapter 8, section 8.3.3). Hence, by altering the specificity of particular factors, interactions of this type are likely to play a crucial role in the complex regulatory networks which allow a relatively small number of transcription factors to control highly complex processes such as development. Such networks are now being progressively elucidated by a combination of experimental and computer-based methods (see Chapter 2, section 2.4.3).

As well as such regulatory interactions between different factors, it has become increasingly clear in recent years that many activating transcription factors need to interact with other factors, such as the mediator complex and/or co-activators in order to stimulate transcription. One of the most important of such co-activators, CBP, was originally characterized as being required for transcriptional activation in response to cyclic AMP treatment, mediated via the CREB transcription factor (see Chapter 5, section 5.4.3). It is also involved, however, in transcriptional activation mediated via a number of other transcription factors, activated by different signalling pathways. In turn, because of the limiting amounts of CBP in the cell, the different transcription factors and signalling pathways compete for CBP, resulting in mutual antagonism between, for example, the signalling pathways mediated by AP1 and the glucocorticoid receptor (see Chapter 6, section 6.6).

Thus, the critical dependence of many activating factors on a specific co-activator can result in a functional link between two different factors which do not themselves interact but which compete for the same co-activator. Moreover, the activity of a transcription factor can be regulated by controlling its ability to interact with its co-activator. For example, in the absence of thyroid hormone, the thyroid hormone receptor has an inhibitory effect on transcription, because it binds co-repressor molecules that act to inhibit transcription. Following exposure to thyroid hormone, however, the receptor undergoes a conformational change which allows it to bind co-activator molecules and hence activate transcription (see Chapter 6, section 6.3.2). Similarly, only the phosphorylated form of CREB can interact with CBP and therefore activate transcription, whereas the non-phosphorylated form does not interact with CBP and is thus inactive (see Chapter 5, section 5.4.3).

This regulation of CREB by phosphorylation is only one example of a plethora of post-translational modifications which have been shown in recent years to modulate transcription factor activity. In addition to phosphorylation, these modifications include methylation, acetylation, ubiquitination and sumoylation. Moreover, they have been shown to modify transcription factor activity via altering processes as diverse as DNA binding, cellular localization and stability, as well as via modulating their interaction with other factors (see Chapter 8, section 8.4).

Moreover, individual factors can be modulated by more than one post-translational modification as seen in the cases of the oncogenic transcription factor β-catenin (see Chapter 9, section 9.4.4) and the anti-oncogenic transcription factors, p53 (see Chapter 9, section 9.4.2) and Rb-1 (see Chapter 9, section 9.4.3). Similarly, these modifications can also target co-activators as well as DNA binding transcription factors. Thus, as discussed in Chapter 8 (section 8.4.4), CBP can be modified by methylation and phosphorylation and this differentially affects its binding to different activating transcription factors. The competition between different activators for binding to CBP can therefore be regulated by modifying CBP itself, as well as by modifying the transcription factors themselves.

Hence, the interaction between activators and co-activators plays a critical role in the activation of transcription and its regulation. Although co-activators are likely to act in some cases by interacting with the basal transcriptional complex (see Chapter 5, section 5.5.2), the finding that many co-activators have histone acetyltransferase activity (see Chapter 5, section 5.5.1) indicates that they may stimulate transcription via altering chromatin structure. Hence, such factors could act by acetylating histones, thereby altering the chromatin structure to a more open structure able to support active transcription (see Chapter 1, section 1.2.3). Similarly, activators and co-activators may also recruit chromatin remodelling complexes such as SWI/SNF, which use ATP-dependent processes to open up the chromatin (see Chapter 1, section 1.2.2 and Chapter 5, section 5.5.1). Conversely, co-repressors which have histone deacetylase activity may act by producing a more closed chromatin structure incompatible with transcription (see Chapter 6, section 6.4.1). Hence, the regulation of chromatin structure by activating and inhibitory transcription factors plays a key role in the regulation of gene expression.

The activation of a target promoter is likely therefore to require the recruitment of activating molecules, histone acetyltransferases and chromatin remodelling complexes, as well as of the basal transcriptional complex. As discussed in Chapter 5 (section 5.7), this is a highly ordered process and at each promoter, a series of events will occur, with the recruitment of each factor facilitating the next stage in the transcription initiation process. Indeed, this process does not end when transcription is initiated since many transcription factors also stimulate or inhibit transcriptional elongation (see Chapter 5, section 5.6 and Chapter 6, section 6.5).

Ultimately, therefore, the understanding of transcription factor function will require a knowledge of the nature and effect of interactions between different transcription factors themselves and with their co-activators and co-repressors, which is as good as that now available for individual factors. Moreover, it will be necessary to establish how such changes modulate transcriptional initiation by the basal transcriptional complex, regulate transcriptional elongation and alter chromatin structure. Clearly, much work remains to be done before this is achieved. The rapid progress since the first edition of this work was published suggests, however, that an eventual understanding in molecular terms of the manner in which transcription factors control highly complex processes such as *Drosophila* and even mammalian development can ultimately be achieved. Similarly, the increasing understanding of the critical role of transcription factors and their involvement in human diseases offers significant potential for the development of novel therapies for a wide range of human diseases (see Chapter 9, section 9.5).

INDEX